SYMMETRIES IN PHYSICS
PHILOSOPHICAL REFLECTIONS

Symmetry considerations dominate modern fundamental physics, both in quantum theory and in relativity. This book presents a collection of new philosophy of physics papers, set in the context of extracts from seminal works by both physicists and philosophers. It covers topical issues such as the significance of gauge symmetry, particle identity in quantum theory, how to make sense of parity violation, the role of symmetry breaking, and the empirical status of symmetry principles. These issues relate directly to more traditional problems in the philosophy of science, including the status of the laws of nature, the relationships between mathematics, physical theory, and the world, and the extent to which mathematics dictates physics. The book is structured into four parts, each with classic texts, review papers, and discussion papers that survey the current situation in the literature and highlight the main issues and controversies. Its aim is to provide a general picture of the current debate, along with a context and framework for future discussion and research in this field. Suitable for courses on the foundations of physics, philosophy of physics, and philosophy of science, it is also a valuable reference for students and researchers in both physics and philosophy.

KATHERINE BRADING is Junior Research Fellow in Philosophy of Science at Wolfson College, University of Oxford.

ELENA CASTELLANI is a Researcher at the Department of Philosophy, University of Florence.

SYMMETRIES IN PHYSICS

Philosophical Reflections

Edited by

KATHERINE BRADING
Wolfson College, University of Oxford
and
ELENA CASTELLANI
University of Florence

CAMBRIDGE
UNIVERSITY PRESS

CAMBRIDGE UNIVERSITY PRESS
Cambridge, New York, Melbourne, Madrid, Cape Town, Singapore,
São Paulo, Delhi, Dubai, Tokyo

Cambridge University Press
The Edinburgh Building, Cambridge CB2 8RU, UK

Published in the United States of America by Cambridge University Press, New York

www.cambridge.org
Information on this title: www.cambridge.org/9780521528894

First published 2003
This digitally printed version 2009

A catalogue record for this publication is available from the British Library

Library of Congress Cataloguing in Publication data
Symmetries in Physics: Philosophical Reflections/edited by Katherine Brading and Elena Castellani.
p. cm.
Includes bibliographical references and index.
ISBN 0 521 82137 1
1. Symmetry (Physics) 2. Physics – Philosophy. I. Brading, Katherine, 1970–
II. Castellani, Elena, 1959–
QC174.17.S9S984 2003
539.7′ – dc21 2003041967

ISBN 978-0-521-82137-7 Hardback
ISBN 978-0-521-52889-4 Paperback

Contents

Contributors

Gordon Belot
Associate Professor of Philosophy, New York University.

Katherine Brading
Junior Research Fellow in Philosophy of Science, Wolfson College, Oxford.

Harvey R. Brown
Reader in Philosophy and University Lecturer in Philosophy of Physics, University of Oxford; Fellow of Wolfson College, Oxford.

Elena Castellani
Researcher in the Department of Philosophy and Lecturer in Foundations of Physics in the Department of Physics, University of Florence.

John Earman
University Professor of History and Philosophy of Physics, University of Pittsburgh.

Steven French
Professor of Philosophy of Science, Division of History and Philosophy of Science, School of Philosophy, University of Leeds.

Nick Huggett
Assistant Professor of Philosophy, University of Illinois at Chicago.

Peter Kosso
Professor of Philosophy, Northern Arizona University.

Jenann Ismael
Assistant Professor of Philosophy, University of Arizona.

Giovanni Jona-Lasinio
Professor, Department of Physics, University of Rome 'La Sapienza'.

Christopher A. Martin
Assistant Professor of History and Philosophy of Science, University of Indiana.

Margaret Morrison
Professor of Philosophy, University of Toronto.

John D. Norton
Professor, Department of History and Philosophy of Science, University of Pittsburgh.

Antigone M. Nounou
Ph.D. in Philosophy of Physics, London School of Economics.

Oliver Pooley
British Academy Postdoctoral Fellow and Fellow of Oriel College, Oxford.

Michael Redhead
Centennial Professor of Philosophy, London School of Economics.

Dean Rickles
Ph.D. student in Philosophy of Science, University of Leeds.

T. A. Ryckman
teaches philosophy at the University of California, Berkeley.

Simon Saunders
University Lecturer in Philosophy of Science, University of Oxford; Fellow of Linacre College, Oxford.

Bas C. van Fraassen
Professor of Philosophy, Princeton University.

David Wallace
D.Phil. in Physics, Oxford; Scholar in Philosophy, Magdalen College, Oxford.

Preface

Symmetry considerations dominate modern fundamental physics, both in quantum theory and in relativity. Philosophers are now beginning to devote increasing attention to such issues as the significance of gauge symmetry, the role of symmetry breaking, the empirical status of symmetry principles, and so forth. These issues relate directly to traditional problems in the philosophy of science, including the status of the laws of nature, the relationships between mathematics, physical theory, and the world, and the extent to which mathematics dictates physics.

In January 2001 the first philosophy of physics workshop on symmetries in physics was held in Oxford. It became clear from the success of the workshop, the enthusiasm and sense of shared work-in-progress, that the time is right for a collection of papers in philosophy of physics on the subject of symmetry. As the organizers of the workshop, we decided to bring together in one book the current philosophical discussions of symmetry in physics, and to do so in a format that would provide a point of entry into the subject for non-experts, including students and philosophers of science in general. As such, the book is intended to be accessible and of interest to a wide audience of physicists and philosophers. It is appropriate for courses in foundations of physics, philosophy of physics, and advanced courses in philosophy of science. Some of the papers in this collection originated from papers presented at the Oxford workshop, but most have been written expressly for this book.

Acknowledgements

Our thanks go to all those who have helped in preparing this volume, first and foremost all of the contributors. We are indebted to all those who gave us invaluable feedback on various papers during the preparation of this volume, including Elaine Landry, Mauricio Suarez, and David Wallace, along with those referees who wish to remain anonymous. A special thank you goes to Ruggero Vaia, for his much appreciated technical support. We are grateful also to our families for the various different kinds of support they have given us; in particular, thank you to Leonardo Castellani for his expertise in theoretical physics, and to Penelope Brading for the willing application of her copy-editing skills in assisting us in the preparation of the manuscript, and for her infinite patience. Finally, we are very grateful to Tamsin van Essen and to the staff of Cambridge University Press.

Copyright acknowledgements

The editors are grateful for permission to include the following copyrighted material.

Extracts from Eugene P. Wigner, 'Invariance in Physical Theory', 'Symmetry and Conservation Laws', 'The Role of Invariance Principles in Natural Philosophy', in *Symmetries and Reflections*, Indiana University Press, 1967. © Martha Wigner Upton. Reprinted by kind permission of Martha Wigner Upton.

Extracts from Hermann Weyl, *Symmetry*. © 1952 Princeton University Press. Reprinted by permission of Princeton University Press.

Extracts from pages 26–7 of *The Leibniz–Clarke Correspondence*, ed. H. G. Alexander. © 1956 by Manchester University Press. Reprinted by permission of Manchester University Press.

Extracts from Immanuel Kant, 'Concerning the ultimate ground of the differentiation of directions in space', in *The Cambridge Edition of the Works of Immanuel Kant: Theoretical Philosophy, 1755–1770*, edited by R. Meerbrote and translated by D. Walford, pp. 365–72. © 1992 by Cambridge University Press. Reprinted with the kind permission of R. Meerbrote, D. Walford, and Cambridge University Press.

Extracts from Max Black, 'The Identity of Indiscernibles', *Mind*, Vol. LXI, no. 242, pp. 155–63. © 1952 by Oxford University Press. Reprinted by permission of Oxford University Press.

Extracts from the translation by J. Rosen and P. Copié of Pierre Curie, 'On symmetry in physical phenomena, symmetry of an electric field and of a magnetic field', in *Symmetry in Physics: Selected Reprints*, ed. J. Rosen, pp. 17–25. Reprinted with permission from American Association of Physics Teachers Publications. Copyright 1982, American Association of Physics Teachers.

Sections II and III of Giovanni Jona-Lasinio, 'Cross Fertilization in Theoretical Physics: the Case of Condensed Matter and Particle Physics', forthcoming in *Highlights in Mathematical Physics*, eds. A. Fokas, J. Halliwell, T. Kibble and B. Zegarlinski, American Mathematical Society. Reprinted with permission from American Mathematical Society.

Michael Redhead, 'The Interpretation of Gauge Symmetry', in *Ontological Aspects of Quantum Field Theory*, eds. M. Kuhlmann, H. Lyre, and A. Wayne, World Scientific, 2002. Reprinted by permission of World Scientific Publishing Co., Inc. The editors are grateful to H. Lyre for his assistance in obtaining this permission.

1
Introduction

KATHERINE BRADING AND ELENA CASTELLANI

This book is about the various symmetries at the heart of modern physics. How should we understand them and the different roles that they play? Before embarking on this investigation, a few words of introduction may be helpful. We begin with a brief description of the historical roots and emergence of the concept of symmetry that is at work in modern physics (section 1). Then, in section 2, we mention the different varieties of symmetry that fall under this general umbrella, outlining the ways in which they were introduced into physics. We also distinguish between two different uses of symmetry: symmetry principles versus symmetry arguments. In section 3 we change tack, stepping back from the details of the various symmetries to make some remarks of a general nature concerning the status and significance of symmetries in physics. Finally, in section 4, we outline the structure of the book and the contents of each part.

1 The meanings of symmetry

Symmetry is an ancient concept. Its history starts with the Greeks, the term συμμετρία deriving from σύν (with, together) and μέτρον (measure) and originally indicating a relation of commensurability (such is the meaning codified in Euclid's *Elements*, for example). But symmetry immediately acquired a further, more general meaning, with commensurability representing a particular case: that of a proportion relation, grounded on (integer) numbers, and with the function of harmonizing the *different* elements into a *unitary whole* (Plato, *Timaeus*, 31c):

> The most beautiful of all links is that which makes, of itself and of the things it connects, the greatest unity possible; and it is the proportion (συμμετρία) which realizes it in the most beautiful way.

From the outset, then, symmetry was closely related to harmony, beauty, and unity, and this was to prove decisive for its role in theories of nature. In Plato's

Timaeus, for example, the regular polyhedra are afforded a central place in the doctrine of natural elements for the proportions they contain and the beauty of their forms: fire has the form of the regular tetrahedron; earth the form of the cube; air the form of the regular octahedron; water the form of the regular icosahedron; while the regular dodecahedron is used for the form of the entire universe. The history of science provides another paradigmatic example of the use of these figures as basic ingredients in physical description: Kepler's 1596 *Mysterium cosmographicum* presents a planetary architecture grounded on the five regular solids.

The regular figures used in Plato's and Kepler's physics for the mathematical proportions and harmonies they contain (and the related properties and beauty of their form) are symmetric in another sense that is not related to proportions. In the language of modern science, the symmetry of geometrical figures – such as the regular polygons and polyhedra – is defined in terms of their invariance under specified groups of rotations and reflections. Where does this definition stem from? Besides the ancient notion of symmetry used by the Greeks and Romans (current until the end of the Renaissance), a different notion of symmetry slowly emerged in the modern era, grounded not on proportions but on an equality relation. More precisely, it is grounded on an equality relation between elements that are opposed, such as the left and right parts of a figure. This notion, explicitly recognized and defined in such terms in a 1673 text by Claude Perrault, is, in fact, nothing other than our reflection symmetry. Reflection symmetry now has a precise definition in terms of invariance under the group of reflections, representing a particular case of the group-theoretic notion of symmetry currently used in modern science.

In moving from Perrault's notion to this abstract group-theoretic notion, the following crucial steps are worth noting. First, we have the interpretation of the *equality of the parts with respect to the whole* in the sense of their *interchangeability* (equal parts can be exchanged with one another, while preserving the whole). Then, we have the introduction of specific mathematical operations, such as reflections, rotations, and translations, that are used to describe with precision how the parts are to be exchanged. As a result, we arrive at a definition of the symmetry of a geometrical figure in terms of its *invariance* when equal component parts are exchanged according to one of the specified operations. Thus, when the two halves of a bilaterally symmetric figure are exchanged by reflection, we recover the original figure, and that figure is said to be invariant under left–right reflections. This is known as the 'crystallographic notion of symmetry', since it was in the context of early developments in crystallography that symmetry was first so defined and applied. The next key step is the generalization of this notion to the *group-theoretic* definition of symmetry, which arose following the nineteenth-century development of the algebraic concept of a group, and the fact that the symmetry operations of a figure were found to satisfy the conditions for forming a group.[1] Finally, as is discussed in more detail later in this volume (see Castellani, Part IV), we have the resulting close connection between the notion of symmetry, equivalence, and group (a symmetry group induces a partition into equivalence classes).

The group-theoretic notion of symmetry is the one that has proved so successful in modern science, and with which the papers of this collection are concerned. Note, however, that symmetry remains linked to beauty (regularity) and unity: by means of the symmetry transformations, distinct (but 'equal' or, more generally, 'equivalent') elements are related to each other and to the whole, thus forming a regular 'unity'. The way in which the regularity of the whole emerges is dictated by the nature of the specified transformation group. Summing up, a *unity of different and equal elements* is always associated with symmetry, in its ancient or modern sense; the way in which this unity is realized on the one hand, and how the equal and different elements are chosen on the other, determines the resulting symmetry and in what exactly it consists.[2]

2 Symmetry in the history of physics

When considering the role of symmetry in physics from a historical point of view, it is worth keeping in mind two preliminary distinctions.

- The first is between implicit and explicit uses of the notion. Symmetry considerations have always been applied to the description of nature, but for a long time in an implicit way only. As we have seen, the scientific notion of symmetry (the one we are interested in here) is a recent one. If we speak about a role of this concept of symmetry in the ancient theories of nature, we must be clear that it was not used explicitly in this sense at that time.
- The second is between the two main ways of using symmetry. First, we may attribute specific symmetry properties to physical situations or phenomena, or to laws (*symmetry principles*). It is the application with respect to *laws*, rather than to objects or phenomena, that has become central to modern physics, as we will see. Second, we may derive specific consequences with regard to particular physical situations or phenomena on the basis of their symmetry properties (*symmetry arguments*).

[1] A group is defined to be a set G, together with a product operation (•), such that: for any two elements g_1 and g_2 of G, $g_1 \cdot g_2$ is again an element of G; the group operation is associative; the group contains the identity element; and for each element there exists an inverse.

[2] Further details of the material in this section can be found in Castellani (2000), chapters 1–3.

2.1 Symmetry principles

Nature offers plenty of examples of (approximate) symmetrical forms: the bilateral symmetry of human (and, in general, of animal) bodies, the pentagonal symmetry frequent in flowers, the hexagonal symmetry of honeycomb cells, the translational symmetry of plant shoots and of animals such as caterpillars, and so on. The natural objects with the richest and most evident symmetry properties are undoubtedly crystals, and so it is not surprising that the systematic study of all possible symmetric configurations – the so-called theory of symmetry – started in connection with the rise of crystallography. The classification of all symmetry properties of crystals, which produced its most notable results in the nineteenth century, in fact marks the first explicit application of the scientific notion of symmetry in science.[3] The real turning point in the use of symmetry in science came, however, with the introduction of the *group* concept and with the ensuing developments in the theory of transformation groups. This is because the group-theoretic definition of symmetry as 'invariance under a specified group of transformations' allowed the concept to be applied much more widely, not only to spatial figures but also to abstract objects such as mathematical expressions – in particular, expressions of physical relevance such as dynamical equations. Moreover, the technical apparatus of group theory could then be transferred and used to great advantage within physical theories.

The first explicit study of the invariance properties of equations in physics is connected with the introduction, in the first half of the nineteenth century, of the transformational approach to the problem of motion in the framework of analytical mechanics. Using the formulation of the dynamical equations of mechanics due to Hamilton (known as the Hamiltonian or canonical formulation), Jacobi developed a procedure for arriving at the solution of the equations of motion based on the strategy of applying transformations of the variables that leave the Hamiltonian equations invariant, thereby transforming step by step the original problem into new ones that are simpler but perfectly equivalent (for further details see Lanczos, 1949).[4] Jacobi's canonical transformation theory, although introduced for the 'merely instrumental' purpose of solving dynamical problems, led to a very important line of research: the general study of physical theories in terms of their transformation properties.

[3] Symmetry considerations were used by Haüy to characterize and classify crystal structure and formation (see his 1801 *Traité de minéralogie*, volume 1), and with this, crystallography emerged as a discipline distinct from mineralogy. From Haüy's work two strands of development led to the 32 point transformation crystal classes and the 14 Bravais lattices, all of which may be defined in terms of discrete groups. These were combined into the 230 space groups by Fedorov and by Schönflies in 1891, and by Barlow in 1894. The theory of discrete groups continues to be fundamental in solid state physics, chemistry, and materials science.

[4] Notice that this is a clear example of a methodological use of symmetry properties: on the basis of the invariance properties of the situation under consideration (in this case, the dynamical problem in classical mechanics), a strategy is applied for deriving determinate consequences. The underlying principle is that equivalent problems have equivalent solutions. This type of symmetry argument (see section 2.2, below) is discussed also by van Fraassen (1989), chapter 10.

Examples of this are the studies of invariants under canonical transformations, such as Poisson brackets or Poincaré's integral invariants; the theory of continuous canonical transformations due to Lie; and, finally, the connection between the study of physical invariants and the algebraic and geometric theory of invariants that flourished in the second half of the nineteenth century, and which laid the foundation for the geometrical approach to dynamical problems. The use of the mathematics of group theory to study physical theories was central to the work, early in the twentieth century in Göttingen, of the group whose central figures were Klein (who earlier collaborated with Lie) and Hilbert, and which included Weyl and later Noether. We will return to Weyl and Noether later.

In the above approach, the equations or expressions of physical interest are already given and the strategy is to study their symmetry properties. There is, however, an alternative way of proceeding, namely the reverse: start with specific symmetries and search for dynamical equations with such properties. In other words, we *postulate* that certain symmetries are physically significant, rather than deriving them from prior dynamical equations. The assumption of certain symmetries in nature is not, of course, a novelty. Although not explicitly expressed as symmetry principles, the homogeneity and isotropy of physical space, and the uniformity of time (forming, together with the invariance under Galilean boosts, 'the older principles of invariance' – see Wigner (1967; this volume, Part IV)), have been assumed as prerequisites in the physical description of the world since the beginning of modern science. Perhaps the most famous early example of the deliberate use of this type of symmetry principle is Galileo's discussion of whether the Earth moves, in his *Dialogue Concerning the Two Chief World Systems* of 1632. Galileo sought to neutralize the standard arguments purporting to show that, simply by looking around us at how things behave locally on Earth – how stones fall, how birds fly – we can conclude that the Earth is at rest rather than rotating, arguing instead that these observations do not enable us to determine the state of motion of the Earth. His approach was to use an analogy with a ship: he urges us to consider the behaviour of objects, both animate and inanimate, inside the cabin of a ship, and claims that no experiments carried out inside the cabin, without reference to anything outside the ship, would enable us to tell whether the ship is at rest or moving smoothly across the surface of the Earth. The *assumption* of a symmetry between rest and a certain kind of motion leads to the prediction of this result, without the need to know the laws governing the experiments on the ship. The 'Galilean principle of relativity' (according to which the laws of physics are invariant under Galilean boosts, where the states of motion considered are now those of uniform velocity) was quickly adopted as an axiom and widely used in the seventeenth century, notably by Huygens in his solution to the problem of colliding bodies and by Newton in his early work on motion. Huygens took the relativity principle as his third hypothesis or axiom, but in Newton's *Principia* it is demoted to a corollary to the laws of motion, its status in Newtonian physics therefore being that of a *consequence* of the laws, even though it remains, in fact, an independent assumption.

Although the spatial and temporal invariance of mechanical laws had been known and used for a long time in physics, and the group of the global spacetime symmetries for electrodynamics was completely derived by Poincaré[5] before Einstein's famous 1905 paper setting out his special theory of relativity, it was not until this work by Einstein that the status of symmetries with respect to the laws was reversed. Wigner (1967; see this volume, Part IV) writes that 'the significance and general validity of these principles were recognized, however, only by Einstein', and that Einstein's work on special relativity marks 'the reversal of a trend: until then, the principles of invariance were derived from the laws of motion . . . It is now natural for us to derive the laws of nature and to test their validity by means of the laws of invariance, rather than to derive the laws of invariance from what we believe to be the laws of nature'. In postulating the universality of the global continuous spacetime symmetries – also known as 'geometrical symmetries' in the terminology introduced by Wigner (1967; see this volume, Part I) – Einstein's construction of his special theory of relativity represents the first turning point in the application of symmetry to twentieth-century physics.[6]

Global spacetime invariance principles are intended to be valid for all the laws of nature. Such a universal character is not shared by the physical symmetries that were next introduced in physics. Most of these were of an entirely new kind, with no roots in the history of science, and in some cases expressly introduced to describe specific forms of interactions – whence the name 'dynamical symmetries' due to Wigner (1967; see this volume, Part I).

The new symmetries were for the most part closely related to specific features of the microscopic world. *Permutation symmetry*, 'discovered' by Heisenberg in 1926 in relation to the indistinguishability of so-called identical quantum particles (see French and Rickles, this volume), was the first non-spatiotemporal symmetry to be introduced into microphysics, and also the first symmetry to be treated with the techniques of group theory in the context of quantum mechanics. The application of the theory of groups and their representations for the exploitation of symmetries in quantum mechanics undoubtedly represents the second turning point in the twentieth-century history of physical symmetries. It is, in fact, in the quantum context that symmetry principles are at their most effective. Wigner and Weyl were among the first to recognize the great relevance of symmetry groups to quantum physics and the first to reflect on the meaning of this. As Wigner

[5] Whence the name 'Poincaré group' introduced later by Wigner, whereas Poincaré himself named the group after Lorentz.

[6] General relativity marks a further important stage in the development, as we will see below. See also Martin, this volume.

emphasized on many occasions, one essential reason for the 'increased effectiveness of invariance principles in quantum theory' (Wigner, 1967, p. 47) is the linear nature of the state space of a quantum physical system, corresponding to the possibility of superposing quantum states. This gives rise to, among other things, the possibility of defining states with particularly simple transformation properties in the presence of symmetries.

In general, if G is a symmetry group of a theory describing a physical system (that is, the fundamental equations of the theory are invariant under the transformations of G), this means that the states of the system transform into each other according to some 'representation' of the group G. In other words, the group transformations are mathematically represented in the state space by operations relating the states to each other. In quantum mechanics, these operations are generally the operators acting on the state space that correspond to the physical observables, and any state of a physical system can be described as a superposition of states of elementary systems, that is, of systems the states of which transform according to the 'irreducible' representations of the symmetry group. Quantum mechanics thus offers a particularly favourable framework for the application of symmetry principles. The observables representing the action of the symmetries of the theory in the state space, and therefore commuting with the Hamiltonian of the system, play the role of the conserved quantities; furthermore, the basis states may be labelled by the irreducible representations of the symmetry group, which accordingly also regulate the transformations from one state to another (state transitions).

But more can be said. Because of the specific properties of the quantum description, symmetries such as spatial reflection symmetry or *parity* (P) and *time reversal* (T) were 'rediscovered' in the quantum context, taking on a new significance.[7] Moreover, new 'quantum symmetries' emerged, such as particle–antiparticle symmetry or *charge conjugation* (C),[8] and the various *internal symmetries* grounded on invariances under phase changes of the quantum states and described in terms of the unitary groups SU(N) (the local versions of which are the *gauge symmetries* at the core of the Standard Model for elementary particles).[9] More recently, new

[7] Parity was introduced in quantum physics in 1927 in a paper by Wigner, where important spectroscopic results were explained for the first time on the basis of a group-theoretic treatment of permutation, rotation, and reflection symmetries. Time reversal invariance appeared in the quantum context, again due to Wigner, in a 1932 paper.

[8] Charge conjugation was introduced in Dirac's famous 1931 paper 'Quantized singularities in the electromagnetic field'. C is a discrete symmetry, connected to the spatial and temporal discrete symmetries P and T by the so-called CPT theorem, demonstrated by Lüders in 1952, which states that the combination of C, P, and T is a general symmetry of physical laws.

[9] The starting point for the idea of internal symmetries was the interpretation of the presence of particles with (approximately) the same value of mass as the components (*states*) of a single physical system, connected to each other by the transformations of an underlying symmetry group. This idea emerged in analogy with what happened in the case of permutation symmetry, and was in fact due to Heisenberg (the discoverer ofpermutation symmetry), who in a 1932 paper introduced the SU(2) symmetry connecting the proton and the neutron (interpreted as the two states of a single system). This symmetry was further studied by Wigner, who in 1937 introduced the term *isotopic spin* (later contracted to *isospin*).

symmetries acquired relevance in theoretical physics, such as *supersymmetry* (the symmetry relating bosons and fermions and leading, when made local, to the theories of *supergravity*), and the various forms of *duality* used in today's superstring theories.

The history of the application of symmetry principles in quantum mechanics and then quantum field theory coincides with the history of the developments of twentieth-century theoretical physics. The salient aspects of this history, from the perspective of the meaning of physical symmetries, are discussed in the contributions to this volume (for details, see section 4, below) and cover four crucial developments.

- The first is the extension of the concept of continuous symmetry from 'global' symmetries (such as the Galilean group of spacetime transformations) to 'local' symmetries, as discussed by Martin (this volume) in his review of continuous symmetries. Einstein was the first to make use of a local symmetry principle in theory construction when developing his General Theory of Relativity (GTR), culminating in 1915.[10] Meanwhile in Göttingen, Klein and Hilbert enlisted the assistance of Noether in their investigations into the status of energy conservation in generally covariant theories of gravitation. This led to Noether's famous 1918 paper containing two theorems, the first of which leads to a connection between global symmetries and conservation laws, and the second of which allows a demonstration of the different status of these conservation laws when the global symmetry group is a subgroup of some local symmetry group of the theory in question (see Brading and Brown, this volume). Prompted by Einstein's work, Weyl's 1918 'unified theory of gravitation and electromagnetism' extended the idea of local symmetries (see Ryckman, this volume), and although this theory is generally deemed to have failed, the theory contains the seeds of later success in the context of quantum theory (see below).
- The second is the extension of the concept of continuous symmetry from spatiotemporal to internal, both global and local. In quantum theory, the phase of the wavefunction encodes internal degrees of freedom. With the requirement that a theory be invariant under *local gauge transformations* involving the phase of the wavefunction, Weyl's ideas found a successful home in quantum theory (see O'Raifeartaigh, 1997). Weyl's new 1929 theory was a theory of electromagnetism coupled to matter. The history of gauge theory is surveyed briefly by Martin (this volume), who highlights various issues surrounding gauge symmetry, in particular the status of the so-called 'gauge principle', first proposed by Weyl. Martin also discusses the ensuing stages in the development of gauge theory, the main steps being the Yang and Mills non-Abelian gauge theory of 1954, and the problems

[10] See Norton (this volume) on the 'Kretschmann objection' to the physical significance of general covariance, and also Martin (this volume, section 2.2) on invariance versus covariance.

and solutions associated with the successful development of gauge theories for the short-range weak and strong interactions.

- The third is the increasing importance of the discrete symmetries of permutation invariance and C, P, and, T mentioned above.
- Finally, the fourth is the introduction in the late 1950s and early 1960s of the concept of spontaneous symmetry breaking in field theory (see Part III of this volume), and the subsequent related results (including the Goldstone 1961 theorem and the 1964 so-called Higgs mechanism), which played a crucial role in the developments of the Standard Model of elementary particles.

2.2 Symmetry arguments

Consider the following cases.

- Buridan's ass: situated between what are, for him, two completely equivalent bundles of hay, he has no reason to choose the one located to his left over the one located to his right, and so he is not able to choose and dies of starvation.
- Archimedes's equilibrium law for the balance: if equal weights are hung at equal distances along the arms of a balance, then it will remain in equilibrium since there is no reason for it to rotate one way or the other about the balance point.
- Anaximander's argument for the immobility of the Earth as reported by Aristotle: the Earth remains at rest since, being at the centre of the spherical cosmos (and in the same relation to the boundary of the cosmos in every direction), there is no reason why it should move in one direction rather than another.

What do these cases have in common?

First, they can all be understood as examples of the application of the Leibnizean Principle of Sufficient Reason (PSR): if there is no sufficient reason for one thing to happen instead of another, the principle says that nothing happens (the initial situation does not change). But there is something more that the above cases have in common: in each of them PSR is applied on the grounds that the initial situation has a given symmetry: in the first two cases, bilateral symmetry; in the third, rotational symmetry. The symmetry of the initial situation implies the complete equivalence between the existing alternatives (the left bundle of hay with respect to the right one, and so on). If the alternatives are completely equivalent, then there is no sufficient reason for choosing between them and the initial situation remains unchanged.

Arguments of the above kind – that is, arguments leading to definite conclusions on the basis of an initial symmetry of the situation plus PSR – have been used in science since antiquity (as Anaximander's argument testifies). The form they most frequently take is the following: a situation with a certain symmetry evolves in such a way that, in the absence of an asymmetric cause, the initial symmetry is preserved. In other words, a breaking of the initial symmetry cannot happen without a reason, or *an asymmetry cannot originate spontaneously*. Van Fraassen (1989) devotes a chapter to considering the way these kinds of symmetry arguments can be used in general problem-solving.

Historically, the first explicit formulation of this kind of argument in terms of symmetry is due to the physicist Pierre Curie towards the end of the nineteenth century. Curie was led to reflect on the question of the relationship between *physical properties* and *symmetry properties* of a physical system by his studies on the thermal, electric, and magnetic properties of crystals, these properties being directly related to the structure, and hence the symmetry, of the crystals studied. More precisely, the question he addressed was the following: in a given physical medium (for example, a crystalline medium) having specified symmetry properties, which physical phenomena (for example, which electric and magnetic phenomena) are allowed to happen? His conclusions, systematically presented in his 1894 work 'Sur la symétrie dans les phénomènes physiques' (see this volume, Part III), can be summarized as follows.

(a) A phenomenon can exist in a medium possessing its characteristic symmetry or that of one of its subgroups. What is needed for its occurrence (i.e. for something rather than nothing to happen) is not the presence, but rather the absence, of certain symmetries: 'Asymmetry is what creates a phenomenon'.
(b) The symmetry elements of the causes must be found in their effects, but the converse is not true; that is, the effects can be more symmetric than the causes.

Conclusion (a) clearly indicates that Curie recognized the important function played by the concept of symmetry breaking in physics (he was indeed one of the first to recognize it). Conclusion (b) is what is usually called 'Curie's principle' in the literature, although notice that (a) and (b) are not independent of one another.

In order for Curie's principle to be applicable, various conditions need to be satisfied: the causal connection must be valid, the cause and effect must be well defined, and the symmetries of both the cause and the effect must also be well defined (this involves both the physical and the geometrical properties of the physical systems considered). Curie's principle then furnishes a *necessary condition* for given phenomena to happen: only those phenomena can happen that are compatible with the symmetry conditions established by the principle.

Curie's principle has thus an important methodological function: on the one side, it furnishes a kind of selection rule (given an initial situation with a specified symmetry, only certain phenomena are allowed to happen); on the other side, it offers a falsification criterion for physical theories (a violation of Curie's principle may indicate that something is wrong in the physical description).[11]

[11] See, for example, Mach's discussion of the Oersted effect in his *Die Mechanik in ihrer Entwickelung historisch – kritisch dargestellt* of 1883.

Such applications of Curie's principle depend, of course, on our accepting its validity, and this is something that has been questioned in the literature, especially in relation to spontaneous symmetry breaking (see this volume, Part III). Different proposals have been offered for justifying the principle. We have presented it here as an example of symmetry considerations based on Leibniz's PSR, while Curie himself seems to have regarded it as a form of causality principle. Chalmers (1970) considers its relation to the invariance properties of physical laws and argues that the principle follows from these in the case of deterministic laws, a point of view taken up again and generalized in Ismael (1997). In this approach, Curie's principle is understood as a condition on the relationship between the symmetries of a problem (an equation) and its solution(s). This has the advantages of avoiding the apparent vagueness of Curie's formulation (the appeal to causality, and so forth) while also extending it to cover symmetries of physical laws. However, trying to generalize Curie's principle as a principle about the link between the symmetries of an equation and its solution(s) is not straightforward and requires further attention (for more on symmetries of laws versus symmetries of solutions, see Belot, this volume, and Castellani, this volume, Part III).

3 Symmetries of modern physics: their status and significance

What is the status and significance of symmetries and symmetry principles in physics? The rich variety of symmetries in modern physics means that such a general question is not easily addressed. Indeed, we might even wonder whether it is well posed, and restrict our questions instead to specific symmetries and the interpretational issues they raise. Much of the recent literature opts for such restrictions on scope, and this is reflected in Parts I–III of this book (see also section 4 of this introduction). However, something can be said in more general terms; here we offer a few remarks in that direction[12] and we refer the reader to Part IV of the book, where general interpretative issues are addressed.

Exploring the roles and meanings of symmetries is deeply intertwined with basic questions regarding physical reality and physical knowledge, along with the methodologies and guiding strategies of contemporary physical inquiry. Thus, in approaching the above question we must take into account the possible ontological, epistemological, and methodological aspects of symmetries. In order to do this, we think that it is helpful to begin by considering the different roles that symmetries play in physics, the main four being, in our opinion, classificatory, normative, unifying, and explanatory.

One of the most important roles played by symmetry is that of *classification* – for example, the classification of crystals using their remarkable and varied symmetry

[12] This section of the introduction is based on Castellani (2002).

properties. In contemporary physics, the best example of this role of symmetry is the classification of elementary particles by means of the irreducible representations of the fundamental physical symmetry groups, a result first obtained by Wigner in his famous paper of 1939 on the unitary representations of the inhomogeneous Lorentz group. If a symmetry classification includes all the necessary properties for characterizing a given type of physical object (for example, all necessary quantum numbers for characterizing a given type of particle), we have the possibility of defining types of entities on the basis of their transformation properties. This has led philosophers of science to explore a structuralist approach to the entities of modern physics, in particular a group-theoretical account of objects (see for example the contributions in Castellani, 1998, part II).

Symmetries also have a *normative* role, being used as constraints on physical theories. The requirement of invariance with respect to a transformation group imposes severe restrictions on the form that a theory may take, limiting the types of quantities that may appear in the theory as well as the form of its fundamental equations. A famous case is Einstein's use of general covariance when searching for his gravitational equations.

The group-theoretical treatment of physical symmetries, with the resulting possibility of unifying different types of symmetries by means of a unification of the corresponding transformation groups, has provided the technical resources for symmetry to play a powerful role in theoretical *unification*. This is best illustrated by the current – dominant – research programme in theoretical physics aimed at arriving at a unified description of all the fundamental forces of nature (gravitational, weak, electromagnetic, and strong) in terms of underlying local symmetry groups.

It is often said that many physical phenomena can be explained as (more or less direct) consequences of symmetry principles or symmetry arguments. In the case of symmetry principles, the *explanatory* role of symmetries arises from their place in the hierarchy of the structure of physical theory, which in turn derives from their generality. For example, an explanatory role for symmetries with respect to conservation laws might be claimed on the basis of Noether's connection between symmetries and conservation laws (see Brading and Brown, this volume), along with the more fundamental status of symmetries in the hierarchy. As Wigner describes the hierarchy (Wigner, 1967; see especially the second extract in Part IV of this volume), symmetries are seen as properties of the laws. Thus, through the requirement that the laws (whatever they may be) must be invariant under certain symmetries, these symmetries place constraints on which events are physically possible (the explanatory role clearly connects to the normative role here). In other words, symmetries may be used to explain (i) the form of the laws and (ii) the occurrence (or non-occurrence) of certain events (this latter in a manner analogous to the way in which the laws explain why certain events occur and not others). Other features of symmetry in physics that are commonly used as an important explanatory basis for physical phenomena are the 'gauge principle' (for the form, or even existence, of the various interactions; see Martin, this volume) and the mechanism of 'spontaneous symmetry breaking' (see this volume, Part III). Finally, insofar as explanatory power may be derived from unification, the unifying role of symmetries also results in an explanatory role.

In the latter case, i.e. that of symmetry arguments, we may, for example, appeal to Curie's principle to explain the occurrence of certain phenomena on the basis of the symmetries (or asymmetries) of the situation, as discussed in section 2.2 above.

From these different roles we can draw some preliminary conclusions about the status of symmetries. It is immediately apparent that symmetries have an important *heuristic* function, indicating a strong *methodological* status. What about the ontological and epistemological status of symmetries?

Adopting an *ontological* view, symmetries are seen as a substantial part of the physical world: the symmetries of theories represent properties existing in nature, or characterize the structure of the physical world. It might be claimed, furthermore, that the ontological status of symmetries provides the reason for the methodological success of symmetries in physics. A concrete example is the use of symmetries to predict the existence of new particles. This can happen via the *classificatory* role, on the grounds of vacant places in symmetry classification schemes, as in the famous case of the 1962 prediction of the particle Ω^- in the context of the hadronic classification scheme known as the 'Eightfold Way'. Or, as in more recent cases, via the *unificatory* role: the paradigmatic example is the prediction of the W and Z particles (experimentally found in 1983) in the context of the Weinberg–Salam gauge theory proposed in 1967 for the unification of the weak and electromagnetic interactions.[13] These impressive cases of the prediction of new phenomena might perhaps be used to argue for an *ontological* status for symmetries, via an inference to the best explanation.

Another reason for attributing symmetries to nature is the so-called geometrical interpretation of spatiotemporal symmetries, according to which the spatiotemporal symmetries of physical laws are interpreted as symmetries of spacetime itself, the 'geometrical structure' of the physical world. Moreover, this way of seeing symmetries can be extended to non-external symmetries, by considering them as properties of other kinds of spaces, usually known as 'internal spaces'. The question of exactly what a realist would be committed to on such a view of internal spaces remains open, and an interesting topic for discussion – in this regard see Nounou, this volume.

One approach to investigating the limits of an ontological stance with respect to symmetries would be to investigate their empirical or observational status: can the

[13] The unificatory role of symmetries in physics is associated with a more general realist metaphysics influential amongst theoretical physicists working towards a unified theory of everything.

symmetries in question be directly observed? Morrison (this volume) raises concerns about a realist approach for the case of spontaneously broken symmetries, and the question can also be tackled for symmetries that are not spontaneously broken. We first have to address what it means for a symmetry to be observable, and indeed whether all (unbroken) symmetries have the same observational status. Kosso (2000) arrives at the conclusion that there are important differences in the empirical status of the different kinds of symmetries. In particular, while global continuous symmetries can be directly observed – via such experiments as the Galilean ship experiment – a local continuous symmetry can have only indirect empirical evidence. Brading and Brown (in press) argue for a different interpretation of Kosso's examples,[14] and hence for a different understanding of why the local symmetries of gauge theory and GTR have an empirical status distinct from that of the familiar global spacetime symmetries. The most fundamental point is this: in theories with local gauge symmetry, the matter fields are embedded in a gauge field, and the local symmetry is a property of both sets of fields *jointly*. Because of this there is, in general, no analogue of the Galilean ship experiment for local symmetry transformations; according to Brading and Brown, the continuous global spacetime symmetries have a special empirical status.[15]

The direct observational status of the familiar global spacetime symmetries leads us to an *epistemological* aspect of symmetries. According to Wigner, the spatiotemporal invariance principles play the role of a prerequisite for the very possibility of discovering the laws of nature: 'if the correlations between events changed from day to day, and would be different for different points of space, it would be impossible to discover them' (Wigner, 1967; see this volume, Part IV). For Wigner, this conception of symmetry principles is essentially related to our ignorance (if we could directly know all the laws of nature, we would not need to use symmetry principles in our search for them). Others, on the contrary, have arrived at a view according to which symmetry principles function as 'transcendental principles' in the Kantian sense (see for instance Mainzer, 1996). It should be noted in this regard that Wigner's starting point, as quoted above, does not imply exact symmetries – all that is needed epistemologically is that the global symmetries hold approximately, for suitable spatiotemporal regions, such that there is sufficient stability and regularity in the events for the laws of nature to be discovered.

There is another reason why symmetries might be seen as being primarily epistemological. As we have mentioned, there is a close connection between the notions of symmetry and equivalence, and this leads also to a notion of irrelevance: the equivalence of space points (translational symmetry) is, for example, understood in the sense of the irrelevance of an absolute position to the physical description; in

[14] Kosso's analysis begins from a set of examples offered by 't Hooft (1980).
[15] See also Brading and Brown, this volume.

the case of local symmetries the irrelevant elements correspond to the presence of 'surplus structure' in the theory.[16] There are two ways that one might interpret the epistemological significance of this: on the one hand, we might say that symmetries are associated with unavoidable redundancy in our *descriptions* of the world, while on the other hand we might maintain that symmetries indicate a limitation of our epistemic access – there are certain properties of objects, such as their absolute positions, that are not observable.

Finally, we would like to mention an aspect of symmetry that might very naturally be used to support either an ontological or an epistemological account. It is widely agreed that there is a close connection between symmetry and objectivity, the starting point once again being provided by spacetime symmetries: the laws by means of which we describe the evolution of physical systems have an objective validity because they are the same for all observers. The old and natural idea that what is objective should not depend upon the particular perspective under which it is taken into consideration is thus reformulated in the following group-theoretical terms: what is objective is what is invariant with respect to the transformation group of reference frames, or, quoting Weyl (1952, p. 132), 'objectivity means invariance with respect to the group of automorphisms [of space-time]'. The link between symmetry and objectivity is one theme of the paper by Kosso in Part IV of this volume.

Summing up, symmetries in physics offer many interpretational possibilities, including ontological, epistemological, and methodological. The position that one takes will depend in part on one's preferred approach to other issues in philosophy of science, including realism, the laws of nature, the relationship between mathematics and physics, the nature of theoretical entities, and so forth. It will also depend on whether one views symmetries as ultimately fundamental or derivative (be that in a methodological sense or, at the other extreme, an ontological sense). How to understand the status and significance of physical symmetries clearly presents a challenge to both physicists and philosophers.

4 Structure of the book

Our aim in this book is to provide a structured picture of the current philosophy of physics debate on symmetry, along with a context and framework for future debate and research in this field. As such, the aim is modest: there is no intention or aspiration to provide a comprehensive discussion of all philosophical issues that might arise from the roles of symmetries in physics. Rather, the content of this book clearly displays the issues that dominate current discussions in philosophy of physics.

[16] See Belot and Castellani, both this volume, Part IV, and also Redhead's 'surplus structure,' this volume, Part I.

We have divided the book into four parts, each of which begins with a selection of classic texts from such authors as Leibniz, Kant, Curie, Weyl, and Wigner. The first three parts of the book concern specific topics falling under the general heading of symmetry in physics: continuous symmetries, discrete symmetries, and symmetry breaking. Each contains a paper that reviews the current situation in the literature and highlights the main issues and controversies. Part IV is devoted to the general interpretational questions arising in connection with symmetries.

Part I. Among the issues raised by Martin in his review of continuous symmetries, the one that dominates the papers that follow is a set of interrelated questions surrounding the interpretation of local symmetries. Martin himself spends some time addressing the status of the so-called 'gauge principle', whose origins in Weyl's 1918 work – and particularly the philosophical background to this work – are the subject of Ryckman's paper. Brading and Brown pick up the historical thread with a discussion of Noether's 1918 theorems and the connection between symmetries and conservation laws. While their paper contains Noether's famous first theorem, concerning global symmetries, they also discuss the more complex case of local symmetries and the question of where the empirical significance of such symmetries lies. A theme common to the papers by Norton, Redhead, Earman, and Wallace is the 'underdetermination problem' associated with theories containing local symmetries. These papers discuss how this problem arises (with respect both to the diffeomorphism freedom of GTR and to gauge theories), what interpretational problems follow, and how these may be tackled. The underdetermination problem is connected to the issue of which quantities in a local gauge theory should be interpreted as real. This problem is made particularly vivid by the Aharonov–Bohm effect; Nounou offers a discussion of this effect, in which she sets out her preferred approach based on the fibre bundle formulation of gauge theories – her paper contains a conceptual introduction to fibre bundles, designed to make the philosophical account accessible.

Part II. Under the general heading of discrete symmetries we find two distinct areas of research, each of which has a large associated literature. The first is permutation symmetry, reviewed by French and Rickles. The second is CPT, or rather, in fact, primarily P. In the philosophy of physics literature, parity (and parity violation) at the level of the fundamental laws has been the focus of attention, the absolute versus relational debate in the philosophy of space and time being the context. This is the topic of Pooley's review paper. Themes arising in these review papers are picked up by both Huggett and Saunders. Huggett's first paper extends the French and Rickles discussion from bosons and fermions to other kinds of quantum particles ('quarticles'), while his second paper is a direct response to the discussion of handedness in Pooley's paper. Saunders advocates a version of Leibniz's Principle of the Identity of Indiscernibles that appeals to 'weak discernibility', which is a natural generalization of Leibniz's law and, he argues, consistent with, and a useful tool with respect to, modern physics.

Part III. Philosophers have come to the topic of symmetry breaking only recently, and hence the main purpose of the review paper by Castellani is to provide an introduction and a framework for further work. We also include an extract from a paper by Jona-Lasinio, offering a first-hand historical account of how the idea of spontaneous symmetry breaking was introduced in particle physics in the early 1960s. In his 'Rough guide' Earman offers an approach to understanding symmetry breaking that makes use of the algebraic formulation of quantum theory, while Morrison's paper raises interpretational questions over the status of spontaneously broken symmetries.

Part IV. The final part contains a selection of papers by Ismael and van Fraassen, Belot, Kosso, and Castellani on general issues of the interpretation of symmetry. They pick up on issues ranging right across the material of the preceding parts, such as those of redundancy and surplus structure, symmetries of laws versus symmetries of solutions, and the relationship between symmetry and objectivity.

Our hope is that this volume will appeal to a wide audience, including philosophers of physics, philosophers of science, and physicists. It offers something for everyone who is curious about symmetries in physics, providing a research tool as well as a point of access into this fascinating area.

References

Brading, K. and Brown, H. R. (in press). 'Are gauge symmetry transformations observable?' *British Journal for the Philosophy of Science*.

Castellani, E., ed. (1998). *Interpreting Bodies: Classical and Quantum Objects in Modern Physics*. Princeton, NJ: Princeton University Press.

Castellani, E. (2000). *Simmetria e Natura*. Roma, Bari: Laterza.

(2002). 'Symmetry, quantum mechanics and beyond'. *Foundations of Science*, **7**, 181–96.

Chalmers, A. F. (1970). 'Curie's principle'. *British Journal for the Philosophy of Science*, **21**, 133–48.

Ismael, J. (1997). 'Curie's principle'. *Synthese*, **110**, 167–90.

Kosso, P. (2000). 'The empirical status of symmetries in physics'. *British Journal for the Philosophy of Science*, **51**, 81–98.

Lanczos, C. (1949). *The Variational Principles of Mechanics*. Toronto: University of Toronto Press.

Mainzer, K. (1996). *Symmetries of Nature*. Berlin: Walter de Gruyter.

O'Raifeartaigh, L. (1997). *The Dawning of Gauge Theory*. Princeton, NJ: Princeton University Press.

't Hooft, G. (1980). 'Gauge theories and the forces between elementary particles'. *Scientific American*, **242**, 90–166.

van Fraassen, B. C. (1989). *Laws and Symmetry*. Oxford, New York: Oxford University Press.
Weyl, H. (1952). *Symmetry*. Princeton, NJ: Princeton University Press.
Wigner, E. (1967). *Symmetries and Reflections*. Bloomington, IN: Indiana University Press.

Part I

Continuous symmetries

2

Classic texts: extracts from Weyl and Wigner

Symmetry

HERMANN WEYL

I think the real situation has to be described as follows. Relative to a complete system of reference not only the points in space but also all physical quantities can be fixed by numbers. Two systems of reference are equally admissible if in both of them all universal geometric and physical laws of nature have the same algebraic expression. The transformations mediating between such equally admissible systems of reference form the group of *physical automorphisms*; the laws of nature are invariant with respect to the transformations of this group. It is a fact that a transformation of this group is uniquely determined by that part of it that concerns the coordinates of space points. Thus we can speak of the physical automorphisms *of space*. Their group does not include the dilatations, because the atomic laws fix an absolute length, but it contains the reflections because no law of nature indicates an intrinsic difference between left and right. Hence the group of physical automorphisms is the group of all proper and improper congruent mappings. If we call two configurations in space congruent provided they are carried over into each other by a transformation of this group, then bodies which are mirror images of each other are congruent. I think it is necessary to substitute this definition of congruence for that depending on the motion of rigid bodies, for reasons similar to those which induce the physicist to substitute the thermodynamical definition of temperature for an ordinary thermometer. Once the group of physical automorphisms = congruent mappings has been established, one may define geometry as the science dealing with the relation of congruence between spatial figures, and then the *geometric automorphisms* would be those transformations of space which carry any two congruent figures into congruent figures – and one need not be surprised, as Kant was, that this group of geometric automorphisms is wider than that of physical automorphisms and includes the dilatations.

Note. Extract from: 1952, Princeton, NJ: Princeton University Press, pp. 129–32.

All these considerations are deficient in one respect: they ignore that physical occurrences happen not only in space but in *space and time*; the world is spread out not as a three- but as a four-dimensional continuum. The symmetry, relativity, or homogeneity of this four-dimensional medium was first correctly described by Einstein. Has the statement, we ask, that two events occur at the same place an objective significance? We are inclined to say yes; but it is clear, if we do so, we understand position as position relative to the earth on which we spend our life. But is it sure that the earth rests? Even our youngsters are now told in school that it rotates and that it moves about in space. Newton wrote his treatise *Philosophiae Naturalis Principia Mathematica* to answer this question, to deduce, as he said, the absolute motion of bodies from their differences, the observable relative motions, and from the forces acting upon the bodies. But although he firmly believed in absolute space, i.e. in the objectivity of the statement that two events occur in the same place, he did not succeed in objectively distinguishing rest of a mass point from all other possible motions, but only motion in a straight line with uniform velocity, the so-called uniform translation, from all other motions. Again, has the statement that two events occur at the same time (but at different places, say here and on Sirius) objective meaning? Until Einstein, people said yes. The basis of this conviction is obviously people's habit of considering an event as happening at the moment when they observe it. But the foundation of this belief was long ago shattered by Olaf Roemer's discovery that light propagates not instantaneously but with finite velocity. Thus one came to realize that in the four-dimensional continuum of space-time only the coincidence of two world points, 'here-now,' or their immediate vicinity has a directly verifiable meaning. But whether a stratification of this four-dimensional continuum in three-dimensional layers of simultaneity and a cross-fibration of one-dimensional fibers, the world-lines of points resting in space, describe objective features of the world's structure became doubtful. What Einstein did was this: without bias he collected all the physical evidence we have about the real structure of the four-dimensional space-time continuum and thus derived its true group of automorphisms. It is called the Lorentz group after the Dutch physicist H. A. Lorentz who, as Einstein's John the Baptist, prepared the way for the gospel of relativity. It turned out that according to this group there are neither invariant layers of simultaneity nor invariant fibers of rest. The light cone, the locus of all world-points in which a light signal given at a definite world-point *0*, 'here-now,' is received, divides the world into future and past, into that part of the world which can still be influenced by what I do at *0* and the part which cannot. This means that no effect travels faster than light, and that the world has an objective causal structure described by these light cones issuing from every world point *0*. Here is not the place to write down the Lorentz transformations and to sketch how special relativity theory with its fixed causal and inertial structure gave way to general relativity where these structures have become flexible by their interaction with matter. I only want to point out that it is the inherent symmetry of the four-dimensional continuum of space and time that relativity deals with.

We found that objectivity means invariance with respect to the group of automorphisms. Reality may not always give a clear answer to the question what the actual group of automorphisms is...

Symmetry and conservation laws

EUGENE P. WIGNER

Symmetry and invariance considerations, and even conservation laws, undoubtedly played an important role in thinking of the early physicists, such as Galileo and Newton, and probably even before them. However, these considerations were not thought to be particularly important and were articulated only rarely....

This situation changed radically, as far as the invariance of the equations is concerned, principally as a result of Einstein's theories. Einstein articulated the postulates about the symmetry of space, that is, the equivalence of directions and of different points of space, eloquently.[1] He also re-established, in a modified form, the equivalence of coordinate systems in motion and at rest. As far as the conservation laws are concerned, their significance became evident when, as a result of the interest in Bohr's atomic model, the angular momentum conservation theorem became all-important. Having lived in those days, I know that there was universal confidence in that law as well as in the other conservation laws. There was much reason for this confidence because Hamel, as early as 1904, established the connection between the conservation laws and the fundamental symmetries of space and time.[2] Although his pioneering work remained practically unknown, at least among physicists, the confidence in the conservation laws was as strong as if it had been known as a matter of course to all. This is yet another example of the greater strength of the physicist's intuition than of his knowledge.

Since the turn of the century, our attitude toward symmetries and conservation laws has turned nearly full circle. Few articles are written nowadays on basic questions of physics which do not refer to invariance postulates, and the connection

Note. Extract from: 1967, *Symmetries and Reflections*, Bloomington, IN: Indiana University Press, pp. 14–27.

[1] See, for instance, his semipopular booklet *Relativitätstheorie*, Braunschweig: Friedr. Vieweg und Sohn, various editions, 1916–1956.

[2] G. Hamel (1904), *Zeitschrift für Angewandte Mathematik und Physik*, **50**, 1; F. Engel (1916), *Königliche Gesellschaft der Wissenschaften zu Göttingen*, p. 270.

between conservation laws and invariance principles has been accepted, perhaps too generally.[3] In addition, the concept of symmetry and invariance has been extended into a new area – an area where its roots are much less close to direct experience and observation than in the classical area of space-time symmetry. It may be useful, therefore, to discuss first the relations of phenomena, laws of nature, and invariance principles to each other. This relation is not quite the same for the classical invariance principles, which will be called geometrical, and the new ones, which will be called dynamical. Finally, I would like to review, from a more elementary point of view than customary, the relation between conservation laws and invariance principles.

Events, laws of nature, invariance principles

The problem of the relation of these concepts is not new; it has occupied people for a long time, first almost subconsciously. It may be of interest to review it in the light of our greater experience and, we hope, more mature understanding.

From a very abstract point of view, there is a great similarity between the relation of the laws of nature to the events on one hand, and the relation of symmetry principles to the laws of nature on the other. Let me begin with the former relation, that of the laws of nature to the events.

If we knew what the position of a planet will be at any given time, there would remain nothing for the laws of physics to tell us about the motion of that planet. This is true also more generally: if we had a complete knowledge of all events in the world, everywhere and at all times, there would be no use for the laws of physics, or, in fact, of any other science. I am making the rather obvious statement that the laws of the natural sciences are useful because without them we would know even less about the world. If we already knew the position of the planet at all times, the mathematical relations between these positions which the planetary laws furnish would not be useful but might still be interesting. They might give us a certain pleasure and perhaps amazement to contemplate, even if they would not furnish us new information. Perhaps also, if someone came who had some different information about the positions of that planet, we would more effectively contradict him if his statements about the positions did not conform with the planetary laws – assuming that we have confidence in the laws of nature which are embodied in the planetary law.

Let us turn now to the relation of symmetry or invariance principles to the laws of nature. If we know a law of nature, such as the equations of electrodynamics, the knowledge of the subtle properties of these equations does not add anything to the

3 See the present writer's article (1954), *Progress in Theoretical Physics*, **11**, 437; also Y. Murai (1954), *Progress in Theoretical Physics*, **11**, 441; and more recently D. M. Greenberg (1963), *Annals of Physics*, **25**, 290.

content of these equations. It may be interesting to note that the correlations between events which the equations predict are the same no matter whether the events are viewed by an observer at rest, or an observer in uniform motion. However, all the correlations between events are already given by the equations themselves, and the aforementioned observation of the invariance of the equations does not augment the number or change the character of the correlations.

More generally, if we knew all the laws of nature, or the ultimate law of nature, the invariance properties of these laws would not furnish us new information. . . .

Geometrical and dynamical principles of invariance

What is the difference between the old and well-established geometrical principles of invariance, and the novel, dynamical ones? The geometrical principles of invariance, though they give a structure to the laws of nature, are formulated in terms of the events themselves. Thus, the time-displacement invariance, properly formulated, is: the correlations between events depend only on the time intervals between the events, not on the time at which the first event takes place. If P_1, P_2, P_3 are positions which the aforementioned planet can assume at times t_1, t_2, t_3, it could assume these positions also at times $t_1 + t, t_2 + t, t_3 + t$, where t is quite arbitrary. On the other hand, the new, dynamical principles of invariance are formulated in terms of the laws of nature. They apply to specific types of interaction, rather than to any correlation between events. Thus, we say that the electromagnetic interaction is gauge invariant, referring to a specific law of nature which regulates the generation of the electromagnetic field by charges, and the influence of the electromagnetic field on the motion of the charges.

It follows that the dynamical types of invariance are based on the existence of specific types of interactions. . . .

Geometrical principles of invariance and conservation laws

Since it is good to stay on *terra cognita* as long as possible, let us first review the geometrical principles of invariance. These were recognized by Poincaré first, and I like to call the group formed by these invariables the Poincaré group.[4] The true meaning and importance of these principles were brought out only by Einstein, in his special theory of relativity. The group contains, first, displacements in space and time. . . .

[4] H. Poincaré (1905), *Comptes Rendus*, **140**, 1504; (1906), *Rendiconti del Circolo Matematico di Palermo*, **21**, 129.

The second symmetry is not at all as obvious as the first one: it postulates the equivalence of all directions....

The last symmetry – the independence of the laws of nature from the state of motion in which it is observed as long as this is uniform – is not at all obvious to the unpreoccupied mind. One of its consequences is that the laws of nature determine not the velocity but the acceleration of a body: the velocity is different in coordinate systems moving with different speeds; the acceleration is the same as long as the motion of the coordinate systems is uniform with respect to each other. Hence, the principle of the equivalence of uniformly moving coordinate systems, and their equivalence with coordinate systems at rest, could not be established before Newton's second law was understood; it was at once recognized then, by Newton himself....

This will conclude the discussion of the geometrical principles of invariance. You will note that reflections which give rise *inter alia* to the concept of parity were not mentioned, nor did I speak about the apparently much more general geometric principle of invariance which forms the foundation of the general theory of relativity. The reason for the former omission is that I will have to consider the reflection operators at the end of this discussion. The reason that I did not speak about the invariance with respect to the general coordinate transformations of the general theory of relativity is that I believe that the underlying invariance is not geometric but dynamic. Let us consider, hence, the dynamic principles of invariance.

Dynamic principles of invariance

When we deal with the dynamic principles of invariance, we are largely on *terra incognita.* Nevertheless, since some of the attempts to develop these principles are both ingenious and successful, and since the subject is at the center of interest, I would like to make a few comments. Let us begin with the case that is best understood, the electromagnetic interaction.

In order to describe the interaction of charges with the electromagnetic field, one first introduces new quantities to describe the electromagnetic field, the so-called electromagnetic potentials. From these, the components of the electromagnetic field can be easily calculated, but not conversely. Furthermore, the potentials are not uniquely determined by the field; several potentials (those differing by a gradient) give the same field. It follows that the potentials cannot be measurable, and, in fact, only such quantities can be measurable which are invariant under the transformations which are arbitrary in the potential. This invariance is, of course, an artificial one, similar to that which we could obtain by introducing into our equations the location of a ghost. The equations then must be invariant with respect to changes of the coordinate of that ghost. One does not see, in fact, what good the introduction of the coordinate of the ghost does.

So it is with the replacement of the fields by the potentials, as long as one leaves everything else unchallenged. One postulates, however, and this is the decisive step, that in order to maintain the same situation, one has to couple a transformation of the matter field with every transition from a set of potentials to another one which gives the same electromagnetic field. The combination of these two transformations, one on the electromagnetic potentials, the other on the matter field, is called a gauge transformation. Since it leaves the physical situation unchanged, every equation must be invariant thereunder. This is not true, for instance, of the unchanged equations of motion, and they would have, if left unchanged, the absurd property that two situations which are completely equivalent at one time would develop, in the course of time, into two distinguishable situations. Hence, the equations of motion have to be modified, and this can be done most easily by a mathematical device called the modification of the Lagrangian. The simplest modification that restores the invariance gives the accepted equations of electrodynamics which are well in accord with all experience.

Let me state next, without giving all the details, that a similar procedure is possible with respect to the gravitational interaction. . . .

We have seen before that the operations of the geometric symmetry group entail conservation laws. The question naturally arises whether this is true also for the operations of the dynamic symmetry groups. Again, there seems to be a difference between the different dynamic invariance groups. It is common opinion that the conservation law for electric charge can be regarded as a consequence of gauge invariance, i.e. of the group of the electromagnetic interaction. On the other hand, one can only speculate about conservation laws which could be attributed to the dynamic group of general relativity. . . .

Events, laws of nature, and invariance principles

EUGENE P. WIGNER

The [geometric] symmetry principles . . . are those of Newtonian mechanics or the special theory of relativity. One may well wonder why the much more general, and apparently geometrical, principles of invariance of the general theory have not been

Note. Extract from: 1967, *Symmetries and Reflections*, Bloomington, IN: Indiana University Press, pp. 38–50.

discussed. The reason is that I believe, in conformity with the views expressed by V. Fock,[5] that the curvilinear coordinate transformations of the general theory of relativity are not invariance transformations in the sense considered here. These were so-called active transformations, replacing events $A, B, C, \ldots$ by events $A', B', C', \ldots$, and unless active transformations are possible, there is no physically meaningful invariance. However, the mere replacement of one curvilinear coordinate system by another is a 'redescription' in the sense of Melvin;[6] it does not change the events and does not represent a structure in the laws of nature. This does not mean that the transformations of general relativity are not useful tools for finding the correct laws of gravitation; they evidently are. However, ... the principle which they serve to formulate is different from the geometrical principles considered here; it is a dynamical invariance principle.

[5] V. A. Fock (1959), *The Theory of Space, Time and Gravitation*, New York: Pergamon Press. The character of the postulate of invariance with respect to general coordinate transformations as a geometrical invariance had already been questioned by E. Kretschman (1917), *Annalen der Physik*, **53**, 575.

[6] M. A. Melvin (1960), *Review of Modern Physics*, **32**, 477.

3

On continuous symmetries and the foundations of modern physics

CHRISTOPHER A. MARTIN

> As far as I see, all a priori statements have their origin in symmetry.
>
> *H. Weyl*

> Symmetry principles have moved to a new level of importance in this century and especially in the last few decades: there are symmetry principles that dictate the very existence of all the known forces of nature.
>
> *S. Weinberg*

1 Introduction

It is difficult to overstate the significance of the concept of symmetry in its many guises to the development of modern physics. Indeed one could reasonably argue that twentieth-century physics with its pillar achievements of successful physical theories of spacetime/gravitation and the electromagnetic and nuclear interactions merits calling that century the 'Century of Symmetry'. For symmetry played a key part in each of these developments. Of particular significance are symmetries described by continuous (Lie) groups of transformations. Such symmetry groups play a central role both in our current understanding of fundamental physics and in various attempts to go beyond this understanding. Today such continuous symmetries are often assigned a, if not *the*, fundamental role in the worldview of modern physics. The precise nature of this role, though, is not entirely unambiguous. Just what significance – physical, philosophical, and otherwise – is to be ascribed to the preeminent role of such symmetry groups in fundamental physics?

This paper aims to provide a brief and necessarily selective survey of the historical development of the place of continuous symmetries in physical theory. I discuss the central role that symmetry came to play with the development of relativity theory at the beginning of the twentieth century. After that, I discuss the central significance of symmetry considerations in quantum theory. My primary focus, though, is on the symmetry-dictates-interaction paradigm characteristic of the gauge field theory programme, which has proved incredibly successful in describing the fundamental physical interactions. It is with this last development that considerations of symmetry have come to occupy the Platonic heights that they are often ascribed in current physics.

In discussing the role of symmetries and symmetry principles in physical theory I touch upon a number of interesting, more 'philosophical' issues. To what heights precisely have symmetry principles really risen, and on the basis of what? Have these come to occupy the realm of the unassailable *a priori* or are they ultimately of mere heuristic value? What, if anything, can be said to be the 'physical content' of the gauge symmetry principle purportedly lying at the heart of successful modern theory? Do such symmetry principles capture/reflect deep ontological features of the world or are they rather artifacts of our particular formal representation? A recurring theme/concern here is whether and, if so, how we are to reconcile the canonical view of gauge invariance as relating to a non-physical, formal redundancy in theory with the general belief that gauge symmetry is in fact of some deep physical significance.

2 The 'Century of Symmetry'

2.1 Groups, invariants, and conservation

Although groups made their appearance in physics at the beginning of the nineteenth century, it was not until the detailed study of group representations in the 1920s which accompanied the developing quantum theory (discussed below) that groups truly made their way into a large part of work-a-day physics. There were, of course, major developments regarding the role of symmetry in physical theory before this. Besides the development of relativity theory, discussed in the next subsection, one development of seminal significance was the formulation of what are now called Noether's theorems.[1] Noether's work incorporated three areas of mathematical physics (Kastrup, 1983, p. 115): (i) algebraic and differential invariant theory; (ii) Riemannian geometry and variational calculus in the contexts of general relativity, mechanics, and field theory; (iii) group theory and, in particular, the methods developed by Lie for solving or reducing differential equations by appeal to their invariance groups. As these areas remain central to modern physics, Noether's theorems have come to play a central role in our current physical worldview.

Noether's theorems grew out of her work while in Göttingen at the invitation of Hilbert and Klein. Klein had many years earlier advanced his influential *Erlangen* programme which concerned the application of group theory to geometry, specifically classification of geometries according to their characteristic invariants under groups of transformations. Both Klein and Hilbert had been interested in the

[1] See Kastrup (1983) for a detailed history of the genesis and reception of Noether's work. See also Rowe (1999).

connection between invariances and conservation laws – Hilbert, in particular, with regard to his work on formulating a relativistic theory of gravitation. Noether's work is now considered as providing the first general treatment of this relationship.

The formal context of Noether's theorems is Lagrangian field theory wherein equations of motion are obtained through a variational procedure (Hamilton's principle) from the action integral, $S = \int \mathcal{L}\, dx$. Noether's theorems are discussed in detail in Earman (2002) and Brading and Brown (this volume). Noether's two theorems concern the invariances of the action integral under continuous groups of transformations of the fields. The first theorem shows that if the integral is invariant under an r-parameter (r finite) Lie group $\mathcal{G}_r$ of transformations then, when the equations of motion are satisfied, there exist r conserved 'currents'. As a familiar example, this theorem relates the invariance of the action under spatial and temporal translations and rotations to the conservation of linear momentum, energy, and angular momentum respectively. The second theorem shows that if the action is invariant under an infinite dimensional (i.e. specified by *r functions*) group $\mathcal{G}_{\infty r}$ of transformations then there will be r identities holding between the Euler–Lagrange equations of motion derived from the action. These identities are commonly called generalized Bianchi identities. In electrodynamics, for example, the generalized Bianchi identities associated with the infinite dimensional group of electromagnetic gauge transformations are just the homogeneous Maxwell field equations. There is another result of Noether's which is perhaps equally as important to understanding the role that invariance plays in physical theory. If the action admits both $\mathcal{G}_{\infty r}$ and, as a rigid subgroup, $\mathcal{G}_r$ as symmetries, then the relations of Noether's first theorem, from which the conservation laws are derived, are consequences of the generalized Bianchi identities associated with the second.

The Bianchi identities are tied to an underdetermination in the dynamical evolution of the field(s), there being fewer independent equations of motion than dynamic variables.[2] This is where gauge symmetry enters the story. The gauge transformations are taken to relate physically equivalent situations – distinct situations evolved from a single set of initial data – and rescue determinism. Lagrangians with such local symmetries, so-called singular Lagrangians, lead, in the canonical Hamiltonian framework, to so-called constrained Hamiltonian systems. Appeal to the constrained Hamiltonian framework is an important part of treating theories with local gauge symmetry, as the Hamiltonian framework is central to (i) the canonical means of formulating quantum field theories and, relatedly, (ii) sorting out the dynamical degrees of freedom. The treatment of gauge theories in the canonical Hamiltonian formalism is discussed in Earman (this volume, Part I) and Castellani (this volume, Part IV).

[2] Note that, at the same time, the equations of motion are overdetermined in the sense that there are constraints on the sets of admissible initial data.

With the advantage of hindsight, we see that the importance of Noether's work is the establishment of systematic links between (i) invariance of the action under groups of transformations and conserved quantities, these commonly being the observables with which physics directly deals (in the quantum context, they in fact generate the symmetry transformations); and (ii) invariance of the action under infinite dimensional groups and certain specific structural properties of the associated theories.[3] This link between invariance(s) of the action and other detailed features of physical theory has come to be one of the defining features of modern physical (gauge) theory.

2.2 Relativity, spacetime and gravitation

Not only is the Special Theory of Relativity (STR) tied directly to considerations of symmetry at the physical and formal levels, but Albert Einstein's formulation of STR is commonly taken to signal an important shift in the overall relation of symmetry to physical theory. When one speaks of symmetry 'principles' this refers to a view of symmetry as lying at the heart of physical theory. Einstein's work on special relativity with its central role for symmetry principles is understood to be the birth of this way of thinking about the significance of symmetry to physical theory.[4] With Einstein's work we see a shift to starting from symmetries and then deducing from these various physical consequences, laws, etc. This is in contrast to taking symmetries to be interesting but after-the-fact features of some known law(s). This elevation of symmetry principles fits squarely with Einstein's well-known regard for theories of principle, such as Euclidean geometry and thermodynamics.

Einstein's principle of relativity, one of the centrepieces of his 1905 formulation of STR, requires that it be impossible to detect one's state of inertial motion. That is, experimentation in one inertial frame must uncover the same laws of physics as experimentation done in any other inertial frame. At base, this has the shape of a symmetry principle – the relevant transformations being those relating inertial frames of reference and the invariant being the (form of the) physical laws. In order to reconcile this invariance requirement with the known constancy of the speed of light, Einstein was forced to consider a new group of transformations relating the space and time coordinates of inertial observers. This group is now termed the inhomogeneous Lorentz group (or Poincaré group, since Poincaré had earlier written down the transformations) .

The work of Minkowski elucidated the nature of the symmetry at the heart of STR. In the spirit of Klein's *Erlangen* programme, he considered the geometry

[3] For further discussion, see Brading and Brown (this volume).

[4] See Yang (1982), Wigner (1991).

associated with the transformations advanced by Einstein. Minkowski showed that these transformations and associated invariants characterized a new and arguably even simpler geometry of the physical world – space and time were now part of a single four-dimensional geometry, (Minkowski) spacetime. With this, symmetry and its mathematical counterpart, group theory, became fixtures in fundamental physics and in its relativistic revamping in the first part of the twentieth century.

As typically construed,[5] the relativity principle at the heart of STR is taken to be grounded in certain regularities in the underlying structure of physical events. Formally, the symmetry transformations relating equivalent inertial observers are taken to be symmetries of the absolute, i.e. non-dynamical, special relativistic Minkowski spacetime – i.e. elements of the symmetry (isometry) group of the Minkowski metric.[6] Some, though, have raised issues with construing the relativity principle in this way. They argue that the common formulation of the principle in terms of global (symmetry) transformations of the entire spacetime (and its contents) lacks significant empirical content and, relatedly, that it overlooks that the basic empirical relativity principle is in fact defined only for isolated subsystems of the universe.[7]

Shortly after the formulation of STR, Einstein turned his attention to developing a relativistic theory of gravitation, and considerations of symmetry and covariance once again figured prominently in his work. Symmetry and covariance are all tied up historically.[8] One of Einstein's primary starting points was the familiar equivalence of gravitational and inertial masses. Einstein was led to advance his principle of equivalence – the equivalence between the effects of homogeneous gravitational fields and uniformly accelerated motion. Towards accommodating this equivalence into a relativistic theory of the gravitational field, Einstein came to consider theories with wider covariance than that characteristic of STR. Through a myriad of interesting twists and turns, Einstein arrived at a principle of general covariance, the form invariance of the laws of nature under arbitrary smooth coordinate transformations.[9] And, with considerations of symmetry playing a significant role, after much struggle Einstein formulated generally covariant gravitational field equations, which relate the distribution of matter and energy in spacetime to the presence of

5 See for example Anderson (1967) and Friedman (1983).

6 Intuitively, the absolute, or non-dynamical objects (fields, etc.) are those that are unaffected by interaction with the other objects of the theory, and are the same in each allowed model of the theory. Dynamical objects, on the other hand, interact with other objects, and vary from model to model.

7 See Brown and Sypel (1995) and Budden (1997), and references therein.

8 The relevant covariance in the context of STR is of course Lorentz (or Poincaré) covariance. I discuss the matter of invariance vs. covariance below. Historically, matters of symmetry/invariance and covariance were sometimes run together.

9 For discussion of the foundations of spacetime theories and, in particular, the place of general covariance and of symmetry generally, see Anderson (1967), Friedman (1983), Earman (1989), Norton (1993 and this volume), and Brown and Sypel (1995).

spacetime curvature.[10] Importantly, the, now, dynamical spacetime metric itself comes to represent/encode the properties of the (dynamical) gravitational field. Relatedly, the familiar equivalence of inertial and gravitational masses is seen to flow from these two in fact being one and the same thing.

With the formulation of the General Theory of Relativity (GTR), symmetry played a key role in developing a new theory of a specific interaction, gravity. Though considerations of symmetry had played a role in the development and understanding of both STR and GTR, it was pointed out early on that there are important differences between these roles. As already mentioned, the symmetry at the heart of STR is grounded in the invariances (isometries) of the Minkowski spacetime. As such, the symmetry applies to all interactions taking place in the spacetime. In contrast to the symmetry or *invariance* requirement in STR, the principle in GTR is most often presented as strictly speaking a *covariance* requirement. General covariance, it is argued, is not tied to any geometrical regularity of the underlying spacetime, but rather to the form invariance (covariance) of laws under arbitrary smooth coordinate transformations. Einstein is at least partly to blame for some initial confusion regarding this distinction since he sometimes said that GTR represented a generalization of STR, that is was a 'more' relativistic or symmetric theory.[11] Suffice it to say that considerations of symmetry/invariance and covariance are historically intertwined. Today we take it that Einstein sometimes spoke of covariance as a symmetry property in ways he perhaps should not have done.

The ascription of *any* physical content to a formal covariance requirement such as general covariance came under fire early on from Kretschmann. There is a long history of agreement and disagreement with Kretschmann's claim. Some of these are discussed in Norton (this volume). Notwithstanding Kretschmann's objection, the place of general covariance in the formulation of GTR, as a (if not *the*) characteristic feature of the theory, is often taken as the first example of the appeal to a new sort of symmetry principle, so-called *local* symmetry principles. More on such principles below. It is also worth mentioning here a related point. The contrasts between the 'symmetries' at the hearts of STR and GTR are an example of what gets codified in Wigner's distinction between 'geometrical-invariances' and 'dynamical-invariances'. I discuss this distinction further in section 4.1.

Weyl (1918a), which contained the first appearance of the concept of local 'gauge' symmetry, is a direct outgrowth of Einstein's work on GTR.[12] Weyl's

[10] The story here is *much* more complicated and interesting. In fact, Einstein wrestled with the implications of general covariance, at one point becoming convinced that he should drop the requirement in founding a relativistic theory of gravity. See Stachel (1989a; 1989b).

[11] cf. note 7, above.

[12] See also Weyl (1918b). Pauli (1921) contains a nice discussion of Weyl's 1918 work.

starting point, ostensibly mathematical, was to consider a sort of generalization of Einstein's work. He contended that the Riemannian geometry of Einstein's GTR retained one vestige of Euclidean geometry-at-a-distance (*ferngeometrisches*). Although parallel transport of vectors was in general non-trivial due to spacetime curvature, the geometry of GTR did allow direct comparison of distant magnitudes. Weyl considered how one might generalize the Riemannian geometry of GTR, removing this methodological inconsistency (*Inkonsequenz*), thereby founding a purely infinitesimal geometry.[13] Specifically, Weyl considered the geometry that results from allowing arbitrary rescalings of the local unit of length – formally rescalings of the spacetime metric – at each point of spacetime. This change of unit length, or change of 'gauge', is the origin of the terminology 'gauge invariance'.[14]

What Weyl found was that the more general geometry resulting from admitting such local changes of 'gauge' apparently described not only gravity but also electromagnetism. In short, the field needed to maintain invariance given the freedom to perform local gauge transformations, part of a generalized affine connection, was formally identical to the field used to represent the electromagnetic potential. Under the local changes of gauge, this field transforms as does the electromagnetic potential under electromagnetic gauge transformations. With this field, specifically a generalization of the Riemannian curvature built from it, Weyl constructed a gauge-invariant action for the unified theory. What Weyl took to be the strongest argument in favour of his new theory, however, was that electric charge conservation followed from the local gauge invariance in precisely the same characteristic way that energy–momentum conservation followed from coordinate invariance.[15] What Weyl refers to is that the conservation laws follow in two distinct ways in theories with local symmetries. This is traceable to the Bianchi identities holding between the coupled equations of motion, which, in turn, are due to the local gauge invariance of the action. Thus, Weyl took it that via the demand of local gauge invariance he was led to the (geometrical) unification of two previously distinct interactions, and that this brought out important similarities between the two theories.

Given Weyl's starting point (see Ryckman, this volume), it is perhaps not surprising that his theory faced immediate challenges from physics. Showing deference to Weyl's mathematical ingenuity, Einstein was quick to point out that nature in fact provides us with standard rods and clocks in the form of atoms with definite spectral lines. This was in contrast to the basic assumption in Weyl's theory that there was no absolute physical significance to be ascribed to the spacetime interval.

[13] See Ryckman (this volume) for further discussion of the origins of local symmetry in Weyl's work.
[14] Originally '*MasstabInvarianz*' and later '*EichInvarianz*'.
[15] See Brading (2002) for a detailed discussion of Weyl's derivation of the conservation laws.

Einstein argued that Weyl's theory could not be correct since it predicts that, owing to the (in general) path-dependent gauge factor, neighbouring identical atoms would show different spectral lines depending upon their respective histories.[16] In the next section I will continue with the historical account of the rise of gauge symmetry.

2.3 Symmetry and quantum theory

In the 1920s, on the back of important work by the likes of Rutherford, Bohr, and many others, physicists began to develop atomic physics, in the process giving birth to a new theory of matter and radiation – quantum theory.[17] A crucial part of the rapid successes in this realm was the application of group-theoretic methods to various central physical problems. The reason for the utility (even necessity) of group theory in the development of quantum mechanics has chiefly to do with the radically different notion of the state of a physical system in the new theory. In classical mechanics, the state space is typically represented by a differentiable manifold, and observables by real-valued functions defined on the manifold. The states of quantum mechanical systems, on the other hand, are represented by vectors in an (infinite dimensional) vector space, and observables are linear mappings of these states onto one another. As such, quantum states can be added to one another yielding new states. The various symmetry groups carve up the quantum state space into subspaces invariant under the action of the associated transformations (alternatively, irreducible representations). The conserved quantities associated with the symmetries in fact 'generate' the associated symmetry transformations. Atomic states can then be characterized by irreducible representations of the various fundamental symmetry groups.

Thus, in the mid to late 1920s group theory was used to show that many empirically determined rules (e.g. spectral formulae, selection rules) were attributable to various symmetries of the physical system under study. For example, certain well-known spectral relations followed directly from the spherical symmetry of the hydrogen atom and the associated invariance of the physics under the rotation group. Hence, group theory provided key insights into the developing atomic theory even before there was a detailed understanding of the dynamics of the electron. This success was so complete that, referring to group theory, Weyl wrote:

> We may well expect that it is just this part of quantum physics which is most certain of a lasting place. . . . *All quantum numbers, with the exception of the so-called principal quantum number, are indices characterizing representations of groups.*[18]

[16] There is more to this story: further argument in favour of his theory from Weyl and further objection by Einstein. See the correspondence and commentary in Einstein (1998).

[17] Pais (1986) provides a detailed account of many of the historical developments discussed here.

[18] Weyl, 1928; Eng. trans. (1950), p. xxi. Emphasis in original.

By the early 1930s group theory had become a staple of the budding physics of fundamental matter. The books by Weyl (1928), Wigner (1931), and van der Waerden (1932) are universally recognized as the canonical works which set the tone and content of much of the development of (quantum) physics in that period.

Towards making contact with our main focus here, note that quantum theory introduced a new treatment of matter and with that a new physical variable to physics. The quantum matter (wave) field was described by a complex-valued function – the field had not only an amplitude but also a *phase*. As I discuss in the next section, this new quantum phase was a key part of a new role for symmetry in fundamental physics.

2.3.1 The ascendancy of symmetry: interaction from symmetry

Weyl (1929) is almost universally considered the origin of modern gauge symmetry principles.[19] While it is sometimes said that Weyl here simply reinterpreted his earlier work, I think it is more correct to see it as substantial modification. The general idea of appealing to some sort of local symmetry requirement in arriving at electromagnetism is preserved, and the formal outline of the derivation remains the same. In his 1929 work, however, the local transformations are imaginary phase transformations of the complex-valued matter field, and the resulting theory is one of the quantum matter field interacting with the electromagnetic field. The demand of invariance under local phase transformations led Weyl to the by then well-known coupling of charged matter to the electromagnetic field.[20] As in his 1918 theory, Weyl attributes a special importance to the fact that the conservation of electric charge followed from the local gauge invariance in the same characteristic way in his theory as does energy momentum conservation from coordinate invariance in GTR. This, again, is related to certain formal features of the theories tied specifically to the local invariance.

Interestingly, Weyl does not stipulate the demand of local gauge (better, phase) invariance as he had in his 1918 theory. Rather he argues that accommodating spinor fields into a general relativistic spacetime requires local phase invariance. In short, Weyl argues that one must use local tetrads in representing spinor matter fields in arbitrary curved spacetimes and that the choice of a local tetrad leaves the spinor field undetermined up to a phase factor. Weyl's reasoning here is now seen to be misguided.[21] As understood today, the phase freedom in quantum electrodynamics

[19] Though Weyl was the first to put all the ingredients together in a nice package, numerous others had drawn attention to the same central idea, among them, London, Fock, Schrödinger, and Dirac. See O'Raifeartaigh (1997) and Vizgin (1994, esp. chapters 3 and 6).

[20] In section 3, I look more closely at the details of this argument.

[21] See O'Raifeartaigh (1997).

(QED) is not related in this way to representing spinors – which, anyway, are not inherently quantum mechanical in the first place – but to the representing of matter via elements of a complex vector space.[22] Despite the questionable motivation for the central demand of local phase invariance, what Weyl does arrive at in his 1929 work is the archetype of the modern gauge symmetry principle: deriving the form of interactions from the demand of invariance under a group of local transformations – in this case, electromagnetism from the demand of invariance under the group of local ($U(1)$) phase transformations.

Jumping ahead in time, after the successful formulation of a fully renormalizable theory of QED, but in the face of many newly discovered particles and the failure of the standard methods for constructing theories of the new particles and their interactions, the 1950s and 1960s saw a low point in the quantum field theory programme – many abandoned its central tenets (even if only temporarily). While some effectively abandoned the appeal to symmetries and group theory, others turned to the further study of symmetries and conservation laws. It was in this context that the idea of local gauge invariance again emerged, in the context of describing the nuclear interactions. The seminal work here is Yang and Mills (1954) – today, the gauge theories central to much of current physics are often simply called Yang–Mills theories.[23] Yang and Mills drew their inspiration from Heisenberg's discussion of the suggestive similarities between the proton and neutron. Heisenberg had hypothesized that the neutron and the proton were two states of the same particle, the nucleon.[24] The idea of the underlying $SU(2)$ symmetry was that, in the absence of electromagnetism, the orientation of the isotopic spin would be of no physical significance – the neutron and the proton would be identical. Such transformations have come to be denoted 'internal' transformations since they are not tied to ('external') spacetime properties but rather to the intrinsic properties of the particles.

The starting point for Yang and Mills was consideration of the $SU(2)$ symmetry of nucleon–nucleon interactions. Yang and Mills took it that this symmetry was subject to an important limitation which should be lifted. This is where local gauge invariance enters. With a *global* $SU(2)$ symmetry, Yang and Mills argue, once one chooses what to call a proton/neutron at one point, one is no longer free to make independent choices at other spacetime points. Yang and Mills took this limitation to be inconsistent with the idea of a *local* field theory. Lifting this limitation through

[22] Consider also that, presumably, the phase freedom is to remain in the case of zero spacetime curvature where the appeal to local tetrads (and, consequently, local phase invariance, as Weyl claims) is no longer necessary.

[23] See O'Raifeartaigh (1997) for discussion of related developments, including the early role of five-dimensional theories, and related and independent work by W. Pauli and R. Shaw.

[24] That the two were identical in this way was suggested by their similar masses and by the near equality of the interaction and self-interaction strengths. The differences were thought to arise from their different electromagnetic properties.

demanding a *local* gauge (or isotopic spin, or $SU(2)$-phase) invariance, they were led to introduce a new field, an $SU(2)$ gauge field. This field bore the same relation to the demand of local $SU(2)$ invariance as did the electromagnetic field to the demand of local $U(1)$ invariance in Weyl's 1929 theory.[25] The appeal to a non-Abelian (i.e. non-commuting) symmetry group has the result that the new field satisfies non-linear equations of motion, because the gauge field carries isotopic spin and is thus self-interacting. Considering the quantization of the gauge field, Yang and Mills indicated a potential problem with putting their considerations to actual use. Though they discuss both, the question as to the mass of the $SU(2)$ gauge field and the related question of the renormalizability of the theory went unanswered in their paper. Such answers were, however, important in determining whether their theory matched experiment. Indeed, these and related matters were to plague the application of gauge theory to the nuclear interactions for some time.

Utiyama, working independently, further generalized this line of thought, cementing a 'general rule for introducing a new field in a definite way when there exists some conservation law . . . or there is a Lie group . . . under which the system is invariant' (Utiyama, 1956, p. 1601) Additionally, Utiyama discussed casting gravitation in this framework.[26] Utiyama's treatment of gauge theories was thus the most general and, as regards physics, the most comprehensive, (O'Raifeartaigh, 1997, p. 208). While the particulars of Yang and Mills' and Utiyama's papers pertaining to the description of the nuclear interactions turned out to be unsuccessful,[27] they represented important generalizations of Weyl's 1929 work. Even after nearly fifty years of intense development in physics, these works still provide the paradigm examples of describing the nuclear interactions through appeal to local symmetry principles.

The foregoing is only a relatively small (though central) part of the rise of gauge symmetry principles and gauge field theories. The generalizations of gauge field theories (GFTs) to non-Abelian symmetry groups by Yang and Mills and Utiyama nevertheless had no immediate home in physics. Most notably, the theories remained unusable due to the lurking 'mass problem': local gauge symmetry apparently required that the gauge field(s) be massless and, hence, long-range, in contradiction to the known short range of the strong interactions. Though there was initial appeal to GFTs given various observed conservations and similarities between certain particle

[25] One can take Yang and Mills' work to effectively generalize Weyl's through adding a matrix index to the electromagnetic (gauge) potential.

[26] Whether and in what way gravitation is a gauge theory in the Yang–Mills sense is a vexed topic that I do not broach here. See Trautman (1980), Ivanenko and Sardanashvily (1983), Weinstein (1999), and Earman (this volume, Part I).

[27] Isospin symmetry is no longer seen as a fundamental symmetry but rather as an incidental symmetry: a consequence of the underlying $SU(3)$ colour symmetry of quantum chromodynamics and the fact that there are two light quarks.

interactions in the rapidly emerging experimental data, numerous other significant developments had to occur before GFTs could be put to work. The formulation of successful (quantum) GFTs occurred during an intense period of development in theoretical physics.[28] Briefly, the key to solving the mass problem proved to be an idea imported from solid-state physics – spontaneous symmetry breaking. The idea was that the gauge particles (bosons) associated with the local symmetry requirement would acquire mass through the so-called 'Higgs mechanism'. This mechanism gives the gauge fields mass without spoiling the gauge invariance of the underlying Lagrangian. Spontaneous symmetry breaking is discussed in Part III of this volume.

One of the most important elements of the development of successful gauge theories of the nuclear interactions was the careful study of the relationship between renormalizability and local gauge symmetry. Appealing to the spontaneous symmetry-breaking mechanism, Weinberg and Salam independently developed a unified theory of the weak and electromagnetic interactions based on the gauge group $SU(2) \times U(1)$. In this theory, what is originally a system of four massless vector bosons becomes, through the spontaneous breakdown of the ground state, a system of one massless particle (the photon), which corresponds to the unbroken $U(1)$ subgroup, and three massive particles (the Z^0 and the $W^{\pm}$) corresponding to the broken part of the group. It was not immediately clear, however, that the theory's renormalizability was not compromised by the spontaneous symmetry breaking.[29] In 1971 't Hooft and Veltman showed that spontaneously broken gauge theories with massless bosons were renormalizable and that, in fact, gauge invariance of the theory was intimately related to the renormalizability of the theories. Renormalizability of the electroweak theory secured, the focus turned to prediction and experimentation. In the following decade, detailed predictions of, for example, the existence and properties of the weak neutral current and the Z^0 and $W^{\pm}$ bosons required by the gauge invariance were confirmed with amazing accuracy.

At the same time, ideas of local symmetry were being applied to developing a theory of the strong interactions – as it turned out, the interaction between particles making up the hadrons (baryons and mesons), quarks. Important here was the determination that non-Abelian gauge theories are asymptotically free, which is directly related to the fact that the effective coupling constant in the theory decreases with energy.[30] And, rather than being a mere peculiarity, this proved promising as a possible

[28] Crease and Mann (1996), Cao (1997; 1999), and Weinberg (1980) contain accounts of these developments.
[29] As Gell-Mann reports (Doncel *et al.*, 1983, p. 551), neither the renormalizability of the massless theory nor of the massive theory had been conclusively settled.
[30] In fact, as was shown in the early to mid-1970s, non-Abelian gauge theories are the only asymptotically free theories in four spacetime dimensions.

explanation for why nucleons, when bombarded with high-energy electrons, behaved as if their constituent particles were essentially free. Moreover, the increase of the coupling strength with distance in such a theory would explain why strongly interacting particles are not produced in isolation – infrared slavery, the other side of asymptotic freedom. These developments culminated in the postulation of an unbroken, internal gauge group, $SU(3)$ (the colour group), relating quark colour multiplets.[31] These quarks interact via the eight massless $SU(3)$ gauge ('gluon') fields – the massless vector fields required by the local $SU(3)$ symmetry (and which also carry colour charge and are, thus, self-interacting) in the Yang–Mills scheme. This led to a successful theory of the strong interactions, quantum chromodynamics (QCD). The developments discussed in this section led to the so-called Standard Model of particle physics based on the gauge group $SU(3) \times SU(2) \times U(1)$, essentially a pasting-together of QCD and the electroweak theory.

Before closing this all-too-brief history let us note that the ascendancy of gauge symmetry as a fundamental fixture of theoretical physics is, historically, to a large extent divorced from consideration of the *a priori* reasonableness, or 'deep physical basis', or 'physical meaning/content' of gauge symmetry principles. Instead, this rise is founded on the hard-fought and impressive success of gauge field theories at the formal and physical level. In the end, the ascendancy of gauge symmetry principles is secured on very pragmatic bases. The eventual successes of gauge field theories of the fundamental interactions cement the heuristic value of local symmetry principles as guides in theorizing.[32] This might seem to constrain any real physical import of gauge principles to the context of discovery. And, along these lines, perhaps we should then count ourselves amazingly fortunate that the 'right' theories just happened to have such a nice structure, i.e. that seen in the theories' tight group-theoretic structure which accompanies the characteristic symmetry/invariance. Notwithstanding this, the 'gauge philosophy' is often elevated and local gauge symmetry principles enshrined. Gauge symmetry principles are regularly invoked in the context of justification, as deep physical principles, fundamental starting points in thinking about why physical theories are the way they are, so to speak. This finds expression, for example, in the prominent current view of symmetry as undergirding our physical worldview in some strong sense. Next, I go on to further consider how literally and in what sense we might understand this undergirding.

[31] Note that $SU(3)$ also characterized the 'eightfold way' of Neéman and Gell-Mann. This was an earlier (and in many ways successful) classification of the hadrons (and resonances) according to an eight-dimensional, non-fundamental representation of $SU(3)$, with the transformations acting on the 'flavour', rather than colour, of the particles.

[32] In the next section, I discuss how the 'reasons' for the success of the gauge field programme might, from our modern standpoint, be construed differently.

3 Getting a grip on the gauge symmetry principle

3.1 Gauge principle – gauge argument – gauge heuristic

The canonical way of understanding the workings or content of local gauge symmetry principles is via the algorithm for producing interacting field theories from the demand of local gauge invariance – the 'gauge argument'.[33] Given the success of the gauge paradigm, some take the argument to reflect the logical order of nature. Symmetry principles as embodied in the argument are then taken to express/reflect deep features of the physical world. In order to make some remarks concerning how literally we can construe the operations of the argument, we first look at a concrete example.[34]

Consider a field Ψ representing electrically charged matter. The free field obeys the Dirac equation which is just the Euler–Lagrange equation(s) for the Lagrangian (density) $\mathcal{L}_{Dirac} = \bar{\Psi}(i\gamma^{\mu}\partial_{\mu} - m)\Psi$.[35] The corresponding action is clearly invariant under so-called 'global' $U(1)$ phase transformations: $\Psi \rightarrow e^{iq\Lambda}\Psi$; $\bar{\Psi} \rightarrow e^{-iq\Lambda}\bar{\Psi}$ with Λ a constant.[36] It follows from Noether's first theorem that when the equations of motion are satisfied there will be a corresponding conserved current.

Consider now 'localizing' these phase transformations, i.e. letting Λ become an arbitrary function of the coordinates $\Lambda(x)$: $\Psi \rightarrow e^{iq\Lambda(x)}\Psi$; $\bar{\Psi} \rightarrow e^{-iq\Lambda(x)}\bar{\Psi}$. As it stands, the free field Lagrangian is clearly not invariant under such transformations, since the derivatives of the arbitrary functions, i.e. $\partial_{\mu}\Lambda(x)$, will not vanish in general. The Lagrangian must be modified if the theory is to admit the local transformations as (variational) symmetries. In particular, we replace the free field Lagrangian with $\mathcal{L}_{interacting} = \bar{\Psi}(i\gamma^{\mu}\partial_{\mu} - m)\Psi - qA_{\mu}\bar{\Psi}\gamma^{\mu}\Psi \equiv \mathcal{L}_{Dirac} - J^{\mu}A_{\mu}$, with $J^{\mu} = q\bar{\Psi}\gamma^{\mu}\Psi$. This current is in fact the conserved current associated with the global $U(1)$ invariance of the interacting theory.[37] Towards securing local invariance we have introduced the field A_{μ}, the *gauge potential*. The particular form of coupling between the matter field and this gauge potential in $\mathcal{L}_{interacting}$ is termed *minimal coupling*.

[33] What I here call the 'gauge argument' is in fact just a convenient label for what is an amalgam of the most common features figuring in many similar such arguments in the physics literature, both popular and technical/text-book. (For an example of the former see Mills (1989) or 't Hooft (1980), and, for the latter, see Aitchison and Hey (1989) or Quigg (1982).) Though there are certainly variations in the way the argument is presented – and, most importantly, in the overall place and significance assigned to the argument relative to other features of gauge theory – what we consider here are the most central elements, the ones to which most presentations make appeal in one way or another.

[34] The following is treated in some more detail in Martin (2002).

[35] Here, I suppress all spinor indices, and $\bar{\Psi}$ is just the Dirac conjugate of Ψ, and γ^{μ} are the usual Dirac matrices.

[36] The action will be invariant, and thus the Euler–Lagrange equations determined from minimizing (extremizing) it covariant, if the Lagrangian is quasi-invariant, i.e. invariant up to an overall divergence, under the transformations.

[37] In general, one must be careful in making this identification since for some types of field the form of the current is modified by the coupling.

This modified Lagrangian is now invariant under the local phase transformations provided that the vector field A_μ is simultaneously transformed according to $A_\mu(x) \to A_\mu(x) - \partial_\mu \Lambda(x)$. This transformation behaviour is, of course, familiar as the covariant analogue of the well-known electromagnetic gauge transformations. This suggests the possibility of viewing the new field A_μ as representing the electromagnetic potential.

Pursuing the idea of viewing the field A_μ as representing the electromagnetic potential, we note that the Lagrangian $\mathcal{L}_{interacting}$ does not yield a fully interacting theory. Varying the Lagrangian with respect to the matter fields yields the latter's equations of motion – the field now being coupled to the A_μ field. But, it remains to add a 'kinetic term' for the A_μ field itself. Such a kinetic term, in effect, imbues the field with its own existence, accounting for the presence of non-zero electromagnetic fields, for the propagation of free photons. The Lagrangian $\mathcal{L}_{kinetic} = \mathcal{L}_{Maxwell} = -\frac{1}{4}F_{\mu\nu}F^{\mu\nu}$, with the *gauge field* $F_{\mu\nu}$ defined as $F_{\mu\nu} := \partial_\mu A_\nu - \partial_\nu A_\mu$, gives (source-free) Maxwell's equations. Putting this all together yields the Lagrangian for the fully interacting theory: $\mathcal{L}_{Maxwell-Dirac} = \mathcal{L}_{interacting} + \mathcal{L}_{kinetic} = \mathcal{L}_{Dirac} - J^\mu A_\mu - \mathcal{L}_{Maxwell}$. The inhomogeneous coupled equations of motion for the gauge field (the electromagnetic field) now follow from varying the full action with respect to A_μ.[38]

Finally, consider that a mass term for the vector field A_μ of standard form $\mathcal{L}_{photon-mass} = \frac{1}{2}m^2 A^\mu A_\mu$ is not gauge invariant. In keeping with the demand of local gauge invariance, the vector field A_μ (i.e. the photon) must then be massless. Local gauge invariance thus necessitates a massless photon.[39]

3.2 *Questioning the gauge logic of nature*

Despite the heuristic success of the gauge argument in its historical context, one must be a little wary of any attempt to read the logical order of nature directly off the argument, through a straightforward, literal reading. Of course one might resist giving it such a reading in the first place (more in section 4.2). Here, though, we proceed along such lines, noting that this argument is often held out as embodying the purportedly deep physical import of local gauge symmetry principles.

[38] The homogeneous field equations follow from the local (gauge) invariance of the action, in fact being just the identities (generalized Bianchi identities) following from Noether's second theorem.

[39] This argument generalizes to fields carrying symmetries associated with arbitrary (in particular, non-Abelian) Lie groups, this yielding further interesting gauge structure. The chief difference is that the non-Abelian gauge group has the result that the gauge field 'generated' in the gauge argument carries its own charge and is thus self-interacting. Specifically, the associated kinetic term necessarily includes self-interactions in the form of a term proportional to the commutator of the (now, matrix-valued) gauge potentials. This commutator, of course, vanishes in the Abelian, electromagnetic case, this corresponding to the photon not carrying electromagnetic charge. The requirement of local gauge invariance similarly determines the form of the (self-)interactions in the non-Abelian case.

First, the initial and all-important demand of local as opposed to global gauge invariance is anything but self-evident, and presumably it must be argued for on some basis. Historically, the arguments surrounding the 'demand' as such are quite thin. The most prevalent form goes back to Yang and Mills' remarks to the effect that 'local' symmetries are more in line with the idea of 'local' field theories.[40] Arguments from a sort of locality, and especially those predicated specifically on the demands of STR (i.e. no communication-at-a-distance),[41] are somewhat suspect, however, and careful treading is needed. Most immediately, the requirement of locality in the STR sense – say, as given by the lightcone structure – does not map cleanly onto to the global/local distinction figuring into the gauge argument – i.e. $\mathcal{G}_r$ vs. $\mathcal{G}_{\infty r}$. Overall, the question of how 'natural', physically, this demand is, is not uncontentious. This is especially so in light of the received view of gauge transformations which maintains that they have no physical significance or counterpart (more below). I will return briefly in the next section to considering possible avenues towards underwriting this initial demand.

For now, let us assume that one can provide some sort of justification for the demand of local gauge invariance. We did not consider above the uniqueness of the minimal modification. This modification is not uniquely dictated by the demand of local gauge invariance. There are infinitely many other gauge-invariant terms that might be added to the Lagrangian if gauge invariance were the only input to the argument. In order to pick out the minimal modification uniquely, we must bring in, besides gauge invariance and knowledge of field theories generally, the requirements of Lorentz invariance, simplicity, and, importantly, renormalizability.[42] The minimal modification is then the simplest, renormalizable, Lorentz and gauge-invariant Lagrangian yielding second-order equations of motion for the coupled system (O'Raifeartaigh, 1979). The point is simply that, in the context of the gauge argument, the requirement of local gauge invariance gets a lot of its formal muscle in combination with other important considerations and requirements.

One might argue that, at the least, other things held fixed, some requirement of formal simplicity selects the minimal modification as the unique gauge-invariant modification. In this way, the demand of local gauge invariance might be seen as going hand-in-hand with a sort of principle of simplicity. While assumptions of simplicity have certainly proved valuable (even necessary) guides in past theorizing, there is no reason to think that they provide unambiguous, let alone infallible, guides in constructing theories and/or in construing any logical order of nature.[43] I suspect that these remarks are not likely to faze a good physicist, who might claim that the

[40] Auyang (1995) contains a more developed argument along these lines.
[41] See for example Ryder (1996, p. 93).
[42] For example, a Pauli term is Lorentz invariant and gauge invariant but not renormalizable.
[43] See Norton (2000) for some critical discussion of considerations of simplicity.

argument requires completion and not critique. My point has been only that the demand of local gauge invariance is not the sole input to the gauge argument, nor is it necessarily the most significant in any strong sense.

Setting aside the issue of the uniqueness of the gauge-invariant minimal coupling, another important point is that, in contrast to how it is often portrayed, one does not strictly speaking 'generate' a new interaction field in running the gauge argument.[44] This gauge field, insofar as it is a physical field, is put in by hand to a large extent. The gauge potentials generated in the gauge argument (in our example, A_μ) form a restricted class of all such A_μ fields – since we start with a free matter field the potentials are of course all gauge transformable to the zero field. Such potentials, however, correspond to zero $F_{\mu\nu}$ fields.[45] An important physical generalization is made then in adding the kinetic term $\mathcal{L}_{Maxwell}$ for the gauge field to the Lagrangian. The generalization is from a *non-physical*, formal coupling of the matter field to trivial gauge fields (since $F_{\mu\nu} \equiv 0$) to the *physical* coupling of the matter field to non-trivial, *dynamical* gauge fields.[46] It is this addition (and the corresponding varying of the full action with respect to the gauge potential(s)) that 'gives physical life' to the field. In making this generalization, one puts in by hand much of the important physics of the fully interacting theory.[47] I believe that this point goes a long way towards explaining the easily acquired illusion of getting more physics out of the gauge argument than one puts in.

All this is not to say that the requirement of local gauge invariance cannot serve as a useful selection criterion in considering modifications of a free theory towards introducing interactions. It can and has served just such a purpose. As we have seen, historically, there was just such a pragmatic basis for the appeal to local symmetry principles. In any event, it is not how a straightforward, literal reading of the gauge argument might portray it: it is not the case that, unaided, the demand of local gauge invariance (i) dictates uniquely the form of the interacting theory or (ii) strictly speaking dictates the existence of, or accounts for, the origin of a new physical gauge field. In order to pick out the correct form of the theory, other considerations must enter. Also, it is not at all clear that these other considerations or requirements are in any sense inferior, conceptually or physically, to that of local gauge invariance.

44 Brown (1999) has drawn attention to these and related issues. See also the discussion in Teller (2000), Lyre (2001), and Healey (1997).

45 Strictly speaking, this is true only locally. There are potential issues in the case of non-trivial global topology as evidenced, for example, in the familiar Aharonov–Bohm effect; see Nounou (this volume).

46 Another issue worth noting is that one can argue that electric charge does not have a firm physical meaning until the physical electromagnetic field is introduced through the kinetic term.

47 I take it that it is in recognition of essentially this same point that Auyang, considering the gauge principle in electrodynamics, remarks: 'It does not stipulate an interaction field but rules against its *a priori* exclusion...' (Auyang, 1995, p. 58). Brown (1999) discusses the step in which the gauge potential is made properly dynamical as being suggested by physical phenomena such as the Aharonov–Bohm effect as well as by an action–reaction principle.

3.3 Other logics of nature

The line of thinking embodied in the gauge argument presupposes a certain logical order to nature in which gauge invariance is privileged in figuring at the very base of our theories as a sort of axiom. There are, however, other ways of thinking about why our theories are the way they are, and some of these have gauge invariance as more of an 'output' than an 'input'. For example, there are arguments to the effect that various consistency requirements, mathematical and/or physical, require theories of, for example, self-interacting spin-one particles to be of Yang–Mills form with its characteristic group properties.[48] Such arguments clearly paint the gauge invariance of physical theory in a different light than does the canonical view.

A more prominent approach that effectively turns the canonical view on its head is that of placing renormalizability (or, alternatively, perturbative unitarity) at the base of fundamental theory. It can be shown that the requirement of renormalizability (respectively, perturbative unitarity) requires that theories have the characteristic form of (spontaneously broken) Yang–Mills gauge theories.[49] As renormalizability can be tied directly to a theory's being well behaved, in the sense that it make sensible predictions for quantities of direct physical interest, one could reasonably argue that gauge invariance is but a feature of the class of well-behaved quantum theories that happen to correctly describe the physics at hand.

Interestingly, renormalizability has itself arguably been superseded in a certain sense. According to the currently prominent effective field theory programme, the familiar renormalizable (quantum) field theories are actually low-energy approximations to some more fundamental underlying theory. Besides the finite number of familiar renormalizable terms, such effective theories (rather, the actions) necessarily contain an infinite number of non-renormalizable terms. However, as long as at high energies there is *some* well-defined underlying local, dynamical theory (e.g. strings, loops, etc.), then at much lower energies these non-renormalizable interactions will be highly suppressed and thus calculationally insignificant, though, strictly speaking, not physically absent.[50] That is, the low-energy 'residue' will in fact 'look like' a renormalizable theory.[51]

[48] See Wald (1986) and Deser (1970; 1987). Weinberg (1995, chapters 5 and 8), in contrast with the flow of the gauge argument, starts from a quantum theory of massless spin-one fields and arrives at the gauge-invariant coupling to matter.

[49] See Froggatt and Nielsen (1991, p. 123), Foerster, Nielsen, and Ninomiya (1980), Cornwall, Levin, and Tiktopoulos (1974), Weinberg (1974a; 1974b), and the references contained therein. See also Weinberg (1965).

[50] Formally, this result goes by the name of the decoupling theorem. Note that, according to this way of thinking, gravity is just such a non-renormalizable interaction – gravity 'shows up' despite being highly suppressed because there are no negative gravitational charges.

[51] Relatedly, Wilczek (2000, p. 4) writes that as only asymptotically free theories make good physical sense, and as the only asymptotically free theories in four spacetime dimensions are Yang–Mills theories, 'the axioms of gauge symmetry and renormalizability are, in a sense, gratuitous. They are implicit in the mere *existence* of non-trivial interacting quantum field theories'.

Such a view would appear to change completely what we take to be 'fundamental.' In light of the effective field theory programme, the appeal of ascribing any deeply fundamental significance to the gauge structure of our theories, especially as resulting from the operations of some deep physical gauge principle, is further diminished.[52] This gauge structure, it could reasonably be argued, is but a direct consequence of (i) the empirical fact that there exist interacting spin-one particles in nature and (ii) the assumption that our theory of such particles is the residue of some more fundamental underlying theory (for example string theory). Under these assumptions, we arrive at the familiar Yang–Mills gauge theories. So, according to this way of thinking, we might avoid altogether any appeal to gauge symmetry principles in describing the shape or content of current theory. Gauge invariance might be taken as an incidental, albeit interesting feature of the only possible theories at current energies.[53, 54]

4 The physical content of the gauge symmetry principle

The history and the nature of relevant physics seems to call out for nothing more than a heuristic reading of the gauge argument and of the role of gauge symmetry principles. Yet one might go on to further consider the question as to the physical content of the demand (or featuring) of local gauge symmetry in successful physical theory. Presumably, the hope is in some way to give or find for gauge symmetry principles some physical 'oomph'. This hope is understandable given the central role often ascribed to these principles and, at the least, their historical role in the development of successful fundamental theories. Also, perhaps with a physically grounded local gauge principle, the gauge argument would be physically underwritten, and possibly construed as something more than a (as it turned out) successful heuristic. In this section, we consider further the received view of gauge invariance and the – I think – associated tension regarding what physical significance to ascribe to gauge symmetry principles. First, I make some brief remarks on some subtleties involved in describing gauge theories.

Aside: On describing gauge theories

Because of the characteristic gauge invariance, describing the basics of formalizing and interpreting gauge theories is a subtle, sometimes tricky business. There are

52 Froggatt and Nielsen (1991, chapter 7) discuss a 'random dynamics' programme wherein one desideratum is the derivation or explanation of (the usually assumed) Lorentz invariance and gauge invariance from underlying randomness at the fundamental level: effectively, a lack of assumption about what is going on.

53 Though symmetry figures prominently into most attempts to go beyond the Standard Model, the search for a new fundamental theory unifying all forces is saddled with the task of determining what, in all likelihood, are radically new ideas and fundamental physical principles governing the theory (see Weinberg, 1999).

54 For further discussion of effective field theories and of the importance of scale considerations, see Castellani (this volume, Part IV).

numerous distinct formalisms for describing gauge theories.[55] Moreover, there are associated interpretational matters to navigate.[56] One branch of the discussion to receive particular attention from philosophers of physics concerns the physical interpretation of the electromagnetic gauge potential.

Reasonable requirements on the interpretation of a gauge theory in a fixed formalism are that it render the theory deterministic and local in some appropriate sense. Yet indeterminism threatens theories with gauge symmetries since, in the canonical framework, these theories do not possess well-posed initial value problems.[57] For further discussion, see Earman (this volume, Part I) and Wallace (this volume). This, though, is where interpretation necessarily enters. For example, if, contrary to the norm, one takes the electromagnetic vector potential to represent a physically real, though unobservable, field, electromagnetism will indeed be indeterministic.[58] The indeterminism owes to the gauge freedom in the electromagnetic potential. We might instead give a gauge-invariant interpretation, taking the physical state as specified completely by the gauge-invariant electric and magnetic field strengths. In this case, electromagnetism is deterministic since the gauge invariance that threatened determinism is in effect washed away from the beginning.

However, in the case of non-trivial spatial topologies, the gauge-invariant interpretation runs into potential complications. The issue is that in this case there are other gauge invariants. So-called holonomies (or their traces, Wilson loops) – the line integral of the gauge potential around closed loops in space – encode physically significant information about the global features of the gauge field. The problem is that these gauge invariants, being ascribed to loops in space, are apparently non-local. But, coming full circle, providing a local description requires appeal to non-gauge-invariant entities such as the electromagnetic potential, whose very reality is in question according to the received understanding.

The context for this discussion is the interpretation of the well-known Aharonov–Bohm (A–B) effect: the manifestation in a particular experimental context of this further gauge-invariant observable, the holonomy, with its apparently non-local nature.[59] The interpretation of the A–B effect is discussed in more detail by Nounou (this volume). Healey (1997) argues that there is an important analogy to be drawn between the analyses of the (quantum) non-locality evidenced in the A–B effect

[55] Creutz (1983), for example, distinguishes four formal notions of gauge theory. See also Earman (this volume, Part I).

[56] Formalism and interpretation are, in some sense, symbiotic concepts, and consequently certain interpretational choices might suggest or be suggested by certain formalisms.

[57] As mentioned above, there are fewer equations of motion than independent variables. See section 2.1.

[58] Belot (1998) lays out and critically discusses the various formal and interpretational options here.

[59] See Anandan (1983) who argues that from both the physical and mathematical points of view, the holonomy contains all the relevant (gauge-invariant) information. Specifically, the connection can be constructed (up to gauge transformation) from a knowledge of the holonomies. Formalizing gauge theories in terms of holonomies associated with (non-local) loops in space appears, though, to require a revamped conception of the notion of a physical field (see Belot, 1998).

and the violation of Bell-type inequalities. Both, Healey takes it, involve possible violations – depending importantly on one's chosen interpretation of quantum mechanics – of one or the other of two distinct conditions, separability and local action (locality). While acknowledging certain disanalogies, Healey (1999) maintains that similarities in the quantum mechanical explanation of the two phenomena highlights a certain non-separability in the physical world. Local accounts, Healey argues, require non-separable electromagnetic and, more generally, gauge fields. Maudlin (1998) contends that the above analogy depends critically on Healey's (gauge-invariant) interpretation of gauge theories. Taking the vector potential to represent a real physical field, properly described by only 'one true gauge', Maudlin argues that we can provide a local and separable account of the A–B effect. One immediate epistemological contention is that nothing in the physics can reveal this one true gauge that properly represents the real gauge potential. Leeds (1999) defends Aharonov and Bohm's original interpretation of the A–B effect in terms of a local interaction of matter and a 'real' gauge potential, the defence underwritten by appeal to and a near-literal reading of fibre bundles.[60]

We have certainly not heard the final word on these matters. What the foregoing discussion does indicate (and what should not be all too surprising) is that assessing the physical content of gauge symmetry and gauge symmetry principles is intertwined with a host of other issues – issues surrounding fundamental matters of formalization, interpretation, possibly even epistemological and metaphysical predilection. Let us now continue to consider the general issue of the physical content of gauge symmetry principles.

4.1 Gauge invariance: profundity, redundancy, or both?

There are numerous ways of classifying the various symmetries figuring in physical theories: continuous vs. discrete; internal vs. external; global vs. local; physical vs. mathematical; geometrical vs. non-geometrical; geometrical vs. dynamical; universal vs. special, etc.[61] The received way of characterizing the domain of gauge symmetries is that gauge symmetry concerns the covariance of the fundamental equations of motion for specific interactions, and that the covariance is tied to a certain descriptive freedom related to the presence of non-physical and, therefore, redundant or 'surplus' quantities in the theory. The basic idea is that in describing the physics we introduce too much, and the symmetry under the covariance group effectively rids the theory of the non-physical excess.[62] This view of gauge

[60] See also Cao (1995) and Auyang (1995) for related views. See also Nounou's discussion of the fibre bundle formalism in this volume.

[61] Note that not all authors classify the same symmetries in the same ways.

[62] This view originated with Dirac. See Castellani (this volume, Part IV).

symmetries can be seen against the backdrop of the more general 'received view' of symmetry and invariance and their relation to physical theory, due to E. Wigner.[63] Not only are Wigner's views lucidly and authoritatively expressed, but they are representative of (even responsible for) the modern view in many respects.

Generally, an invariance principle, according to Wigner, expresses that two systems with the same initial conditions relative to two putatively equivalent coordinate systems develop in the same way, i.e. according to the same physical laws. Understanding the precise import of invariance principles is, however, a trickier business than this. For, as Wigner (and many to follow in his footsteps) notes, there are different types of symmetries and associated invariance principles. This in turn is reflected in a more intricate relationship between invariances, laws, and the physical world. In particular, certain sorts of invariances, such as gauge invariance, do not, according to Wigner, fit this general characterization.

Wigner maintains that there are two qualitatively and fundamentally different sorts of symmetries and associated invariance principles: geometrical and dynamical. At a general level, the primary difference is that the former concern the invariance of *all* the laws of nature under geometric transformations tied to regularities of the underlying spacetime, while the latter concern the form invariance (i.e. covariance) of the laws governing *particular* interactions under groups of transformations not tied to spacetime.[64] Both of these types of symmetries posit/embody a certain structure to some set of physical laws in placing restrictions on their possible forms. However, Wigner maintains that the geometrical principles are properly construed as concerning directly the physical events subsumed under these laws. For this reason, he takes it that geometrical invariances are properly physical invariance principles, their being a direct statement about physical events. For example, the geometrical invariance of physical laws under spatial translations is, Wigner says, properly viewed as a statement to the effect that correlations amongst events (i.e. laws) depend only on the relative distances between these events and not on any 'absolute position' of the events in the spacetime. The physical features of spacetime (the regularities of its defining structure in fact) 'undergird' geometric invariances.

Dynamic invariance principles, on the contrary, concern directly the laws of theory and cannot be formulated directly in terms of physical events.[65] Thus, for

[63] See Wigner (1967b) and the collected works, Wigner (1992); see also Houtappel, Van Dam, and Wigner (1965) and the introduction to this volume.

[64] Redhead (1975) continues this symmetry taxonomy, further dividing the dynamical symmetries into those having heuristic potential and those that are merely accidental. The former refers to symmetries used as constraints on theory building and the latter to dynamical accidents having no fundamental physical significance. As Redhead notes, this division is apparently not fixed, there being examples of changes in status – gauge symmetry he takes to be an accidental symmetry which came to be a heuristic symmetry. Kosso (2000) further discusses the distinction between fundamental and accidental symmetries.

[65] *Pace* Wigner, one can bring dynamical invariance principles closer to geometrical invariances in this regard by making appeal to the mathematics of principal fibre bundles. Gauge transformations are then automorphisms of an enlarged geometrical space of sorts. This, though, might be taken to require that one take the bundle structure quite literally at an ontological level. See Leeds (1999) and Cao (1995).

example, the (dynamic) gauge invariance characteristic of electrodynamics is a statement about the specific *laws* governing electromagnetic interactions and nothing else, Wigner takes it. Briefly stated, there is no regularity in physical events that undergirds such symmetries. The geometrical invariances, concerning as they do directly the physical world, are thus physical in a way that dynamical invariances simply are not.

At root, the above distinction rests on what Wigner takes to be a fundamental difference in the nature of the respective symmetry transformations. The crux of the issue is Wigner's holding that proper invariance principles (i.e. physically meaningful ones) concern transformations which it is, at least in principle, possible to conceive of *actively* (Wigner, 1967a, p. 45). In particular, physically meaningful invariance transformations for Wigner are those transformations that can be taken as relating (equivalent) physical observers. This, Wigner takes it, constrains the physically meaningful invariance principles to be the familiar geometric ones associated with the (improper) Lorentz group. Wigner maintains that the transformations figuring into dynamical invariance principles cannot be viewed actively, but rather must be conceived of *passively*, as mere changes of description.

For Wigner, dynamical symmetry principles are not properly physical since the associated transformations cannot be taken to relate equivalent physical observers and (relatedly) do not characterize any regularity of physical events *per se*. Rather than relating different physical situations (taken to be equivalent from the point of view of the two observers), they relate redescriptions or re-coordinatizations of one and the same physical situation. So, for example, Wigner (1967a, p. 22) likens appealing to the vector potential and associated gauge invariance to introducing the coordinates of a ghost into our physics and then noting that nothing physical depends on transforming the coordinates of the ghost – a descriptive freedom of no real physical consequence. In the end, Wigner stresses that we should speak only of dynamical 'invariance' (or better, perhaps, 'covariance') and not dynamical 'symmetry': 'I do not like the idea of gauge invariance being a symmetry principle' (Wigner, 1984, p. 729).

The basis of this view is straightforward. Wigner's theory of theories quite sensibly takes observables, specifically probability functions, as fundamental. Laws are in effect nothing but convenient ways of encompassing the various probability distributions for observable outcomes. It immediately follows that dynamical symmetry principles are not physical. The associated transformations change nothing physical since they correspond to the identity transformation on observables. And, since they affect nothing physical, there is no question of viewing the transformations actively, as relating equivalent physical observers.

Redhead (1975) provides a somewhat similar account. Key to the account is the notion of 'surplus structure,' essentially mathematics used in the formulation of some theory that outstrips the physics of the theory. In effect, Redhead gives Wigner's ghost a name. Redhead considers the familiar distinction between physical and mathematical symmetries, the former a proper subset of the latter. The need for such a distinction arises from the fact that symmetries are not formulation independent. The idea is that mathematical symmetries are features of some particular choice of formal description, whereas physical symmetries are present in all formalisms and hence characterize the underlying physics itself. Within this framework, Redhead characterizes gauge symmetries as non-trivial mathematical symmetries that are, however, physically trivial. Gauge transformations affect only surplus structure and are the identity transformation on that part of the mathematics necessary in representing the actual physics at hand. Redhead (this volume) further discusses these matters, providing further insight into the interesting role of gauge invariance in physical theory.

4.1.1 Received view, received tension

The received view, in effect, presupposes a gauge-invariant formulation. It is not surprising then that gauge transformations are physically impotent, since any potential physical significance of the characteristic gauge symmetry has been washed away from the start. This way of carving out the domain of gauge symmetries, though, seems to pose the following question: how is it that symmetries having to do with a non-physical, formal/descriptive freedom come to have any substantive physical import as it is generally thought gauge symmetries do? In this way, the received view can be taken to sow the seeds that, left unattended, can grow into a sort of tension – a tension between what we might call the 'redundancy of gauge' and the 'profundity of gauge'. On the one hand, as just discussed, we have the view that dynamical symmetry principles are void of physical content, the symmetry transformations being purely mathematical changes of description. On the other hand, as we have also discussed, gauge invariance is often invoked as a supremely powerful, beautiful, deeply physical, even undeniably necessary feature of current fundamental physical theory.[66] Redhead (this volume) calls attention to just this tension, remarking that we may be led to ascribe a 'mysterious, even mystical, Platonist-Pythagorean role for purely mathematical considerations'.

4.1.2 A generalized Kretschmannian objection

This tension is already familiar, at least in outline, from the wranglings over the physical content of general covariance and its place in the foundations of GTR. I mentioned above Kretschmann's objection to the ascription of physical content to

[66] This tension was once lucidly (even if unknowingly) indicated to me by a working particle physicist when in discussion he likened gauge invariance to 'a phantom that is somehow real'.

general covariance (see Norton, this volume). The tension just discussed might be construed as resulting from what is, in effect, a generalized Kretschmannian objection. The general challenge is to say how it is that the mathematical requirement of covariance can have any deep physical significance. Any theory of appropriate type, the generalized objection might go, can be made gauge invariant (read covariant) if one is willing to add whatever formal/mathematical machinery is necessary to cancel out the effects of the local gauge transformations. Securing gauge invariance is then, it would seem, a mere formal manoeuvre of no physical content. This analogy is perhaps not unexpected since, as already mentioned, general covariance is taken to be the first example of appeal to a local gauge symmetry principle. In the next section, I consider some possible avenues toward addressing this objection and any corresponding tension.

4.2 How gauge principles got their oomph – a Just So Story?

One way to respond to the above tension is simply to deny that there is one. We might argue that we can accept both the redundancy and the profundity aspects of gauge symmetry without worrying about any resulting tension. One might take the historical, heuristic success to be where the only real physical oomph of gauge invariance lies. As discussed above, the role of local symmetry requirements as an attractive selection criterion in arriving at (as it turned out, successful) physical theories should not to be understated. Relatedly, there are the many important affiliations between local gauge invariance and other important, even fundamental, features of physical theory (e.g. renormalizability and asymptotic freedom). Moreover, we might, as is often done, ascribe substantial pragmatic and aesthetic value to the 'simplicity' of the gauge paradigm, in that relatively few inputs are required to specify full theories.[67] The fact that all non-gravitational interactions fit into the gauge framework then lends this simplicity to a large part of fundamental physics.[68] This 'minimalist view' is not incompatible with taking gauge symmetry as having no direct physical counterpart – as concerning mere formal redundancy. The physical import of these principles is not tied to their (or, rather, the associated transformations') having direct physical correlates. And, as already discussed, one can reasonably argue that this view is all that the relevant history and physics support.

Perhaps, though, we take it that there is, or that there must be, or that we simply want more than this. One might hold out hope of locating some sort of interesting *physical* counterpart to the featuring of local gauge invariance, perhaps even a sense

[67] Weinberg (1993) discusses the appeal of the structural rigidity of gauge theories.
[68] How precisely and to what extent the gauge paradigm provides a true unification of the various interactions has been the topic of some philosophical discussion. See Maudlin (1996) and Morrison (2000).

in which gauge invariance can, contra Wigner, properly be said to be a symmetry. Some have claimed, for example, that gauge transformations can be or, even, must be viewed actively lest the associated forces (interactions) be relegated to fictitious forces (Rosen, 1990). The intuition here is clear enough. It is that physically meaningful symmetries, in agreement with Wigner, must have some sort of physical realization or physical counterpart if they are to be associated with physically meaningful invariance principles.

One possibility already mentioned in the context of assessing the gauge argument was seeing the demand of a local (rather than a global) gauge invariance as accompanying a sort of a 'locality' requirement. As I said, though, making precise the connection between the global/local distinction as it figures in classifying gauge transformations and as it does in the spacetime sense is non-trivial in the first place. The fields and transformations ostensibly live in configuration space and need to be 'brought down' to spacetime.[69] Also, the local vs. global distinction for gauge transformations as commonly employed is itself, in general, dependent on a choice of gauge. This stems from the fact that in order to construe the fields (and transformations) as living in spacetime, one must fix a gauge. This is made clear in treatments of GFTs in the principal fibre bundle framework, where gauge transformations are bundle automorphisms and the typical global/local distinction for gauge transformations is seen to be of little use (Bleecker, 1981).

In any event, it would seem that such a local-gauge-as-locality argument, if it can be mounted, must rely on some sort of active reading of the local gauge transformations. The force of the argument, presumably, derives from the untenability of actually performing the 'same' transformation everywhere at the same time (i.e. a global transformation), this violating some notion of locality. But if, as the received view holds, the transformations are taken to be merely passive coordinate relabellings – even 'ultra-passive' in the sense that there is no change of physical reference frame but only a change in mathematical description – then it seems that global transformations cannot possibly pose any threat to locality. This is because, by stipulation, there is nothing physical that gets changed under the global transformation. Thus, I do not see how one can mount any argument for local gauge symmetry in the name of locality if one ascribes to the received view of gauge symmetry/invariance.

Perhaps we could take it that the physical significance of the requirement of local gauge invariance lies specifically in that the 'required' gauge potential(s) be *dynamical*. The common definition of what it is to be dynamical in the context of field theory is for the field in question to appear in the action and for it to be varied in deriving equations of motion, including its own. Weinberg, for

[69] See Wald and Lee (1990) for further discussion.

example, argues that the real force of the demand of local gauge invariance is carried by the specific requirement that the gauge potential (or the metric connection for gravity) be dynamical in this way (Weinberg, 1996, p. 1). Similarly, Mack (1981) cashes out his construal of the demand of local gauge invariance, which he presents as a principle of no information at a distance (*Naheinformationprinzip*), in terms of its requiring specifically a dynamical connection/gauge potential to compare (generalized) phases at different spacetime points.

Locating any potential 'oomph' of gauge principles in the dynamical nature of the associated gauge fields in this way is consistent with what I indicated above in discussing the gauge argument. There we saw that the physically loaded step in the argument was the investing of the gauge potential with physical life through the addition to the Lagrangian of a kinetic term describing the free electromagnetic field and the subsequent variation of the gauge potential. Recall, though, that this 'dynamization' was in large part effected by hand, rather than necessitated by the local invariance requirement. Thus, any argument to the effect that the *dynamical* gauge field is associated with a, therefore, *physical* local invariance requirement faces challenges. As I discussed, one can reasonably argue that the local gauge symmetry is rather a *by-product* or accompaniment of the specific dynamical interaction field(s) in question.

In considering further the ascription of any physical content to gauge invariance, one might seek inspiration from discussions of the physical content and place of general covariance in the foundations of GTR. Why might one do this? I have already mentioned that many take GTR to be a (in fact, the first) gauge field theory, in *some* sense. More generally, some take it that GTR and gauge field theories are, at the very least, 'constructed' in roughly analogous ways according to similar principles of local invariance. Whether there are more disanalogies than analogies here, either in the respective historical contexts or according to current understanding, is debatable.[70]

Certainly, one appeal to seeing some such analogy here is that it might bring a certain unity or economy of fundamental principles to physical theories which, arguably, are already close in formalism.[71] A further promise of this analogy lies in that many have argued that, in the case of GTR, there are avenues available towards ascribing to general covariance a non-trivial physical content, thereby addressing the Kretschmannian objection.[72] The hope might be that we could provide some sort of analogous rejoinder to a generalized Kretschmannian objection facing the gauge principle in non-gravitational gauge theories.

70 See note 26, above.

71 Moriyasu (1978), Mack (1981), and Lyre (2001) speak of 'gauge principles of equivalence' – essentially, the ability to chose a gauge (cf. coordinate system) such that the gauge potential (cf. metric connection) vanishes in some neighbourhood (generally, it vanishes only at a point). Mack (1981) further discusses gauge invariance as a generalized principle of general covariance.

72 See the references in note 9, above. See also Rovelli (1991; 1997).

Taking such a tack is somewhat complicated by the fact that there is no universally accepted view concerning what physical content, if any, is to be ascribed to the demand of general covariance. Typically, physical content is ascribed to the (active) diffeomorphism invariance of GTR rather than to the (passive) general coordinate invariance of the theory. Though the two types of transformations can be identified with one another mathematically, symmetry under the respective transformations can have different physical significances. Most immediately, active diffeomorphism invariance precludes absolute spacetime objects, such as, for example, the non-dynamical Minkowski metric of STR. As we discussed above, the metric is dynamical in GTR, encoding, besides the metrical properties of spacetime, also the features of the gravitational field itself.

It is clear that, in its historical context, coming to terms with this active diffeomorphism invariance of GTR engendered a major revision in our physical conception of spacetime: it was not a fixed, non-dynamical background arena as was previously thought. Relatedly, spacetime reference systems are, themselves, thoroughly dynamical, part of the physics, and spacetime points are individuated only via the *dynamical*, physical fields (including, importantly, the metric). These profound realizations about the nature of spacetime (and/or gravitation) go hand-in-hand with identifying and understanding a certain redundancy in the theory: the redundancy that is washed away by the active diffeomorphism invariance characteristic of GTR. Ridding the theory of this redundancy, through physically identifying models related by active diffeomorphisms, is crucial to securing a deterministic theory.[73] And this identification, at least, in historical context, enjoins a revamped notion of spacetime. Such invariance is non-trivial physically: the profound realization concerns the physical consequences of this redundancy (and corresponding (active) invariance) for what spacetime (and spacetime reference systems) are *not*, namely absolute, or non-dynamical. In this way, redundancy and profundity go hand-in-hand in the case of diffeomorphism invariance and its role (particularly, its historical role) in the formulation of GTR.

There are clearly many historical disanalogies between the two cases, and attempting to construct some sort of analogous account about the figuring of gauge invariance in gauge field theories is strained. Even if one can (contra Wigner) in some way sensibly speak of invariance under active gauge transformations – say, for example, as automorphisms of a reified bundle space – it is not clear that this invariance could necessarily be parleyed into anything interesting physically, as in the case of diffeomorphism invariance and GTR. Though not part of the historical discussion, one possible avenue is to view the physical content of local gauge

[73] This is the central issue in the well-known 'hole argument'. This identification is often denoted, separately, as 'Leibniz equivalence'. See Earman and Norton (1987).

symmetry principles in a manner similar to the way some view the physical content of diffeomorphism invariance of GTR: as underwriting the non-existence of certain absolute (non-dynamical) elements in the theory.

In assessing the physical content of gauge symmetry principles, any analogy with the 'conceptual foundations' of GTR (or of spacetime theories generally) is perhaps more trouble than it is worth. In addition to general risks faced in attempting to provide such 'principled' stories about theories in the first place, I take it that, in the end, any such hypothetical analogy must be navigated very carefully, if for no other reason than the entirely different historical and theoretical contexts of general relativity and of gauge theories respectively. In the end, how far any such analogy can (or should) be taken, even forgetting the obvious historical disanalogies, is not immediately clear. The promise of such an analogy, though, is presumably that it may yield insight into seeing *both* profundity and redundancy in the figuring of gauge invariance in successful physical theory, thereby alleviating a certain nagging tension.

References

Aitchison, I. J. R., and A. J. G. Hey (1989). *Gauge Theories in Particle Physics*. Bristol: IOP Publishing.

Anandan, J. (1983). 'Holonomy groups in gravity and gauge fields'. In *Conference on differential geometric methods in theoretical physics*, Trieste, ed. G. Denardo and H. D. Doebner, pp. 211–16. Singapore: World Scientific.

Anderson, R. (1967). *Principles of Relativity Physics*. New York: Academic Press.

Auyang, S. (1995). *How is Quantum Field Theory Possible?* Oxford: Oxford University Press.

Belot, G. (1998). 'Understanding electromagnetism'. *British Journal for the Philosophy of Science*, **49**, 531–55.

Bleecker, D. (1981). *Gauge Theory and Variational Principles*. Reading, MA: Addison-Wesley.

Brading, K. A. (2002). 'Which symmetry? Noether, Weyl, and conservation of electric charge'. *Studies in History and Philosophy of Modern Physics*, **33**, 3–22.

Brown, H. (1999). 'Aspects of objectivity in quantum mechanics'. In *From Physics to Philosophy*, ed. J. Butterfield and C. Pagonis. Cambridge: Cambridge University Press.

Brown, H., and Sypel, R. (1995). 'On the meaning of the relativity principle and other symmetries'. *International Studies in the Philosophy of Science*, **9**, 233–51.

Budden, T. (1997). 'Galileo's ship and spacetime symmetry'. *British Journal for the Philosophy of Science*, **48**, 483–516.

Cao, T. Y. (1995). 'Group theory and the geometrization of fundamental physics'. In *Philosophical Foundations of Quantum Field Theory*, ed. H. Brown and R. Harré, pp. 155–75. Cambridge: Cambridge University Press.

(1997). *The Conceptual Development of Twentieth Century Field Theories*. Cambridge: Cambridge University Press.

Cao, T. Y., ed. (1999). *Conceptual Foundations of Quantum Field Theory*. Cambridge: Cambridge University Press.

Cornwall, J., Levin, D., and Tiktopoulos, G. (1974). 'Derivation of gauge invariance from high-energy unitarity bounds on the S-matrix'. *Physical Review D*, **10**, 1145–67.
Crease, R. P., and Mann, C. (1996). *The Second Creation*. New Brunswick: Rutgers University Press.
Creutz, M. (1983). *Quarks, Gluons and Lattices*. Cambridge: Cambridge University Press.
Deser, S. (1970). 'Self-interaction and gauge invariance'. *General Relativity and Gravitation*, **1**, 9–18.
(1987). 'Gravity from self-interaction in a curved background'. *Classical and Quantum Gravity*, **4**, L99–L105.
Doncel, M., Hermann, A., Michel, L., and Pais, A., eds. (1983). *Symmetries in physics (1600–1980): Proceedings of the 1st international meeting on the history of scientific ideas*, Barcelona. Barcelona: Servei de Publicacions.
Earman, J. (1989). *World Enough and Space-time*. Cambridge, MA: MIT Press.
(2002). 'Gauge matters'. *Philosophy of Science*, **69**, S209–20.
Earman, J., and Norton, J. D. (1987). 'What price spacetime substantivalism: the hole story'. *British Journal for the Philosophy of Science*, **38**, 515–25.
Einstein, A. (1998). *The Collected Papers of Albert Einstein*, Vol. 8: *The Berlin Years: Correspondence 1914–1918*, ed. R. Schulmann, A. J. Kox, M. Janssen, and J. Illy. Princeton: Princeton University Press.
Foerster, D., Nielsen, H., and Ninomiya, M. (1980). 'Dynamical stability of local gauge symmetry'. *Physics Letters B*, **94**, 135–40.
Friedman, M. (1983). *Foundations of Space-time Theories*. Princeton, NJ: Princeton University Press.
Froggatt, C. D., and Nielsen, H. B., eds. (1991). *Origin of Symmetries*. Singapore: World Scientific.
Healey, R. (1997). 'Nonlocality and the Aharonov–Bohm effect'. *Philosophy of Science*, **64**, 18–41.
(1999). 'Quantum analogies: a reply to Maudlin'. *Philosophy of Science*, **66**, 440–7.
Houtappel, R., Van Dam, H., and Wigner, E. (1965). 'The conceptual basis and use of the geometric invariance principles'. *Reviews of Modern Physics*, **37**, 595–632.
Ivanenko, D., and Sardanashvily, G. (1983). 'The gauge treatment of gravity'. *Physics Reports*, **94**, 1–45.
Kastrup, H. A. (1983). 'The contributions of Sophus Lie, Felix Klein, and Emmy Noether to the concept of symmetry in physics'. In *Symmetries in physics (1600–1980): Proceedings of the 1st international meeting on the history of scientific ideas*, Barcelona, ed. M. G. Doncel, pp. 115–58. Barcelona: Servei de Publicacions.
Kosso, P. (2000). 'Fundamental and accidental symmetries'. *International Studies in the Philosophy of Science*, **14**, 109–21.
Leeds, S. (1999). 'Gauges: Aharonov, Bohm, Yang, Healey'. *Philosophy of Science*, **66**, 606–27.
Lyre, H. (2001). 'The principles of gauging'. *Philosophy of Science*, **68**, S371–81.
Mack, G. (1981). 'Physical principles, geometrical aspects, and locality properties of gauge field theories'. *Fortschritte der Physik*, **29**, 135–85.
Martin, C. (2002). 'Gauge principles, gauge arguments and the logic of nature'. *Philosophy of Science*, **69**, S221–34.
Maudlin, T. (1996). 'On the unification of physics'. *Journal of Philosophy*, **93**, 129–44.
(1998). 'Discussion: Healey on the Aharonov–Bohm effect'. *Philosophy of Science*, **65**, 361–8.
Mills, R. (1989). 'Gauge fields'. *American Journal of Physics*, **57**, 493–507.

Moriyasu, K. (1978). 'Gauge invariance rediscovered'. *American Journal of Physics*, **46**, 274–8.

Morrison, M. (2000). *Unifying Scientific Theories: Physical Concepts and Mathematical Structures*. Cambridge: Cambridge University Press.

Norton, J. D. (1993). 'General covariance and the foundations of general relativity: eight decades of dispute'. *Reports on Progress in Physics*, **56**, 791–858.

Norton, J. (2000). ' "Nature is the realisation of the simplest conceivable mathematical ideas": Einstein and the canon of mathematical simplicity'. *Studies in History and Philosophy of Modern Physics* **31**, 135–70.

O'Raifeartaigh, L. (1979). 'Hidden gauge symmetry'. *Reports on Progress in Physics*, **42**, 159–223.

(1997). *The Dawning of Gauge Theory*. Cambridge: Cambridge University Press.

Pais, A. (1986). *Inward Bound*. Oxford: Oxford University Press.

Pauli, W. (1921). *Theory of Relativity*. New York: Dover.

Quigg, C. (1982). *Gauge Theories of the Strong, Weak, and Electromagnetic Interactions*. Reading, MA: Addison-Wesley.

Redhead, M. (1975). 'Symmetry in intertheory relations'. *Synthese*, **32**, 77–112.

Rosen, J. (1990). 'Fundamental manifestations of symmetry in physics'. *Foundations of Physics*, **20**, 282–308.

Rovelli, C. (1991). 'What is observable in classical and quantum gravity'. *Classical and Quantum Gravity*, **8**.

(1997). *Halfway Through the Woods: Contemporary Research on Space and Time*, pp. 180–223. Pittsburgh: University of Pittsburgh Press.

Rowe, D. (1999). 'The Göttingen response to general relativity and Emmy Noether's theorems'. In *The Symbolic Universe*, ed. J. Gray. Oxford: Oxford University Press.

Ryder, L. H. (1996). *Quantum Field Theory*. Cambridge: Cambridge University Press.

Stachel, J. (1989a). 'Einstein's search for general covariance'. In *Einstein and the History of General Relativity*, ed. D. Howard and J. Stachel, pp. 63–100, Vol. 1 of *Einstein Studies*. Boston: Birkhäuser.

(1989b). 'How Einstein found his field equations: 1912–1915'. In *Einstein and the History of General Relativity*, ed. D. Howard and J. Stachel, pp. 101–59, Vol. 1 of *Einstein Studies*. Boston: Birkhäuser.

't Hooft, G. (1980). 'Gauge theories of the forces between elementary particles'. *Scientific American*, **242**, 104–38.

Teller, P. (2000). 'The gauge argument'. *Philosophy of Science*, **67**, S466–81.

Trautman, A. (1980). 'Yang–Mills theory and gravitation: a comparison'. In *Geometric Techniques in Gauge Theories*, ed. R. Martini and E. M. Jager, pp. 179–88. New York: Springer-Verlag.

Utiyama, R. (1956). 'Invariant theoretical interpretation of interaction'. *Physical Review*, **101**, 1597–607.

van der Waerden, B. (1932). *Die Gruppentheoretische Methoden in der Quantenmechanik*. Berlin: Springer-Verlag.

Vizgin, V. (1994). *Unified Field Theories in the First Half of the Twentieth Century*. Translated from the Russian original by J. Barbour. Basel, Boston, Berlin: Birkhäuser Verlag.

Wald, R. (1986). 'Spin-two fields and general covariance'. *Physical Review D*, **33**, 3613–25.

Wald, R., and Lee, J. (1990). 'Local symmetries and constraints'. *Journal of Mathematical Physics*, **31**, 725–43.

Weinberg, S. (1965). 'Photons and gravitons in perturbation theory: derivation of Maxwell's and Einstein's equations'. *Physical Review*, **138**, 988–1002.
(1974a). 'Gauge and global symmetries at high temperature'. *Physical Review D*, **9**, 3357–78.
(1974b). 'Recent progress in gauge theories of the weak, electromagnetic and strong interactions'. *Reviews of Modern Physics*, **46**, 255–77.
(1980). 'Conceptual foundations of the unified theory of weak and electromagnetic interactions'. *Reviews of Modern Physics*, **52**, 515–23.
(1993). *Dreams of a Final Theory*. London: Vintage.
(1995). *The Quantum Theory of Fields, Vol. 1: Foundations*. Cambridge: Cambridge University Press.
(1996). *The Quantum Theory of Fields, Vol. 2: Modern applications*. Cambridge: Cambridge University Press.
(1999). 'A unified physics by 2050?' *Scientific American*, **281**, 68–75.
Weinstein, S. (1999). 'Gravity and gauge theory'. *Philosophy of Science*, **66**, S146–55.
Weyl, H. (1918a). 'Gravitation und Elektrizität'. *Preussische Akademie der Wissenschaften (Berlin) Sitzungsberichte. Physikalisch-Mathematische Klasse*, pp. 465–80. Translation in O'Raifeartaigh (1997), pp. 24–37.
(1918b). 'Reine Infinitesimalgeometrie'. *Mathematische Zeitschrift*, **2**, 384–411.
(1928). *Gruppentheorie und Quantenmechanik*. Translated by H. R. Robertson as *The Theory of Groups and Quantum Mechanics*. London: Methuen. Translation reprinted (1950). New York: Dover.
(1929). 'Elektron und Gravitation'. *Zeitschrift für Physik*, **56**, 330–52. Translation in O'Raifeartaigh (1997), pp. 121–44.
Wigner, E. P. (1931). *Gruppentheorie und ihre Anwendung auf die Quantenmechanik der Atomspektren*. Translated by J. J. Griffin as *Group Theory and its Application to the Quantum Mechanics of Atomic Spectra* (1959). New York: Academic Press.
(1967a). 'Events, laws of nature and invariance principles'. In Wigner (1967b), pp. 38–50.
(1967b). *Symmetries and Reflections*. Bloomington, IN: Indiana University Press.
(1967c). 'Symmetry and conservation laws'. In Wigner (1967b), pp. 14–27.
(1984). 'The meaning of symmetry'. In *Gauge interactions theory and experiment, Proceedings of the 20th international school of subnuclear physics*, Erice, Sicily, ed. A. Zichichi, pp. 729–33. New York: Plenum Press.
(1991). 'The role and value of the symmetry principles and Einstein's contribution to their recognition'. In Froggatt and Nielsen (1991), pp. 193–201.
(1992). *The Collected Works of Eugene Paul Wigner*. Berlin: Springer-Verlag.
Wilczek, F. (2000). 'What QCD tells us about nature – and why we should listen'. *Nuclear Physics, A* **663–4**, 3–20.
Yang, C. N. (1982). 'Einstein and the physics of the second half of the twentieth century'. In *Proceedings of the second Marcel Grossmann meeting on general relativity*, Amsterdam, ed. R. Ruffini, pp. 7–14. Amsterdam: North-Holland.
Yang, C. N., and Mills, R. (1954). 'Isotopic spin conservation and a generalized gauge invariance'. *Physical Review*, **96**, 191.

4

The philosophical roots of the gauge principle: Weyl and transcendental phenomenological idealism

T. A. RYCKMAN

> As far as I see, all *a priori* statements in physics have their origin in symmetry.
> (*H. Weyl, 1952, p. 126*)

> The most important lesson that we have learned in this century is that the secret of nature is symmetry.
> (*D. Gross, 1999, p. 57*)

1 Introduction

Most readers of this volume will know that the ancestry of gauge field theories extends back to Hermann Weyl's 1918 theory of 'gravitation and electricity'. Since papers of Yang and others in the 1970s recovered this lineage from obscurity, considerable interest has been shown, and a few years ago a new English translation of Weyl's original paper appeared.[1] The broad outline of the story is now common currency and need not be rehearsed in any detail here. Yet there are several largely unacknowledged aspects that arguably have more than incidental interest for those interested in philosophical issues of the significance of local symmetries. First of all, Weyl did not start out with the objective of unifying gravitation and electromagnetism, but sought to remedy a perceived blemish in Riemannian 'infinitesimal' geometry. The resulting 'unification' was, as it were, serendipitous. Then there is the not insignificant matter of Einstein's 'pre-history' objection, commonly affirmed to have sealed the unhappy fate of Weyl's theory. Just several weeks after hailing Weyl's theory as 'a stroke of genius of the first magnitude', Einstein reasoned that it was in blatant contradiction with the known constancy of the spectral lines of the chemical elements.[2] For if two atoms, say of hydrogen, at one spacetime location are separated and then brought together again via two different routes, then in

[1] In O'Raifertaigh (1997), pp. 24–37.

[2] Einstein to Weyl, 6 May 1918 and 15 June 1918; nos. 498 and 507 in Einstein (1998). See also Einstein to W. Dällenbach (after 15 June 1918), no. 565.

general, according to Weyl's theory, they should have distinct spectral frequencies, corresponding to the differing electromagnetic field strengths through which each has passed. Elaborating Einstein's argument, Pauli (see section 4.3 below) showed that this difference could become arbitrarily large with proper times. Empirically refuted in its original form of a local scale invariance of spacetime geometry, Weyl's idea was reborn as the modern concept of gauge (phase) invariance some ten years later, with Weyl's assistance and in several intermediate steps.[3] So the story goes.[4]

However, this familiar narrative ignores that Weyl dismissed the prehistory objection as *irrelevant*, stating that the 'ideal process of congruent displacement' (of a vector's magnitude) at the basis of his 'purely infinitesimal geometry' had, in itself, 'nothing to do' with the actual behaviour of rods and clocks (or atoms).[5] Rather, the salient issue lay in the fact that Einstein accepted, while Weyl did not, the postulates of 'practically rigid rods' and 'perfect clocks' as necessary to ground the empirical meaning of the invariant line element ds. Although a 'natural gauge of the world' is tacitly assumed in these postulates of Einstein's general relativity, Weyl objected that its existence was a datum to be explained, not posited. To that end, he argued that spacetime measurement disguised the scale arbitrariness of an *Aether-geometrie*, the geometry of a matter-free state of the world, its *local* degrees of freedom effaced by dynamical interaction with matter. His hypothetical outline of such a mechanism, wherein atoms are regarded as in a continuous equilibrium of 'adjustment' (*Einstellung*) to the curvature scalar of a purely infinitesimal spacetime geometry, was only intended as a placeholder pending a completed theory of matter.[6] However, such an account was not to be expected from a 'unified field theory'. Already in 1920, Weyl declared that atomic phenomena are irreducibly stochastic, and so outside the deterministic explanatory frame of classical field theory. Despite its fourth-order field equations, Pauli and Weyl could demonstrate that the theory recovered (up to terms beyond the limit of observation) both the Einstein and the Maxwell field equations.[7] But in uniting them in spacetime geometry, Weyl

[3] Weyl (1929) argued that accommodating spinor fields in curved spacetimes required the use of local tetrads that left these fields determined only up to an arbitrary factor of phase, hence the matter field equations are locally phase invariant. Of course, we would now say that phase invariance of the matter field is independent of the presence of curved spacetime.

[4] See Yang (1977), Mielke and Hehl (1988), Bergia (1993), Vizgin (1994), Cao (1997), Scholz (1994; 2001), O'Raifertaigh (1997), O'Raifertaigh and Straumann (2000), and Straumann (1987 and 2001).

[5] Weyl (1919a, p. 67): 'The functioning of these instruments of measurement however is a physical occurrence whose course is determined through laws of nature and which has as such nothing to do with the ideal process of congruent displacement of world tracts (*Verpflanzung von Weltstrecken*).' Here, and throughout, the translations are mine, unless otherwise indicated.

[6] Weyl (1921a), p. 230; (1923a), pp. 299–300. In coordinates giving the 'naturally measured volume of the world', the curvature scalar F of Weyl's geometry is chosen as the invariant of an action integral, and is normalized $F = -1$, the unit of length obtained by averaging the radii of all sectional curvatures through a given point P. The 'natural gauge of the world' is then of the order of magnitude of the 'curvature radius of the world'; its reciprocal corresponds to measurements with rods and clocks. See Ryckman (forthcoming) for details.

[7] Pauli (1919) and Weyl (1921b), pp. 477–8.

maintained that the explanatory template just sketched furnished a preferable, and in principle more consistent, understanding of Einstein's gravitational theory, a view reiterated as late as 1949.[8] The virtues were apparent: there is no longer a need to glue electromagnetism on to a prior spacetime geometry; moreover, the ideal of explanatory closure is upheld by regarding measuring apparatus as not exempt from the great field laws of gravitation and electromagnetism. As Weyl's research in the mid-1920s returned to pure mathematics, it was left to Eddington to place this explanatory ideal at the centre of his own (1923) treatment of general relativity, in a book Einstein surprisingly praised, much later, as 'the best presentation of the subject in any language'.[9]

It is the aim of this paper to consider this hidden history further by shedding illumination on the philosophical motivations underlying Weyl's 1918 *a priori* posit of the concept of local gauge symmetry and his subsequent prolonged defence of it. To contemporary philosophers of physics that context is highly unfamiliar, if not fantastic: the transcendental phenomenological idealism of Edmund Husserl. We shall show that Weyl's *a priori* grounds in introducing this concept were rooted nominally in 'mathematical speculation', but organically in a phenomenologically guided 'essential analysis' of the fundamental concepts of Riemannian infinitesimal geometry. Emphasizing the epistemological priority of what is immediately surveyable within the localized space of phenomenological intuition, Weyl considered parallel displacement of a tangent vector along a curve, rather than the Riemannian metric, as the paradigm comparison relation for any genuinely infinitesimal geometry. This led to the construction of a 'pure infinitesimal geometry'. Consistent with the prohibition against 'action at a distance' that is the hallmark of field theories, Weyl affirmed that, analogous to comparisons of vector direction, direct magnitude comparisons can also be made only at infinitesimally neighbouring points. Within the resulting geometry, structures identical to the vector and tensor fields of electromagnetism 'naturally' emerged, and in this way, electromagnetism as well as gravitation was brought within the spacetime metric. These mathematical constructions were accompanied by a passionate belief in his geometry's '*a priori* reasonableness'. To Einstein, Weyl proclaimed that his geometry must be the true local geometry (*wahre Nahegeometrie*), and that he should accuse the 'dear Lord' of an inconsistency were it not so.[10]

In order to assess just how ideas of Husserlian phenomenology guided Weyl's development and justification of a 'pure infinitesimal' geometry underlying field physics, an expository outline of the salient concepts of transcendental

[8] Weyl's comment is in a section that first appears in the 1949 translation of Weyl (1926), p. 288.

[9] See Douglas (1956), p. 100.

[10] Weyl to Einstein, 19 May 1918, no. 544 in Einstein (1998). Einstein's reply, that the 'dear Lord' might, with the same consistency, have foregone the conformal invariance of Weyl's geometry (31 May 1918, no. 551 in Einstein, 1998), made no discernible impression on Weyl.

phenomenological idealism, its goals and its methodology, is unavoidable. This will be undertaken in section 3, while we begin in section 2 with abundant evidence that Husserl was the operative and significant influence governing the hardly suppressed 'philosophical spirit' of Weyl's classic *Raum-Zeit-Materie*, a work already in press in the spring of 1918 as Weyl hit upon the idea of local gauge symmetry. Section 4 then considers the distinct stages of Weyl's 'essential analysis' of the foundational concepts of Riemannian geometry that issued in a 'pure infinitesimal geometry'. Employed as a 'world geometry', a conceptual space of 'essential possibilities', Weyl sought to distinguish the 'actual world', *viz.*, the field laws of gravitation and electromagnetism, via a uniquely determined 'rational action function' of the field variables. In conclusion, we suggest that revisiting this unknown 'context of discovery' of Weyl's theory may yield hitherto unexploited resources for addressing perplexing ontological and epistemological issues raised by posit of unobserved local symmetries in fundamental physical theories.

2 The 'Introduction' to *Raum-Zeit-Materie*

The first nine pages of Weyl's *Raum-Zeit-Materie* (*RZM*) surely come as a shock to readers who believe they hold in their hands what the journal *Nature* (in the person of A. S. Eddington) described, on its appearance in English, as 'the standard treatise on the general theory of relativity'.[11] Few readers prepared to take on a work of this kind could possibly be expected to also possess the background requisite to successfully grapple with the 'several philosophical discussions' these pages contain. Within them are found the following passages, whose meanings, even if but dimly perceived, seem quite remote indeed from the kind of 'philosophy' customarily found in technical treatises.[12]

> The real world (*wirkliche Welt*), each of its components and all their determinations, is, and can only be, given as intentional objects of conscious acts. The conscious experiences that I have are absolutely given – just as I have them.
>
> The immanent is absolute, that is, it is exactly what it is as I have it and I can eventually bring this, its essence (*Wesen*) to givenness (*Gegebenheit*) before me in acts of reflection.
>
> The given to consciousness (*Bewußtseins-Gegebene*) is the starting point in which we must place ourselves in order to comprehend the sense and the justification of the posit of reality (*Wirklichkeitsetzung*).
>
> 'Pure consciousness' is the seat of the philosophical *a priori*.[13]

[11] Eddington (1922), p. 634. The review is signed 'A.S.E.'.
[12] Weyl (1918b), pp. 3–4; (1923a), pp. 3–4.
[13] Weyl (1921a), p. 4; cf. Eng. trans. (1953), p. 5.

To the *cognoscenti*, Weyl's orientation to Husserlian transcendental phenomenology is immediately apparent in the language employed; to others, it is announced in the book's first footnote.[14]

To those who know Weyl's book only in the English translation of the fourth (1921) edition by the Australian physicist H. L. Brose, the difficulty of comprehending these mysterious incantations is considerably compounded. With painful consistency, their translations are so garbled that little comprehensible meaning can be gleaned, or else their intended meaning is completely subverted, with the ensuing obscurities and inconsistency widely, and unjustly, attributed to Weyl. Yet judging by the extant literature on Weyl, even those who read him in the original German have found these discussions too murky or idiosyncratic to merit serious attention and further investigation. Still, the expressed philosophical alignment is not a momentary infatuation. Despite many changes in the various editions in other parts of the book, these passages remain unmodified, while additional evidence of this attachment accumulates through the texts of the successive editions. No doubt correctly, Weyl observed that his goal of 'attaining insight into the essence (*Wesen*) of space, time and matter, insofar as these participate in the construction (*Aufbau*) of objective reality' is all too readily obscured by 'the Flood (*Sintflut*) of formulas and indices deluging the guiding ideas of infinitesimal geometry'.[15] Admittedly, the book's mathematical and physical content can be extracted and independently studied, a fact explaining the longevity of its publishing history. Yet doing so ignores an express declaration that 'philosophical clarification' of science, though a task of a completely different kind than that of the individual sciences, 'remains a great responsibility'. For, Weyl noted, 'as things stand today', the individual sciences must be allowed to proceed unhindered along the fresh paths opened up by revision of fundamental principles and newly emerged ideas. In this they are guided by a 'reasonable motive of good faith' in the domain of competence of their particular methods, but, by the same token, they 'proceed in this sense, dogmatically (*in diesem Sinne dogmatisch zu verfahren*)'. It is precisely for this reason that philosophical elucidation, while not obstructing the forward steps of the special sciences and respecting the difficulty of the problems they face, is needed.[16] The incursion of philosophy into his technical treatise is deemed a necessity, not mere window-dressing, for the mathematician's methods are useless in the 'darker depths' belonging to the origins of cognition. Eloquent testimony to this effect concludes the 'Introduction':

[14] Weyl (1918b), p. 4: 'The precise wording of these thoughts is closely modelled upon Husserl, *Ideen zu einer reinen Phänomenologischen Philosophie (Jahrbuch f(ür) Philos(ophie) u(nd) phänomenol(ogische) Forschung*, Vol. I, Halle, 1913).'

[15] Weyl (1919b), p. 123; (1921a), p. 124; (1923a), p. 136.

[16] Weyl (1918b), p. 2; (1923a), p. 2.

All beginnings are obscure.... from time to time the mathematician, above all, must be reminded that origins (*die Ursprüngen*) lie in depths darker than he is capable of grasping with his methods. Beyond all the individual sciences, the task of comprehending (*zu begreifen*) remains. In spite of philosophy's endless swinging from system to system, to and fro, we may not renounce it altogether, else knowledge be transformed into a meaningless chaos.[17]

Much of *RZM*, together with related papers of 1918–1923 on Weyl's theory, manifests this striving to attain philosophical comprehension of the new physical knowledge brought by the General Theory of Relativity. On recognition of the many clearly visible intimations of Weyl's transcendental-phenomenological idealist leanings, these texts evince a remarkably sustained attempt to probe the 'darker depths' of the 'origins' of the objective physical world portrayed in classical field physics through mathematical construction guided by the phenomenological method of 'essential analysis'. To be sure, Weyl himself judged his treatment of the epistemological questions raised by general relativity theory as preliminary and tentative, lamenting he had not been able to provide such answers as would salve his scientific conscience.[18] Even so, there can be little doubt that his attempt to cast illumination in these dim regions was carried out in agreement with the fundamental thesis of transcendental phenomenological idealism, as stated at the end of section 49 of Husserl's *Ideen I* (1913):

the whole spatiotemporal world, which includes man himself and the human Ego as subordinate single realities is, according to its sense, a merely intentional being, thus one having the merely secondary, relative sense of a being for a consciousness. It is a being which consciousness in its experiences posits that, in principle, is only determined and intuited as something identical by motivated manifolds of phenomena: beyond that it is nothing.

Understood as guided by this beacon, Weyl's 1918 re-setting of general relativity within the new framework of a 'pure infinitesimal geometry' was an effort to demonstrate that the vector and tensor fields of gravitation and electromagnetism likewise have the 'sense of an intentional being', of 'a being for a consciousness'. That classical field physics admits of a complete differential geometrical representation is absolutely crucial to this endeavour. For as purely infinitesimal relations of comparison of vectors and tensors are the axiomatic basis of identical mathematical structures, such relations – 'given to consciousness' in localized phenomenological intuition – can be deemed the starting point from which may be understood 'the sense and the justification' of the posit of reality accorded to such objects. Accordingly, Weyl's theory of gravitation and electricity appears as an epistemologically motivated mathematical construction whose principal aim is to show that

[17] Weyl (1918b), p. 9; (1923a), p. 9.
[18] Weyl (1918b), p. 2; (1923a), p. 2.

the theoretical objects of classical field theory are constituted as objects of thought, and so have the sense of *intentional objects*.[19] Before turning to the details of this construction in section 4, it is necessary first to briefly review the epistemological theory that guided this construction.

3 The transcendental phenomenological context

The following is a crude road map of the salient methodological aspects of transcendental phenomenological idealism providing epistemological motivation for Weyl's 'purely infinitesimal' geometry and its use as a 'world geometry' for gravitation and electromagnetism. These simply expository remarks are intended to serve only that purpose and can in no way be considered an introduction to this incomplete, even fragmentary, nexus of ideas. Indeed, as Husserl acknowledged, Weyl was less an acolyte than a philosophically inclined mathematician who creatively adapted phenomenological methodology to his own pursuits and interests.[20]

3.1 'The natural attitude', 'transcendental subjectivity'

It is first to be observed that a transcendental–phenomenological elucidation of a particular physical science does not put in question the correctness of its accepted theories; here there is, Husserl sternly warned, 'no room for private 'opinions', 'intuitions', or 'points of view'.[21] So if not scepticism of science, or an attempt to refer physics to the higher cognitive authority of a pet metaphysical theory or worldview, what is the point of such an endeavour? A hint of an answer has already been seen in Weyl's remark cited above, that the individual sciences, operating under the 'reasonable motive of good faith' in the domain of competence of their particular methods, nonetheless proceed 'dogmatically' when considered from an epistemological point of view.

Accordingly, it is necessary to begin with the dogmatic 'natural attitude', common both to working scientists and to everyday life, regarding a pre-given world of material objects, located in space and time, having properties entirely independent of human perception and cognition. But in accepting nature as simply 'there', the natural sciences display, to use Husserl's term, an 'immortal naiveté', exhibited each time recourse is made to experience. This is not to say that natural science does not have, in its own way, a 'very critical' account of experience, placing little stock in isolated experiences. It is rather that the methods of natural science do not put experience itself in question. For it is simply not the office of natural science

[19] The full case for this claim is contained in Ryckman (forthcoming).
[20] Husserl to Weyl, 10 April 1918 and 5 June 1920; see Husserl (1994), pp. 287–90.
[21] Husserl (1911), p. 290; Eng. trans., p. 74.

to consider 'how experience as consciousness can give or contact an object', an explicitly epistemological issue whose adequate treatment is regarded as outside the competency of the methods of natural science. In virtue of this necessary omission, 'all natural science is naive in regard to its point of departure'.[22]

Yet phenomenology's critical animus is directed not to the 'pre-epistemological' realism of the 'natural attitude' but rather to its hypostatization in the philosophical naturalism that 'dominates the age', posing a 'growing danger for our culture'.[23] On account of the latter's '*philosophical* absolutizing' of the natural world, an interpretation far outstripping conclusions following from the experiential methods of the natural sciences, it is deemed 'completely alien' to the natural attitude.[24] Maintaining that a genuinely scientific philosophy must itself be based upon physical natural science and its methods, it nonetheless has both epistemological and metaphysical components. Husserl (and it may be conjectured, Weyl as well) considered each not only wrong, but 'counter-sensical' – absurd. Epistemologically, naturalism affirms the *Ur*-thesis of scientific realism, that cognition of 'nature in itself' is indeed possible through the methods of the natural sciences, and even, in certain areas, wholly or partly achieved. To Husserl, such a claim is literally 'counter-sensical' for the very sense of objectivity, of what it means to be an object of a rational proposition, can only be made evident or understandable within consciousness itself.[25] On the other hand, as a metaphysics, naturalism 'sees only nature, and primarily, physical nature'. Thus it maintains that 'whatever is, belongs to psychophysical nature', *viz.*, is either physical, belonging to the unified totality of physical nature, or else is derivatively mental, variably depending on the physical. The mental is, then, 'at best, a secondary "parallel accompaniment"' to the physical, while nature is regarded 'as a unity of spatio-temporal being subject to exact laws'. Consciousness, values, reason, logic, and mathematics are thus completely naturalized; these are socio-psychological-biological phenomena to be accounted for in empirically attested psycho-physical causal laws. As already argued at length in the *Logical Investigations*, Husserl regarded this metaphysics as self-refuting; in particular, its supposition that the 'exemplary index of ideality, formal logic' is rooted in 'natural laws of thinking' leads to absurdity.[26]

Phenomenology's antidote to naturalism stems from the discovery of 'transcendental subjectivity' presupposed in all experience. Identified through a 'phenomenological reduction' that notoriously brackets or suspends the unquestioned 'posit of reality' endemic to the natural attitude, it hypothetically 'puts out of action' all existential posits regarding a world of objects transcendent to consciousness.

[22] Husserl (1911), p. 298; Eng. trans., p. 85.
[23] Husserl (1911), p. 293; Eng. trans., p. 78.
[24] Husserl (1913), section 55.
[25] Husserl (1911), p. 301; Eng. trans., p. 90. See also Husserl (1913), section 52.
[26] Husserl (1911), p. 295; Eng. trans., pp. 79–80. See also Husserl (1900), *passim*.

The intent is not to deny the existence of extra-mental objects, but rather to uncover the 'ultimate origin of all sense-bestowal' in 'pure consciousness'.[27] Phenomenology's epistemological task, broadly construed, is then to exhibit 'the origins of objectivity in transcendental subjectivity'. The aim is to show that the 'sense' of objects of a given theoretical domain (as indeed of all objects, insofar as they are objects of thought) is that of 'a being for consciousness'. Its critique is directed not against the natural attitude but against its philosophical hypostatization that is philosophical naturalism, or, as we should presently say, scientific realism, and the appraisal of theoretical entities as having 'nothing to do with consciousness or the Ego pertaining to consciousness'.[28] It is not a subjective idealism *à la* Berkeley, but rather the removal of an interpretation of physical theory that both Husserl and Weyl considered counter-sensical – a contradiction in terms.

3.2 Evidenz

Well-known difficulties beset the attempt to find a suitable English equivalent for the core phenomenological notion of *Evidenz*. The obvious candidate, *evidence*, while not quite a false cognate, wrongly suggests commonly manifested proof or grounds for belief. But certainly this manner of intersubjectivity, obviously taken for granted within 'the natural attitude' of ordinary life and the working scientist, cannot be fundamental in a phenomenology concerned to show the origins of objectivity in transcendental subjectivity. Rather it must be established later on, somewhat as Carnap, in sections 148–149 of the *Aufbau*, constituted an intersubjective world from the quasi-phenomenological standpoint of 'methodological solipsism'.[29] Neither is *self-evidence* completely accurate, for it lacks the connotation of intentional achievement stemming from the coincidence of the object as presented in a 'fulfilling act' with the object as intended. It is not at all a feeling causally appended to a true judgement. Rather, the particular phenomenological content of such an act is entirely distinctive, an 'immediate insight' containing 'the "experience" of truth':

> *Evidenz* is rather nothing but the 'experience' of truth (*das 'Erlebnis' der Wahrheit*).[30]

It is noteworthy that Weyl reverts to this phrase *verbatim* in his 1918 monograph *Das Kontinuum*, in a passage where he takes issue with Dedekind's injunction that, in science, belief should be accorded only to what is actually proven.[31]

[27] Husserl (1913), section 55.
[28] Husserl (1913), section 47.
[29] Carnap (1928).
[30] Husserl (1900), p. 190; Eng. trans., p. 194.
[31] Weyl (1918a), p. 11, footnote: 'As if such an indirect collocation of reasons as we designate as "proof" is capable of arousing any belief without our securing the justification of each single step in immediate insight (*unmittelbarer Einsicht*). This (and not the proof) generally remains the ultimate source of justification of knowledge; it is the "experience of truth" (*das "Erlebnis der Wahrheit"*).'

In the years between 1918 and 1923, when working on and defending his 'pure infinitesimal geometry' as a basis for field physics, Weyl maintained this phenomenological foundation for mathematical and physical cognition. Thus, in a section of *RZM* entitled 'Report on non-Euclidean geometry', noting the warning of Proclus against the 'misuse' made of appeals to self-evidence in the justification of Euclid's fifth (parallel) postulate, he remarked:

But one also must not grow weary of emphasizing that, despite its many misuses, evidence (*die Evidenz*) is the ultimate anchoring ground of all knowledge, even of empirical knowledge.[32]

By 1926, Weyl showed many signs of relinquishing an exclusive evidential basis for cognition in intuition, in what is 'immediately given to consciousness', a foundation to some extent common to both phenomenology and to Brouwer's intuitionism in mathematics, 'idealism in mathematics thought through to the end'.[33] But consideration of this matter goes beyond the concerns of the present paper.[34]

3.3 *Essence* (Wesen/Eidos), *essential seeing* (Wesenerschauung)

A particularly compressed statement of the nature and method of transcendental phenomenology, one that at once serves also to illustrate the difficulties of entering into the semantical web of Husserlian neologisms, occurs in section 66 of *Ideen I* (Husserl, 1913):

In phenomenology, which is to be nothing other than a doctrine of essences (*Wesenslehre*) within pure intuition (*reiner Intuition*), we therefore carry out (acts of) immediate essential seeing (*unmittelbare Wesenserschauungen*) on exemplarily givens (*exemplarischen Gegebenheiten*) of transcendentally pure consciousness, and fix them *conceptually*, or else terminologically.

The crucial methodological pieces of phenomenological context requisite to understanding Weyl's idea of a 'world geometry' concern the notions of *essential seeing* or *eidetic intuition* and *eidetic analysis.* Both pertain to the basic phenomenological method for exploration and investigation of the 'pure possibilities of consciousness' within 'transcendental subjectivity'. The fundamental notion is that of *essence*, admittedly a 'new kind of object' that, as the datum of 'eidetic intuition' or 'essential seeing', is analogous to the individual object that is the datum of empirical intuition.[35]

Obviously, the method of eidetic intuition is closely modelled on, and indeed considered an extension of, 'the mathematical style of thinking' directed to the

[32] Weyl (1918b), p. 69; (1923a), p. 71; cf. Eng. trans. (1953) of Weyl (1921a), p. 77.
[33] Weyl (1925), pp. 23–4.
[34] See Mancosu and Ryckman (2002).
[35] Husserl (1913), section 3: '*Essential seeing* (*Wesenerschauung*) *is also precisely intuition* (*Anschauung*).' (Emphasis in original.)

categorical relationships between individuality and universality, actuality and possibility, the contingent and the necessary, experience and imagination. In the phenomenological account of intentionality, an individual object of attention or, more particularly, of outer perception, is a never-repeatable 'this-here' (*Dies-da*), immediately apprehended or 'seized upon' as a bare particular. Yet this mode of apprehension does not give the object 'in itself', as a 'thus and so'. As so qualified, the object has a 'specific character' determinable by attending to its given accompanying background, its 'horizon' of more or less unthematized possibilities of further meanings and connections.[36] According therefore to the *sense* of what a contingent being *is*, every factual contingency has an essence. By directing attention within the 'horizon', these objects of pure regard or reflection can be 'brought into "the sharply illuminated circle of perfect givenness" '.[37] Collectively considered, they comprise an object's field of 'eidetic possibilities'.[38] Within this field lies the essence or *eidos* of each individual object, 'its stock of essential predicables which must belong to it making it *necessarily* the thing that it is', and so determining what 'other, secondary, contingent determinations can belong to it'.[39]

In short, its essence or *eidos* makes a given individual object rational. Essence is the intuited structure of necessity against which the contingency of the given is contrasted, determining the sense in which any particular matter of fact 'could have been otherwise'. This is the basis of the claim that in transcendental phenomenology, knowledge of actuality (*Wirklichkeit*), presupposes knowledge of the space of possibilities that is bounded by the eidetic structures of experience.[40] 'Correctly understood', i.e. in terms of transcendental idealism, this means that the 'constitution' of objects of knowledge, in particular those of the empirical sciences, must presuppose eidetic structures of experience, the space of possibilities bounding the sense of any actual object. The aggregate of all eidetic possibilities (*a priori* objects of possible experience) and of the actualized subset encountered in empirical cognition comprises 'the world'.

> The world is the 'sum-total' of objects of possible experience together with objects of experiential cognition, objects of actual experiences cognized in correct theoretical thinking.[41]

[36] Husserl (1913), section 2. The simplest cases of eidetic intuition arise from within perceptual experience, as may be see from Weyl's discussion of the perception of a chair. This example, also from the 'Introduction' to *RZM*, briefly recapitulates the argument for transcendental idealism in sections 33–55 of the *Ideen I*.

[37] Husserl (1913), section 69.

[38] Husserl (1913), section 83.

[39] Husserl (1913), section 2.

[40] Husserl (1913), section 79: 'The old ontological doctrine that knowledge of "possibilities" must precede knowledge of actualities (*Wirklichkeiten*) is, in my opinion, so far as it is correctly understood and made use of in the right ways, a great truth.'

[41] Husserl (1913), section 1: '*Die Welt ist der Gesamtinbegriff von Gegenständen möglicher Erfahrung und Erfahrungserkenntnis, von Gegenständen, die auf Grund aktueller Erfahrungen in richtigen theoretishen Denken erkennbar sind.*'

The terminology is instructive. In Weyl's sense, a 'world geometry' will represent, through mathematical construction from an axiomatic/epistemologically privileged basis, a space of eidetic possibilities sufficient to comprise the laws of gravitation and electromagnetism, constituting what is phenomenologically understood to be the unity of sense of that world.

3.4 *Essential analysis* (Wesensanalyse)

Essential seeing displays the characteristic *noetic–noematic* structure of intentional acts, a subjective structure as a particular intending act (*noesis*), but also an objective structure that is its correlated imminent meaning (*noema*). As new intentional objects may appear with each act of reflection or eidetic insight, an 'essential analysis' (*Wesensanalyse*) is called for to conceptually fix or otherwise determine the levels of structure of each successive act (*noesis*) and of each act's object, its *noema.* In view of many misunderstandings on this point, it may be helpful to note that in taking essences as its objects of investigation, transcendental phenomenology proclaims itself a 'descriptive science of essence' (*deskriptive Wesenswissenschaft*) and proclaims metaphysical neutrality, neither asserting nor denying the non-immanent reality of essences. In this regard, one of several salient analogies, Husserl's phenomenological method bears comparison with the metaphilosophy of the *Aufbau*, as Carnap indeed recognized.[42]

3.5 Eidetic science; regional ontology

Husserl's conception of phenomenology as a 'doctrine of essence' derives, clearly, from his attempt, as *Exmathematicus*, to model phenomenological inquiry on the modes of thinking characteristic of the 'pure eidetic sciences' of pure mathematics: geometry, arithmetic, analysis, and so on.[43] 'Pure phenomenology' is then itself a science purified of any assertions about empirical actuality. The pure phenomenologist, just as the pure mathematician, abstains in principle from judgements concerning the actual, dealing only with ideal possibilities and their related laws. Through a method, later termed 'eidetic variation', the *eidos* or essence that alone is necessary of the object is 'grasped' or 'seized upon' (*Wesenserfassung*). Higher order acts of phenomenological reflection upon these ideal possibilities (belonging to any intentional object) are then to determine what is universal, *invariant*, and pervasively identical in every imaginable particularization constituted as a *possible* actuality. It is this 'mathematical style of thinking', Husserl wrote, that 'orient(ed)

[42] Husserl (1913), sections 52, 55; Carnap (1928), section 3, pp. 64, 65.

[43] Husserl received his doctorate in Vienna in 1882 with a dissertation on the calculus of variations, and was briefly *Assistant* to Weierstrass in Berlin in 1883 before turning to philosophy. From Weierstrass, Husserl claimed to receive 'the ethos of his scientific aspiration' ('*das Ethos seines wissenschaftlichen Strebens*') (Bernet, Kern, and Marhach, 1989, p. 217).

our concept of the *a priori*'. Thus, the pure geometer (to take a favourite example), while employing particular figures or constructions as examples, abstracts from the particulars of the given case to turn it into an arbitrary example. From a few basic concepts, such as point, line, plane, angle, and so on, the geometer can 'derive purely deductively *all* the spatial shapes "existing"', that is, all ideally possible shapes in space and their relationships. This is to 'imagine them in a world of fantasy', thematizing 'pure spatial structures' according to their 'ideal possibilities'. Accordingly, the grounding act of geometrical (and pure mathematical) cognition lies not in observation and experiment but in '*the seeing of essence*'.[44]

The eidetic or *a priori* disciplines fall within two broad categories: so-called 'formal ontology' and 'regional ontology'. In each case, 'ontology' signifies the field of pure *a priori* possibility necessarily within which all objects are constituted in 'transcendental subjectivity'. Inspired by the Leibnizian idea of a *mathesis universalis*, 'formal ontology' is the eidetic science of 'objectivity as such', comprising the disciplines of formal logic, mathematical logic and set theory, whose ideal goal is the development of a 'theory of possible forms of theory'. Here Husserl had an explicit analogue in mind, 'a theory of manifolds in general', such as, somewhat later, is given in Veblen and Whitehead (1932). Cognitions pertaining to formal ontology are analytic and hold of all objects whatsoever. A 'regional ontology', in contrast, is the eidetic science of a particular material discipline comprising *a priori* suppositions underlying the objects of a given domain of inquiry, such as (parts of) physics or chemistry. So, to the various sciences of matter of fact, of actuality, correspond different eidetic sciences or regional ontologies. Each eidetic science of this kind is oriented towards, and explores, the interdependent connection between individual object and essence in its domain, a necessary dependence in both directions.[45] Such a regional essence determines synthetic (as not belonging to formal ontology) eidetic truths; the set of these comprise the regional ontology. Regional axioms ('Kant's synthetic cognitions *a priori*') in turn determine the set of regional categories.[46] These concepts express what is particular to the regional essence or express what must belong, *a priori* and synthetically, to an individual object within the extension of the region. Specified through regional axioms and concepts, a regional ontology is the locus of synthetic *a priori* cognitions concerning the region in question.

In this way a regional ontology is an ideal 'rationalization' of factual existence, corresponding to the 'idea of a completely rationalized science of experience'. Indeed, the very idea of a regional ontology captures the ideal of a complete mathematization of nature, posed at the birth of modern science in the seventeenth

44 Husserl (1913), section 7: 'But for the *geometer* who explores not actualities but "ideal possibilities", not preciatively formed actuality-complexes but predicatively formed eidetic affair-complexes, *the ultimate grounding act* is not experience but rather *the seeing of essences*.'

45 Husserl (1913), section 7.

46 Husserl (1913), section 16.

century.[47] Such a complete rationalization of nature is intended to explicitly recognize that cognition of possibility proceeds cognition of actuality, though the notion of possible object can be established only through phenomenological investigation of the empirical researches of the particular sciences themselves. Hence, this use of eidetic analysis takes empirical science as given. As will be seen, Weyl's 'purely infinitesimal geometry' is characterized in terms showing it is conceived as a 'world geometry' for gravitation and electromagnetism, an ideal regional ontology for field physics.

4 Constituting field physics from 'purely infinitesimal' comparison relations

The broad currents of transcendental phenomenology can now be illuminated within Weyl's attempt (as noted in the 'Introduction' to *RZM*) to 'comprehend the sense and justification of the posit of reality' belonging to field physics *c.* 1918 through a geometry that is 'purely infinitesimal'. As noted, an elucidation of this kind puts in question the fundamental correctness of neither Maxwell's theory nor Einstein's theory of gravitation. Its purpose lies elsewhere. For the supposition that these theories literally portray an objective mind-independent nature is unobjectionable for the purposes of physics, in its exclusive task of the investigation of nature. Yet as a philosophical attitude it is 'dogmatic' in the light of epistemological reflection on the 'sense and justification' of this supposition. Such reflection reminds us that it is also true of the physical world comprising the mathematical structures of field theory that (again in the language of *RZM*'s 'Introduction') 'each of its parts, and all determinations in them, are, and can only be given as intentional objects of acts of consciousness'. In consequence, it will be seen that even this objective world has the primary sense of a 'being for a consciousness'.

How can this possibly be done? The general procedure for a transcendental phenomenological constitution of nature is, according to Weyl, the 'construction of objective reality out of the material of immediate experience'.[48] In itself, this appears to be the epistemological hurdle faced by early twentieth-century sense-datum theorists such as Russell. However, as 'objective reality' is a physical world that is neither space nor time, but field functions of four spacetime variables, the task appears altogether insurmountable. How can epistemological reconstruction get off the ground at all when general relativity has shown that 'nothing of the essence of space and time, given to us in intuition, enters into the mathematically constructed physical world'?[49] How might any such construction show that the generically

47 Husserl (1913), section 9.
48 Weyl (1920), p. 738; (1968), Vol. 2, p. 113.
49 Weyl (1918b); (1923a), p. 3. See also (1918b), p. 172; (1923a), p. 218, and Eng. trans. (1953) of Weyl (1921a), p. 217.

inhomogeneous spacetimes of general relativity arise from what is 'given to consciousness' in space and time, *forms of intuition*, that are necessarily homogeneous?[50] Weyl's approach to a solution to these difficulties was twofold. First, he upheld the phenomenological thesis of 'the general form of consciousness' as a permeation of 'being' and 'essence', understanding of which is 'the key to all philosophy':

The world comes into our consciousness only in the general form of consciousness, a mutual penetration of being (*Seins*) and of essence (*Wesens*), of the 'this' ('*Dies*') and 'thus' ('*So*'). (The profound understanding of this penetration is, incidentally, in my opinion the key to all philosophy.) In acts of reflection we are capable of bringing the essence (*Wesen*), the being-thus (*So-sein*) of phenomena into prominence, to be noticed for itself, without *de facto* being able to loosen it from the individual existence (*Sein*) of that intuitive given in which it appears. Here is the origin of concepts.[51]

Now the 'origin of concepts' lies in locating, through 'essential insight', the hierarchy of essential genera, categories and relations, the 'essence' or set of 'essential predicables', of each contingent phenomenologically given individual 'this'. In the interpenetration of contingency and necessity that is the general form of consciousness also lies the 'origin' of the 'mathematical contemplation of actuality' of theoretical physics, in which 'the attempt is made to represent in the absoluteness of pure being the world given to consciousness'.[52] Mathematical representation of 'actuality' in terms of the 'absoluteness of pure being' recasts the contingently given physical world solely in terms of the purely immanent objects of ideal mathematical possibility. Such a representation carries nothing of the intuitive essence of space and time, possessing no sensory or perceptual qualities at all. Yet according to the thesis of transcendental phenomenological idealism, the representation's sense as an intentional being follows from the mutual penetration of contingent fact and necessary essence, the form of consciousness that 'phenomenologically, one cannot get beyond'.[53] Re-fashioning existing fundamental physical theory from materials provided within the *a priori* subsisting eidetic possibilities of a world geometry is to cast a description of the actual world upon a background canvas of ideal geometric possibilities. It is, ultimately, in furnishing this space of possibilities (in 1918, the four-dimensional spacetime continuum) for the design of the

50 Weyl (1918b); (1923a), pp. 5, 7; (1923b), pp. 24, 44ff.

51 Weyl (1920), p. 738; (1968), Vol. 2, p. 114: '*Die Welt kommt uns nur zum Bewußtseins, welche da ist: eine Durchdringung des Seins und Wesens, des "Dies" und "So". (Das innige Verständnis dieser Durchdringung ist, nebenbei bemerkt, meiner Ueberzeugung nach der Schlüssel zu aller Philosophie.) In Akten der Reflexion sind wir imstande, des* Wesen, *das* So-Sein *der Phänomene zur Abhebung zu bringen, für sich zu bemerken, ohne es doch von dem einzelnen Sein des jeweils anschaulich Gegegenen,* in dem *es erscheint, de facto lösen zu können. Hier der Ursprung der* Begriffe!' (Emphasis in original.)

52 Weyl (1921d), pp. 57–8; reprinted (1968), Vol. 2, p. 159. Eng. trans., p. 100: 'It may be said that, through the mathematical contemplation (*Betrachtung*) of actuality, the attempt is made to represent, in the absoluteness of pure being (*reinen Seins*), the world that is given to consciousness in its more general form of a penetration of being (*Sein*) and essence (*Wesen*) (of the "this" ("*dies*") and the "thus" ("*so*").'

53 Weyl (1926), p. 93; Eng. trans., p. 135.

theoretical description of the physical world that epitomizes the role of geometry in physics.[54]

Secondly, Weyl restricted the homogeneous space of phenomenological intuition, the locus of phenomenological *Evidenz*, to what is given at, or neighbouring, the cognizing ego:

> Only the spatio-temporally coinciding and the immediate spatial-temporal neighbourhood have a directly clear meaning exhibited in intuition.... The philosophers may have been correct that our space of intuition bears a Euclidean structure, regardless of what physical experience says. I only insist, though, that to this space of intuition belongs the ego-centre (*Ich Zentrum*) and that the coincidences, the relations of the space of intuition to that of physics, becomes vaguer the further one distances oneself from the ego-centre.[55]

Guided by the phenomenological methods of 'eidetic insight' and 'eidetic analysis', the epistemologically privileged purely infinitesimal geometric relations of *parallel transport* of a vector, and the *congruent displacement* of vector magnitude, will be the foundation stones of Weyl's reconstruction. The task of comprehending 'the sense and justification' of the mathematical structures of classical field theory is accordingly to be addressed through a construction or *constitution* of the latter within a world geometry generated from these basic geometrical relations immediately evident within a purely infinitesimal space of intuition. This wholly *epistemological* project coincides with the explicitly *metaphysical* aspirations of Leibniz and Riemann to 'understand the world from its behaviour in the infinitesimally small'.[56]

By delimiting what Husserl termed 'the sharply illuminated circle of perfect givenness', the domain of 'eidetic vision', to the infinitely small homogeneous space of intuition surrounding the 'ego-centre', Weyl could limit his attention to linear relations, since only these need be considered in passing to the tangent space of a point in a manifold. Linearity, in turn, gave the expectation of 'uniform elementary laws'.[57] Thus Weyl initially restricted the concept of a coordinate

[54] Weyl (1924), p. 81: 'That for the purpose of its theoretical description we must set the actual (*das Wirkliche*) upon the background of the possible (*des Möglichen*) (of the spacetime continuum with its field structure) signifies, when all is said and done, the appearance of geometry in physics.' This passage occurs in a section of *Erlänterungen und Zusätze* that appeared first in the monographic publication. See also Weyl (1926), p. 94; cf. Eng. trans., p. 131: 'It is rooted in the double nature of the actual (*des Wirklichen*) that we can only design a theoretical image (*Bild*) of the existing (*des Seienden*) upon the background of the possible. Thus the four-dimensional continuum of space and time is, above all else, the field of *a priori* subsisting possibilities of coincidences. Accordingly, Leibniz named the 'abstract space the order of all positions (*Stellen*) assumed possible' and adds: 'consequently it is something ideal (*etwas Ideales*)' (Leibniz's Fifth Letter to Clarke; cf. Leibniz and Clarke (1956, paragraph 104)).

[55] Weyl (1931), pp. 49, 52. The quotation continues: 'That is mirrored in theoretical construction in the relation between the curved surface and its tangent plane at the point P: both cover the immediate surroundings of the centre P, but the further one proceeds from P, the more arbitrary becomes the continuation of an unambiguous correspondence of the covering relation between surface and plane.' Cf. Weyl (1926), p. 98; Eng. trans., p. 135: 'a space of intuition whose metrical structure on essential grounds (*aus Wesensgründen*) fulfills the Euclidean laws does not contradict physics in so far as it clings to the Euclidean character of the infinitely small region of a point O (at which I momentarily find myself).'

[56] Weyl (1918b), p. 82; (1923a), p. 86. Also Weyl (1926), p. 61; Eng. trans., p. 86.

[57] Weyl (1926), p. 61; Eng. trans., p. 86.

system to the tangent space covering each manifold point P, essentially assuming a four-dimensional manifold that is Hausdorff, simply connected, and differentiable. Imposition of a local coordinate system is regarded as the original constitutive act of 'a pure, sense-giving ego'. Required by the operations of differential calculus on a manifold, a coordinate system always bears an indelible mark of transcendental subjectivity, it is 'the unavoidable residue of the ego's annihilation in that geometrico-physical world which reason sifts from the given under the norm of "objectivity" '.[58] In this necessary presupposition of any differential structure, Weyl recognized an intimation of the phenomenological postulate that 'existence is only given and *can* only be given as the intentional content of the conscious experience of a pure, sense-giving ego'.[59]

The next steps concern the immediately evident 'purely infinitesimal' relations of comparison of direction and magnitude that depend on a specific choice of coordinates and unit of scale. The construction of pure infinitesimal geometry is laid out as taking place in three distinct stages of 'connection': topological manifold or 'continuous connection' (*stetiger Zusammenhang*), affine connection, and 'metric (or, length) connection'.[60] The construction itself, 'in which each step is executed in full naturalness, visualizability and necessity' (*in voller Natürlichkeit, Anschaulichkeit, und Notwendigkeit*), is 'in all essential parts the final result' of the renewed investigation of the mathematical foundations of Riemannian geometry opened up by Levi-Civita's discovery of the concept of infinitesimal parallel displacement.[61] The physical world is then to be distinguished within this 'world geometry' through the unique choice of a gauge invariant action function $S(g_{\mu\nu}, \varphi_\mu)$, where $g_{\mu\nu}$ is the (only conformally invariant) metric tensor and φ_μ is the electromagnetic four potential.[62] However, to Weyl's dismay, it soon became apparent that a number of such functions could be constructed, choice among them being essentially arbitrary.[63]

4.1 First stage: continuous connection (topology)

Weyl's several discussions of topology in the context of his geometry add little to topology *per se* but take over the modern topological concepts of 'point' and 'neighbourhood', first clarified in his own 1913 book on Riemann surfaces. There

58 Weyl (1918a), p. 72; Eng. trans., p. 93: 'The coordinate system is the unavoidable residuum of the ego's annihilation (*das unvermeidliche Residuum der Ich-Vernichtung*) in that geometrico-physical world which reason sifts from the given under the norm of "objectivity" – a final scanty token in this objective sphere that existence (*Dasein*) is only given and *can* only be given as the intentional content of the conscious experience of a pure, sense-giving ego.' See also Weyl (1921a), p. 8; Eng. trans., p. 8.

59 Weyl (1918a), p. 72; Eng. trans., p. 94.

60 Weyl (1918d); (1919b); (1921a); (1923b).

61 *'Vorwort zur dritten Auflage'*, in Weyl (1919b), p. vi.

62 Weyl (1918d), p. 385; (1968), Vol. 2, p. 2.

63 See Weitzenböck (1920).

is a clearly identified reason for his reticence in extending phenomenological constitution to the manifold, and so to the concept of 'continuous connection' itself. With reference to his discussion in *Das Kontinuum* of the 'deep chasm' separating the intuitive and the mathematical continuum, Weyl observed that a 'fully satisfactory analysis of the concept of the *n-dimensional manifold* is not possible today' in view of the 'difficulty of grasping the intuitive essence (*anschauliche Wesen*) of continuous connection through a purely logical construction'.[64] Setting that task aside, Weyl simply assumed that in the tangent space covering each manifold point P, there is an affine linear space of vectors centred on P in that line elements dx radiating from P are infinitely small vectors. In this way, functions at P and in its neighbourhood (in particular, the displacement functions – see below) transform linearly and homogeneously. Weyl's attention then concentrated on the manifold's *Strukturfeld*, its metric, affine (and conformal, and projective) structures, originating the now familiar machinery of connections in a specifically philosophical context.

4.2 Second stage: affine connected manifold

The concept of parallel transport of a tangent vector in a Riemannian manifold M was first developed in 1917 by Levi-Civita (and independently by Schouten in 1918). It provided a geometric interpretation – as the parallel displacement of a vector along a path connecting a point P to another point P' in the infinitesimal neighbourhood (tangent space T_P) of $P(T_{P'}M = T_P M)$ – to the hitherto purely analytical Christoffel symbols (of the second kind) of covariant differentiation. This enabled covariant differentiation to be understood as a means of comparing infinitesimal changes in vector or tensor fields at neighbouring points with respect to a parallel transported vector. Parallel transport is purely infinitesimal in the sense that directional comparison of vectors at finitely distant points P and Q can be made only by specifying a path of displacement from P to Q and 'transporting' from point to point along it a comparison vector defined at Q as 'parallel to' the original vector at P. In general, parallel displacement is not integrable, i.e. the new vector arising at Q will depend upon the path taken between P and Q.

In Weyl's assessment, Levi-Civita's concept marked a significant advance of 'simplicity and visualizability (*Anschaulichkeit*) in the construction of Riemannian infinitesimal geometry'.[65] But whereas Levi-Civita had employed an auxiliary construction, embedding M in a Euclidean space where parallel transport was defined, and then projecting it into the tangent space of M, Weyl gave the first intrinsic characterization in terms of bilinear functions $\Gamma(A^{\mu}, dx)$ since known as the

[64] Weyl (1918d), p. 386; (1968), Vol. 2, p. 3.
[65] Weyl (1923b), p. 11.

components of a (symmetrical) affine connection.[66] In general, the change δA^{μ} in a given vector A^{μ} displaced from P to $P'_{(x^{\nu}+dx^{\nu})}$ is defined

$$\delta A^{\mu} = -\Gamma^{\mu}_{\alpha\beta}\, A^{\alpha}\, dx^{\beta}, \tag{1}$$

while the *covariant derivative* of A^{μ} (a tensor, and so of objective significance) is defined

$$A^{\mu}_{\alpha} = \frac{\partial A^{\mu}}{\partial x^{\alpha}} + \Gamma^{\mu}_{\beta\alpha}\, A^{\beta}. \tag{2}$$

Parallel transport occurs when the components of the affine connection vanish. Next followed the concept of a manifold with an affine connection. A point P is affinely connected with its immediate neighbourhood just in case it is determined, for every vector at P, the vector at P' to which it gives rise under parallel transport from P to P'. If it is possible to single out a unique affine connection, among all the possible ones at each point P, then M is called a *manifold with an affine connection.* This is essentially a conception of space as stitched together in linear fashion from infinitely small homogeneous patches. To Weyl, parallel transport was the paradigm comparison relation of infinitesimal geometry for it satisfied the epistemological demand that all integral (and so, not immediately surveyable) relations between finitely separated points cannot be posited but must be constructed from a specified infinitesimal displacement along a given curve connecting them. He also introduced the idea of the curvature of a connection $R(\Gamma)$, a (1, 3) tensor analogous to the Riemann–Christoffel tensor of Riemannian geometry, and showed that the calculus of tensors could be developed on the basis of the concept of infinitesimal parallel transport, without any reliance on a metric.[67] However, it was Eddington, not Weyl, who first fully exploited this idea in physics.[68]

The 'essence of parallel displacement' (*das Wesen der Parallelverschiebung*) is expressed in that, in a given coordinate system covering P and its neighbourhood, the components of an arbitrary vector A^{μ} do not change when A^{μ} is parallel-displaced from P to a neighbouring point P'.[69] Unaltered displacement accordingly depends on a particular 'geodesic' (at P) coordinate system, proleptically referring to the fact that at P the $g_{\mu\nu}$ have stationary values, $\partial g_{\mu\nu}/\partial x^{\sigma} = 0$, and so the components of the affine connection vanish. According to the principle of equivalence, such geodesic coordinates at a point always exist. In this dependence on a particular coordinate system, parallel displacement of a vector or tensor without 'absolute change' is not an invariant or 'objective' relation. But a specifically epistemological and non-conventional meaning is intended for the

66 Following Cartan, such a connection is called 'without torsion'.
67 Weyl (1923b), p. 17. A metric tensor is needed only to raise or lower indices.
68 Eddington (1921); for discussion, see Ryckman (forthcoming).
69 Weyl (1923a), p. 113; (1921c), p. 542, reprinted in (1968), Vol. 2, p. 238. See also Scholz (1994).

statement that some vector at P' is 'the same' as a given vector at P. Namely from the original vector at P, a new vector arises at P' that, in the purely local comparison made, as it were, by a particularly situated consciousness, is affirmed to be 'without change'. Despite the subjectivity of the 'experienced' condition $\partial g_{\mu\nu}/\partial x^\sigma = 0$ required by this construction, such comparison is nonetheless the basis for the invariant relation of covariant differentiation. Obviously, the idea is an analogy formed from Einstein's theory in which the non-tensorial gravitational field strengths $\Gamma^\sigma_{\mu\nu}$ (in Weyl's suggestive terminology, the 'guiding field' (*Führungsfeld*)) can be locally, but not generally, 'transformed away', an observer-dependent 'disappearance' of a gravitational field. At the same time, invariant space-time curvatures are derived from the $\Gamma^\sigma_{\mu\nu}$ that have an objective significance for all observers.

4.3 Third stage: metrically connected manifold

In Weyl's eyes,

> *a truly infinitesimal geometry* (*wahrhafte Nahegeometrie*) *should know only a principle of displacement* (*Übertragung*) *of a length from one point to another infinitely close by.*[70]

As the 'essence of space' is metric, the fundamental metrical concept, congruence, also must be conceived 'purely infinitesimally'.[71] Enshrined as 'the epistemological principle of relativity of magnitude', a postulate is laid down that direct comparison of vector magnitudes can be immediately made only at a given point P or at infinitesimally nearby points P' ($P' - P = \overrightarrow{P'P} \in T(M_P)$). Just as an affine connection governs direct infinitesimal comparisons of orientation, or parallelism, so a *length* or *metric connection* is required to determine infinitesimal comparisons of congruence. This also requires a vector to be displaced from P to P' and, in general, the (square of the) 'length' l of the vector is altered. Thus if l is the (squared) length of a vector A^μ at $P_{(x)}$, $l_{P_{(x)}}(A^\mu) = ds^2 = g_{\mu\nu}A^\mu A^\nu$, then on being displaced to P', the change of length is defined to be a definite fraction of l,

$$\frac{dl}{dx^\mu} := -l\frac{d\varphi}{dx^\mu}, \tag{3}$$

where $d\varphi = \sum_\mu \varphi_\mu \, dx^\mu$ is a homogeneous function of the coordinate differentials. The new vector at P', corresponding to A^μ at P, accordingly has the length

$$l_{P'_{(x+dx)}} = (1 - d\varphi)(g_{\mu\nu} + dg_{\mu\nu})A^\mu A^\nu, \tag{4}$$

[70] Weyl (1918c), p. 466; (1968), Vol. 2, p. 30. Emphasis in original.
[71] Weyl (1923b), p. 47.

where $(1 - d\varphi)$ is a proportionality factor, arbitrarily close to 1. In analogy to (1), the change in length of A^μ is defined as

$$\delta l := \frac{\partial l}{\partial x^\mu} dx^\mu + l\, d\varphi. \tag{5}$$

Then, just as the vanishing of its covariant derivative means that a vector has been parallel-displaced from P to P' without 'absolute change', so here the vanishing of δl indicates that A^μ has been *congruently displaced* from P to P':

$$\delta l = 0 \Leftrightarrow dl = \frac{\partial l}{\partial x^\mu} dx^\mu = -l\, d\varphi. \tag{6}$$

Congruent displacement involves the assumption of a particular 'gauge', a unit of scale Weyl called 'geodetic' at P. Re-calibrating the unit of length at P through multiplication by λ, an always positive function of the coordinates, multiplies the length $l_{P(x)}$ by λ, $l' = \lambda l$. Then the change in length at P', dl', corresponds to a transformation of the 'length connection' $d\varphi$,

$$\begin{aligned} dl' &= d(\lambda l) = l\, d\lambda + \lambda\, dl = l\, d\lambda - \lambda l\, d\varphi \\ &= -\lambda l \left(d\varphi - \frac{d\lambda}{\lambda} \right) = -l' d\varphi', \end{aligned} \tag{7}$$

$(d\lambda/\lambda = d \log \lambda)$. A *metrically connected manifold* is then one in which each point P is metrically connected to every point P' is its immediate neighbourhood through a *length connection* $\phi_\mu dx^\mu$. In general, length is not integrable for (8) follows from (5) by integration,

$$\log l\big|_P^Q = -\int_P^Q \varphi_\mu\, dx^\mu, \ \text{and so}\ \ l_Q = l_P^{\,-\int_P^Q \varphi_\mu\, dx^\mu}. \tag{8}$$

As Pauli demonstrated, displacement of a vector along different paths between finitely separated points P and Q will lead to arbitrarily different results at Q.[72] But when the linear form φ_μ vanishes, the magnitude of a vector is independent of the path along which it is displaced, which is just the case of Riemannian geometry. The necessary and sufficient condition for this is the disappearance of the 'length curvature' (*Streckenkrümmung*) of Weyl's geometry

$$F_{\mu\nu} = \frac{\partial \varphi_\nu}{\partial x^\mu} - \frac{\partial \varphi_\mu}{\partial x^\nu}, \tag{9}$$

just as the vanishing of the Riemann tensor is the necessary and sufficient condition for flat space.

Implementation of the local comparison condition means that the fundamental tensor $g_{\mu\nu}$ of Riemannian geometry induces only a local conformal structure on the manifold. There is then an immediate meaning given to the angle between two

[72] Pauli (1921); Eng. trans., pp. 195–6.

vectors at a point, or to the ratio of their lengths there, but not to their absolute lengths. These transform at a point x as $g'_{\mu\nu}(x) = \lambda\, g_{\mu\nu}(x)$ where $\lambda(x) = e^{\phi_\mu(x)}$ is the gauge function. This weakening of the metrical structure has two important consequences. Such a metric no longer determines a unique linear (affine) connection, but only an equivalence class of connections. Yet Weyl required, as the 'fundamental fact' of infinitesimal geometry, that there be unique affine compatibility in the sense that the transport of tangent vectors along curves associated with the connection, i.e. affine geodesics, leave the vectors congruent with themselves with respect to the metric. Weyl showed that a unique connection, coupled to given choice of a metric tensor, is found by incorporating into its definition the linear differential form φ_μ of his length connection. Then, when the components of φ_μ vanish identically at a point, the connection becomes identical to the 'Levi-Civita' connection, as can be seen from comparison of the definitions of the two connections in components:

$$\text{Levi-Civita: } \Gamma^\sigma_{\mu\nu} = \frac{1}{2} g^{\sigma\tau} \left(\frac{\partial g_{\tau\mu}}{\partial x^\nu} + \frac{\partial g_{\nu\tau}}{\partial x^\mu} - \frac{\partial g_{\mu\nu}}{\partial x^\tau} \right); \tag{10}$$

$$\text{Weyl: } \Gamma^\sigma_{\mu\nu} = \frac{1}{2} g^{\sigma\tau} \left(\frac{\partial g_{\tau\mu}}{\partial x^\nu} + \frac{\partial g_{\nu\tau}}{\partial x^\mu} - \frac{\partial g_{\mu\nu}}{\partial x^\tau} \right) + \frac{1}{2}(g_{\mu\sigma}\varphi^\nu + g_{\nu\sigma}\varphi^\mu + g_{\mu\nu}\varphi^\sigma). \tag{11}$$

Given this 'Weyl connection', it is possible to speak of a manifold with an affine connection where, as in the Riemannian case, there is a unique determination of parallel displacement of a vector at every point. Only in the case of 'congruent displacement' (*kongruente Verpflanzung*), or displacement without alteration of length, is parallel displacement possible, and so it is that infinitesimal length or 'tract displacement' (*Streckenübertragung*), the 'foundational principle of metric geometry', brings along also directional displacement (*Richtungsübertragung*). This is to say, according to Weyl, that '*according to its nature, a metric space bears an affine connection.*'[73]

Justification of his 'essential analysis' of infinitesimal geometry culminated in Weyl's purely mathematical group-theoretical proof of Riemann's posit of an 'infinitesimal Pythagorean (Euclidean) metric', the capstone of his efforts to show that the supposition of the purely infinitesimal character of the geometry underlying field physics was not arbitrary.[74] Writing to Husserl on 26 March 1921, Weyl could report that he had finally captured the '*a priori* essence of space (*apriorische Wesen des Raumes*), through a notable deepening of its mathematical foundations (*eine merkliche Tieferlegung der Fundamente*).'[75] This philosophical linkage was

[73] Weyl (1923a), p. 124: '*ein metrischer Raum tragt von Natur einen affinen Zusammenhang*'. Laugwitz (1958) proved Weyl's conjecture, showing that this condition singles out infinitesimal Euclidean metrics from the wider class of Finsler metrics.

[74] Weyl (1921b), p. 497; (1968), Vol. 2, p. 235; the full treatment appears in Weyl (1923b).

[75] Husserl (1994), p. 291.

publicly announced in a newly appended fourth (1921) edition of *RZM*. There Weyl declared that his 'investigations concerning space . . . appear to me to be a good example of the essential analysis (*Wesenanalyse*) striven for by phenomenological philosophy (Husserl)'.[76] Referring to this passage some thirty years later, Weyl observed that he still essentially held to its implicit characterization of the relation between cognition and reflection underlying his method of investigation, one that combined experimentally supported experience, analysis of essence (*Wesensanalyse*) and mathematical construction.[77]

4.4 The transition to physics[78]

Just as Einstein required the invariance of physical laws under arbitrary continuous transformation of the coordinates (general covariance), Weyl additionally demanded their invariance under the 'gauge transformations'

$$g \Rightarrow g'_{\mu\nu} = \lambda\, g_{\mu\nu}, \quad \text{and} \quad \varphi \Rightarrow \varphi'_{\mu} = \varphi_{\mu} - \frac{1}{\lambda}\frac{\partial \lambda}{\partial x^{\mu}}. \tag{12}$$

And since the first system of Maxwell's equations

$$\frac{\partial F_{\mu\nu}}{\partial x^{\sigma}} + \frac{\partial F_{\nu\sigma}}{\partial x^{\mu}} + \frac{\partial F_{\sigma\mu}}{\partial x^{\nu}} = 0 \tag{13}$$

follows immediately from (9) on purely formal grounds, Weyl made the obvious identifications of his length curvature $F_{\mu\nu}$ with the already gauge-invariant electromagnetic field tensor (of 'gauge weight 0'), and his metric connection φ_{μ} with the spacetime four potential. As a mathematical consequence of his geometry, equations (13) are held to express 'the essence of electricity'; they are an 'essential law' (*Wesensgesetze*) whose validity is completely independent of the actual laws of nature.[79] Furthermore, Weyl could show that a vector density and contravariant second rank tensor density follow from the *general* form of a hypothetical action function invariant under local changes of gauge $\lambda = 1 + \pi$, where π is an arbitrarily

[76] Weyl (1921a), p. 133; cf. Eng. trans. (1953), p. 148: 'The investigations made concerning space in chapter two appear to me to be a good example of the essential analysis (*Wesenanalyse*) striven for by phenomenological philosophy (Husserl), an example that is typical for such cases where a non-immanent essence is dealt with. We see in the historical development of the problem of space, how difficult it is for us reality-prejudiced humans to hit upon what is decisive. A long mathematical development, the great unfolding of geometrical studies from Euclid to Riemann, the physical exploration of nature and its laws since Galileo, together with all its incessant boosts from empirical data, finally, the genius of singularly great minds – Newton, Gauss, Riemann, Einstein – all were required to tear us loose from the accidental, non-essential characteristics to which we at first remain captive. Certainly, once the true standpoint has been attained, reason (*Vernunft*) is flooded with light, recognizing and accepting what is understandable out-of-itself (*das ihr aus-sich-selbst Verständliche*)'. 'The example of space', Weyl continued, 'is most instructive for that question of phenomenology that seems to me particularly decisive: to what extent the delimitation of the essentialities (*Wesenheiten*) rising up to consciousness express a characteristic structure of the domain of the given itself and to what extent mere convention participates in it.'

[77] Weyl (1955), p. 161.

[78] Weyl (1918c); (1919a); (1919b).

[79] Weyl (1919b), p. 244.

specified infinitesimal scalar field. These are respectively identified with the four current density $\mathbf{j}^\mu$ and the electromagnetic field density $\mathbf{h}^{\mu\nu}$, through the relation

$$\frac{\partial \mathbf{h}^{\mu\nu}}{\partial x^\nu} = \mathbf{j}^\mu, \tag{14}$$

i.e. the second system of Maxwell equations. Thus Weyl claimed that, without having to specify a particular action function, 'the entire structure of the Maxwell theory could be read off of gauge invariance'.[80] Again, using only the general form of such a function, he demonstrated that conservation of energy–momentum and of charge follow from the field laws in two *distinct* ways.[81] Accordingly he asserted that, just as the Einstein theory had shown that the agreement of inertial and gravitational mass was 'essentially necessary' (*wesensnotwendig*), his theory did so in regard to the facts finding expression in the structure of the Maxwell equations, and in the conservation laws. This appeared to him to be 'an extraordinarily strong support' for the 'hypothesis of the essence of electricity' (*Wesen der Elektrizität*).[82] The domain of validity of Einstein's theory of gravitation, with its assumption of a global unit of scale, was originally held to correspond to $F_{\mu\nu} = 0$, the vanishing of the electromagnetic field tensor. By 1919, Weyl substituted his own 'dynamical' account of the origin of 'the natural gauge of the world' noted above. These details of Weyl's theory will suffice for present purposes; further discussion is available elsewhere.[83] We will only reiterate that Weyl gave a sustained and detailed response to Einstein's well-known criticism, that the theory is incompatible with the observed sharp spectral lines of the chemical elements, replying that his theory, but not Einstein's, provided the possibility of accounting for this 'natural gauge of the world'. In the fifth (1923) edition of *Raum-Zeit-Materie*, the last to appear in his lifetime, the theory was defended not so much as a physical hypothesis, but as 'a theoretically very satisfying amalgamation and interpretation of our whole knowledge of field physics.'[84]

5 Concluding remarks

Dirac's celebrated 1931 paper on magnetic monopoles is prefaced with several remarks that have the tenor of announcing a sea-change in the methodology of theoretical physics. Anticipating that drastic revision of fundamental concepts may well be required to address the current problems of theoretical physics, Dirac cautioned that such a transformation in outlook is likely be beyond the power of human

80 Weyl (1919b), p. 251.
81 Weyl (1919b), pp. 251–2; (1923a), pp. 314–15; for discussion see Brading (2002).
82 Weyl (1919b), p. 253.
83 See Ryckman (forthcoming), chapter 4.
84 Weyl (1923a), p. 308.

intelligence to immediately form the required new ideas. In the face of these cognitive limitations, a more indirect approach is suggested, wherein 'the most powerful method of advance' would be

> to perfect and generalize the mathematical formalism that forms the existing basis of theoretical physics, and *after* each success in this direction, to try to interpret the new mathematical features in terms of physical entities (by a process like Eddington's Principle of Identification).[85]

Now this principle, as Eddington himself makes clear,[86] was inspired by Weyl's mathematical identification of the vector and tensor structures of his purely infinitesimal world geometry with those of gravitation and electromagnetism. That being the case, Weyl in 1918 can be justly regarded as the modern revival of the method of *a priori* mathematical conjecture in fundamental physical theory. That this method can ever be fruitful in according physical reality to mathematical entities has long appeared a mystery, for which extreme solutions (such as Platonism) have been seriously proposed. But in revisiting the philosophical context of origin of the concept of local symmetry we see that less desperate measures may have been overlooked. For if we recall, with Einstein, that the 'truly valuable' core of Kant's doctrine is expressed in the sentence 'the real is not given to us but posed as a problem (*nicht gegeben, sondern aufgegeben*)',[87] a more intelligible meaning accrues to Weyl's notable statement that 'all *a priori* statements in physics have their origin in symmetry'. Despite the difficult nexus of transcendental phenomenology through which Weyl himself arrived at this conclusion, his example in 1918 is a canonical demonstration of how and why *a priori* constraints of reasonableness can be imposed on nature without proudly presuming them to be inherent in nature itself.

References

Bergia, S. (1993). 'The fate of Weyl's theory of 1918'. In *History of Physics in Europe in the Nineteenth and Twentieth Centuries*, ed. F. Bevilacqua (European Physical Society Conference, Como, 2–3 September 1992), pp. 185–93. Bologna: Italian Physical Society.

Bernet, R., Kern, I., and Marhach, E. (1989). *Edmund Husserl: Darstellung seines Denkens*. Hamburg: Felix Meiner Verlag.

Brading, K. A. (2002). 'Which symmetry? Noether, Weyl, and the conservation of charge'. *Studies in History and Philosophy of Modern Physics*, **33**, 3–22.

Cao, T. N. (1997). *Conceptual Developments of Twentieth Century Field Theories*. Cambridge: Cambridge University Press.

85 Dirac (1931), p. 60.
86 Eddington (1923), section 96; for discussion see Ryckman (forthcoming).
87 Einstein (1949), p. 680.

Carnap, R. (1928). *Der logische Aufbau der Welt*. Berlin: Weltkreis Verlag. Translated by R. A. George as *The Logical Structure of the World* (1969). Berkeley and Los Angeles: University of California Press.
Dirac, P. A. M. (1931). 'Quantised singularities in the electromagnetic field'. *Proceedings of the Royal Society of London A*, **133**, 60–72.
Douglas, A. V. (1956). 'Forty minutes with Einstein'. *Journal of the Royal Astronomical Society of Canada*, **50**, 99–102.
Eddington, A. S. (1921). 'A generalization of Weyl's theory of gravitation and electromagnetism'. *Proceedings of the Royal Society of London A*, **99**, 104–22.
(1922). 'Review of *Space-time-matter*'. *Nature*, **109**, 634–6.
(1923). *The Mathematical Theory of Relativity*. Cambridge: Cambridge University Press.
Einstein, A. (1949). 'Reply to criticisms'. In *Albert Einstein, Philosopher–Scientist*, ed. P. A. Schilpp, pp. 663–88. Evanston, IL: Northwestern University Press.
(1998). *The Collected Papers of Albert Einstein*, Vol. 8, Part B, ed. R. Schulmann, A. J. Kox, M. Janssen, and J. Illy. Princeton, NJ: Princeton University Press.
Gross, D. (1999). 'The triumph and the limitations of quantum field theory'. In *Conceptual Foundations of Quantum Field theory*, ed. T. Y. Cao, pp. 56–67. New York and Cambridge: Cambridge University Press.
Husserl, E. (1900). *Logische Untersuchungen, Vol. 1: Prologomena zur reinen Logik*. Halle: Max Niemeyer. Pagination according to 1922 reprint of 1913 second edition. Translated by J. N. Findlay as *Logical Investigations, Vol. 1: Prolegomena to Pure Logic* (1970). London: Routledge and Kegan Paul.
(1911). 'Philosophie als strenge Wissenschaft'. *Logos*, **1**, 289–341. Translated by Q. Lauer as 'Philosophy as rigorous science' (1965). In E. Husserl, *Phenomenology and the Crisis of Philosophy*. New York: Harper and Row.
(1913). *Ideen zu einer reinen phänomenologischen Philosophie (Jahrbuch für Philosophie und phänomenologische Forschung)*, Vol. 1, pp. 1–323. Halle a.d. Saale: Max Niemeyer. Translated by F. Kersten as *Ideas Pertaining to a Pure Phenomenology and to a Phenomenological Philosophy. First book* (1983). The Hague: Martinus Nijhoff.
(1994). *Edmund Husserl Briefwechsel*, Vol. 7, ed. K. and E. Schuhmann. Dordrecht, Boston, London: Kluwer Academic Publishers.
Laugwitz, D. (1958). 'Über eine Vermutung von Hermann Weyl zum Raumproblem'. *Archiv der Mathematik*, **9**, 128–33.
Leibniz, G. W., and Clarke, S. (1956). *The Leibniz–Clarke Correspondence*, ed. H. G. Alexander. Manchester: Manchester University Press.
Mancosu, P., and Ryckman, T. A. (2002). 'Mathematics and phenomenology: the correspondence between O. Becker and H. Weyl.' *Philosophia Mathematica*, **10**, 130–202.
Mielke, E., and Hehl, F. (1988). 'Die Entwicklung der Eichtheorien: Marginalien zu deren Wissenschaftsgeschichte'. In *Exact sciences and their philosophical foundations: Proceedings of the International Hermann Weyl Congress, Kiel 1985*, ed. W. Deppert *et al.* pp. 191–231. Frankfurt, Bern, New York: Peter Lang Verlag.
O'Raifeartaigh, L. (1997). *The Dawning of Gauge Theory*. Princeton, NJ: Princeton University Press.
O'Raifeartaigh, L., and Straumann, N. (2000). 'Gauge theory: historical origins and some modern developments'. *Reviews of Modern Physics*, **72**, 1–23.
Pauli, W. (1919). 'Merkurperihelbewegung und Strahlenablenkung in Weyls Gravitationstheorie'. *Verhandlungen der Deutschen Physikalischen Gesellschaft*, **21**, 742.

(1921). 'Relativitätstheorie'. In *Encyklopädie der mathematischen Wissenschaften*, Vol. 19. Leipzig: Teubner. Translated as *The Theory of Relativity* (1958). New York: Pergamon Press.
Ryckman, T. A. (forthcoming). *The Reign of Relativity: Philosophy in Physics, 1915–1925.*
Scholz, E. (1994). 'Hermann Weyl's contribution to geometry, 1917–1923'. In *The Intersection of History and Mathematics*, ed. S. Chikara, S. Mitsuo, and J. W. Dauben, pp. 203–30. Basel, Boston, Berlin: Birkhäuser Verlag.
(2001). 'Weyls Infinitesimalgeometrie'. In *Hermann Weyl's 'Raum-Zeit-Materie' and a General Introduction to his Scientific Work*, ed. E. Scholz, pp. 48–104. Basel, Boston, Berlin: Birkhäuser Verlag.
Straumann, N. (1987). 'Zum Ursprung der Eichtheorien bei Hermann Weyl'. *Physikalische Blätter*, **43**, 414–21.
(2001). 'Ursprünge der Eichtheorien'. In *Hermann Weyl's 'Raum-Zeit-Materie' and a General Introduction to his Scientific Work*, ed. E. Scholz, pp. 138–55. Basel, Boston, Berlin: Birkhäuser Verlag.
Veblen, O., and Whitehead, J. H. C. (1932). *The Foundations of Differential Geometry*. London: Cambridge University Press.
Vizgin, V. (1994). *Unified Field Theories in the First Third of the Twentieth Century.* Translated from the Russian original by J. Barbour. Basel, Boston, Berlin: Birkhäuser Verlag.
Weitzenböck, R. (1920). 'Über die Wirkungsfunktion in der Weyl'schen Physik'. *Akademie der Wissenschaften in Wien. Sitzungsberichte, Abteilung IIa. Mathematisch-naturwissenschaftliche Klasse*, **129**, 683–708.
Weyl, H. (1913). *Die Idee der Riemannschen Fläche*. Leipzig: Teubner.
(1918a). *Das Kontinnum: Kritische Untersuchungen über die Grundlagen der Analysis.* Leipzig: Veit. Translated by S. Pollard and T. Bowl as *The Continuum: A Critical Examination of the Foundations of Analysis* (1994). New York: Dover.
(1918b). *Raum-Zeit-Materie*. Berlin: Springer-Verlag. The second edition (1919) is a reprint of this edition.
(1918c). 'Gravitation und Elektrizität'. *Preußische Akademie der Wissenschaften (Berlin) Sitzungsberichte. Physikalisch-Mathematische Klasse*, pp. 465–80. Reprinted in Weyl (1968), Vol. 2, pp. 29–42. Translation in O'Raifeartaigh (1997), pp. 24–37.
(1918d). 'Reine Infinitesimalgeometrie'. *Mathematische Zeitschrift*, **2**, 384–411. Reprinted in Weyl (1968), Vol. 2, pp. 1–28.
(1919a). 'Eine neue Erweiterung der Relativitätstheorie'. *Annalen der Physik*, **21**, 649–81. Reprinted in Weyl (1968), Vol. 2, pp. 55–87.
(1919b). *Raum-Zeit-Materie*, 3rd edn. Berlin: Springer-Verlag.
(1920). 'Das Verhältnis der kausalen zur statischen Betrachtungsweise in der Physik'. *Schweiserische Medizinische Wochenschrift*, 737–41. Reprinted in Weyl (1968), Vol. 2, pp. 113–22.
(1921a). *Raum-Zeit-Materie*, 4th edn. Berlin: Springer-Verlag. Translated by H. L. Brose as *Space-Time-Matter* (1922). London: Methuen. Translation reprinted (1953). Dover: New York.
(1921b). 'Über die physikalischen Grundlagen der erweiterten Relativitätstheorie'. *Physikalische Zeitschrift*, **22**, 473–80. Reprinted in Weyl (1968), Vol. 2, pp. 99–112.
(1921c). 'Feld und Materie'. *Annalen der Physik*, **65**, 541–63. Reprinted in Weyl (1968), Vol. 2, pp. 237–59.

(1921d). 'Über die neue Grundlagenkrise der Mathematik'. *Mathematische Zeitschrift*, **10**, 39–79. Reprinted in Weyl (1968), Vol. 2, pp. 143–79. Translated by B. Müller as 'On the new foundational crisis of mathematics', in P. Mancosu, *From Brouwer to Hilbert: The Debate on the Foundations of Mathematics in the 1920s* (1998), pp. 86–122. New York and Oxford: Oxford University Press.

(1923a). *Raum-Zeit-Materie*, 5th edn. Berlin: Springer-Verlag.

(1923b). *Mathematische Analyse des Raumproblems: Vorlesungen gehalten in Barcelona und Madrid*. Berlin: Springer-Verlag.

(1924). 'Was ist Materie?' *Die Naturwissenschaften*, **12**, 561–68; 585–93; 604–11. Separately issued as *Was ist Materie? Zwei Aufsätze zur Naturphilosophie*. Berlin: Springer-Verlag. Reprinted in Weyl (1968), Vol. 2, pp. 486–510.

(1925). 'Die Heutige Erkenntnislage in der Mathematik'. *Symposium*, **1**, 1–32. Reprinted in Weyl (1968), Vol. 2, pp. 511–42.

(1926). 'Philosophie der Mathematik und Naturwissenschaft.' In *Handbuch der Philosophie*, Part 2, ed. A. Baeumler and M. Schröter. Munich and Berlin: Oldenbourg. Translated by O. Helmer as *Philosophy of Mathematics and Natural Science* (1949), a revised and augmented English edition. Princeton, NJ: Princeton University Press.

(1929). 'Elektron und Gravitation'. *Zeitschrift für Physik*, **56**, 330–52. Reprinted in Weyl (1968), Vol. 3, pp. 245–67. Translation in O'Raifeartaigh (1997), pp. 121–44.

(1931). 'Geometrie und Physik'. *Die Naturwissenschaften*, **19**, 49–58. Reprinted in Weyl (1968), Vol. 3, pp. 336–45.

(1952). *Symmetry*. Princeton, NJ: Princeton University Press.

(1955). 'Erkenntnis und Besinnung' (Vortrag, gehalten an der Universität Lausanne im Mai 1954). *Studia Philosophia* (Jahrbuch der Schweizerischen Philosophischen Gesellschaft), **15**, 153–71. Reprinted in Weyl (1968), Vol. 4, pp. 631–49.

(1968). *Gesammelte Abhandlungen*, Vols. 1–4, ed. K. Chandrasekharan. Berlin, Heidelberg, New York: Springer-Verlag.

Yang, C. N. (1977). 'Magnetic monopoles, fibre bundles, and gauge fields'. *Annals of the New York Academy of Sciences*, **294**, 86–97.

5

Symmetries and Noether's theorems

KATHERINE BRADING AND HARVEY R. BROWN

1 Introduction

Emmy Noether's greatest contributions to science were in algebra, but for physicists her name will always be remembered for her paper of 1918 on an invariance problem in the calculus of variations.[1] The most celebrated part of this work, associated with her 'first theorem', has to do with the connection between continuous (global) symmetries in Lagrangian dynamics and conservation principles, though the main focus of the paper was the relationship between this and the second part of her paper, where she gives a systematic treatment of the more subtle and general case of continuous *local* symmetries (symmetries depending on arbitrary functions of the spacetime coordinates).

The connection between global or 'rigid' symmetries and conservation principles in classical mechanics was hardly news in 1918. As Kastrup (1987) discusses in his historical review, it had been appreciated in the previous century by Lagrange, Hamilton, Jacobi, and Poincaré, and an anticipation of Noether's first theorem in the special cases of the 10-parameter Lorentz and Galilean groups had been given by Herglotz in 1911 and Engel in 1916, respectively. Noether's own contribution is often praised for its degree of generality, and not without reason. But interestingly it does not cover the cases in which the symmetry transformation preserves the Lagrangian or Lagrangian density only up to a divergence term. It does not therefore cover such cases as the boost symmetry in classical pre-relativistic dynamics, although modern treatments of Noether's first theorem commonly rectify this defect.

In this connection it is worth noting that the problem originally addressed by Noether involves specifying the conditions under which a specific class of infinitesimal transformations of the independent and/or dependent variables leaves

[1] Noether's paper was presented to the Royal Society of the Sciences in Göttingen on 7 July 1918 by Felix Klein. See Kastrup (1987), who also discusses the historical background to the Noether paper.

the action invariant. There was no explicit attempt on Noether's part to relate this issue to that of dynamical symmetries. Indeed, the word 'symmetry' never appears in the 1918 paper. Though Noether was well aware that the Euler–Lagrange equations of motion – obtained by application of Hamilton's stationarity principle – are unaffected by the addition of a divergence term to the Lagrangian, this simple fact complicates somewhat the connection between the existence of *symmetry transformations* and the *variational* properties of the action. Even today, texts on general relativity, for instance, sometimes give the impression that the gravitational action, whether or not it contains variables related to matter fields, *must* be invariant under arbitrary coordinate transformations, i.e. that the Lagrangian density must be a scalar density – despite the counterexample offered by Einstein as early as 1916![2] However, it has long been a commonplace in the specialist literature on Noether-type theorems that symmetries are not always connected with transformations which leave the action strictly invariant. What is still perhaps open to discussion is how this fact should be understood. In this respect a recent detailed account of Noether's first theorem in the excellent textbook on Lagrangian dynamics by Doughty (1990), which follows Hill's classic 1951 analysis, is quite at odds with the 1962 treatment by Trautman, for example. In our discussion of the connection between the existence of symmetries and Noether's variational problem (see sections 2 and 3), we will come down on the side of Trautman and others who have defended the form-invariance rather than the 'scalarity' of the Lagrangian in the case of symmetries.

In sections 4 and 5 we state and comment on the three theorems that follow from the Noether variational problem. The main point that we wish to emphasize is this: the three theorems are mathematical tools that enable us to explore and extract the structural properties of our theories that are associated with symmetries. The results may be interesting, even surprising perhaps, but there is no quick route to physics here: any physical significance attaching to the results obtained is tied entirely to the physical significance of the assumptions that are put in, and this significance is something to which the theorems themselves are entirely blind. In the section on Noether's first theorem we discuss how it is that gauge symmetry, sometimes thought of as a freedom in description that is 'merely mathematical',[3] can be connected to a physically significant result such as conservation of electric charge. Noether's second theorem shows how underdetermination will arise for any theory with a local symmetry, while the third theorem (the Boundary theorem) shows in detail the tight restrictions on the possible form of any theory with a local symmetry. These theorems are mathematical tools that may be brought to bear on

[2] For such general relativity texts, see Stephani (1990, p. 95), and Misner, Thorne, and Wheeler (1973, p. 503). The 1916 Einstein counterexample is discussed below.

[3] See Martin, section 4.1, this volume.

interpretational problems; our aim here is to present the theorems in an accessible form, and to offer some brief interpretational remarks along the way.

The final aspect of this paper concerns the historical origins of these theorems. The stimulus to Noether's paper, and to a related paper by Klein, had to do with concerns that the Göttingen mathematicians Hilbert and Klein had surrounding the significance of energy conservation in generally covariant gravitational theories based on Einstein's idea of a dynamical metric field in spacetime. Their claim is that such conservation laws lack physical content (in contrast to the situation in classical mechanics, for example), something Einstein contested. In section 5 of this paper we show how the work done by Noether and Klein bears on the formulation of this issue, and how it should be resolved.

2 Noether's variational problem

From the point of view of physics, Noether's paper concerns theories that can be given a Lagrangian formulation. This covers a wide range of theories, from particle mechanics to general relativity. In the case of relativistic field theories it is standard to make use of Lagrangian densities; these depend on the independent variables x^μ ($\mu = 0, 1, 2, 3$) and the dependent variables (the fields) $\varphi_i(x)$ ($i = 1, \ldots, N$) and their derivatives:[4]

$$L = L(\varphi_i, \partial_\mu \varphi_i, x^\mu). \tag{1}$$

The action, S, is given in terms of the Lagrangian density by:

$$S = \int_R L \, d^4x, \tag{2}$$

where the compact spacetime region R is bounded by initial and final space-like surfaces. The equations of motion for a given field φ_k are the Euler–Lagrange equations; these are derived from the action using *Hamilton's Principle*, according to which the first-order functional variation in the action, resulting from arbitrary infinitesimal transformations of φ_k, vanishes for arbitrary regions of integration, where the variations in φ_k are stipulated to vanish on the boundary of the region of integration. The necessary condition for satisfaction of this variational principle is that φ_k satisfies Euler–Lagrange equations:

$$\frac{\partial L}{\partial \varphi_k} - \partial_\mu \frac{\partial L}{\partial(\partial_\mu \varphi_k)} = 0. \tag{3}$$

In this and what follows we use the Einstein convention to sum over Greek indices, all other summations being shown explicitly. We will refer to the left-hand side of (3)

[4] The restriction to the first derivative of the fields is for convenience; nothing of principle hangs on it and the generalization to higher derivatives is straightforward.

as the 'Euler expression'. (Notice that not all the fields on which a given Lagrangian density depends may satisfy this principle – such fields are not 'dynamical' in the sense that they do not satisfy Euler–Lagrange equations.)

The variational problem addressed by Noether in her 1918 paper, on the other hand, may be posed as follows. If the first-order functional variation of the action – involving a specific smooth infinitesimal transformation of the independent and/or dependent variables (a transformation which may not vanish on the boundary) – vanishes for an arbitrary region of integration, what general conditions must the variables satisfy?

The first-order functional variation of the action takes the following form:[5]

$$\delta S = \int_R \sum_i \left\{ \left(\frac{\partial L}{\partial \varphi_i} - \partial_\mu \frac{\partial L}{\partial(\partial_\mu \varphi_i)} \right) \delta_0 \varphi_i + \partial_\mu \left(\frac{\partial L}{\partial(\partial_\mu \varphi_i)} \delta_0 \varphi_i + L \delta x^\mu \right) \right\} d^4 x, \tag{4}$$

where the first term on the right-hand side is an 'interior' term arising from the 'bulk' contributions to the variation, and the second term may be thought of as a 'boundary' or 'surface' term since it may be converted into a surface term using Gauss's theorem.[6] The variation $\delta_0 \varphi_i$ is the Lie drag $\delta_0 \varphi_i = \varphi_i'(x) - \varphi_i(x)$.

Hence, setting $\delta S = 0$ (and recalling that the region of integration is arbitrary), the solution to Noether's 1918 variational problem is:

$$\sum_i \left(\frac{\partial L}{\partial \varphi_i} - \partial_\mu \frac{\partial L}{\partial(\partial_\mu \varphi_i)} \right) \delta_0 \varphi_i = - \sum_i \partial_\mu \left(\frac{\partial L}{\partial(\partial_\mu \varphi_i)} \delta_0 \varphi_i + L \delta x^\mu \right). \tag{5}$$

This expresses a restriction on the form of the Lagrangian density that must be satisfied *if* the first-order variation of the action vanishes. The above expression (5) is the crucial result from which Noether's 1918 theorems follow. We will see in section 3 below that when considering the theorems in the context of physics, however, this result is subject to an important generalization in which the strict invariance of the action is relaxed.[7] The theorems that we state in sections 4 and 5 take account of this.

Thus, the variational problem addressed by Noether differs from Hamilton's Principle in four crucial respects. First, it is a *problem* not a *principle*: it asks what follows *if* a given condition is satisfied, rather than requiring that a certain condition be satisfied. Second, the variation considered is a *specific* infinitesimal transformation of the variables, not an arbitrary transformation; and third, the variation considered may involve more than one of the variables, including the independent

[5] For derivations see, for example, Barbashov and Nesterenko (1983), Doughty (1990), and Trautman (1962).

[6] See Schutz (1990, p. 111).

[7] Remarks made by Noether (1918; p. 194 of the English translation) and Klein (1918, section 3 and p. 181, comments (d) and (e)) show that they were aware of the possibility of relaxing the strict invariance of the action integral, but had not yet fully understood the significance of this move.

variables. Finally, the variations are not required to vanish on the boundary of the region of integration.

3 Transformations, invariance of the action, and symmetries in physics

Noether's variational problem is a conditional, where the antecedent is the assumption that the action is invariant under some variation of the dependent and/or independent variables. In this and the following sections we discuss the significance of this assumption and how one should understand it in the context of symmetry transformations in physics. Our target is to link invariance of the action under a variation of the dependent and/or independent variables with infinitesimal symmetry transformations in physics. We begin with general point transformations, and then move to the case of symmetry transformations. The outcome is a revised version of the variational problem addressed by Noether in 1918, appropriate to symmetries in physics, along with the general solution to this problem.

Let us consider a *change* in the action defined by

$$\begin{aligned}\Delta S &= S'[\varphi'_i, \partial'_\mu \varphi'_i, x'^\mu] - S[\varphi_i, \partial_\mu \varphi_i, x^\mu] \\ &= \int_{R'} L'(\varphi'_i, \partial'_\mu \varphi'_i, x'^\mu)\, d^4x' - \int_R L(\varphi_i, \partial_\mu \varphi_i, x^\mu)\, d^4x. \qquad (6)\end{aligned}$$

Following such a transformation we require that our new Lagrangian density, expressed as a function of the new variables, satisfies Euler–Lagrange equations. In other words, we require that if $S[\varphi_i, \partial_\mu \varphi_i, x^\mu]$ satisfies Hamilton's Principle, then $S'[\varphi'_i, \partial'_\mu \varphi'_i, x'^\mu]$ also satisfies Hamilton's principle. A sufficient condition for this requirement to be met is that the transformed and untransformed actions differ by at most the integral of a total divergence,

$$\Delta S = \int_R \partial_\mu \Lambda^\mu \, d^4x, \qquad (7)$$

since the Euler expression associated with a total divergence vanishes identically.[8] If the transformation is continuous and infinitesimal, then (7) holds when the function Λ is replaced with the infinitesimal function $\Delta\Lambda$.[9] Hence,

$$\Delta S = \int_R \partial_\mu (\Delta \Lambda^\mu)\, d^4x. \qquad (8)$$

We now turn our attention to those special transformations that are symmetry transformations.

[8] This is a point noted in passing by Noether (1918; p. 194 of the English translation), but not included in her derivation and statement of her theorems. Allowing for a divergence term gives the most familiar generalization of the results proved by Noether herself, but notice that even this generalization expresses a sufficient *but not necessary* condition on the change in the action; we shall have more to say on this below.

[9] See Doughty (1990, p. 196).

For the purposes of classical theories including field theories prior to second quantization (such as non-relativistic Schrödinger quantum mechanics and relativistic field theories associated with the Klein–Gordon and Dirac equations), the appropriate concept of symmetry in physics is a transformation of the independent and/or dependent variables that leaves the *explicit* form of the equations of motion unchanged. We are concerned with Lagrangian theories, and with the subset of point transformations that leave the explicit form of Euler–Lagrange equations unaffected. A sufficient condition for this is that the transformed Lagrangian density has *the same functional form* as the untransformed Lagrangian density:

$$L'(\varphi_i, \partial_\mu \varphi_i, x^\mu) = L(\varphi_i, \partial_\mu \varphi_i, x^\mu). \tag{9}$$

More generally, the Euler–Lagrange equations will be unaffected if the form of the new Lagrangian density differs from the form of the old by a divergence term:

$$L'(\varphi_i, \partial_\mu \varphi_i, x^\mu) = L(\varphi_i, \partial_\mu \varphi_i, x^\mu) + \partial_\mu \Theta^\mu. \tag{10}$$

This is the second place in the derivation in which the freedom to pick up a divergence term arises, and the reason is the same: the Euler expression associated with a total derivative vanishes identically.[10] In principle, we could permit the existence of both divergence terms, but in the literature the options seem to be limited to the following two: (a) treat the Lagrangian density as strictly form-invariant, but allow for a non-zero change in the action as indicated in equation (7) above, or (b) treat the action as strictly numerically invariant (so that the surface term in equation (7) vanishes), but allow for the possibility that the Lagrangian density be form-invariant only up to a divergence term, as in equation (10).

In most cases, it is possible to avoid divergence terms altogether. For example, in general relativity the standard electromagnetic action and (Hilbert) gravitational action are each numerically invariant under arbitrary coordinate transformations, with the Lagrangian densities taken to be form-invariant. But this is not possible for the free particle action in classical non-relativistic mechanics when the coordinate transformations include boosts, nor for the 1916 Einstein '$\Gamma\Gamma$' gravitational action[11] – which depends on only first derivatives of the metric – when the coordinate transformations are non-linear. In such cases, it is surely appropriate to

[10] There are some very confusing remarks concerning this point in the literature. Both Hill (1951) and Doughty (1990) restrict the dependence of Θ^μ on the independent and dependent variables to exclude the derivatives of the dependent variables. As they remark, this ensures that the new Lagrangian is of the same order of derivatives of the dependent variable as the original Lagrangian. However, since the resulting Euler–Lagrange equations are independent of any such restriction, it is not clear what the motivation is for imposing this restriction. Furthermore, Hill (1951, p. 257, note 14) indicates that his restriction is not just a sufficient but also a necessary condition for a symmetry transformation. For our purposes, we require only that we have a sufficient condition for a symmetry transformation. (We are already restricting ourselves to theories that can be given a Lagrangian formulation, and to continuous symmetries; the treatment given here is not intended to cover all possible cases of symmetry transformations in physics, but rather to address what follows if certain specified general conditions are satisfied.)

[11] See Dirac (1996, section 26), and Brown and Brading (2002).

choose option (a) above. To choose option (b) strictly means specifying a coordinate system before specifying the form of the Lagrangian (density), something which in practice is never done. In opting for (a) we side with Trautman (1962) and Anderson (1967, p. 91), for example, in opposition to Hill (1952) and Doughty (1990, pp. 190–91) who go for (b).

We are now in a position to connect the sufficient conditions for our transformation to be a symmetry transformation with the *variation* in the action.

Consider the *variation* in the action:

$$\begin{aligned}\delta S &= S[\varphi_i', \partial_\mu \varphi_i', x'^\mu] - S[\varphi_i, \partial_\mu \varphi_i, x^\mu] \\ &= \int_{R'} L(\varphi_i', \partial_\mu \varphi_i', x'^\mu)\, d^4x' - \int_R L(\varphi_i, \partial_\mu \varphi_i, x^\mu)\, d^4x. \qquad (11)\end{aligned}$$

First, we add the requirement (9), form-invariance of the Lagrangian density, and we compare this with the change in the action under a general infinitesimal point transformation ΔS (see equation (6)). We see that, under this condition,

$$\Delta S = \delta S. \qquad (12)$$

Then, using (8), we conclude that, for an infinitesimal symmetry transformation,

$$\delta S = \int_R \partial_\mu (\Delta \Lambda^\mu)\, d^4x, \qquad (13)$$

where the condition (8) is now necessary for the ensuing derivation to proceed.[12]

The next step is to equate (13) with the expression for δS derived above (see equation (4)), giving

$$\begin{aligned}&\int_R \sum_i \left(\frac{\partial L}{\partial \varphi_i} - \partial_\mu \frac{\partial L}{\partial(\partial_\mu \varphi_i)} \right) \delta_0 \varphi_i\, d^4x \\ &= \int_R \sum_i \partial_\mu \left(\frac{\partial L}{\partial(\partial_\mu \varphi_i)} \delta_0 \varphi_i + L \delta x^\mu - \Delta \Lambda^\mu \right) d^4x, \qquad (14)\end{aligned}$$

where the left-hand side is the interior contribution and on the right-hand side we have collected all the boundary contributions. Since the region of integration is arbitrary, we therefore arrive at the following solution to the Noether variational problem for symmetries in physics.

'Noether general expression'

$$\sum_i \left(\frac{\partial L}{\partial \varphi_i} - \partial_\mu \frac{\partial L}{\partial(\partial_\mu \varphi_i)} \right) \delta_0 \varphi_i = - \sum_i \partial_\mu \left(\frac{\partial L}{\partial(\partial_\mu \varphi_i)} \delta_0 \varphi_i + L \delta x^\mu - \Delta \Lambda^\mu \right) \qquad (15)$$

[12] This allows us to prove the most familiar generalization of Noether's own results, as given in what follows. Further generalization is possible by relaxing the condition given in equation (7), leading to more complicated and less familiar results; see below for further discussion of this point, and an illustration from electromagnetism.

This differs from the general solution to Noether's original variational problem (5) by the additional term in Λ^μ appearing on the right-hand side. The connection to symmetries in physics has been made by imposing condition (7) on the action and requiring form-invariance of the Lagrangian density – notice that while these conditions are *sufficient* for the explicit form-invariance of the Euler–Lagrange equations, they are *necessary* for the derivation of the 'Noether general expression' derived here, and in what follows we refer to the class of symmetries satisfying these conditions as 'Noether symmetries'. However, this excludes familiar cases such as the gauge symmetry of Maxwell electromagnetism with a background current, the Lagrangian for which is:

$$L_{EM} = \frac{1}{4} F^{\mu\nu} F_{\mu\nu} - J^\mu A_\mu, \tag{16}$$

where $F^{\mu\nu}$ is defined in terms of A^μ, A^μ being the electromagnetic 4-potential, and J^μ is the 4-current. The Euler–Lagrange equations for the electromagnetic potential are invariant under a gauge transformation,

$$A_\mu \to A_\mu + \partial_\mu \theta, \tag{17}$$

where θ is an arbitrary function of the coordinates. Nevertheless, the variation in the action arising from a gauge transformation is:

$$\delta S = -\int J^\mu \partial_\mu \theta \, d^4x, \tag{18}$$

which does not satisfy condition (7), and hence the 'Noether general expression' (15), derived via (13), is not applicable.[13] Generalized results that include such cases may be derived by considering the *necessary* and sufficient conditions for the explicit form-invariance of the Euler–Lagrange equations.[14] However, this further generalization introduces additional complications, and leads to results that differ from those usually associated with the title 'Noether's theorems'. In our presentation here we give the more familiar results which, in fact, cover almost all cases of interest in physics whilst also displaying clearly the features of Noether's theorems that are of central conceptual interest.

Three theorems are derivable from the 'Noether general expression', two of which are generalized versions of those presented by Noether in her 1918 paper and the third of which derives from Noether's discussions with Klein and his presentation in the context of general relativity in Klein (1918).

[13] For further discussion of this example, in conjunction with Noether's first theorem, see Brading (2002, appendix).

[14] See K. Brading and H. R. Brown, 'Noether's theorems, gauge symmetries and general covariance', unpublished manuscript.

4 Noether's first theorem: global symmetries and conservation laws

The first theorem is the most famous of the three theorems. This theorem applies to 'global' symmetries, where by 'global' we mean, in this context, symmetries depending on constant parameters. Examples of such symmetries are the spacetime symmetries of spatial translations and rotations, temporal translations, and boosts, familiar in Galilean or Lorentzian spacetimes. In sloganized form, Noether's first theorem says that for every continuous global symmetry there exists a conservation law. In fact, the specialization to global symmetries allows us to derive the following result from the Noether general expression (15).

Noether's first theorem

If a continuous group of transformations depending smoothly on ρ constant parameters ω_k ($k = 1, 2, \ldots, \rho$) is a Noether symmetry group of the Euler–Lagrange equations associated with $L(\varphi_i, \partial_\mu \varphi_i, x^\mu)$, then the following ρ relations are satisfied, one for every parameter on which the symmetry group depends:

$$\sum_i \left(\frac{\partial L}{\partial \varphi_i} - \partial_\mu \frac{\partial L}{\partial(\partial_\mu \varphi_i)} \right) \frac{\partial(\delta_0 \varphi_i)}{\partial(\Delta\omega_k)} = \partial_\mu j^\mu_{k(Noether)}, \tag{19}$$

where $\Delta\omega_k$ indicates that we are taking infinitesimal symmetry transformations,

$$\delta_0 \varphi_i = \frac{\partial(\delta_0 \varphi_i)}{\partial(\Delta\omega_k)} \Delta\omega_k, \tag{20}$$

and where $j^\mu_{k(Noether)}$ is the Noether current associated with the kth parameter,

$$j^\mu_{k(Noether)} := -\sum_i \left\{ \frac{\partial L}{\partial(\partial_\mu \varphi_i)} \frac{\partial(\delta_0 \varphi_i)}{\partial(\Delta\omega_k)} + L \frac{\partial(\delta x^\mu)}{\partial(\Delta\omega_k)} - \frac{\partial(\Delta\Lambda^\mu)}{\partial(\Delta\omega_k)} \right\}. \tag{21}$$

Then, if the left-hand side of (19) vanishes we get:

$$\partial_\mu j^\mu_{k(Noether)} = 0, \tag{22}$$

which is a continuity equation for the Noether current. Integrating over an entire space-like surface we obtain, subject to suitable boundary conditions, conservation of the associated Noether charge.[15]

The most familiar applications of Noether's first theorem are in classical particle mechanics, where we link spatial translation, spatial rotation, time translation, and boost symmetries to conservation of linear momentum, angular momentum, energy, and centre-of-mass motion respectively.[16] Perhaps more puzzling is the

[15] For a more detailed discussion of the derivation of Noether's first theorem, see K. Brading and H. R. Brown (*op. cit.*).

[16] For details see, for example, Landau and Lifshitz (1976, chapter 2) and Doughty (1990, chapter 9).

connection (in relativistic field theory for a complex scalar field coupled to an electromagnetic field)[17] between global gauge symmetry and conservation of electric charge. On one possible account, a state and its gauge-transformed counterpart are mere redescriptions of the same physical situation,[18] from which it might seem that gauge freedom is nothing more than a mathematical freedom of description with no physical significance. And yet conservation of electric charge is surely a physically significant result. At first sight it might seem that Noether's first theorem gives us a very strong result, allowing us to derive a physically significant conservation law from a 'merely mathematical' freedom in how we choose to describe a physical situation. Can it be the case that we are getting physically significant results from 'mere' mathematical inputs?

In fact, there is no mystery here, but the reasons why this puzzle dissolves are informative. First, we need to re-emphasize the point that in order to make the connection between a certain symmetry and an associated conservation law we must ensure that the left-hand side of (19) vanishes, and this will involve using dynamically significant information or assumptions, such as the assumption that all the fields appearing in the theory satisfy Euler–Lagrange equations of motion. If we want to search for the underlying basis of the existence of the general connection between symmetries and conservation laws, then the appropriate place to look is at these assumptions and at their nature – the structure of the Euler–Lagrange equations and the conditions under which these equations are satisfied, for example. For any given Noether symmetry and its associated conservation law, the one can be obtained from the other directly from the Euler–Lagrange equations without the use of Noether's first theorem: the power of Noether's first theorem lies in the fact that it gives us a simple and general mathematical recipe for extracting a Noether conserved quantity from a Noether symmetry,[19] and vice versa, *subject to* the left-hand side of (19) vanishing. Thus, when we use Noether's first theorem to connect a symmetry with a conservation law we have to put in the relevant dynamical information. Nevertheless, it might still seem puzzling that a 'mere freedom in description' should be connected to a physically significant feature, even via the equations of motion. We need to think more carefully about the empirical content of the symmetries themselves.

One way of getting our hands on the empirical significance of a symmetry is through 'Galilean ship' type experiments.[20] Here, we take an effectively isolated subsystem of the universe, transform it (in the case of Galileo's ship we go from the ship being at rest to the ship being in uniform motion), and observe that the

[17] For details see for example Ryder (1996, section 3.3), and K. Brading and H. R. Brown (*op. cit.*)
[18] The interpretation of gauge transformations is discussed in detail in various papers in Part I of this volume (see Earman, Martin, Norton, Nounou, Redhead, and Wallace).
[19] See Kastrup (1987, p. 140), and also the quotation from Klein given by Pais (1987, p. 610).
[20] See Brown and Sypel (1995) and Budden (1997).

two states of the subsystem are empirically indistinguishable except in relation to (parts of) the rest of the universe. Thus, in the case of Galileo's ship, no experiments carried out inside the cabin of the ship, and without reference to anything outside the ship, enable us to tell whether the ship is at rest or moving uniformly. The two states of motion are empirically indistinguishable except by looking out of the porthole. The very fact that we can get on with reading, writing, eating, dancing, playing tennis, and so forth, while on a cruise-liner is evidence of the symmetry between rest and uniform motion. Similarly, the fact that our bodies, TVs, watches, and computers work the same way whether we are in London or New York is evidence of translational symmetry. Given this very direct empirical significance of the global spacetime symmetries, perhaps it doesn't seem mysterious that something as empirically significant as a conservation law can be derived from a symmetry of the dynamical laws (given that those laws are satisfied).

This is, perhaps, a tempting line of thought, drawing on an important feature of global spacetime symmetries – their active interpretation. On an active interpretation, as understood here, we apply the symmetry transformation to an effectively isolated *sub*system of the universe, yielding two *empirically distinct* scenarios[21] across which the internal evolutions of the subsystem are empirically indistinguishable. (The Galilean ship experiment involves the stronger requirement that we must be able to implement the two distinct scenarios in practice.) However, in the case of global gauge symmetry, this approach doesn't work. Gauge transformations have no analogue to the 'Galilean ship',[22] they have no active interpretation.

While it is true that global gauge symmetry does not have the same direct empirical significance, arising from an active interpretation, that global spacetime symmetries have, this does not imply that global gauge symmetry is without empirical content. *The very fact that a global gauge transformation does not lead to empirically distinct predictions is itself non-trivial.* In other words, the freedom in our descriptions is no 'mere' mathematical freedom – it is a consequence of a physically significant structural feature of the theory. The same is true in the case of global spacetime symmetries: the fact that the equations of motion are invariant under translations, for example, is empirically significant. This is independent of whether there is an active interpretation of the symmetry in question. The imposition of a symmetry on a theory places a restriction on the possible form of the equations of motion of that theory, and insofar as this restriction has empirical significance then so too does the symmetry itself. *This* is the proper place to look when analysing the empirical significance of a given Noether symmetry.

Given a Lagrangian theory, Noether's first theorem applies to any isolated system; thus, it applies to an isolated subsystem of the universe, such as Galileo's ship,

[21] By 'empirically distinct' we mean observationally distinct in principle.
[22] See K. Brading and H. R. Brown (in press).

but it also applies to the universe as a whole. Thinking about Noether's first theorem for the case of the universe as a whole makes vivid the point that the empirical significance of the theorem is not connected to active interpretations of Noether symmetries. For the universe as a whole there is no active interpretation of a symmetry transformation. A Noether symmetry, however, is a property of the equations of motion; thus, insofar as these are well defined for any isolated system, *including the universe as a whole*, all results of the form (19), and indeed the move from (19) to (22) – i.e. linking the symmetry to the continuity equation via satisfaction of the equations of motion – are meaningful. Although the active interpretation of global spacetime symmetries is an important feature of these symmetries, it is a red herring when thinking about Noether's theorems.

In sum, it would be a mistake to think that Noether's first theorem, and indeed the Noether results more generally, allow us to pull physically significant conclusions out of 'merely' mathematical hats: symmetries place restrictions on the form of the equations of motion, and thereby derive much – and in some cases all – of any empirical significance they may have from that of the equations of motion themselves. As we will see in the next section, the second and third theorems derivable from the Noether variational problem show that local gauge symmetry is an even stronger restriction than global gauge symmetry on the possible form of the equations of motion.

5 The Boundary theorem and Noether's second theorem

Turning our attention to local symmetries (i.e. symmetries depending on arbitrary functions of space and time), we can derive two theorems. This is because we can separate the interior and boundary contributions to (14) by noticing that we must allow for the possibility that the arbitrary functions vanish on the boundary,[23] and so we require that the interior and boundary contributions vanish.

5.1 The Boundary theorem

On the basis of the vanishing of the boundary contribution we derive the Boundary theorem from the 'Noether general expression' (15).

The Boundary theorem

If a continuous group of transformations depending smoothly on ρ arbitrary functions of time and space $p_k(x)$ $(k = 1, 2, \ldots, \rho)$ and their first derivatives is a Noether symmetry group of the Euler–Lagrange equations associated with $L(\varphi_i, \partial_\mu \varphi_i, x^\mu)$,

[23] For a more detailed derivation of Noether's second theorem and the Boundary theorem, see K. Brading and H. R. Brown, 'Noether's theorems, gauge symmetries and general covariance', unpublished manuscript.

then the following three sets of ρ relations are satisfied, one for every parameter on which the symmetry group depends:

$$\sum_i \partial_\mu \left\{ \left(\frac{\partial L}{\partial \varphi_i} - \partial_\nu \frac{\partial L}{\partial(\partial_\nu \varphi_i)} \right) b^\mu_{ki} \right\} = \partial_\mu j^\mu_{k(Noether)} \tag{23}$$

$$\sum_i \left(\frac{\partial L}{\partial \varphi_i} - \partial_\nu \frac{\partial L}{\partial(\partial_\nu \varphi_i)} \right) b^\mu_{ki} = j^\mu_{k(Noether)} - \sum_i \left\{ \partial_\nu \left(\frac{\partial L}{\partial(\partial_\nu \varphi_i)} b^\mu_{ki} + \frac{\partial(\Delta\Lambda^\mu)}{\partial(\partial_\nu \Delta p_k)} \right) \right\} \tag{24}$$

$$\sum_i \left\{ \left(\frac{\partial L}{\partial(\partial_\mu \varphi_i)} b^\nu_{ki} + \frac{\partial(\Delta\Lambda^\nu)}{\partial(\partial_\mu \Delta p_k)} \right) + \left(\frac{\partial L}{\partial(\partial_\nu \varphi_i)} b^\mu_{ki} + \frac{\partial(\Delta\Lambda^\mu)}{\partial(\partial_\nu \Delta p_k)} \right) \right\} = 0 \tag{25}$$

where the infinitesimal transformation $\delta_0 \varphi_i$ is given by

$$\delta_0 \varphi_i = \sum_k \left\{ a_{ki}(\varphi_i, \partial_\mu \varphi_i, x) \Delta p_k(x) + b^\nu_{ki}(\varphi_i, \partial_\mu \varphi_i, x) \partial_\nu \Delta p_k(x) \right\}, \tag{26}$$

where Δp_k indicates that we are considering infinitesimal transformations, and $j^\mu_{k(Noether)}$ is the Noether current once again, see (21) above, in this case that associated with the kth arbitrary function.

These three identities, along with that of Noether's second theorem (see section 5.2, below), are not independent of one another, but we present all four here since this makes it easier to see their origin and their connection to the related results found in the literature.[24]

The significance of the Boundary theorem is best seen by means of examples, as demonstrated in section 5.3, below. One point is worth making at this stage, however, since it will be needed in the following section in the discussion of Noether's second theorem.[25] The roots of this point are historical, going back to Klein's analysis of Hilbert's energy conservation theorem (see Klein, 1917), and his correspondence with Einstein in March 1918 on the status of energy conservation in general relativity.[26] Indeed, the reasoning leading to the Boundary theorem was first given by Klein (1918) in the context of general relativity, work done in close co-operation with Noether.

Re-arranging the first identity of the Boundary theorem, equation (23), we get:

$$\partial_\mu \left\{ j^\mu_{k(Noether)} - \sum_i \left(\frac{\partial L}{\partial \varphi_i} - \partial_\nu \frac{\partial L}{\partial(\partial_\nu \varphi_i)} \right) b^\mu_{ki} \right\} = 0. \tag{27}$$

[24] Related results can be found in: Barbashov and Nesterenko (1983), de Wet (1947), Govearts (1991), Heller (1951), Julia and Silva, 'Currents and superpotentials in classical gauge invariant theories. I: Local results with applications to perfect fluids and general relativity', E-print gr-qc/9804029 v2 (1998), Klein (1918), Utiyama (1956; 1959), and Weyl (1919).

[25] This material was first presented at the Sixth International Conference on the History of General Relativity.

[26] Einstein (1998).

Hence, defining

$$\Theta^{\mu}_{k} := j^{\mu}_{k(Noether)} - \sum\nolimits_{i} \left(\frac{\partial L}{\partial \varphi_i} - \partial_{\nu} \frac{\partial L}{\partial(\partial_{\nu}\varphi_i)} \right) b^{\mu}_{ki}, \tag{28}$$

we have that

$$\partial_{\mu}\Theta^{\mu}_{k} = 0 \tag{29}$$

holds identically. From this, we infer the existence of the so-called 'superpotentials' $U^{\mu\nu}_{k}$, such that

$$\Theta^{\mu}_{k} = \partial_{\nu} U^{\mu\nu}_{k}, \tag{30}$$

where

$$\partial_{\mu}\partial_{\nu} U^{\mu\nu}_{k} = 0 \tag{31}$$

holds identically. All we have done here is some mathematical manoeuvring, allowing us to re-write the Noether current in the following form:

$$j^{\mu}_{k(Noether)} = \sum\nolimits_{i} \left(\frac{\partial L}{\partial \varphi_i} - \partial_{\nu} \frac{\partial L}{\partial(\partial_{\nu}\varphi_i)} \right) b^{\mu}_{ki} + \partial_{\nu} U^{\mu\nu}_{k}. \tag{32}$$

In other words, the Noether current can be expressed as consisting of a term which vanishes on-shell (i.e. when the Euler–Lagrange equations are satisfied), and a term whose divergence vanishes identically.

Now consider the conservation law

$$\partial_{\mu} j^{\mu}_{k(Noether)} = 0. \tag{33}$$

Given that the Noether current can be re-written in the form (32), we see that (33) can be understood as the vanishing of the divergence of two contributions. The first contribution vanishes on-shell without any need to take the divergence; the divergence of the second contribution vanishes identically. We can therefore raise a query over what physical significance the vanishing of the divergence of the Noether current can possibly have. This is the basis of the concerns that Klein raised over the status of Einstein's conservation law. At least a part of Einstein's response seems to be that (33) holds only when the field equations are satisfied, and that we are therefore making use of physically significant information in order to move from (32) to (33). This is true, but it doesn't address the full weight of the problem: the term of the Noether current involving the Euler–Lagrange equations vanishes on-shell *without* any need to take the divergence of the Noether current. Taking the divergence plays a role only with respect to the second term, and there the divergence vanishes identically. We are back to the question: wherein lies the physical content in taking the divergence of the Noether current and finding that it vanishes?

Something more can, and should, be said at this point. We have shown that whenever we have a local symmetry the associated Noether current can be re-written in the form (32), such that on-shell

$$j^{\mu}_{k(Noether)} = \partial_{\nu} U^{\mu\nu}_{k}. \tag{34}$$

Part of the Klein worry is that the associated continuity equation for $j^{\mu}_{k(Noether)}$ lacks physical content because of (31). But notice: while it is true that we can always write an expression of the form (34) on-shell, there remains the question of whether, and if so when, this equation expresses a physically significant relation. So far in doing the re-writing all we have done is mathematics, and only mathematics. The relation (34) gains *physical* significance only when it holds 'not as an identity or definition, but as a field equation postulated to relate two separate systems' (Deser, 1972, p. 1082). Consider, for example, the Maxwell field equations

$$J^{\mu} = \partial_{\nu} F^{\mu\nu}. \tag{35}$$

These equations are of the form (34), and

$$\partial_{\mu}\partial_{\nu} F^{\mu\nu} = 0 \tag{36}$$

holds simply in virtue of the antisymmetry of $F^{\mu\nu}$. Nevertheless, we do not say that conservation of electric charge is a mathematical identity without physical significance. This is because the equations (35) are not a mere mathematical re-expression of the current J^{μ}; they express a physically significant relation between two different types of field: on the left-hand side we have a current, J^{μ}, depending on the matter fields carrying the electric charge, and on the right-hand side we have an expression depending the electromagnetic fields, $F^{\mu\nu}$. Thus, the current conservation law follows, via (35) and (36), and since (35) is physically significant, so is the current conservation law.

Similarly in the case of general relativity, the re-expression of energy–momentum through a relation of the form (34) has physical content because it gives a relation between the behaviour of the metric and the matter fields, it is a field equation with physical content, and hence the conservation law that follows from it (via an identity for the right-hand side) also has physical content. This is perhaps what Einstein was alluding to when he said in a letter to Klein (24 March 1918) that his continuity equation contains 'a part of the content of the field equations'.[27]

5.2 Noether's second theorem

On the basis of the vanishing of the interior contribution we derive Noether's second theorem from the 'Noether general expression' (15).

[27] Einstein (1998, document 492).

Noether's second theorem

If a continuous group of transformations depending smoothly on ρ arbitrary functions of time and space $p_k(x)$ $(k = 1, 2, \ldots, \rho)$ and their first derivatives is a Noether symmetry group of the Euler–Lagrange equations associated with $L(\varphi_i, \partial_\mu \varphi_i, x^\mu)$, then the following ρ relations are satisfied, one for every parameter on which the symmetry group depends:

$$\sum_i \left(\frac{\partial L}{\partial \varphi_i} - \partial_\mu \frac{\partial L}{\partial(\partial_\mu \varphi_i)} \right) a_{ki} = \sum_i \partial_\nu \left\{ b^\nu_{ki} \left(\frac{\partial L}{\partial \varphi_i} - \partial_\mu \frac{\partial L}{\partial(\partial_\mu \varphi_i)} \right) \right\} \tag{37}$$

where the infinitesimal transformation $\delta_0 \varphi_i$ is given by (26), above.

Thus, we have a dependency between the Euler expressions for the various fields appearing in the theory. Noether's second theorem tells us that in any theory with a local Noether symmetry there is always a *prima facie* case of underdetermination: more unknowns than there are independent equations of motion.[28] This central feature of gauge theories is discussed elsewhere in Part I of this volume; see especially Earman, Norton, and Wallace. The underdetermination means that there are, in general, as many identities involving the fields as there are arbitrary functions involved in defining the local symmetry transformations. In the case of general relativity, for example, inserting the specific form of the Lagrangian density into the second theorem identities associated with general covariance leads to the twice contracted Bianchi identities. These, and their analogues for other symmetries in other theories, are the results most usually associated with Noether's second theorem (sometimes referred to collectively as the 'generalized Bianchi identities').

Noether's main objective in her paper was not this, however. In an exchange with Klein mentioned above (see section 5.1, above, and Klein (1917)), Hilbert conjectured that the difference between generally covariant theories such as general relativity, and earlier theories such as classical mechanics, can be characterized by the differing status of energy conservation: in generally covariant theories the energy conservation law can be re-written, using the Euler–Lagrange equations, such that it holds 'identically'. We have discussed the status of conservation laws in theories with local symmetries in section 5.1, above. What Noether did was to show that this is indeed characteristic of generally covariant theories. Using both her first and second theorems, she showed that the Noether current associated with a *global* symmetry can be re-written in the form (32) *only* when that global symmetry is a special case of a local symmetry. In classical mechanics (for example), the global

[28] For further discussion of exactly how this underdetermination arises, and its connection to the Cauchy problem and to the 'Bianchi identities' (see below), see Anderson (1967, pp. 95–101) and K. Brading and H. R. Brown, 'Noether's theorems, gauge symmetries and general covariance', unpublished manuscript.

space and time symmetry group is *not* a subgroup of a local symmetry group; so, the energy conservation law (associated with global time translations) cannot be rewritten in the form (32). Such a conservation law is, according to Hilbert, a 'proper' conservation law, as distinct from the energy conservation laws associated with generally covariant theories.[29] In this way, Noether proved Hilbert's conjecture, and generalized it beyond the case of energy conservation to all continuous global and local symmetry groups.

5.3 Illustrations

What is the general significance of these results? The four identities together place powerful restrictions on the possible form that a theory can take. This is best seen by looking at an example. We choose here to use Weyl's 1918 theory – the original 'gauge' theory – for the purposes of illustration,[30] beginning from Weyl's assumption that, whatever the detailed form of his theory, the following condition must be satisfied (in our notation):[31]

$$\delta S = \int \partial_\mu B^\mu \, dx + \int (W^{\mu\nu} \delta g_{\mu\nu} + \omega^\mu \delta A_\mu) \, dx = 0 \tag{38}$$

where

- $\int \partial_\mu B^\mu \, dx$ represents the boundary contribution to the variation in the action;
- $W^{\mu\nu} = 0$ and $\omega^\mu = 0$ are the gravitational and electromagnetic Euler–Lagrange equations, respectively, whose explicit form is unknown;
- $\delta g_{\mu\nu} = g_{\mu\nu} \Delta\rho$ is an infinitesimal scale transformation involving the metric $g_{\mu\nu}$;
- $\delta A_\mu = \partial_\mu(\Delta\rho)$ is an infinitesimal transformation of the electromagnetic 4-potential A_μ;
- $\rho(x)$ is an arbitrary function parameterizing the local 'gauge' transformations of Weyl's 1918 theory.

From the vanishing of the interior contribution, Weyl derives (Weyl, 1918, p. 32)

$$W^\mu_\mu = \partial_\mu \omega^\mu, \tag{39}$$

which is the result that follows from Noether's second theorem (although Weyl appears to have derived it independently of Noether).[32]

The result (39) expresses a dependence between the Euler expressions associated with the gravitational fields and the electromagnetic fields: this theory has the characteristic underdetermination problem.

[29] See also K. Brading and H. R. Brown (*op. cit.*), and Trautman (1962, p. 179).
[30] See Ryckman and Martin (end of section 2.2), both this volume.
[31] Weyl (1922, pp. 286–9).
[32] See Brading (2002).

For the vanishing of the boundary contribution, Weyl (1922, pp. 286–9) makes explicit use of Klein (1918) to derive the following three sets of equations (in our notation):

$$\begin{aligned} \partial_\mu J^\mu &= \partial_\mu \omega^\mu \\ J^\mu + \partial_\nu F^{\nu\mu} &= \omega^\mu \\ F^{\mu\nu} + F^{\nu\mu} &= 0. \end{aligned} \tag{40}$$

Using the third of these equations, we may re-write the second in the familiar form of the Maxwell Euler expression, $\partial_\nu F^{\mu\nu} - J^\mu$, associated with the Maxwell equations (see Weyl's equation (82)),[33] and current conservation follows when these equations are assumed to be satisfied. These restrictions on the form of Weyl's theory arise from the imposition of local 'gauge' symmetry as specified by the transformations of the metric and vector potential given above, and show how Weyl 'recovered' the structure of Maxwell electromagnetism in his 1918 theory. They follow straightforwardly from the Boundary theorem.

Other illustrations of the power of Noether's second theorem and the Boundary theorem can be found in Utiyama (1959; see also 1956) and the review paper by Barbashov and Nesterenko (1983), along with Brown and Brading (2002), for example.[34] In Brown and Brading (2002) we discuss the case of general covariance and general relativity, where we look at the way in which adding assumptions to 'mere' general covariance gives the results associated with the Noether variational problem increasingly significant bite.[35] We see, for example, the important role of the assumption that the gravitational part of the action be independently invariant (up to a surface term). Also, if the matter part of the Lagrangian density depends on the metric but none of its derivatives, and on the first derivatives only of a vector field, then the gauge structure of those vector fields is dictated. These remarks serve only to hint at the power of the results following from the Noether variational problem. For more details see the papers listed above and the references therein. The results following from the Noether variational problem obviously cannot dictate physical theory to us – they cannot tell us which objects in our theories are physically significant, or which symmetries to impose. Nevertheless, the conjunction of opting to impose a certain symmetry on our physical theory, along with placing restrictions on how that symmetry is to be realized by the mathematical objects appearing in

33 Weyl uses (39) to arrive at conservation of electric charge in his 1918 theory, $\partial_\mu J^\mu = 0$, via satisfaction of his gravitational field equations (see Brading, 2002). In this derivation he appears to 'help himself' to the Maxwell Euler expression and the antisymmetry of $F^{\mu\nu}$. The second and third equations of (40) are what make this legitimate. We are grateful to Tom Ryckman for bringing the relevant section of Weyl (1922) to our attention.

34 See also K. Brading and H. R. Brown, 'Noether's theorems, gauge symmetries and general covariance', unpublished manuscript.

35 This is connected to the 'Kretschmann objection' to Einstein; see Earman (Part I) and Norton (this volume).

the theory, leads to some very strong consequences and restrictions in terms of the possible form that this theory can take.[36]

6 Conclusion

We have explored two aspects of the results relating to Noether's seminal 1918 paper. The first is the relationship between symmetries of the Euler–Lagrange equations and the variational results obtained by Noether and Klein. Their results must be modified to allow for the non-scalarity of the Lagrangian, this being needed for such cases as boosts in Galilean particle mechanics and the arbitrary coordinate freedom of Einstein's '$\Gamma\Gamma$' Lagrangian.

The second is the significance of the three theorems derivable from the Noether variational problem. Here, we have emphasized the power of the theorems, especially with respect to the structure of gauge theories, while also insisting on the point that the results are mathematics, and their significance in the context of a particular physical theory depends upon interpretational steps that go beyond what the theorems alone can tell us.

Finally, we have also made some remarks concerning the historical origins of the Noether and Klein papers, concerning the issue of whether conservation of energy in general relativity is physically meaningful, and whether this is characteristic of conservation laws connected to local symmetries. Our conclusion is that such conservation laws can indeed be re-written in a characteristic form (this result following mathematically), but that this does not necessarily imply that they are physically meaningless (this issue hinging on further interpretational steps).

References

Anderson, J. L. (1967). *Principles of Relativity Physics*. New York and London: Academic Press.

Barbashov, B. M., and Nesterenko, V. V. (1983). 'Continuous symmetries in field theory'. *Fortschritte der Physik*, **31**, 535–67.

Brading, K. A. (2002). 'Which symmetry? Noether, Weyl, and conservation of electric charge'. *Studies in History and Philosophy of Modern Physics*, **33**, 3–22.

Brading, K., and Brown, H. R. (in press). 'Are gauge symmetry transformations observable?' *British Journal for the Philosophy of Science*.

Brown, H. R., and Brading, K. A. (2002). 'General covariance from the perspective of Noether's theorems'. *Diálogos*, **79**, 59–86.

Brown, H. R., and Sypel, R. (1995). 'On the meaning of the relativity principle and other symmetries'. *International Studies in the Philosophy of Science*, **9**, 233–51.

[36] Various interpretational issues associated with the structure of theories possessing local symmetries are discussed in papers elsewhere in Part I of this volume (see especially Earman, Norton, Nounou, Redhead, and Wallace).

Budden, T. (1997). 'Galileo's ship and spacetime symmetries'. *British Journal for the Philosophy of Science*, **48**, 483–516.

de Wet, J. S. (1947). 'Symmetric energy–momentum tensors in relativistic field theories'. *Proceedings of the Cambridge Philosophical Society*, **43**, 511–20.

Deser, S. (1972). 'Note on current conservation, charge, and flux integrals'. *American Journal of Physics*, **40**, 1082–4.

Dirac, P. A. M. (1996). *General Theory of Relativity*. Princeton, NJ: Princeton University Press.

Doughty, N. A. (1990). *Lagrangian Interaction*. Reading, MA: Addison-Wesley.

Einstein, A. (1998). *The Collected Papers of Albert Einstein*, Vol. 8, ed. R. Schulmann, A. J. Kox, M. Janssen, and J. Illy. Princeton, NJ: Princeton University Press.

Govearts, J. (1991). *Hamiltonian Quantization and Constrained Dynamics*. Leuven notes in mathematical and theoretical physics, Vol. 4. Leuven: Leuven University Press.

Heller, J. (1951). 'Covariant transformation law for the field equations'. *Physical Review*, **81**, 946–8.

Hill, E. L. (1951). 'Hamilton's principle and the conservation theorems of mathematical physics'. *Reviews of Modern Physics*, **23**, 253–60.

Kastrup, H. A. (1987). 'The contributions of Emmy Noether, Felix Klein and Sophus Lie to the modern concept of symmetries in physical systems'. In *Symmetries in Physics (1600–1980): 1st international meeting on the history of scientific ideas*, Catalonia, Spain, 20–26 September 1983, pp. 113–63. Barcelona: Universitat Autonoma de Barcelona.

Klein, F. (1917). 'Zu Hilberts erster Note über die Grundlagen der Physik'. *Königliche Gesellschaft der Wissenschaften zu Göttingen. Mathematisch-physikalische Klasse. Nachrichten*, pp. 469–82.

(1918). 'Über die Differentialgesetze für die Erhaltung von Impuls und Energie in der Einsteinschen Gravitationstheorie'. *Königliche Gesellschaft der Wissenschaften zu Göttingen. Mathematisch-physikalische Klasse. Nachrichten*, pp. 171–89. Translated by J. Barbour as 'On the differential laws for conservation of momentum and energy in Einstein's theory of gravitation' (unpublished manuscript).

Landau, L. D., and Lifshitz, E. M. (1976). *Mechanics*, 3rd edn. Oxford, New York, Seoul, Tokyo: Pergamon.

Misner, C. W., Thorne, K. S., and Wheeler, J. A. (1973). *Gravitation*. San Francisco: Freeman.

Noether, E. (1918). 'Invariante Variationsprobleme'. *Königliche Gesellschaft der Wissenschaften zu Göttingen. Mathematisch-Physikalische Klasse. Nachrichten*, pp. 235–57. Page numbers refer to the English translation: Tavel (1971).

Pais, A. (1987). 'Conservation of energy'. In *Symmetries in physics (1600–1980): 1st international meeting on the history of scientific ideas*, Catalonia, Spain, 20–26 September 1983, pp. 361–75. Barcelona: Universitat Autonoma de Barcelona.

Ryder, L. H. (1996). *Quantum Field Theory*, 2nd edn (1st edn, 1985). Cambridge: Cambridge University Press.

Sarlet, W., and Cantrijn, F. (1981). 'Generalizations of Noether's theorem in classical mechanics'. *Siam Review*, **23**, 467–94.

Schutz, B. F. (1990). *A First Course in General Relativity*, Cambridge: Cambridge University Press.

Stephani, H. (1990). *General Relativity*, 2nd edn (1st edn, 1982). Cambridge: Cambridge University Press.

Tavel, M. A. (1971). 'Noether's theorem'. *Transport Theory and Statistical Physics*, **1**, 183–207.

Trautman, A. J. (1962). 'Conservation laws in general relativity'. In *Gravitation: An Introduction to Current Research*, ed. L. Witten, pp. 169–98. New York: Wiley.

Utiyama, R. (1956). 'Invariant theoretical interpretation of interaction'. *Physical Review*, **101**, 1597–607.

(1959). 'Theory of invariant variation and the generalized canonical dynamics'. *Progress of Theoretical Physics Supplement*, **9**, 19–44.

Weyl, H. (1918). 'Gravitation und Elektrizität.' *Preußische Akademie der Wissenschaften (Berlin) Sitzungsberichte. Physikalisch-Mathematische Klasse*, pp. 465–80. Translated as 'Gravitation and electricity', in L. O'Raifeartaigh (1997), *The Dawning of Gauge Theory*, pp. 24–37. Princeton, NJ: Princeton University Press.

(1919). *Raum-Zeit-Materie*, 3rd edn (1st edn, 1918). Berlin: Springer-Verlag. See Weyl (1922) for the English translation of the 4th edn.

(1922). *Space-Time-Matter*, translation by H. L. Brose of the 4th edn (1921) of *Raum-Zeit-Materie*. London: Methuen. Page numbers refer to the Dover edition (an unaltered republication), 1952. New York: Dover.

6

General covariance, gauge theories, and the Kretschmann objection

JOHN D. NORTON

1 Introduction

Two views . . .

When Einstein formulated his General Theory of Relativity, he presented it as the culmination of his search for a generally covariant theory. That this was the signal achievement of the theory rapidly became the orthodox conception. A dissident view, however, tracing back at least to objections raised by Erich Kretschmann in 1917, holds that there is no physical content in Einstein's demand for general covariance. That dissident view has grown into the mainstream. Many accounts of general relativity no longer even mention a principle or requirement of general covariance.

What is unsettling for this shift in opinion is the newer characterization of general relativity as a gauge theory of gravitation, with general covariance expressing a gauge freedom. The recognition of this gauge freedom has proved central to the physical interpretation of the theory. That freedom precludes certain otherwise natural sorts of background spacetimes; it complicates identification of the theory's observables, since they must be gauge invariant; and it is now recognized as presenting special problems for the project of quantizing of gravitation.

. . . that we need not choose between

It would seem unavoidable that we can choose at most *one* of these two views: the vacuity of a requirement of general covariance or the central importance of general covariance as a gauge freedom of general relativity. I will urge here that this is not so; we may choose *both*, once we recognize the differing contexts in which they arise. Kretschmann's claim of vacuity arises when we have some body of physical fact to represent and we are given free rein in devising the formalism that will capture it. He urges, correctly I believe, that we will always succeed in finding a generally covariant formulation. Now take a different context. The theory – general relativity – is fixed both in its formalism and physical interpretation. Each formal property of the theory will have some meaning. That holds for its general covariance which turns out to express an important gauge freedom.

To come

In section 4 I will lay out this reconciliation in greater detail. As preparation, in sections 2 and 3, I will briefly review the two viewpoints. Finally in section 5 I will relate the reconciliation to the fertile 'gauge principle' used in recent particle physics. Appendix 1 discusses the difficulty of making good on Kretschmann's claim that generally covariant reformulations are possible for any spacetime theory.

2 Einstein and Kretschmann's objection

Einstein...

In November 1915 an exhausted and exhilarated Einstein presented the gravitational field equations of his General Theory of Relativity to the Prussian Academy of Science. These equations were generally covariant; they retained their form under arbitrary transformation of the spacetime coordinate system. This event marked the end of a seven-year quest, with the final three years of greatest intensity, as Einstein struggled to see that a generally covariant theory was physically admissible.[1]

Einstein had several bases for general covariance. He believed that the general covariance of his theory embodied an extension of the principle of relativity to acceleration. This conclusion seemed automatic to Einstein, just as the Lorentz covariance of his 1905 formulation of special relativity expressed its satisfaction of the principle of relativity of inertial motion.[2] He also advanced what we now call the 'point-coincidence' argument. The physical content of a theory is exhausted by a catalogue of coincidences, such as the coincidence of a pointer with a scale, or, if the world consisted of nothing but particles in motion, the meetings of their worldlines. These coincidences are preserved under arbitrary coordinate transformations; all we do in the transformations is relabel the spacetime coordinates assigned to each coincidence. Therefore a physical theory should be generally covariant. Any less covariance restricts our freedom to relabel the spacetime coordinates of the coincidences and that restriction can be based in no physical fact.

[1] Over the last two decades there has been extensive historical work on this episode. Earlier works include Stachel (1980) and Norton (1984); the definitive work will be Renn *et al.* (forthcoming).

[2] The analogy proved difficult to sustain and has been the subject of extensive debate. See Norton (1993).

... and Kretschmann

Shortly after, Erich Kretschmann (1917) announced that Einstein had profoundly mistaken the character of his achievement. In demanding general covariance, Kretschmann asserted, Einstein had placed no constraint on the physical content of his theory. He had merely challenged his mathematical ingenuity. For, Kretschmann urged, any spacetime theory could be given a generally covariant formulation as long as we are prepared to put sufficient energy into the task of reformulating it. In arriving at general relativity, Einstein had used the 'absolute differential calculus' of Ricci and Levi-Civita (now called 'tensor calculus'). Kretschmann pointed to this calculus as a tool that made the task of finding generally covariant formulations of theories tractable.[3]

Kretschmann's argument was slightly more subtle than the above remarks. Kretschmann actually embraced Einstein's point-coincidence argument and turned it to his own ends. In his objection, he agreed that the physical content of spacetime theories is exhausted by the catalogue of spacetime coincidences; this is no peculiarity of general relativity. For this very reason all spacetime theories can be given generally covariant formulations.[4]

Kretschmann's objection does seem sustainable. For example, using Ricci and Levi-Civita's methods it is quite easy to give special relativity a generally covariant formulation. In its standard Lorentz covariant formulation, using the standard spacetime coordinates (t, x, y, z), special relativity is the theory of a Minkowski spacetime whose geometry is given by the invariant line element

$$ds^2 = c^2\,dt^2 - dx^2 - dy^2 - dz^2. \tag{1}$$

Free fall trajectories (and other 'straights' of the geometry) are given by

$$d^2x/dt^2 = d^2y/dt^2 = d^2z/dt^2 = 0. \tag{2}$$

We introduce arbitrary spacetime coordinates x^i, for $i = 0, \ldots, 3$ and the invariant line element becomes

$$ds^2 = g_{ik}\,dx^i\,dx^k, \tag{3a}$$

where the matrix of coefficients g^{ik} is subject to a field equation

$$R_{iklm} = 0 \tag{3b}$$

3 For further discussion of Kretschmann's objection, of Einstein's response, and of the still-active debate that followed, see Norton (1993) and Rynasiewicz (1999).

4 Rhetorically, Kretschmann's argument was brilliant. To deny it, Einstein may have had to deny his own point-coincidence argument. However, a persistent ambiguity remains in Einstein's original argument. Just what is a point-coincidence? Einstein gives no general definition. He gives only a list of illustrations and many pitfalls await those who want to make the argument more precise. For example, see Howard (1999).

with R_{iklm} the Riemann–Christoffel curvature tensor. The free falls are now governed by

$$d^2x^i/ds^2 + \{^{\,i}_{km}\}\, dx^k/ds\, dx^m/ds = 0, \tag{4}$$

where $\{^{\,i}_{km}\}$ are the Christoffel symbols of the second kind.

Examples such as this suggest that Kretschmann was right to urge that generally covariant reformulations are possible for all spacetime theories. While the suggestion is plausible it is certainly not proven by the examples and any final decision must await clarification of some ambiguities. For further discussion see appendix 1.

3 The gauge freedom of general relativity

Active general covariance

Einstein spoke of general covariance as the invariance of form of a theory's equations when the spacetime coordinates are transformed. It is usually coupled with a so-called 'passive' reading of general covariance: if we have some system of fields, we can change our spacetime coordinate system as we please and the new descriptions of the fields in the new coordinate systems will still solve the theory's equations. Einstein's form invariance of the theory's equations also licenses a second version, the so-called 'active' general covariance. It involves no transformation of the spacetime coordinate system. Rather, active general covariance licenses the generation of many new solutions of the equations of the theory *in the same coordinate system* once one solution has been given.

For example, assume the equations of some generally covariant theory admit a scalar field $\varphi(x^i)$ as a solution. Then general covariance allows us to generate arbitrarily many more solutions by, metaphorically speaking, spreading the scalar field differently over the spacetime manifold of events. We need a smooth mapping on the events – a diffeomorphism – to effect the redistribution. For example, assume we have such a map that sends the event at coordinate x^i to the event at coordinate x'^i in the same coordinate system. Such a map might be a uniform doubling, so that x^i is mapped to $x'^i = 2x^i$. To define the redistributed field φ', we assign to the event at x'^i the value of the original field φ at the event with coordinate x^i.[5] If the field is not a scalar field, the transformation rule is slightly more complicated. For further details of the scalar case see appendix 2.

[5] To visualize this redistribution in the two-dimensional case, imagine that the original field is represented by numbers written on a flat rubber membrane. If we now uniformly stretch the rubber membrane so it doubles in size, we have the new field.

Why it is a gauge freedom

The fields $\varphi(x^i)$ and $\varphi'(x^i)$ are mathematically distinct. But do they represent physically distinct fields? The standard view is to assume that they do not, so that they are related by a gauge transformation, that is, one that relates mathematically distinct representations of the same physical reality. That this is so cannot be decided purely by the mathematics. It is a matter of physics and must be settled by physical argumentation.

A vivid way to lay out the physical arguments is through Einstein's 'hole argument'.[6] The transformation on the manifold of events can be set up so that it is the identity everywhere outside some nominated neighbourhood of spacetime ('the hole') and comes smoothly to differ within. We now use the transformation to duplicate diffeomorphically all the fields of some generally covariant theory. Do the new fields represent the same physical reality as the old? It would be very odd if they did not. Both systems of fields agree completely in all invariants; they are just spread differently on the manifold. Since observables are given by invariants, they agree in everything observable. Moreover, the two systems of field will agree everywhere outside the hole, but they differ only within. This means that, in a generally covariant theory, fixing all fields outside this neighbourhood fails to fix the fields within. This is a violation of determinism. In short, if we assume the two systems of fields differ in some physical way we must insist upon a difference that transcends both observation and the determining power of the theory. The ready solution is that these differences are purely ones of mathematical representation and that the two systems of fields represent the same physical reality.

Its physical consequences

Accepting this gauge freedom has important consequences for the physical interpretation of a theory such as general relativity.[7] The theory is developed by positing a manifold of spacetime events which is then endowed with metric properties by means of a metric tensor field g_{ik}. The natural default is to take the manifold of events as supplying some kind of independent background spacetime in which physical processes can unfold. The gauge freedom makes it very difficult to retain this view. For, when we apply a diffeomorphism to the field and spread the metrical properties differently over events, the transformation is purely gauge and we end up changing nothing physical. So now the same events are endowed with different properties, yet nothing physical has changed. The simplest, and perhaps only, way to make sense of this is to give up the idea of an independent existence of the events

6 See Earman and Norton (1987), Norton (1999).

7 For further discussion of these and related issues and their import for the quantization of gravity see Rovelli (1997).

of the manifold. Insofar as we can associate an event of the manifold with real events in the world, that association must change in concert with our redistribution of the metrical field over the manifold.

Our notion of what is observable is affected by similar considerations. What is observable is a subset of the physically real and that in turn is expressed by the invariants of a theory. Might an observable result consist of the assertion that an invariant of some field has such and such a value at some event of the manifold? No. The invariance must also include invariance under the gauge transformation and the assertion would fail to be invariant under the gauge transformation. In redistributing the fields, the transformation might relocate that invariant with that value at quite another event of the manifold. If some result is eradicated by a gauge transformation, it cannot have been a result expressing physical fact since the gauge transformation alters nothing physical. We must resort to more refined ways of representing observables. For example, they may be expressed by an assertion that two invariants are equal. The event at which the equality resided may vary under gauge transformation; but the transformation will preserve the equality asserted.

4 Reconciliation

The context in which Kretschmann's objection succeeds

Kretschmann's objection succeeds because he allows us every freedom in reformulating and reinterpreting terms within a theory. Thus we easily transformed special relativity from its Lorentz covariant formulation (1), (2) to a generally covariant formulation (3a), (3b), and (4). In doing so, we introduced new variables not originally present. These are the coefficients of the metric tensor g_{ik} and the Christoffel symbols $\{{}^{\,i}_{km}\}$.

With this amount of freedom, it is plausible that we can arrive at formulations of any theory that have any designated formal property.[8] Imagine, for example, that we wanted a formulation of Newtonian particle mechanics in which the string of symbols '$E = mc^2$' appears. (This is a purely formal property since we place no conditions on what the string might mean.) Here is one way we can generate it. We take the usual expression for the kinetic energy K of a particle of mass m moving at velocity v, $K = (1/2)mv^2$. We introduce a new quantity E, defined by $E = 2K$,

[8] I am distinguishing the formalism of the theory (and its formal properties) from its interpretation. The formalism of a theory would be the actual words used, if the theory consisted of an English language description, independently of their meanings. Formal properties would include such things as the choice of English and the number of words. More commonly, physical theories use mathematical structures in place of words. These structures can be considered quite independently of what we take them to represent in the world. The properties we then consider are the purely formal properties. A real-valued field on some manifold is just a mathematical structure until we specify what it may represent in the world. That specification is the job of the interpretation. See footnote 9.

and also a new label 'c' for velocity v, so that $c = v$. Once we substitute these new variables into the expression for kinetic energy, our reformulated theory contains the string '$E = mc^2$'.

The physical vacuity arises because we are demanding the formal property of general covariance (or some other formal property) without placing further restrictions that would preclude it always being achievable. The vacuity would persist even if we demanded a fixed physical content; we must simply be careful not to alter our initial physical content as we adjust its formal clothing. In the case of the discovery of general relativity, Einstein did not keep the physical content fixed. It became fully fixed only after he found a generally covariant formulation that satisfied a number of restrictive physical limitations.

The context in which the diffeomorphism gauge freedom has physical content

Matters are quite different if we fix the formalism of the theory and its interpretation. So we might be given general relativity in its standard interpretation.[9] If a theory has any content at all, we must be able to ascribe some physical meaning to its assertions. *A fortiori* there must some physical meaning in the general covariance of general relativity. It may be trivial or it may not.[10] Consulting the theory, as we did in section 2 above, reveals that the content is not trivial.

Things are just the same in our toy example of forcing the string '$E = mc^2$' into a formulation of Newtonian particle mechanics. Let us fix the formulation to be the doctored one above. We had forced the string '$E = mc^2$' into it. But now that we have done it, the string uses symbols that have a meaning and, when we decode what it says about them, we discover that the string expresses something physical, the original statement that kinetic energy is half mass $\times$ (velocity)^2. Mimicking Kretschmann, we would insist that, given Newtonian particle mechanics or any other theory, some reformulation with the string is assuredly possible; so the demand for it places no restriction on the physically possible. But, once we have the reformulation, that string will express something.

The analogous circumstance arises in the generally covariant reformulation of special relativity. The existence of the reformulation is assured. Once we have it, its general covariance does express something. In this case, it is a gauge freedom

[9] By 'interpretation' I just mean the rules that tell us how to connect the various terms or mathematical structures of the theory with things in the physical world. These rules can vary from formulation to formulation and theory to theory. So, in ordinary formulations of special relativity, 'c' refers to the speed of light. In thermodynamics 'c' would refer to specific heat.

[10] Indeed the assertion may prove to be a logical truth, that is, it would be true by the definition of the terms it invokes or it may amount to the definition of term. While their truth is assured, such assertions need not be trivial. For example in a formulation of special relativity we may assert that the coefficients of the metric tensor are linear functions of the coordinates. This turns out to place no physical restriction on the theory; it merely restricts us to particular coordinate systems. It is what is known as a coordinate condition that defines the restricted class of coordinate systems in which the formulation holds.

of the geometric structure just like that of general relativity. The Lorentz covariant formulation of equations (1) and (2) admits preferred coordinate systems. In effect, some of the physical content of the theory is encoded in them. They specify, for example, which are the inertial motions; a body moves inertially only if there is a coordinate system in which its spatial coordinates do not change with the time coordinate. In the transition to the generally covariant formulation, this content is stripped out of the coordinate systems. We can no longer use constancy of spatial coordinates to discern which points move inertially. This content is relocated in the Christoffel symbols, which, via equation (4), determine whether a particular motion is inertial. The general covariance of (3a), (3b), and (4) leaves a gauge freedom in how the metric g_{ik} and the Christoffel symbols $\{{}_{km}^{\ i}\}$ may be spread over some coordinate system. In one coordinate system, they may be spread in many mathematically distinct but physically equivalent ways.

To summarize

There is no restriction on physical content in saying that *there exists* a formulation of the theory that has some formal property (general covariance, the presence of the string of symbols '$E = mc^2$', etc.). But once we fix a *particular* formulation and interpretation, that very same formal property will express something physical, although there is no assurance that it will be something interesting.

5 Gauge theories in particle physics

This summary generates a new puzzle. One of the most fertile strategies in recent decades in particle physics has been to extend the gauge symmetries of non-interacting particles and thereby infer to new gauge fields that mediate the interaction between the particles. Most simply, the electromagnetic field can be generated as the gauge field that mediates interactions of electrons. This power has earned the strategy the label of the 'gauge principle'. How can this strategy succeed if Kretschmann is right and there is no physical content in our being able to arrive at a reformulation of expanded covariance? In the particle context, this corresponds to a reformulation of expanded gauge freedom. So why doesn't Kretschmann's objection also tell us that the strategy of the gauge principle is physically vacuous?

The solution lies in the essential antecedent condition of Kretschmann's objection. The physical vacuity arises since there are no restrictions placed on how we might reformulate a theory in seeking generally covariance. It has long been recognized that the assured achievement of general covariance can be blocked by some sort of additional restriction on how the reformulation may be achieved. Many additional conditions have been suggested, including demands for simplicity and restrictions on which extra variables may introduced. (For a survey, see Norton, 1993, section 5, and Norton, 1995, section 4.) The analogous solution is what gives the gauge principle its content. In generating gauge fields, we are most definitely *not* at liberty to expand the gauge freedom of some non-interacting particle field in any way we please. There is a quite precise recipe that must be followed: we must promote a global symmetry of the original particle field to a local symmetry, using the exemplar of the electron and the Maxwell field, and the new field arises from the connection introduced to preserve gauge equivalence.[11]

There is considerably more that should be said about the details of the recipe and the way in which new physical content arises. The recipe is standardly presented as merely expanding the gauge freedom of the non-interacting particles, which should mean that the realm of physical possibility is unaltered; we merely have more gauge-equivalent representations of the same physical situations. So how can physically new particle fields emerge? This question is currently under detailed and profitable scrutiny.[12]

Appendix 1: Is a generally covariant reformulation always possible?

As Earman[13] has pointed out, it is not entirely clear whether a generally covariant reformulation is always possible for any spacetime theory. The problem lies in ambiguities in the question. Just what counts as 'any' spacetime theory? Just what are we expecting from a generally covariant reformulation? Let me rehearse some of the difficulties and suggest that, for most reasonable answers to these questions, generally covariant reformulation will be possible, though not necessarily pretty.

The substitution trick...

Let us imagine that we are given a spacetime theory in a formulation of restricted covariance. It is given in just one spacetime coordinate system X^i. Let us imagine that the laws of the theory happen to be given by n equations in the $2n$ quantities A^k, B^k

$$A^k(X^i) = B^k(X^i) \tag{5a}$$

where $k = 1, \ldots, n$ and the A^k and B^k are functions of the coordinates as indicated. Consider an arbitrary coordinate system x^i to which we transform by means of the

[11] The transition from special relativity in (1) and (2) to the generally covariant formulation (3a), (3b), and (4) can be extended by one step. We replace the flatness condition (3a) by a weaker condition, a natural relaxation, $R_{ik} = g^{lm} R_{ilmk} = 0$. The result is general relativity in the source-free case. Arbitrary, source-free gravitational fields now appear in the generalized connection $\{^{\,i}_{km}\}$. We have what amounts to the earliest example of the use of the gauge recipe to generate new fields. The analogy to more traditional examples in particle physics is obvious.

[12] See Martin (2002a,b), and contributions to this volume.

[13] 'Once more general covariance,' unpublished manuscript, section 3.

transformation law

$$x^i = x^i(X^m). \tag{6}$$

We can replace the n equations (5a) by equations that hold in the arbitrary coordinate system by the simple expedient of inverting the transformation of (6) to recover the expression for the X^m as a function of the x^i, that is $X^m = X^m(x^i)$. Substituting these expressions for X^m into (5a), we recover a version of (5a) that holds in the arbitrary coordinate system

$$A^k(X^i(x^m)) = B^k(X^i(x^m)). \tag{5b}$$

We seem to have achieved a generally covariant reformulation of (5a) by the most direct application of the intuition that coordinate systems are merely labels and we can relabel spacetime events as we please.

...yields geometric objects

While equation (5b) is generally covariant, we may not be happy with the form of the general covariance achieved – one of the ambiguities mentioned above. We might, as Earman (*ibid.*) suggests, want to demand that (5b) be expressed in terms of geometric object fields. The standard definition of a geometric object field is that it is an n-tuple valued field of components on the manifold, with one field for each coordinate system, and that the transformation rule that associates the components of different coordinate systems has the usual group properties.

While this definition may appear demanding, it turns out to be sufficiently permissive to characterize each side of (5b) as a geometric object field. For example, in each coordinate system x^m, the geometric object field A has components $A^k(X^i(x^m))$, which I now write as $A^k(x^m)$. The transformation rule between the components is induced by the rule for coordinate transformations. That is, under the transformation x^m to $y^r(x^m)$, $A^k(x^m)$ transforms to $A^k(x^m(y^r))$, where $x^m(y^r)$ is the inverse of the coordinate transformation. With this definition of the transformation law for A^k, the components will inherit as much group structure as the coordinate transformations themselves have; that is, it will be as much of a geometric object field as we can demand.[14] For example, assume the transformations of coordinate systems z^p to y^r and y^r to x^m conform to transitivity. Then this same transitivity will be inherited by A. We will have $A^k(x^m(z^p)) = A^k(x^m(y^r(z^p)))$ since the transitivity of the coordinate transformation yields $x^m(z^p) = x^m(y^r(z^p))$.

[14] Why the hedged 'as much group structure as the coordinate transformations themselves have'? These general coordinate transformations may not have all the group properties if the manifold patches covered by the coordinate systems do not all coincide.

But are they the geometric objects we expect?

While the components A^k turn out to be geometric object fields, they are probably not the ones we expected. In brief, the reason is that the transformation rule induced by the substitution trick does not allow any mixing of the components. That precludes it yielding vectors or tensors or like structures; it turns everything into scalar fields. To see how odd this is, take a very simple case. Imagine that we have special relativity restricted to just one coordinate system X^i. Our law might be the law governing the motion of a body of unit mass, $F^i = A^i$, where F^i is the 4-force and A^i the 4-acceleration. Under a Lorentz transformation

$$Y^0 = \gamma(X^0 - vX^1); \quad Y^1 = \gamma(X^1 - vX^0); \quad Y^3 = X^3; \quad Y^4 = X^4$$

with velocity v in the X^1 direction, $c = 1$ and $\gamma = (1 - v^2)^{-1/2}$. The usual Lorentz transformation for the components A^i of the four acceleration would be

$$A'^0 = \gamma(A^0 - vA^1); \quad A'^1 = \gamma(A^1 - vA^0); \quad A'^3 = A^3; \quad A'^4 = A^4 \tag{7a}$$

Note that the transformed A'^0 and A'^1 are linear sums of terms in A^0 and A^1. For this same transformation, the substitution trick merely gives us

$$A'^i = A^i(X^m(Y^r)) \tag{7b}$$

That is, A'^0 is a function of A^0 only and A'^1 is a function of A^1 only.

This oddity becomes a disaster if we apply the substitution rule in a natural way. Instead of starting with A^i in one fixed coordinate system X^i, we might start with the full set of all components of A^i in all coordinate systems related by a Lorentz transformation to X^i. If we now try and make this bigger object generally covariant by the substitution trick, we will end up with two incompatible transformation laws for the transformation X^i to Y^i when we try to transform the components A^i – law (7a) and law (7b). We no longer have a geometric object field since we no longer have a unique transformation law for the components.

The escape from this last problem is to separate the two transformation groups. We consider A^i in coordinate system X^i and A'^i in coordinate system Y^i separately and convert them into distinct geometric object fields by the substitution trick. As geometric object fields they have become, in effect, scalar fields. The Lorentz transformation then reappears as a transformation between these geometric objects.

The coordinates as scalars trick

If this is our final goal, then another general trick for generating generally covariant reformulations could have brought us there much faster. We return to $A^k(X^i)$ of equation (5a). We can conceive the X^i as scalar fields on the manifold – that is really all they are.[15] Scalar fields are geometric object fields already. The A^i are functions of X^i, that is, functions of scalar fields. Therefore they are also geometric objects. So we can conceive of the entire structure $A^k(X^i)$ as a geometric object field. We have got general covariance on the cheap. We cannot avoid a cost elsewhere in the theory, however. Our reformulation is overloaded with structure, one geometric object field for each of what was originally a component. There is clearly far more mathematical structure present than has physical significance. So the theory will need a careful system for discerning just which parts of all this structure has physical significance.

Temptations resisted

These devices for inducing general covariance are clumsy but they do fall within the few rules discussed. We might be tempted to demand that we admit generally covariant formulations only if their various parts fall together into nice compact geometric objects. But what basis do we have for demanding this? Are we to preclude the possibility that the theory we started with is just a complicated mess that can only admit an even more complicated mess when given generally covariant reformulations? (Newtonian theory has been accused of this!) And if we are to demand only nice and elegant reformulations, just how do we define 'nice and elegant'?

My conclusion is that generally covariant reformulations are possible under the few rules discussed and that efforts to impose further rules to block the more clumsy ones will cause more trouble than they are worth elsewhere.

Appendix 2: From passive to active covariance

As above, assume the equations of some generally covariant theory admit a scalar field $\varphi(x^i)$ as a solution. We can transform to a new coordinate system by merely relabelling the events of spacetime; x^i is relabelled x'^i, where the x'^i are smooth functions of the x^i. The field $\varphi(x^i)$ transforms to field $\varphi'(x'^i)$ by the simple rule $\varphi'(x'^i) = \varphi(x^i)$. Since the equations of the theory hold in the new coordinate system, the new field $\varphi'(x'^i)$ will still be a solution. The two fields $\varphi(x^i)$ and $\varphi'(x'^i)$ are just representations of the same physical field in different spacetime coordinate systems.

This is the passive view of general covariance. It can be readily transmogrified into an active view, a transition that Einstein had already undertaken with his 1914

[15] Ask, what is the X^0 coordinate in coordinate system X^i of some event p? The answer will be the same number if we ask it from any other coordinate system y^i as long as we are careful to ask it of the original coordinate system X^i. That is, each coordinate can be treated as a scalar field.

statements of the 'hole argument'. What makes $\varphi'(x'^i)$ a solution of the theory under discussion is nothing special about the coordinate system x'^i. It is merely the particular function that φ' happens to be. It is a function that happens to satisfy the equations of the theory. We could take that very same function and use it in the original coordinate system, x^i. That is, we could form a new field $\varphi'(x^i)$. Since this new field uses the very same function, it retains every property except the mention of the primed coordinate system x'^i. Thus it is also a solution of the equations of the theory.

In short, the passive general covariance of the theory has delivered us two fields, $\varphi(x^i)$ and $\varphi'(x^i)$. They are not merely two representations of the same field in different coordinate systems. They are defined in the *same* coordinate system and are mathematically distinct fields, insofar as their values at given events will (in general) be different. Active general covariance allows the generation of the field $\varphi'(x^i)$ from $\varphi(x^i)$ by the transformation x^i to x'^i.

Acknowledgements

I thank Carlo Rovelli, John Earman, Elena Castellani, and Chris Martin for their discussion and for forcing me to think this through. I am also grateful for discussion by the participants in the 'International Workshop: General covariance and the quantum: where do we stand?' Department of Physics, University of Parma, 21–23 June 2001, organized by Massimo Pauri.

References

Earman, J., and Norton, J. D. (1987). 'What price spacetime substantivalism? The hole argument'. *British Journal for the Philosophy of Science*, **38**, 515–25.

Howard, D. (1999). 'Point coincidences and pointer coincidences: Einstein on the invariant content of space–time theories'. In *The Expanding Worlds of General Relativity*, Einstein Studies, Vol. 7, ed. H. Goenner, J. Renn, J. Ritter, and T. Sauer, pp. 463–500. Boston: Birkhäuser.

Kretschmann, E. (1917). 'Über den physikalischen Sinn der Relativitätspostulat, A. Einsteins neue und seine ursprüngliche Relativitätstheorie'. *Annalen der Physik*, **53**, 575–614.

Martin, C. (2002a). 'Gauge principles, gauge arguments and the logic of nature'. *Philosophy of Science*, **69**, S221–34.

(2002b). 'Gauging gauge: remarks on the conceptual foundations of gauge symmetry'. Ph.D. dissertation, University of Pittsburgh.

Norton, J. D. (1984). 'How Einstein found his field equations: 1912–1915'. *Historical Studies in the Physical Sciences*, **14**, 253–316. Reprinted in *Einstein and the History of General Relativity*, Einstein Studies, Vol. 1 (1989), ed. D. Howard and J. Stachel, pp. 101–59. Boston: Birkhäuser.

(1993). 'General covariance and the foundations of general relativity: eight decades of dispute'. *Reports on Progress in Physics*, **56**, 791–858.

(1995). 'Did Einstein stumble? The debate over general covariance'. *Erkenntnis*, **42**, 223–45.
(1999). 'The hole argument'. *Stanford Encyclopedia of Philosophy.* http://plato.stanford.edu/entries/spacetime-holearg.
Renn, J., Sauer, T., Janssen, M., Norton, J. D., and Stachel, J. (forthcoming). *General Relativity in the Making: Einstein's Zurich Notebook.*
Rovelli, C. (1997). 'Halfway through the woods: contemporary research on space and time'. In *The Cosmos of Science: Essays of Exploration*, ed. J. Earman and J. D. Norton, pp. 180–223. Pittsburgh: University of Pittsburgh Press.
Rynasiewicz, R. (1999). 'Kretschmann's analysis of general covariance and relativity principles'. In *The Expanding Worlds of General Relativity*, Einstein Studies, Vol. 7, ed. H. Goenner, J. Renn, J. Ritter, and T. Sauer, pp. 431–62. Boston: Birkhäuser.
Stachel, J. (1980). 'Einstein's search for general covariance'. Paper read at the Ninth International Conference on General Relativity and Gravitation, Jena. Reprinted in *Einstein and the History of General Relativity*, Einstein Studies, Vol. 1 (1989), ed. D. Howard and J. Stachel, pp. 63–100. Boston: Birkhäuser.

7

The interpretation of gauge symmetry

MICHAEL REDHEAD

1 Introduction

The term 'gauge' refers in its most general everyday connotation to a system of measuring physical quantities, for example by comparing a physical magnitude with a standard or 'unit'. Changing the gauge would then refer to changing the standard. The original idea of a gauge as introduced by Weyl in his (1918) in an attempt to provide a geometrical interpretation of the electromagnetic field was to consider the possibility of changing the standard of 'length' in a four-dimensional generalization of Riemannian geometry in an arbitrary local manner, so that the invariants of the new geometry were specified not just by general coordinate transformations but also by symmetry under conformal rescaling of the metric. The result was, in general, a non-integrability or path dependence of the notion of length which could be identified with the presence of an electromagnetic field. In relativistic terms this meant that, unacceptably, the frequencies of spectral lines would depend on the path of an atom through an electromagnetic field, as was pointed out by Einstein.

With the development of wave mechanics the notion of gauge invariance was revived by Weyl himself (1929) following earlier suggestions by Fock and by London, so as to apply to the non-integrability of the *phase* of the Schrödinger wave function, effectively replacing a scale transformation $e^{\alpha(x)}$ by a phase transformation $e^{i\alpha(x)}$. Invariance under these local phase transformations, referred to as gauge transformations of the second kind (as contrasted with constant global phase transformations of the first kind), necessitated the introduction of an interaction field which could be identified with the electromagnetic potential, a point of view which was particularly stressed by Pauli (1941). The extension of this idea to other sorts of interaction was introduced by Yang and Mills in their article (1954) (although mention should be made of the independent work of Shaw (1954) and the proposals made in an unpublished lecture by Oskar Klein in 1938). The extension to a gauge theory of gravitation was considered by Utiyama (1956). The great advantage of gauge theories was that they offered the possibility of renormalizability, but this was offset by the fact that the interactions described by gauge fields were carried by massless quanta and so seemed inappropriate to the case of the short-range weak and strong interactions of nuclear physics. In the case of the weak interactions this defect was remedied by noticing that renormalizability survived the process of spontaneous symmetry breaking that would generate effective mass for the gauge quanta, while the key to understanding strong interactions as a gauge theory lay in the development of the idea of 'asymptotic freedom', expressing roughly the idea that strong interactions were actually weak at very short distances, effectively increasing rather than decreasing with distance.

With this brief historical introduction we turn to consider the fundamental conceptual issues involved in gauge freedom and the closely associated idea of gauge symmetry.

2 The ambiguity of mathematical representation

As we have seen, the term gauge refers in a primitive sense to the measurement of physical magnitudes, i.e. of associating physical magnitudes with mathematical entities such as numbers. Of course the numerical measure is not unique, varying indeed inversely with the magnitude of the unit chosen. Both the unit and the measure can, with some confusion, be referred to as the gauge of the quantity, in everyday parlance.

We now want to generalize this usage by referring to the mathematical representation of any physical structure as a gauge for that structure. By narrowing down this very general definition we shall focus in on more standard definitions of gauge in theoretical physics, such as the gauge freedom of constrained Hamiltonian systems and Yang–Mills gauge symmetries.

But let us start with the most general concept.[1] Consider a physical structure P consisting of a set of physical entities and their relations, and a mathematical structure M consisting of a set of mathematical entities and their relations, which represents P in the sense that M and P share the same abstract structure, i.e. there exists a one–one structure-preserving map between P and M, what mathematicians call an isomorphism. In the old-fashioned statement view of theories, P and M could be regarded as models for an uninterpreted calculus C, as illustrated in figure 1. On the more modern semantic view, theories are of course identified directly with a collection of models such as P. We do not need to take sides in this debate. For our purposes we need merely to note that P does not refer directly to the world, but typically to a 'stripped-down', emasculated, idealized version of the world.

[1] The following account leans heavily on Redhead (2001).

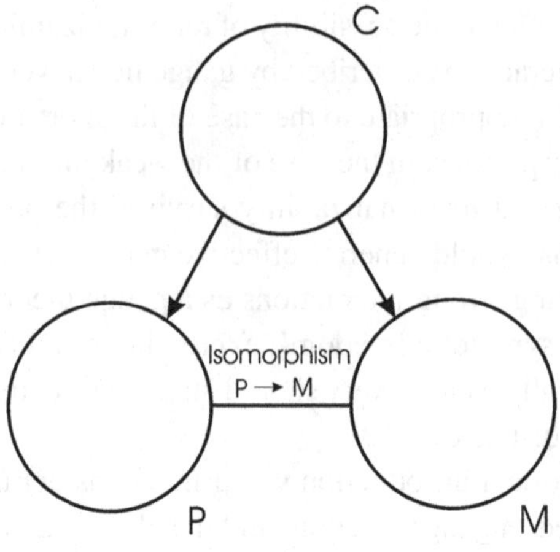

Figure 1. A physical structure P and a mathematical structure M are isomorphic models of an uninterpreted calculus C.

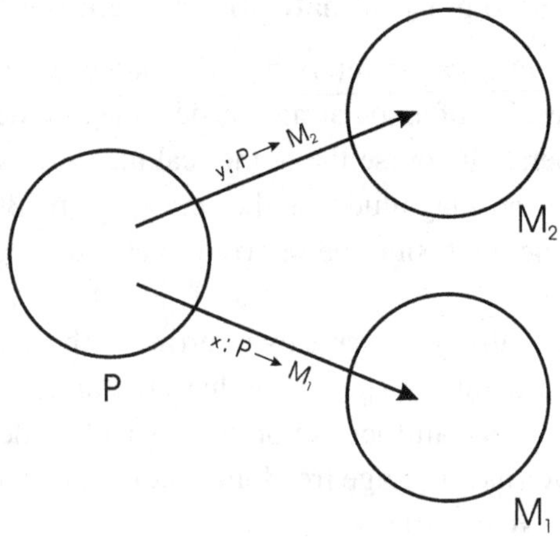

Figure 2. Ambiguity of gauge. M_1 and M_2 are distinct mathematical structures each of which represents P via isomorphisms x and y respectively.

(Only in the case of a genuine Theory of Everything would there be a proposed isomorphism between the *world* and a mathematical structure.)

In our new terminology we shall call M a gauge for P (another way of expressing the relationship between P and M, would be to say that M 'coordinatizes' P in a general sense).

In general there will be many different gauges for P. Consider, as a very elementary example, the ordinal scale provided by Moh's scale of hardness. Minerals are arranged in order of 'scratchability' on a scale of 1 to 10, i.e. the physical structure involved in ordering the hardness of minerals is mapped isomorphically onto the finite segment of the arithmetical ordinals running from 1 to 10. But of course we might just as well have used the ordinals from 2 to 11 or 21 to 30 or whatever. The general situation is sketched in figure 2, which shows two maps x and y which are isomorphisms between P and distinct mathematical structures M_1 and M_2. Of course M_1 and M_2 are also isomorphically related via the map $y \circ x^{-1} : M_1 \to M_2$ and its inverse $x \circ y^{-1} : M_2 \to M_1$.

But how can the conventional choice between M_1 and M_2 as gauges for P have any *physical* significance? To begin to answer this question we introduce the notion of a symmetry of P and its connection with the gauge freedom in the generalized sense we have been discussing.

3 Symmetry

Consider now the case where the ambiguity of representation (the gauge freedom) arises within a *single* mathematical structure M. Thus we consider two distinct isomorphisms $x : P \to M$ and $y : P \to M$, as illustrated in figure 3.

Clearly the composite map $y^{-1} \circ x : P \to P$ is an automorphism of P. This is referred to by a mathematician as a point transformation of P and by physicists as an *active* symmetry of P. The composite map $y \circ x^{-1} : M \to M$ is a 'coordinate' transformation or what physicists call a *passive* symmetry of P. It is easy to show that *every* automorphism of P or M can be factorized in terms of pairs of isomorphic maps between P and M in the way described. It is, of course, not at all surprising that the automorphisms of P and M are themselves in one–one correspondence. After all, since P and M are isomorphically related, they share the same abstract structure, so the structural properties of P represented by the symmetries of P can be simply read off from the corresponding symmetries of M.

Now the symmetries of P express very important structural properties of P, and we can see how they are related to the gauge freedom in this very important special case where the ambiguity of representation is within a *single* mathematical structure M.

The gauge freedom represented in figure 2 does not, in general, have physical repercussions related to symmetry. For example, in the case of Moh's scale of hardness, there simply are no non-trivial automorphisms of a finite ordinal scale.

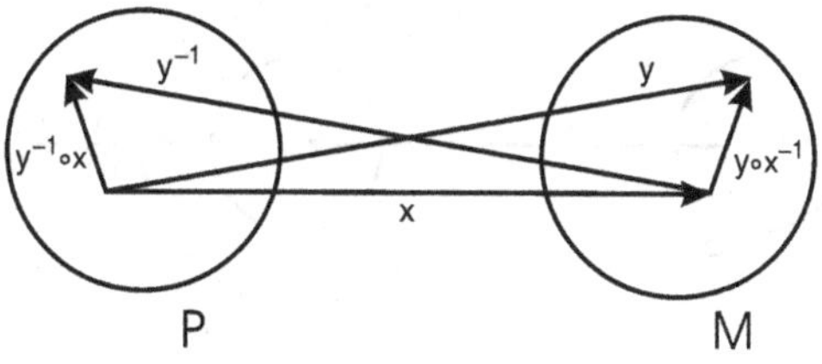

Figure 3. x and y are two distinct isomorphisms between P and M. Then $y^{-1} \circ x : P \to P$ is an automorphism of P and $y \circ x^{-1} : M \to M$ is an associated automorphism of M.

We now want to extend our discussion to a more general situation, which frequently arises in theoretical physics and which we introduce via a notion we call 'surplus structure'.

4 Surplus structure

We consider now the situation where the physical structure P is *embedded* in a larger structure M' by means of an isomorphic map between P and a substructure M of M'. This case is illustrated in figure 4.

The relative complement of M in M' comprises elements of what we shall call the surplus structure in the representation of P by means of M'. Considered as a structure rather than just as a set of elements, the surplus structure involves both relations among the surplus elements and relations between these elements and elements of M.

A simple example of this surplus structure would arise in the familiar use of complex currents and impedances in alternating current theory, where the physical quantities are embedded in the wider mathematical structure of complex numbers.

Another example is the so-called S-matrix theory of the elementary particles that was popular in the 1960s, in which scattering amplitudes considered as functions of real-valued energy and momentum transfer were continued analytically into the complex plane and axioms introduced concerning the location of singularities of these functions in the complex plane were used to set up systems of equations controlling the behaviour of scattering amplitudes considered as functions of the real physical variables. This is an extreme example of the role of surplus structure in formulating a physical theory, where there was no question of identifying any physical correlate with the surplus structure.

In other examples the situation is not so clear. What starts as surplus structure may come to be seen as invested with physical reality. A striking example is the case of energy in nineteenth-century physics. The sum of kinetic and potential energy was originally introduced into mechanics as an auxiliary, purely mathematical entity,

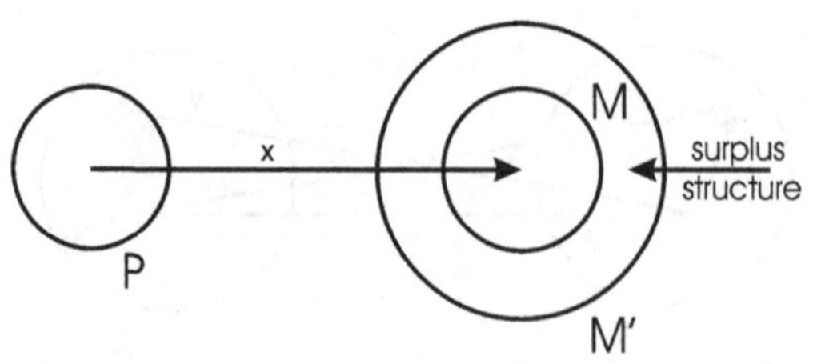

Figure 4. $x : P \to M$ is an embedding of P in the larger structure M'.

arising as a first integral of the Newtonian equations of motion for systems subject to conservative forces. But as a result of the formulation of the general principle of the conservation of energy and its incorporation in the science of thermodynamics (the First Law) it came to be regarded as possessing ontological significance in its own right. So the sharp boundary between M and the surplus structure as illustrated in figure 4 may become blurred, with entities in the surplus structure moving over time into M. Another example would be Dirac's hole theory of the positron, allowing a physical interpretation for the negative-energy solutions of the Dirac equation.

Ambiguities in representation, i.e. gauge freedom, can now arise via automorphisms of M' that reduce to the identity on M, i.e. the transformations of representation act non-trivially only on the surplus structure. Nevertheless such transformations can have repercussions in controlling the substructure M and hence the physical structure P. This is the situation that arises in Yang–Mills theories which we shall describe in section 6. But first we shall make a short digression to discuss the example of constrained Hamiltonian systems, of which free-field electromagnetism is a very important special case.

5 Constrained Hamiltonian systems[2]

The idea of surplus structure describes a situation in which the number of degrees of freedom used in the mathematical representation of a physical system exceeds the number of degrees of freedom associated with the physical system itself. A familiar example is the case of a constrained Hamiltonian system in classical mechanics. Here the Legendre transformation from the Lagrangian to the Hamiltonian variables is singular (non-invertible). As a result the Hamiltonian variables are not all independent, but satisfy identities known as constraints. This in turn means that the Hamiltonian equations underdetermine the time-evolution of the Hamiltonian variables, leading to a gauge freedom in the description of the time-evolution, which means in other words a breakdown of determinism for the evolution of the state of the system as specified by the Hamiltonian variables.

More formally the arena for describing a constrained Hamiltonian system is what mathematicians call a *presymplectic manifold.* This is effectively a phase space equipped with a degenerate symplectic two-form ω. By degenerate one means that the equation $\omega(X) = 0$, where X is a tangent vector field, has non-trivial solutions, the integral curves of which we shall refer to as null curves on the phase space. The equations of motion are given in the usual Hamiltonian form as $\omega(X) = dH$, where H is the Hamiltonian function. The integral curves derived from this equation

[2] The treatment of this topic broadly follows the excellent account in Belot (1998).

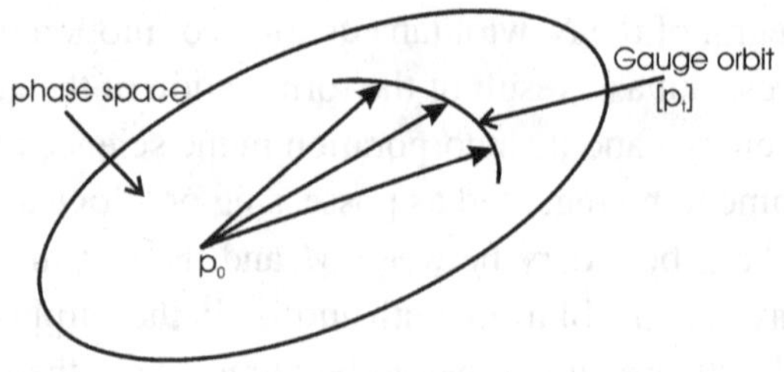

Figure 5. The indeterministic time-evolution of a constrained Hamiltonian system.

represent the dynamical trajectories in the phase space. But in the case we are considering there are many trajectories issuing from some initial point p_0, at time t_0. At a later time t the possible solutions of the Hamiltonian equations all lie on a *gauge orbit* in the phase space which is what we may call a null subspace of the phase space, in the sense that any two points on the orbit can be joined by a null curve as we have defined it. The situation is illustrated schematically in figure 5.

Instead of the initial phase point p_0 developing into a unique state p_t at a later time t as in the case of an unconstrained Hamiltonian system, we now have an indeterministic time-evolution, with a unique p_t replaced by a gauge orbit, which we denote by $[p_t]$ in figure 5. Effectively what is happening here is that the 'physical' degrees of freedom at time t are being multiply represented by points on the gauge orbit $[p_t]$ at time t in terms of the 'unphysical' degrees of freedom.

A familiar example of a constrained Hamiltonian system is the case of electromagnetism described by Maxwell's equations in vacuo. Here the Hamiltonian variables may be taken as the magnetic vector potential $\vec{A}$ and the electric field $\vec{E}$ subject to the constraint $div\ E = 0$. On a gauge orbit $\vec{E}$ is constant but $\vec{A}$ is specified only up to the gradient of a scalar function. The magnetic induction $\vec{B}$ defined by $\vec{B} = curl\ \vec{A}$ is then also gauge-invariant, i.e. constant on a gauge orbit. So $\vec{A}$ involves unphysical degrees of freedom, whose time-evolution is not uniquely determined. It is only for the physical degrees of freedom represented by $\vec{E}$ and $\vec{B}$ that determinism is restored. The gauge freedom in $\vec{A}$ belongs to surplus structure in the terminology of section 4.

6 Yang–Mills gauge theories

We turn now to a still more restricted sense of gauge symmetry associated with Yang–Mills gauge theories of particle interactions. To bring out the main idea we shall consider the simplest case of non-relativistic (first-quantized) Schrödinger field. The field amplitude $\psi(x)$ (for simplicity we consider just one spatial dimension for the time being) is a complex number, but quantities like the charge density

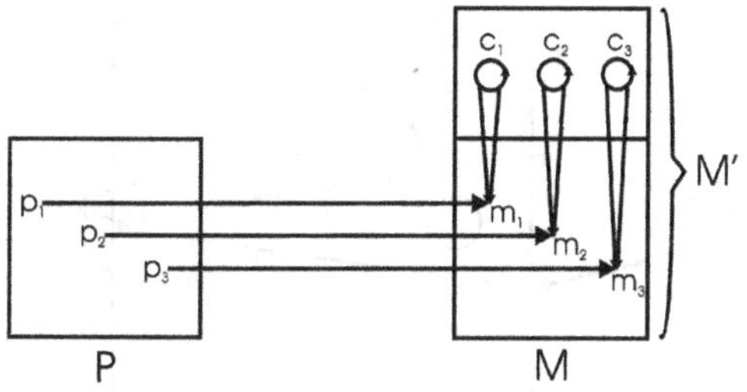

Figure 6. Gauge transformations and surplus structure.

$\phi = e\psi^*\psi$ and the current density $j = \frac{1}{2}ie\big(\psi^*\frac{d}{dx}\psi - \psi\frac{d}{dx}\psi^*\big)$ are real quantities and can represent physical magnitudes. Consider now phase transformations of the form $\psi \to \psi e^{i\alpha}$. These are known as global gauge transformations since the phase factor α does not depend on x. If we now demand invariance of physical magnitudes under such gauge transformations, then ϕ and j satisfy this requirement. But suppose we impose *local* gauge invariance, i.e. allow the phase factor α to be a function $\alpha(x)$ of x. ϕ remains invariant but j does not. In order to obtain a gauge-invariant current we introduce the following device. Replace $\frac{d}{dx}$ by a new sort of derivative $\frac{d}{dx} - iA(x)$ where A transforms according to $A \to A + \frac{d}{dx}\alpha(x)$. Then the modified current $j(x) = \frac{1}{2}ie\big(\psi^*\big(\frac{d}{dx} - iA\big)\psi - \psi\big(\frac{d}{dx} - iA\big)\psi^*\big)$ *is* gauge-invariant. But this has been achieved by introducing a new field $A(x)$ as a necessary concomitant of the original field $\psi(x)$. Reverting to three spatial dimensions, the $\vec{A}$ field can be identified (modulo the electronic charge e) with the magnetic vector potential, and the transformation law for $\vec{A}$ is exactly that described for the vector potential in the last section. The requirement of local gauge-invariance can be seen as requiring the introduction of a magnetic interaction for the ψ field.

Again we have an example here of physical structure being controlled by requirements imposed on surplus mathematical structure. The situation is illustrated schematically in figure 6. p_1, p_2, p_3 are three physical magnitudes, for example the charge or current at three different spatial locations. They are mapped onto m_1, m_2, m_3 in the mathematical structure M which is a substructure in the larger structure M'. The circles c_1, c_2, c_3 in the surplus structure represent possible phase angles associated with m_1, m_2, m_3 in a many–one fashion as represented by the arrows projecting c_1, c_2, c_3 onto m_1, m_2, m_3. Local gauge transformations represented by the arrows on the circles act independently at different spatial locations. They correspond to identity transformations on M and correlatively on P.

The $\vec{A}$ field establishes what mathematicians call a connection, correlating phases on the different circles c_1, c_2, c_3. The gauge transformations alter the connection as well as the individual phases in such a way as to maintain the gauge-invariance of the corrected 'derivative' $\vec{\nabla} - i\vec{A}$.

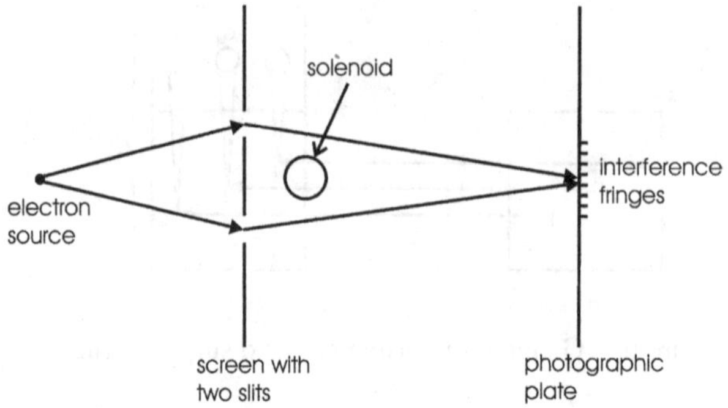

Figure 7. The Aharonov–Bohm experiment.

Two ways of dealing with the surplus structure inherent in gauge theories suggest themselves. Firstly, we might just fix the gauge by some arbitrary convention,[3] but then we have lost the possibility of expressing gauge transformations which lead from one gauge to another. Alternatively, we might try to formulate the theory in terms of gauge-invariant quantities, which are the physically 'real' quantities in the theory. Thus instead of the gauge potential, the $\vec{A}$ field in electromagnetism, we should employ the magnetic induction $\vec{B}$, specified by the equation $\vec{B} = curl\, \vec{A}$.

However, this manoeuvre has the serious disadvantage of rendering theory non-local! This is most clearly seen in the Aharonov–Bohm effect[4] in which a phase shift occurs between electron waves propagating above and below a long (in principle infinitely long) solenoid. The experiment is illustrated schematically in figure 7.

The magnetic induction is, of course, confined within the solenoid, so if it is regarded as responsible for the phase shift, it must be regarded as acting non-locally. On the other hand the vector potential extends everywhere outside the solenoid, so if invested with physical reality its effect on the electron phases can be understood as occurring locally. This is an argument for extending physical reality to elements which originated as elements of surplus structure.

However, just as in the case of free electromagnetism discussed in the previous section, the time-evolution of the vector potential is indeterministic since it is only specified up to the unfolding of a, in general, time-dependent gauge transformation. To restore determinism we must regard the gauge as being determined by additional 'hidden variables' which pick out the One True Gauge; this seems a highly *ad hoc* way of proceeding as a remedy for restoring determinism. This is indeed a quite general feature of Yang–Mills gauge theories.[5]

[3] In some pathological cases this may not be consistently possible, a phenomenon known in the trade as the Gribov obstruction.

[4] The interpretation of the Aharonov–Bohm effect has occasioned considerable controversy in the philosophical literature. See, in particular, Healey (1997), Belot (1998), and Leeds (1999).

[5] For a detailed discussion see a paper presented by Lyre at the Fifth International Conference on the History and Foundations of General Relativity, 8–11 July 1999, University of Notre Dame, Notre Dame, Indiana (E-Print: gr-qc/9904036).

7 The case of general relativity

The general arena for Yang–Mills gauge theories is provided by the notion of a fibre bundle. Speaking crudely a fibre bundle can be thought of as being constructed by attaching one sort of space, the fibre, to each point of a second sort of space, the base space, so that *locally* the structure is just the familiar Cartesian product.

We can effectively redraw figure 6 in a way that brings out the bundle structure, as illustrated in figure 8.

The local gauge group changes the phases according to the action of the $U(1)$ group. A cross-section of 'parallel' or constant phase is specified by the connection field, i.e. the gauge potential.

In the case of general relativity (GR) we are dealing with the bundle of tangent spaces at each point of the spacetime manifold, or more appositely the frame bundle, specifying the basis (or frame) for the tangent space at every point. The gauge group is now the group of general 4-dimensional frame transformations, usually denoted by $GL(4, \mathbb{R})$. If consideration is restricted to Lorentzian frames the gauge group reduces to the familiar Lorentz group $SO(1, 3)$ (or one might want to consider $SL(2, \mathbb{C})$, the covering group of $SO(1, 3)$, if spinor fields are to be introduced). There are now two ways to go. Stick with the Lorentz group, and introduce a connection field to define parallel transport of frames from one point of spacetime to another. This was the original approach of Utiyama (1956). But it has been claimed repeatedly in the literature that if one wants to generalize classical relativity, so as to allow for torsion in the spacetime manifold, it is necessary to introduce an affine structure into the fibres (to be sharply distinguished from an affine connection on the bundle), so the local symmetry group becomes the *inhomogeneous* Lorentz group, i.e. the Poincaré group. Of course, this can be done from a purely mathematical point of view, but does not really make any *physical* sense at all. The translation subgroup effectively changes the origin, i.e. the point of attachment of the tangent space to the spacetime manifold, so inhomogeneous frame transformations correspond picturesquely to sliding the tangent space over the base space, but that is *not* what

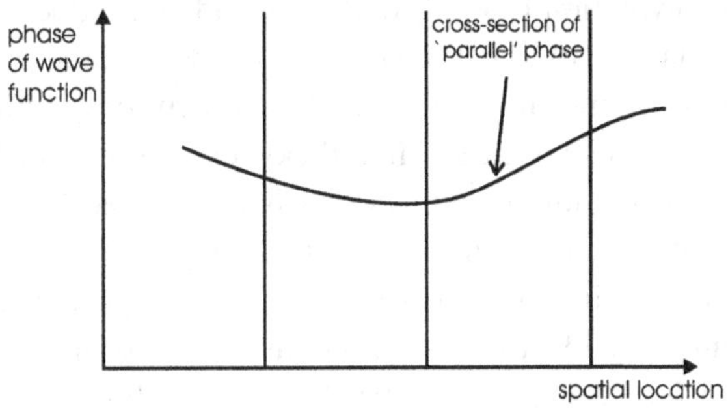

Figure 8. Fibre bundle structure of Yang–Mills gauge theory corresponding to figure 6.

local gauge transformation are supposed to do – they move points around in the fibre at a *fixed* point on the base space. I refer the reader to Ivanenko and Sardanshvily (1983) or Göckeler and Schücker (1987), who support, in my view correctly, the view that we do not need an affine bundle at all in order to extend GR to the Einstein–Cartan $\mathbb{U}_4$ theory incorporating spin and torsion.

So, there is considerable confusion between the Lorentz group and the Poincaré group as to the appropriate Yang–Mills gauge group for GR and its generalizations, but it is also often claimed that general coordinate transformations (the subject of general covariance) provide the gauge group of GR! The following comments are intended to clarify what is going on here. Firstly, it should be noted that general coordinate transformations do not in general constitute a group from the global point of view, since in general they cannot be defined globally. But there is a globally defined symmetry group, which is an invariance group of GR, namely the diffeomorphism group, *diff*, which from the local point of view is the active version of local coordinate transformations. From the bundle point of view described above, elements of *diff* move points around in the base space, which is just the spacetime manifold. This is not directly connected with gauge freedom in the more specialized sense we have defined, that is to say either in the Yang–Mills sense or as arising in the theory of constrained Hamiltonian systems as described in section 5 above. To link up with the latter notion, we need to exhibit GR in a canonical formulation, sometimes referred to as the $(3+1)$ approach to GR as compared with the 4-dimensional approach of the more familiar covariant formulation. In the $(3+1)$ approach the configuration variables are the 3-geometries on a spatial slice at a given coordinate time. (The collection of all possible 3-geometries is what is often referred to as *superspace*.) The Hamiltonian (canonical) variables satisfy constraints, indeed the Hamiltonian itself vanishes identically. The gauge freedom arises essentially as a *manifestation* of the diffeomorphism invariance of the 4-dimensional covariant formulation, in the $(3+1)$ setting. In this setting there are two sorts of gauge motion, one sort acting in the spatial slices and corresponding to diffeomorphisms of the 3-geometries, the other acting in time-like directions and corresponding to time-evolution of the 3-geometries.

The fact that time-evolution is a gauge motion, and hence does not correspond to any change at all in the 'physical' degrees of freedom in the theory, produces the famous 'problem of time' in canonical GR! Crudely this is often referred to under the slogan 'time does not exist!' In a Pickwickian sense the indeterminism problem for constrained Hamiltonian systems is solved because time-evolution itself lies in a gauge orbit rather than cutting across gauge orbits, as in figure 5. The solution of the problem of time (which plagues attempts to quantize canonical GR), must involve in some way identifying some combination of the *physical* degrees of freedom with an *internal* time variable. But exactly how to do this remains a matter of controversy among the experts in canonical approaches to quantum gravity.[6]

8 BRST symmetry

In the path integral approach to general (non-Abelian) gauge theories, a naive approach would involve integrating over paths which are connected by gauge transformations. To make physical sense of the theory, the obvious move is to 'fix the gauge', so that each path intersects each gauge orbit in just one point. However, early attempts to derive Feynman rules for expanding the gauge-fixed path integral in a perturbation expansion led to an unexpected breakdown of unitarity.[7] This was dealt with in an *ad hoc* fashion by introducing fictitious fields, later termed ghost fields, which only circulated on internal lines of the Feynman diagrams in such a way as to cure the unitarity problem, but could never occur as real quanta propagating along the external lines of the diagrams. So getting rid of one sort of surplus structure, the unphysical gauge freedom, seemed to involve one in a new sort of surplus structure associated with the ghost fields.

The whole situation was greatly clarified by the work of Fadeev and Popov (1967) who pointed out that when fixing the gauge in the path integral careful consideration must be given to transforming the measure over the paths appropriately. The transformation of the measure was expressed in a purely mathematical manoeuvre as an integral over scalar Grassmann (i.e. anticommuting) fields which were none other than the ghost (and antighost) fields!

The effective Lagrangian density could now be written as the sum of three terms, $\mathcal{L}_{eff} = \mathcal{L}_{gi} + \mathcal{L}_{gf} + \mathcal{L}_{ghost}$, where $\mathcal{L}_{gi}$ is a gauge-invariant part, $\mathcal{L}_{gf}$ is a non-gauge-invariant part arising from the gauge fixing, and $\mathcal{L}_{ghost}$ is the contribution from the ghost fields.

$\mathcal{L}_{eff}$ no longer, of course, has the property of gauge-invariance, but it was discovered by Becchi, Rouet, and Stora (1975) and independently by Tyutin (1975) that $\mathcal{L}_{eff}$ does exhibit a kind of generalized gauge symmetry, now known as BRST symmetry, in which the non-invariance of $\mathcal{L}_{gf}$ is compensated by a suitable transformation of the ghost fields contributing to $\mathcal{L}_{ghost}$.

To see how this comes about we consider the simplest (Abelian) case of scalar electrodynamics. The matter field ψ satisfies the familiar Klein–Gordon equation. Under the local gauge transformation $\psi \to \psi e^{i\alpha(x)}$, where x now stands for the 4-dimensional spacetime location x^μ, the gauge-invariance of the Lagrangian for the free field is restored by using the corrected derivative $\partial \to \partial_\mu - iA_\mu$, where

[6] For a comprehensive account of canonical quantum gravity and the 'problem of time' reference may be made to Isham (1993).

[7] Cf. Feynman (1963).

the gauge potential A_μ can be identified, modulo the electronic charge, with the electromagnetic 4-potential. A_μ transforms as $A_\mu \to A_\mu + \partial_\mu \alpha(x)$. The field strength $F_{\nu\mu} = A_{\nu,\mu} - A_{\mu,\nu}$ is gauge-invariant and measures the curvature of the connection field A_μ in the geometrical fibre bundle language. All that we have done here is just a relativistic generalization of the discussion already given in section 6.

To formulate the BRST transformation we consider a 5-component object

$$\Phi = \begin{pmatrix} \psi \\ A_\mu \\ \eta \\ \omega \\ b \end{pmatrix}$$

where ψ is the matter field, A_μ the gauge potential which we have already introduced above, η is the ghost field, ω the antighost field, and b is what is usually termed a Nakanishi–Lautrup field.

η and ω are anticommuting (Grassmann) scalar fields. The fact that they violate the spin-statistic theorem, which would associate scalar fields with commuting variables, emphasizes the unphysical character of the ghosts and antighosts.

We have then

$$\omega^2 = \eta^2 = 0.$$

The BRST symmetry is defined by

$$\Phi \to \Phi + \epsilon s \Phi$$

where ϵ is an infinitesimal Grassmann parameter and

$$s\Phi = \begin{pmatrix} i\eta\psi \\ \partial_\mu \eta \\ 0 \\ b \\ 0 \end{pmatrix}.$$

The first two components of $s\Phi$ comprise just the infinitesimal version of a gauge transformation with the arbitrary spacetime function $\alpha(x)$ replaced by the ghost field η. But ϵ is a constant so the BRST transformation is a curious hybrid. It is in essence a non-linear rigid fermionic transformation, which contains within itself, so to speak, a local gauge transformation specified by a dynamical field, namely the ghost field.

What is the role of the Nakanishi–Lautrup field? By incorporating this field the transformation is rendered nilpotent,[8] i.e. it is easily checked that $s^2\Phi = 0$. But this means that s behaves like an exterior derivative on the extended space of fields.

[8] The original BRST transformation failed to be nilpotent on the antighost sector.

This in turn leads to a beautiful generalized de Rham cohomology theory in terms of which delicate properties of gauge fields, such as the presence of anomalies, the violation of a classically imposed symmetry in the quantized version of the theory, can be given an elegant geometrical interpretation.[9]

But now we can go further. Instead of arriving at the BRST symmetry via the Fadeev–Popov formalism, we can forget all about gauge symmetry in the original Yang–Mills sense, and impose BRST symmetry directly as the fundamental symmetry principle. It turns out that this is all that is required to prove the renormalizability of anomaly-free gauge theories such as those considered in the standard model of the strong and electroweak interactions of the elementary particles.

But we may note in passing that for still more recondite gauge theories, further generalizations have had to be introduced.[10]

1. In a sense the ghosts compensate for the unphysical degrees of freedom in the original gauge theories. But in some cases the ghosts can 'overcompensate' and this has to be corrected by introducing ghosts of ghosts, and indeed ghosts of ghosts of ghosts etc!
2. For the more general actions contemplated in string and membrane theories the so-called Batalin–Vilkovisky antifield formalism has been developed. This introduces partners (antifields) for all the fields, but the antifield of a ghost is not an antighost and the anti (antighost) is not a ghost!

9 Conclusion

As we have seen, there are three main approaches to interpreting the gauge potentials.

The first is to try and invest them with physical reality, i.e. to move them across the boundary from surplus structure to M in the language of figure 4. The advantage is that we may then be able to tell a local story as to how the gauge potentials bring about the relative phase shifts between the electron wave functions in the Aharonov–Bohm effect, but the disadvantage is that the theory becomes indeterministic unless we introduce *ad hoc* hidden variables that pick out the One True Gauge.

The second approach is to try and reformulate the whole theory in terms of gauge-invariant quantities. But then the theory becomes non-local. In the case of the Aharonov–Bohm effect this can be seen in two ways. If the phase shift is attributed to the gauge-invariant magnetic induction this is confined *within* the solenoid whereas the experiment is designed so that the electron waves propagate *outside* the solenoid. Alternatively we might try to interpret the effect not in terms

9 See Fine and Fine (1997) for an excellent account of these developments.

10 Weinberg (1996, chapter 15) may be consulted for further information on these matters.

of the $\vec{A}$ field itself which of course is not gauge-invariant but in terms of the gauge-invariant holonomy integral $\oint \vec{A} \cdot d\vec{l}$ taken round a closed curve C encircling the solenoid. (This by Stokes theorem is of course just equal to the flux of magnetic induction through the solenoid.) But if the fundamental physical quantities are holonomies, then the theory is again clearly 'non-local', since these holonomies are functions defined on a space of loops, rather than a space of points.

Furthermore, with this second approach, the principle of gauge invariance cannot even be formulated since gauge transformations are defined by their action on non-gauge-invariant quantities such as gauge potentials, and in the approach we are now considering the idea is to eschew the introduction of non-gauge-invariant quantities altogether!

So this leaves us with the third approach. Allow non-gauge-invariant quantities to enter the theory via surplus structure. And then develop the theory by introducing still more surplus structure, such as ghost fields, antifields and so on. This is the route that has actually been followed in the practical development of the concept of gauge symmetry as we have described in the previous section.

But this leaves us with a mysterious, even mystical, Platonist-Pythagorean role for purely mathematical considerations in theoretical physics. This is a situation which is quite congenial to most practising physicists. But it is something which philosophers have probably not paid sufficient attention to in discussing the foundations of physics. The gauge principle is generally regarded as the most fundamental cornerstone of modern theoretical physics. In my view its elucidation is the most pressing problem in current philosophy of physics. The aim of the present paper has been, not so much to provide solutions, but rather to lay out the options that need to be discussed, in as clear a fashion as possible.

References

Becchi, C., Rouet, A., and Stora, R. (1975). 'Renormalization of the Abelian Higgs–Kibble model'. *Communications in Mathematical Physics*, **42**, 127–62.

Belot, G. (1998). 'Understanding electromagnetism'. *The British Journal for the Philosophy of Science*, **49**, 531–55.

Fadeev, L. D., and Popov, V. N. (1967). 'Feynman diagrams for the Yang–Mills field'. *Physics Letters B*, **25**, 29–30.

Feynman, R. P. (1963). 'Quantum theory of gravity'. *Acta Physica Polonica*, **24**, 697–722.

Fine, D., and Fine, A. (1997). 'Gauge theory, anomalies and global geometry: the interplay of physics and mathematics'. *Studies in History and Philosophy of Modern Physics*, **28**, 307–23.

Göckeler, M., and Schücker, T. (1987). *Differential Geometry, Gauge Theories, and Gravity*. Cambridge: Cambridge University Press.

Healey, R. (1997). 'Nonlocality and the Aharonov–Bohm effect'. *Philosophy of Science*, **64**, 18–40.

Isham, C. J. (1993). 'Canonical quantum gravity and the problem of time'. In *Integrable Systems, Quantum Groups, and Quantum Field Theories*, ed. L. A. Ibort and M. A. Rodriguez, pp. 157–287. Dordrecht: Kluwer.
Ivanenko, D., and Sardanshvily, G. (1983). 'The gauge treatment of gravity'. *Physics Reports*, **94**, 1–45.
Leeds, S. (1999). 'Gauges: Aharonov, Bohm, Yang, Healey'. *Philosophy of Science*, **66**, 606–27.
Pauli, W. (1941). 'Relativistic field theories of elementary particles'. *Reviews of Modern Physics*, **13**, 203–32.
Redhead, M. L. G. (2001). 'The intelligibility of the universe'. In *Philosophy at the New Millennium*, ed. A. O'Hear. Cambridge: Cambridge University Press.
Shaw, R. (1954). *The problem of particle types and other contributions to the theory of elementary particles*. Ph.D. thesis, Cambridge University.
Tyutin, I. V. (1975). 'Gauge invariance in field theory and statistical mechanics'. *Lebedev Institute Reprint N 39.*
Utiyama, R. (1956). 'Invariant theoretical interpretation of interaction'. *Physical Review*, **101**, 1597–607.
Weinberg, S. (1996). *The Quantum Theory of Fields, Vol. 2: Modern Applications*. Cambridge: Cambridge University Press.
Weyl, H. (1918). 'Gravitation und Elektrizität'. *Sitzungsberichte der Preussischen Akademie der Wissenschaften*, pp. 465–80.
(1929). 'Elektron und Gravitation'. *Zeitschrift für Physik*, **56**, 330–52.
Yang, C. N., and Mills, R. L. (1954). 'Conservation of isotopic spin and isotopic gauge invariance'. *Physical Review*, **96**, 191–5.

8

Tracking down gauge: an ode to the constrained Hamiltonian formalism

JOHN EARMAN

1 Introduction

Like moths attracted to a bright light, philosophers are drawn to glitz. So in discussing the notions of 'gauge', 'gauge freedom', and 'gauge theories', they have tended to focus on examples such as Yang–Mills theories and on the mathematical apparatus of fibre bundles. But while Yang–Mills theories are crucial to modern elementary particle physics, they are only a special case of a much broader class of gauge theories. And while the fibre bundle apparatus turned out, in retrospect, to be the right formalism to illuminate the structure of Yang–Mills theories, the strength of this apparatus is also its weakness: the fibre bundle formalism is very flexible and general, and, as such, fibre bundles can be seen lurking under, over, and around every bush. What is needed is an explanation of what the relevant bundle structure is and how it arises, especially for theories that are not initially formulated in fibre bundle language.

Here I will describe an approach that grows out of the conviction that, at least for theories that can be written in Lagrangian/Hamiltonian form, gauge freedom arises precisely when there are Lagrangian/Hamiltonian constraints of an appropriate character. This conviction is shared, if only tacitly, by that segment of the physics community that works on constrained Hamiltonian systems.[1] The approach taps into one of the root notions of gauge transformations – namely, that of transformations that connect equivalent descriptions of the same state or history of a physical system – and one of the key motivations for seeking gauge freedom – namely, to take up the slack that would otherwise constitute a failure of determinism.

[1] Here is one explicit expression of that conviction: 'It is well known that all the theories containing gauge transformations are described by constrained systems' (Gomis, Henneaux, and Pons, 1990, p. 1089). The conviction is all but explicit in Henneaux and Teitelboim (1992), the standard reference on the quantization of gauge systems, which opens with a long and detailed treatment of constrained Hamiltonian systems. In section 7 below I will indicate a way in which this conviction can be challenged.

2 Noether's theorems, constrained Hamiltonian systems, and all that

The literature on gauge theories is filled with talk about 'global' and 'local' symmetries, talk which is annoying both because it is often unaccompanied by any attempt to make it precise and because it is potentially very misleading (e.g. a global mapping of the spacetime onto itself can count as a 'local' symmetry in the relevant sense of corresponding to a gauge transformation). One way to get a grip on the global vs. local distinction is to place it in the context of Noether's two theorems.[2] Both theorems apply to theories whose equations of motion or field equations are derivable from an action principle and, thus, are in the form of (generalized) Euler–Lagrange (EL) equations. And both concern *variational symmetries*, that is, a group $\mathcal{G}$ of transformations that leave the action $\mathcal{A} = \int_\Omega L(\mathbf{x}, \mathbf{u}, \mathbf{u}^{(n)})\, d^p\mathbf{x}$ invariant, where $\mathbf{x} = (x^1, \ldots, x^p)$ stands for the independent variables, $\mathbf{u} = (u^1, \ldots, u^r)$ are the dependent variables, and the $\mathbf{u}^{(n)}$ are derivatives of the dependent variables up to some finite order n with respect to the x^i.[3] Every such variational symmetry is a symmetry of the EL equations $L_A = 0$, $A = 1, 2, \ldots, r$; that is, a variational symmetry carries solutions of the EL equations into solutions. The converse, however, is not guaranteed to hold, e.g. it is often the case that scaling transformations are symmetries of the EL equations but are not variational symmetries.

Noether's first theorem concerns the case of an s-parameter Lie group $\mathcal{G}_s$, which I take to be the explication of the (badly chosen) term 'global symmetry'. The theorem states that the action admits such a group $\mathcal{G}_s$ of variational symmetries iff there are s linearly independent combinations of the EL expressions L_A which are divergences, i.e. there are s p-tuples $\mathbf{P}_j = (P_j^1, \ldots, P_j^p)$, $j = 1, 2, \ldots, s$, where the P_j^i are functions of $\mathbf{x}$, $\mathbf{u}$, and $\mathbf{u}^{(n)}$ such that

$$Div\,(\mathbf{P}_j) = \sum_A c_j^A L_A, \qquad j = 1, 2, \ldots, s \tag{1}$$

where the c_j^A are constants, $Div\,(\mathbf{P}_j)$ stands for $\sum_{i=1}^{p} D_i P_j^i$, and D_i is the total derivative with respect to x^i. Thus, as a consequence of the EL equations there are s *conservation laws*

$$Div\,(\mathbf{P}_j) = 0, \qquad j = 1, 2, \ldots, s. \tag{2}$$

[2] These theorems were presented in Noether (1918). For relevant historical information, see Kastrup (1987) and Byers (1999). For a modern presentation of the Noether theorems, see Olver (1993), Brading (2002), and Brading and Brown (this volume).

[3] The transformations $\mathcal{G} \ni g : (\mathbf{x}, \mathbf{u}) \to (\mathbf{x}', \mathbf{u}')$ may depend on both the independent variables $\mathbf{x}$ and the dependent variables $\mathbf{u}$. Noether's theorems can be generalized to handle transformations that depend on the $\mathbf{u}^{(n)}$ as well (see Olver, 1993), but these generalized transformations will play no role here. But what is relevant here is the fact that Noether's original theorems can be generalized to handle so-called divergence (variational) symmetries that leave the action invariant only up to a term of the form $\int_{\partial\Omega} Div\,(\mathbf{B})$, where the variation of $\mathbf{B}$ vanishes on the boundary $\partial\Omega$. For example, the Galilean velocity boosts do not leave the familiar action for Newtonian particle mechanics invariant, but these boosts are divergence (variational) symmetries. This is crucial in deriving the constancy of the velocity of the centre of mass of the system. Noether does not seem to have been aware of this fact. From here on when I speak of variational symmetries I will mean divergence (variational) symmetries.

The $\mathbf{P}_j$ are called the *conserved currents*. In some cases these conservation laws can be written in the form $D_t T + div(X) = 0$, where div is the spatial divergence. Then if the flux density X vanishes on the spatial boundary of the system, the spatial integral of the density T is a constant of the motion. It is well to note that there are equations of motion (or field equations) which cannot be derived from an action principle, and in such cases there is no guarantee that a symmetry of the equations of motion will give rise to a corresponding conserved quantity.[4]

A concrete application of Noether's first theorem is provided by interacting point masses in Newtonian mechanics, provided that the equations of motion follow from an action principle. The requirement that the inhomogeneous Galilean group is a variational symmetry entails the conservation of energy, angular and linear momentum, and the uniform motion of the centre of mass. Conversely, the existence of these conservation laws entails that the action admits a 10-parameter Lie group of variational symmetries.

Noether's second theorem is concerned with the case of an infinite-dimensional Lie group $\mathcal{G}_{\infty s}$ depending on s arbitrary functions $h_j(\mathbf{x})$, $j = 1, 2, \ldots, s$, of all of the independent variables. This I take to be the explication of the (badly chosen) term 'local symmetries'. The theorem states that the action admits such a group $\mathcal{G}_{\infty s}$ of variational symmetries iff there are s dependencies among the EL equations, in the form of linear combinations of the L_A and their derivatives, which vanish identically (also known as 'generalized Bianchi identities'). Since the EL equations are not independent, we have a case of underdetermination, and as a result the solutions of these equations contain arbitrary functions of the independent variables – an apparent violation of determinism since the initial data do not seem to fix a unique solution of the EL equations.

Now comes an insight of Noether's that is important enough to be labelled the third Noether theorem. Suppose that $\mathcal{G}_{\infty s}$ possesses a rigid Lie subgroup $\mathcal{G}_s$ that arises from fixing $h_j(\mathbf{x}) = \alpha_j = const$. Then each of the conserved currents $\mathbf{P}_j$, corresponding to the invariance under $\mathcal{G}_s$ by Noether's first theorem, can be written as a linear combination of the EL expressions L_A plus the divergence of an antisymmetric quantity: $P_j^i = \sum_{A=1}^{r} L_A \xi_j^{Ai} + \sum_{k=1}^{p} D_k X_j^{ik}$, where the 'superpotentials' X_j^{ik} are functions of $\mathbf{x}$, $\mathbf{u}$, and $\mathbf{u}^{(n)}$, and $X_j^{ik} = -X_j^{ki}$. Thus, on any solution to the EL equations $Div\,(\mathbf{P}_j) = \sum_{i,k=1}^{p} D_i D_k X_j^{ik} \equiv 0$. In this case the conservation laws (2) were dubbed 'improper' by Noether. If the independent variables are spacetime coordinates and x^4 is the time coordinate, the charges Q_j associated with the conserved currents of improper conservation laws are defined by the spatial volume

[4] See section 7 below for a discussion of the issue of what equations of motion can be derived from an action principle.

integrals $\int_{V_3} P_j^4 \, d^3x = \int_{V_3} \sum_{k=1}^{4} D_k X_j^{4k} \, d^3x = \int_{V_3} \sum_{k=1}^{3} D_k X_j^{4k} \, d^3x$, which are equal to $\int_\sigma \sum_{k=1}^{3} n_k X_j^{4k} \, d\sigma$, where σ is the 2-surface bounding the 3-volume V_3, and n_k is the unit normal to the surface. In electromagnetism this relation expresses the charge in a spatial volume as the flux of the electric field through the bounding surface. The upshot is that the content of improper conservation laws amounts to a Gauss-type relation.[5] The conservation law associated with the conserved current $\mathbf{P}_j$ will be completely trivial if $\mathbf{P}_j$ vanishes on any solution of the EL equations; this happens just in case the total divergence of the superpotential for $\mathbf{P}_j$ vanishes. Finally, it is worth mentioning that there is a converse for Noether's third theorem; namely, if corresponding to a non-trivial variational symmetry there is a conserved quantity that can be written as a linear combination of the EL expressions and the divergence of a superpotential, then there is an infinite dimensional Lie group of variational symmetries and the EL equations are underdetermined (see Olver, 1993, section 5.3).

In cases where Noether's second theorem applies there are also what physicists call 'strong conservation laws' which hold 'off shell', i.e. regardless of whether the EL equations hold or not. These strong laws, which are consequences of the Bianchi identities, should not be confused with improper conservation laws.

To return to the main theme: the underdetermination encountered in Noether's second theorem points to one of the principal roots of the notions of gauge and gauge transformation. In one of its main uses, a 'gauge transformation' is supposed to be a transformation that connects what are to be regarded as equivalent descriptions of the same state or history of a physical system. And one key motivation for seeking gauge freedom is to take up the slack that would otherwise constitute a breakdown of determinism: taken at face value, a theory which admits 'local' gauge symmetries is indeterministic because the initial value problem does not have a unique solution; but the apparent breakdown is to be regarded as merely apparent because the allegedly different solutions for the same initial data are to be regarded as merely different ways of describing the same evolution. Putting the point in different terminology, the evolution of the genuine or gauge-invariant quantities (or 'observables') is manifestly deterministic.

Now the obvious danger here is that determinism will be trivialized if, whenever it is threatened by non-uniqueness, we stand willing to sop up the non-uniqueness in temporal evolution with what we regard as gauge freedom to describe the evolution in different ways. Is there then some non-question begging and systematic way

[5] For an interesting discussion of how the distinction between proper and improper conservation laws illuminates Hermann Weyl's work on gauge theories, see Brading (2002).

to identify gauge freedom and to characterize the observables? The answer is yes, but specifying the details involves a switch from the Lagrangian to the constrained Hamiltonian formalism. To motivate that switch, let me note that, subject to some technical provisos, if one is in the domain of Noether's second theorem (i.e. the action admits 'local' symmetries – a group $\mathcal{G}_{\infty s}$ of variational symmetries), as I have been assuming is the case for gauge theories, then the Lagrangian density (more properly, its Hessian) is singular (see Wipf, 1994), and the Legendre transformation which defines the canonical momenta shows that these momenta are not all independent but must satisfy a family of constraints. Hence, one is in the domain of the constrained Hamiltonian theories. Following Dirac (1950; 1951; 1964) one then identifies gauge transformations as mappings of the phase space that are generated by a subset of the constraints.[6] To illustrate what this means and to underscore the point that the key ideas about gauge arise in the humblest settings, I will concentrate in the following section on finite-dimensional systems.

3 Gauge transformations and first-class Hamiltonian constraints[7]

In this section I confine attention to systems where $\dim(Q) = N < \infty$, Q being the configuration space. I further restrict attention to first-order Lagrangians $L(q^i, \dot{q}^i)$, $i = 1, 2, \ldots, N$, $\dot{q}^i := \frac{dq^i}{dt}$, and assume in addition that the Lagrangians are independent of the time t. In this setting the EL equations assume their familiar form

$$\frac{d}{dt}\left(\frac{\partial L}{\partial \dot{q}^n}\right) - \frac{\partial L}{\partial q^n} = 0, \qquad n = 1, 2, \ldots, N \tag{3}$$

These equations can be rewritten as

$$\ddot{q}^m \left(\frac{\partial^2 L}{\partial \dot{q}^m \partial \dot{q}^n}\right) = \frac{\partial L}{\partial q^n} - \dot{q}^m \left(\frac{\partial^2 L}{\partial q^m \partial \dot{q}^n}\right). \tag{4}$$

When the Hessian matrix $W_{ij} := (\frac{\partial^2 L}{\partial \dot{q}^i \partial \dot{q}^j})$ is singular, one cannot solve for $\ddot{q}^m$ in terms of the positions and velocities, and determinism (apparently) fails because arbitrary functions of time appear in the solutions and, thus, initial data will not single out a unique solution. A way to recoup the fortunes of determinism appears in the Hamiltonian treatment. But before turning to that treatment, I need to define another structure.

Setting $v^k := \dot{q}^k$ and $\alpha_i := \frac{\partial L}{\partial q^i} - \frac{\partial^2 L}{\partial v^i \partial q^k} v^k$, the EL equations (4) can be rewritten as

$$\ddot{q}^m W_{mn} = \alpha_i. \tag{5}$$

[6] The constrained Hamiltonian formalism was developed simultaneously by Peter Bergmann and co-workers; see, for example, Bergmann (1949) and Bergmann and Brunings (1949).

[7] The general reference for this section is Henneaux and Teitelboim (1992).

So if the null vector fields of the Hessian are $V_\rho := \gamma^i_\rho \frac{\partial}{\partial v^i}$, the first-generation Lagrangian constraints are

$$\chi^{(1)}_\rho := \alpha_i \gamma^i_\rho = 0, \qquad \rho = 1, 2, \ldots, R < N. \tag{6}$$

Requiring that these constraints be preserved by the motion produces a second generation of Lagrangian constraints, etc. Eventually this process terminates. The final Lagrangian constraint manifold $\mathcal{C}^\ell_F$ is then the submanifold of the Lagrangian velocity phase space $\Gamma(q, v) := T(Q)$ where all of the constraints hold.

Passing to the Hamiltonian phase space $\Gamma(q, p) := T^*(Q)$ is accomplished by the Legendre transformation $FL : \Gamma(q, v) \to \Gamma(q, p)$, where $FL(q^i, v^i) = (q^i, p_i)$ with the canonical momenta given by $p_i := \frac{\partial L}{\partial v^i}$. When the Hessian is singular, the canonical momenta are not all independent but must satisfy *primary constraints*

$$\phi^{(0)}_\mu(q, p) = 0, \qquad \mu = 1, 2, \ldots, J < N \tag{7}$$

that follow from the definitions of the momenta. These equations define the primary Hamiltonian constraint manifold $\mathcal{C}^h_o$. The Hamilton–Dirac (HD) equations of motion take the form

$$\dot{q}^i \underset{\mathcal{C}^h_o}{=} \{q^i, H_T\} \tag{8a}$$

$$\dot{p}_i \underset{\mathcal{C}^h_o}{=} \{p_i, H_T\}. \tag{8b}$$

Here $\underset{\mathcal{C}^h_o}{=}$ means equality on $\mathcal{C}^h_o$ and $\{,\}$ is the usual Poisson bracket. The total Hamiltonian H_T is $H_c(q, p) + \lambda^\mu \phi^{(0)}_\mu(q, p)$, where the canonical Hamiltonian H_c is any function of (q, p) satisfying $H_c(q, p) = \frac{\partial L}{\partial \dot{q}^i}\dot{q}^i - L$. The Lagrange multipliers λ^μ appearing in the total Hamiltonian can be arbitrary functions of time. Requiring that the primary constraints be preserved by the motion can lead to *secondary constraints*. Requiring that the secondary constraints be preserved by the motion can produce *tertiary constraints*, etc. The final Hamiltonian constraint manifold $\mathcal{C}^h_F$, which is reached after a finite number of steps, is the submanifold of $T^*(Q)$ where all the constraints are satisfied. A constraint is said to be *(final) first class* just in case its Poisson bracket with any constraint 'vanishes weakly', i.e. is zero when evaluated on $\mathcal{C}^h_F$. On $\mathcal{C}^h_F$ where (8a) and (8b) have solutions, the total Hamiltonian can be written as $H_T = H^F_c + u^{\mu_f}\psi^{(0)}_{\mu_f}$, where $\psi^{(0)}_{\mu_f}$, $\mu_f = 1, 2, \ldots, K$, are the final primary first-class constraints and H^F_c is a particular canonical Hamiltonian which happens to be a first-class function, i.e. its Poisson bracket with every constraint vanishes weakly.

Under the regularity conditions that the Legendre map has constant rank and that the rank of the Poisson bracket of constraints is constant, the Hamiltonian and Lagrangian treatments of constrained systems are equivalent in the sense that for any solution $q^i(t)$ of the EL equations, there is a solution $(q^i(t), p_i(t))$ of the HD equations, and vice versa (see Batlle *et al.*, 1986).

We are now in a position to be more specific about what counts as a gauge transformation in the Hamiltonian formalism. The first version treats a gauge transformation as a point transformation on the Hamiltonian phase space $\Gamma(q, p)$.

Def. 1. (Hamiltonian version). $(q_1, p_1), (q_2, p_2) \in C_F^h$ are Hamiltonian gauge equivalent iff there is a $(q_0, p_0) \in C_F^h$ such that (q_1, p_1) and (q_2, p_2) are both obtained from (q_0, p_0) as solutions to the HD equations in the same lapse time Δt.

The motivation for this definition starts with the conviction that the physical state of the system at any given time is completely specified by a point in phase space, and then proceeds to the realization that, if the EL/HD equations for a constrained system are to determine the physical state at later times, the physical-state-to-phase-space correspondence must be one–many.[8] The slack of the 'many' side is identified as gauge freedom. Using Def. 1 we can proceed to relate the gauge freedom to the constraints. Note first that any (final) first-class constraint can be obtained from an iterative procedure that starts with the primary first-class constraints and successively takes the Poisson bracket of the preceding stage with H_c^F. It is natural to conjecture that (i) all (final) first-class constraints generate Hamiltonian (point) gauge transformations, and (ii) any (point) Hamiltonian gauge transformation is generated by the (final) first-class constraints. This will be true provided that the ancestry of all the first-class constraints is untainted in the sense that the iterative procedure which generates the first-class constraints does not pass through an *ineffective constraint*, i.e. a constraint whose gradient vanishes weakly (see Cabo and Louis-Martinez, 1990).[9] Geometrically the gauge orbits on C_F^h are the integral curves of the vector field formed by linear combinations of the gradients of the first-class constraints. This vector field is null with respect to the pre-symplectic form induced on C_F^h by the symplectic form for $\Gamma(q, p)$.[10]

With the proviso that ineffective constraints are absent, the results (i) and (ii) justify the Dirac slogan 'The gauge transformations are those transformations

[8] As Henneaux and Teitelboim put it: '[A]lthough the physical state is uniquely defined once the set of p's and q's is given, the converse is not true – i.e., there is more than one set of values of the canonical variables representing a given physical state. To see how this conclusion comes about, we notice that if we give an initial set of canonical variables at time t_1 and thereby completely define the physical state at that time, we expect the equations to *fully determine the physical state at other times*. Thus, by definition any ambiguity in the value of the canonical variables at $t_2 \neq t_1$ should be a physically irrelevant ambiguity' (1992, pp. 16–17).

[9] For examples of what can go wrong when ineffective constraints are present, see Henneaux and Teitelboim (1992, pp. 19–20) and Gotay and Nester (1984).

[10] The mathematically precise way to formulate Hamiltonian mechanics uses a *symplectic form* ω that is a non-degenerate two-form on the phase space $T^*(Q)$. The Poisson bracket for phase functions is defined by $\{f, g\} := \omega(df, dg)$. Locally, coordinates (q^i, p_i), $i = 1, 2, \ldots, N$, can be chosen so that $\{f, g\} = \sum_i \left(\frac{\partial f}{\partial q^i} \frac{\partial g}{\partial p_i} - \frac{\partial f}{\partial p_i} \frac{\partial g}{\partial q^i} \right)$.

generated by the first-class constraints.' Adding non-primary first-class constraints with their Lagrange multipliers to the total Hamiltonian results in what is called the *extended Hamiltonian*. This extension is harmless (and pointless) when ineffective constraints are absent. But if ineffective constraints are present the extension can be pernicious in that it can break the equivalence of the Lagrangian and Hamiltonian treatments.[11]

There is, of course, a Lagrangian version of Def. 1. The equivalence of the Lagrangian and Hamiltonian approaches means that the Lagrangian and Hamiltonian (point) gauge transformations have a natural correspondence. So if there are no ineffective Hamiltonian constraints, every first-class Hamiltonian constraint generates a Hamiltonian (point) gauge transformation which has a corresponding Lagrangian (point) gauge transformation, and every Lagrangian (point) gauge transformation has a corresponding Hamiltonian (point) gauge transformation generated by a first-class constraint.

Given these nice equivalences of the Lagrangian and Hamiltonian treatments, it is not surprising that, despite the fact that the final Lagrangian and Hamiltonian constraint manifolds can have different dimensions, the two treatments give the same counts for the number of physical degrees of freedom. Count these degrees of freedom in the Hamiltonian and Lagrangian approaches respectively by the dimensions of $\mathcal{C}_G^h \subset \mathcal{C}_F^h$ and $\mathcal{C}_G^\ell \subset \mathcal{C}_F^\ell$, where the subscript G indicates the submanifold obtained by killing the gauge freedom by gauge fixing conditions. If there are no ineffective constraints, $\dim(\mathcal{C}_G^h) = \dim(\mathcal{C}_G^\ell) = 2N - M - P$, where N is the dimension of the configuration space, M is the number of Hamiltonian constraints, and P is the number of first-class Hamiltonian constraints (see Gràcia and Pons, 1988).

It remains to link the notion of gauge transformation detailed above with the Noether transformations. There appears to be a mismatch since the former are point transformations while the latter are mappings of solutions onto solutions. To bridge the gap, one can introduce a notion of gauge transformation connecting solutions. The Hamiltonian version is given in:

Def. 2. Two solutions $(q(t), p(t))$ and $(q'(t), p'(t))$ of the HD equations are gauge equivalent just in case for each t the points of phase space they determine are gauge equivalent in the sense of Def. 1.

A mapping $G : T(Q)^* \times \mathbb{R} \to \mathbb{R}$ of the form $G(q, p, t) = \sum_{k \geq 1} \varepsilon^k(t) G_k(q, p)$,

[11] Thus, without the restriction of no ineffective constraints, 'Dirac's conjecture' (as it is called in the literature) that all first-class constraints generate gauge transformations is false.

where $\varepsilon^k(t)$ is the kth time derivative of the gauge generator $\varepsilon(t)$, is a solution gauge transformation provided that the G_k satisfy the chain condition

$$\begin{aligned} G_1 &\equiv PFC \\ \{PFC, G_k\} &\equiv PFC \\ \{G_k, H\} + G_{k+1} &\equiv PFC \end{aligned} \tag{9}$$

where 'PFC' stands for a primary first-class constraint and H is a first-class Hamiltonian, and provided that the maximum value of k is the number of steps in the Dirac consistency algorithm (see Gràcia and Pons, 1988; Gomis *et al.*, 1990).[12]

Corresponding to this transformation there will be a Lagrangian gauge transformation that maps any solution to the EL equations to another solution that is gauge equivalent to the first in the sense analogous to Def. 2. However, there is no guarantee that this Lagrangian gauge transformation is a Noether transformation in the sense that it leaves the action invariant (up to a total divergence). But it is known that the latter will hold if all of the Hamiltonian constraints are primary first class. In the opposite direction, a pair of gauge equivalent solutions to the EL equations has a corresponding pair of gauge equivalent solutions to the HD equations where the solution gauge transformation connecting them is of the form given above.

The final concept I will introduce for the Hamiltonian formalism is that of an *observable* or gauge independent quantity. There are two equivalent ways to make this precise: an observable can be defined to be a function from the Hamiltonian phase space $\Gamma(q, p)$ to $\mathbb{R}$ which has weakly vanishing Poisson brackets with the first-class constraints or, equivalently, which is constant along the gauge orbits. An observable in this sense corresponds to a function on the reduced phase space obtained by quotienting $\Gamma(q, p)$ by the gauge orbits.

The reader who is encountering the apparatus of constrained systems for the first time may be aghast at the seeming complexity. I have no comfort to offer. For not only have I slurred over some of the complexities in my overly brief presentation above, but I have additional bad news as well. Many of the systems one wants to study in physics have infinite-dimensional configuration spaces, and the relatively simple results reported above for finite-dimensional systems do not always carry over to the infinite-dimensional case. Nevertheless, in what follows I will simply forge ahead under the assumption that in constrained systems gauge is generated by the first-class Hamiltonian constraints.[13]

In closing this section I want to underscore the point that the use of the recommended apparatus for getting a fix on gauge takes the overcoming of an apparent

[12] The restriction to the case of no ineffective constraints is also in force here.

[13] As far as I am aware there are no proofs in the literature of the infinite-dimensional analogues of all of the basic results quoted above for finite dimensional systems. Belot (2002) provides a detailed discussion of a large class of gauge theories that arises when the conserved quantities of a Hamiltonian system are set equal to zero. This class contains Yang–Mills type theories but not all constrained Hamiltonian systems.

underdetermination to be the key motivation for recognizing gauge freedom. It is well to note, however, that other motivations can operate, and these motivations may produce a different verdict on gauge. Suppose, for example, that you want to be a relationist about space and time, and also that you want to acknowledge the striking success in the use of Minkowski spacetime for formulating the theories of modern physics. You could reconcile these two desires by saying that the relational spatiotemporal structure of physical events conforms to those prescribed by Minkowski spacetime while at the same time denying that physical events are in any literal sense located in a spacetime container. But to make such a stance consistent requires treating a Poincaré boost of the matter fields on Minkowski spacetime as a gauge transformation in the sense that it produces not a different physical situation but a different representation of the same physical situation. By contrast, an application of the constraint apparatus to Maxwell electromagnetic theory and other standard special relativistic theories does not produce the verdict that the Poincaré group is a gauge group. The difference between the two motivations for seeing gauge freedom is brought out by the contrasting attitudes towards what counts as an observable. A non-relationist who is guided by the constrained Hamiltonian formalism sees no interesting gauge freedom in the familiar special relativistic treatment of electromagnetism and, therefore, will treat the electromagnetic field as an observable. By contrast, the relationist is (on my reading) committed to denying that the electromagnetic field is a genuine physical observable since, by her lights, this field is not gauge invariant; and by her lights, only non-local quantities, such as the spacetime volume integral of the field energy, will pass muster. Lest one think that this result is a *reductio* of relationism, the relationist can note that worse is to follow in the context of the General Theory of Relativity (GTR). But in contrast to special relativistic theories, there is no choice to be made in GTR where the constraint structure of the theory dictates that local fields are not observables (see section 5).

4 A toy example

Since the discussion thus far has been both abstract and complicated, it may assist the reader to work through a concrete toy example that illustrates some of the above concepts. Those of you who have read Maxwell's (1877) *Matter and Motion* may have been puzzled by his apparently contradictory claim that acceleration is *relative* even though rotation is *absolute* (see sections 32–35 and 104–105). Maxwell is consistent if we take him to be proposing that physics be done in the setting of what I have dubbed *Maxwellian spacetime*. Like Newtonian spacetime, Maxwellian spacetime has absolute simultaneity, the $\mathbb{E}^3$ structure of the instantaneous space, and a time metric, but it eschews the full inertial structure in favour of a family of relatively non-rotating rigid frames.[14] In terms of coordinate systems adapted to the absolute simultaneity, the $\mathbb{E}^3$ structure, and the privileged non-rotating frames, the symmetry transformations of Maxwellian spacetime are

$$\mathbf{x} \to \mathbf{x}' = \mathbf{R}\mathbf{x} + \mathbf{a}(t) \qquad \text{(Max)}$$
$$t \to t' + const.$$

where $\mathbf{R}$ is a constant rotation matrix and $\mathbf{a}(t)$ is an arbitrary smooth function of time. In such a setting it seems hopeless to have determinism if, as ordinarily assumed, positions and velocities of particles are regarded as observables. For we can choose $\mathbf{a}(t)$ such that it is zero for all $t \leq 0$ and non-zero for $t > 0$. Since a symmetry of the spacetime should be a symmetry of the equations that specify the permitted particle motions, the application of (Max) to a solution of the equations of motion will produce another solution that agrees on the particle trajectories of the first solution for all past time but disagrees with it in the future – an apparent violation of even the weakest form of determinism.

Now let's see how this example gets reinterpreted when cranked through the Lagrangian/constrained Hamiltonian formalism. An appropriate Lagrangian invariant under (Max) is

$$L = \sum\sum\nolimits_{j<k} \frac{m_j m_k}{2M}(\dot{\mathbf{r}}_j - \dot{\mathbf{r}}_k)^2 - V(|\mathbf{r}_j - \mathbf{r}_k|), \qquad M := \sum_i m_i. \quad (10)$$

The transformations (Max) are global mappings of Maxwellian spacetime onto itself, but they are 'local' in that Noether's second theorem applies since (Max) contains an infinite-dimensional group $\mathcal{G}_{\infty 3}$ whose parameters are arbitrary functions of t, the only independent variable in the action $\int L\, dt$. It is easy to verify that the Hessian matrix for (10) is singular. The EL equations are:

$$\frac{d}{dt}\left(m_i\left(\dot{\mathbf{r}}_i - \frac{1}{M}\sum_k m_k\dot{\mathbf{r}}_k\right)\right) = \frac{\partial V}{\partial \dot{\mathbf{r}}_i}. \quad (11)$$

These equations do not determine the evolution of the particle positions uniquely: if $\mathbf{r}_i(t)$ is a solution, so is $\mathbf{r}'_i(t) = \mathbf{r}_i(t) + \mathbf{f}(t)$, for arbitrary $\mathbf{f}(t)$, confirming the intuitive argument given above for the apparent breakdown of determinism. Determinism can be restored by regarding the transformation $\mathbf{r}_i(t) \to \mathbf{r}_i(t) + \mathbf{f}(t)$ as a gauge transformation.

Now let's switch to the Hamiltonian formalism and find the constraints. The canonical momenta are:

$$\mathbf{p}_i := \frac{\partial L}{\partial \dot{\mathbf{r}}_i} = \frac{m_i}{M}\sum_k m_k(\dot{\mathbf{r}}_i - \dot{\mathbf{r}}_k) = m_i\dot{\mathbf{r}}_i - \frac{m_i}{M}\sum_k m_k\dot{\mathbf{r}}_k. \quad (12)$$

[14] For details, see Earman (1989, chapter 2, section 3).

These momenta are not independent but must satisfy three primary constraints, which require the vanishing of the x, y, and z components of the total momentum:

$$\phi_\alpha = \sum_i p_i^\alpha = 0, \qquad \alpha = 1, 2, 3. \tag{13}$$

These primary constraints are the only constraints – there are no secondary constraints – and they are first class. These constraints generate in each configuration variable $\mathbf{r}_i$ the same gauge freedom; namely, a Euclidean shift given by the same arbitrary function of time. The gauge-invariant quantities include such things as relative particle positions and relative particle velocities.

In this toy example there is simple connection between the senses of gauge freedom derived from the Lagrangian and Hamiltonian approaches. In more complicated cases, however, the connection between the two approaches is not so transparent.

5 The General Theory of Relativity as a gauge theory

The Hilbert action for the source-free Einstein gravitational field equations reads $\int_{\mathcal{M}} R\sqrt{-g}\, d^4x$, where R is the Ricci scalar and $g := \det(g_{ij})$, g_{ij} being the coordinate components of the spacetime metric. The diffeomorphism group *diff*$(\mathcal{M})$ of the spacetime manifold $\mathcal{M}$, which contains arbitrary functions of the independent variables in the action (here the spacetime coordinates x^i, $i = 1, 2, 3, 4$), is a variational symmetry. Thus, Noether's second theorem applies, telling us that we have a case of underdetermination. We expect that in the Hamiltonian formulation there will be constraints that generate non-trivial gauge. Our expectations are not disappointed. The configuration variables (the qs) are Riemann 3-metrics, interpreted as giving the intrinsic geometry of a 3-manifold that is to be embedded as a time slice of spacetime, and the conjugate momentum variables (the ps) are tensor fields related to the exterior curvature of the 3-manifold. When the crank of the Dirac algorithm is turned, it is found that the constraints are all primary first class. That sounds nice, but in fact when the details are unpacked three surprises/puzzles are revealed.

There are two families of constraints: the momentum constraints and the Hamiltonian constraints.[15] When the Poisson bracket algebra of these constraints is computed, it is found that it does not close, so this algebra is not a Lie algebra. This means that a defining feature of Yang–Mills theories is missing from the most natural formulation of GTR as a Hamiltonian theory. Now some writers want to reserve the label 'gauge theory' for Yang–Mills theories. This seems to me to be a merely terminological matter – if you do not wish to call GTR a gauge theory

[15] 'Families' because for each point of space there is a momentum constraint and a Hamiltonian constraint.

because it is not Yang–Mills, that is fine with me; but please be aware that the constrained Hamiltonian formalism provides a perfectly respectable sense in which the standard textbook formulation of GTR using tensor fields on differentiable manifolds does contain gauge freedom. What goes beyond label-mongering is the issue of why GTR fails to be a Yang–Mills theory and, more generally, what features separate constrained Hamiltonian theories that are Yang–Mills from those which are not. Some important results of these matters have been obtained by Lee and Wald (1990), but these results are too technical to review here.

A second puzzle is how to find, among the Dirac gauge transformations on the Hamiltonian phase space of GTR, a counterpart of the action of the diffeomorphism group of spacetime. The most obvious way to identify a counterpart by finding a homomorphic copy of the Lie algebra of $diff(\mathcal{M})$ in the constraint algebra is blocked by the fact that the latter isn't a Lie algebra. One resolution of the puzzle is given by Isham and Kuchař (1986a; 1986b) who show that if the embedding variables (which describe how a 3-manifold is embedded as an initial value hypersurface of spacetime) and their conjugate momenta variables are adjoined to the phase space of GTR, there is a natural homomorphism of the Lie algebra of spacetime diffeomorphisms into the Poisson bracket algebra of constraints on the extended phase space. An alternative approach is taken by Ashtekar and Bombelli (1991), who show that Hamiltonian mechanics for general relativity does not require a $(3+1)$-cotangent bundle structure. Instead of taking the phase space of the theory to be the space $\Gamma(q, p)$ of instantaneous states, they work with the space $\hat{\Gamma}$ of entire histories or solutions to the Einstein field equations, which implies that dynamics is implemented not by a mapping from one state to another state in the same solution but as a mapping from one solution to another solution. The space $\hat{\Gamma}$ has a presymplectic structure given by a degenerate 2-form $\hat{\omega}$. There is no constraint surface, as in the $(3+1)$ formulation; rather, the gauge directions Y are given directly by the null vectors of $\hat{\omega}$. It turns out that two solutions lie on the same gauge orbit (i.e. integral curve of the gauge field Y) iff they are diffeomorphically related.

The third and most contentious puzzle arises from the fact that since the Hamiltonian constraints generate the motion, motion is pure gauge, and the observables of the theory are constants of the motion in the sense that they are constants along the gauge orbits. Taken at face value, the gauge interpretation of GTR implies a *truly* frozen universe: not just the 'block universe' that philosophers endlessly carp about – that is, a universe stripped of A-series change or shifting 'nowness' – but a universe stripped of its B-series change in that no genuine physical magnitude (= gauge-invariant quantity) changes its value with time. Philosophers of science have generally ignored this puzzle. But it deserves a resolution, either by showing how the 'no change' consequence can be avoided or by showing how the consequence can be reconciled with our perceptions of a world filled with (B-series) change. Since I have given my own resolution at length elsewhere, I will not repeat it here (see Earman, 2002). But lest one think that the problem depends on peculiarities of infinite-dimensional systems or of general relativity, I will mention that an analogue arises for a subclass of the humble theories discussed in the preceding sections. These are so-called reparameterization invariant Lagrangian theories where the action is invariant under $t \rightarrow f(t)$ for arbitrary $f(t)$. For such theories the canonical Hamiltonian, if non-zero, is a first-class constraint so that motion is pure gauge.

Finally, I will mention that the analysis of gauge recommended here helps to illuminate some of the issues surrounding the never-ending debate about the nature and status of the requirement of general covariance. What makes the issues so difficult to disentangle is that as the debate unfolded, every confusion it was possible to make was in fact made – coordinate transformations were confused with point transformations, relativity principles were confused with gauge principles, etc.[16] This is not the place to attempt a disentanglement, and I will have to be content with noting that much of the confusion is swept away by distinguishing two forms of the requirement of general covariance. The *weak requirement* demands that the equations of motion of a theory are valid in an arbitrary coordinate system (or, equivalently, that the equations should be covariant under an arbitrary coordinate transformation). Assuming that nature can be fully described by geometric object fields, this requirement is a restriction on the form rather than the content of a theory. The *strong requirement* demands that the spacetime diffeomorphism group is a gauge group of the theory. If the recommended Hamiltonian constraint apparatus is used to detect gauge freedom, then it is obvious that a theory can satisfy the weak requirement without at the same time satisfying the strong requirement. However, one can wonder whether the strong requirement, like its weak sister, is also a matter of form rather than content. That is, can a theory that satisfies the weak but not the strong requirement always be rewritten in a form that conforms to the latter? The application of the constraint formalism reveals that monkeying with a theory so as to make it satisfy strong general covariance may change the constraint structure of the theory and, thus, what counts as an observable (= gauge-invariant quantity). Arguably, such a change amounts to a change in the content rather than the form of the theory. It then becomes an empirical question as to whether nature is best described in terms of the observables of the original or of the new theory.[17]

[16] For a historical review of the debate, see Norton (1993).
[17] See Earman, 'Once more general covariance', unpublished manuscript.

6 The fibre bundle apparatus

How does the relevant fibre bundle structure arise for theories that are not presented in bundle-theoretic language? The constrained Hamiltonian formalism provides the basis of an answer. One would like it to be the case that when the reduced phase space $\tilde{\Gamma}$ of a constrained Hamiltonian system is formed by quotienting the constraint surface $\mathcal{C}$ by the gauge orbits, these orbits are the fibres of a bundle with base space $\tilde{\Gamma}$. Suppose that the first-class constraints form a Lie algebra and that this algebra exponentiates to give a Lie group $\mathcal{G}$ which acts freely on $\mathcal{C}$. And suppose that the quotient $\mathcal{C}/\mathcal{G}$ is a manifold. Then our desire is satisfied in that $\mathcal{C}/\mathcal{G}$ will be the base space of a $\mathcal{G}$-bundle whose fibres are the orbits of the group action. In practice, however, some of the stated suppositions may fail. For instance, although $\tilde{\Gamma}$ always exists as a topological space, it may lack a manifold structure – GTR is a particular example.[18] If one believes that the fibre bundle apparatus captures an essential feature of nature, then one could posit that the emergence of the appropriate bundle structure is a necessary condition for genuine physical possibility. This is an interesting idea, but obviously it requires critical examination.

Even for paradigm cases of gauge theories that wear their fibre bundle structure on their sleeves – e.g. Yang–Mills theories – understanding the geometry of constraints is crucial to quantization, as we will see in section 8.

7 The reach of the constraint apparatus

In order to produce simple and understandable examples I have emphasized applications where, from the Lagrangian point of view, the gauge transformations are purely transformations of the independent variables[19] of the action and where these variables are identified with spatiotemporal variables. But nothing in the constraint formalism depends on these simplifying assumptions, and the formalism serves to identify the gauge freedom even when these assumptions do not hold.

However, there is one obvious and absolute limitation of the apparatus: it does not apply to equations of motion that are not derivable from an action principle. This might seem to be a mild limitation because almost all of the candidates for fundamental equations of motion in modern physics are derivable from an action principle. But appearances can be deceptive; in particular, the apparent ubiquity of equations of motion that are derivable from an action principle might represent a selection effect deriving from the facts that modern physicists always have

[18] In this case, however, the reduced phase space is not badly behaved since it is the disjoint union of manifolds; see Fischer and Moncrief (1996).

[19] These are sometimes referred to as 'gauge transformations of the first kind', whereas transformations depending only on the dependent variables are referred to as 'gauge transformations of the second kind'.

quantization in the back of their minds and that the standard cookbook procedures for producing a quantization start from a Hamiltonian formulation.

To get a feel for just how strong the limitation really is we would need to know what the necessary and sufficient conditions are for equations of motion to be derivable from an action principle. A special case of this problem is what is known as the *Helmholtz problem*: Consider a system of Newtonian second-order ordinary differential equations: $\Delta^i = 0$, $\Delta^i := \ddot{q}^i - F^i(q^i, \dot{q}^i, t)$; under what conditions does there exist a Lagrangian $L(q^i, \dot{q}^i, t)$ and a non-singular matrix A^{ik} such that $\Delta^i = A^{ik} EL_k$, where $EL_k = 0$ are the EL equations of L? Helmholtz found a set of necessary conditions which were later proved to be sufficient as well. Darboux proved that for $n = 1$ the Helmholtz conditions can always be satisfied. The case of $n = 2$ has also been solved. But the general problem remains unsolved (see, for example, Sarlet, 1982). Even less is known when more complicated equations of motion are considered and when singular Lagrangians are permitted.

The issue under discussion is also intimately linked with the status of determinism. Determinism becomes a trivial doctrine if whenever cracks appear in the doctrine we stand ready to paper them over by seeking gauge freedom. I gave the impression that the trivialization is halted by providing a principled way to detect gauge freedom. This impression is badly misleading if the means of detecting gauge freedom is that of Dirac. Start with any theory whose equations of motion are derivable from an action principle, and suppose that the EL equations do not suffer from overdetermination but do suffer from underdetermination – they fail to determine a unique solution from initial data because arbitrary functions of time appear in the solutions. A cure for this form of indeterminism is always at hand in that in the constrained Hamiltonian formalism the gauge transformations, as identified by Dirac's prescription, are sufficient to sop up the underdetermination. (Of course, it may happen that the 'cure' takes the drastic form of freezing the dynamics, as in the case of GTR or time reparameterization theories in general.) Thus, to decide just how *a priori* or contingent determinism is, it is crucial to know how strong is the demand that the equations of motion admit a (possibly singular) Lagrangian formulation – the stronger (respectively, weaker) the demand is, the more contingent (respectively, *a priori*) determinism is.

In the absence of any convincing argument to the effect that acceptable equations of motion must be derivable from an action principle, it seems necessary to confront the issue of how to identify gauge freedom and observables for equations of motion that are not derivable from an action principle. I know of no systematic approach to this issue.

Finally, I want to indicate a way in which the conviction that theories containing gauge freedom are described by constraints can be challenged. Independently of the desire to save determinism, a motivation for seeing gauge freedom at work comes from considerations of observability. Suppose, for example, that one has reasons for thinking that a complex-valued scalar field φ is not observable whereas combinations such as $\varphi^*\varphi$ are. If the action is constructed from such combinations it will be invariant under the group of transformations of the form $\varphi \to \varphi' = \exp(i\alpha)\varphi$, $\partial_\mu \alpha = 0$, which might be taken as gauge transformations even though the constraint apparatus does not apply since the group parameters do not involve arbitrary functions of the independent variables. In the case of quantum field theory this line on gauge leads to what I find to be unacceptable consequences, such as that unitarily inequivalent representations of the algebra of field operators are gauge equivalent and, thus, are to be regarded as merely different ways of describing the same physical situation (see Earman, this volume, Part III). And apart from considerations of quantum field theory, the present motivation for seeing gauge freedom is potentially confused. That the complex-valued scalar field φ is not observable or measurable is a necessary but not sufficient condition for consigning φ to the category of quantities which are not to be regarded as genuine physical magnitudes because their values can be changed without changing the real physical situation. Neverthless, it is important to recognize that there can be a number of different motivations for seeing gauge at work.

8 The quantization of gauge theories

There are at least four extant approaches to quantization of gauge theories.[20] The first is *gauge fixing*: fix a gauge and quantize in that gauge. But when one tries to do this for Yang–Mills theories using the analogues of familiar gauge conditions (e.g. Lorentz gauge) the procedure may break down. The difficulty is explained by the fact that the gauge condition may fail to define a global transversal in the constraint surface, i.e. a hypersurface that meets each of the gauge orbits exactly once.

A second approach is *reduced phase space quantization*. Quotient out the gauge orbits to produce the reduced phase space. If this procedure goes smoothly (see section 10 below) the normal method of quantization can be applied to the resulting unconstrained Hamiltonian system. This approach faces the practical difficulty of having to solve the constraints, and even if one overcomes this difficulty one may find that the reduced phase space has features that complicate the quantization (see section 10 below).

The third approach is called *Dirac constraint quantization*. Here the procedure is to promote the first-class constraints to operators on a Hilbert space and then require that the vectors in the physical sector of this Hilbert space be annihilated by the constraint operators. Of course, the forming of the constraint operators is subject

[20] The standard reference on this topic is Henneaux and Teitelboim (1992).

to operator ordering ambiguities. But even modulo such ambiguities, it can happen that the resulting Dirac quantization is inequivalent to that obtained by reduced phase space quantization. In such a case, which is the correct quantization? And how would one tell? I will return to these matters below.

The fourth approach is called *BRST quantization* after Becchi, Rouet, Stora, and Tyutin. The idea is to mirror the original gauge symmetry by a symmetry transformation on an extended phase space obtained by adding auxiliary variables. The additional phase space variables are chosen so that the BRST symmetry has a simple form that facilitates quantization.

Apart from technical issues, the point to keep firmly in mind is that it is, presumably, the observables – in the sense of gauge-invariant quantities – of a constrained system that get promoted to quantum observables – in the sense of self-adjoint operators on the appropriate Hilbert space. Thus, it is hardly surprising that one of the key issues in the search for a quantum theory of gravity is what to take as the appropriate set of observables of classical GTR (see Isham, 1992).

9 The magical gauge argument

In the physics literature there is something called the 'gauge argument' that goes like this. Start with a free field which admits a 'global symmetry' and obeys (by Noether's first theorem) a 'global conservation law'. An appeal to relativity theory and locality is then used to motivate a move from the 'global' to a 'local symmetry'. But this move necessitates the introduction of a new field that interacts with the original field (and, perhaps, with itself) in a prescribed way. The success of the gauge argument in capturing some of the most fundamental interactions in nature has been taken to indicate that the argument reveals an important strand of the logic of nature.

I am in agreement with Martin (2002a,b; this volume) who finds the 'getting something from nothing' character of the gauge argument too good to be true. In particular, a careful look at applications of this argument reveals that a unique theory of the interacting field results only if some meaty restrictions on the form of the final Lagrangian are implicitly in operation; and furthermore, the kind of locality needed for the move from the 'global symmetry' (invoking Noether's first theorem) to the 'local symmetry' (invoking Noether's second theorem) is not justified by an appeal to the no-action-at-a-distance sense of locality supported by relativity theory. Not only is there no magic to be found in the gauge argument, but the 'gauge principle' that prescribes a move from global to local symmetries for interacting fields can be viewed as output rather than input: for example, it can be viewed as the product of a self-consistency requirement (see, for example, Wald, 1986) or as a consequence of the requirement of renormalizability (see, for example, Weinberg, 1974).

What is missing is an explanation of the heuristic power of the gauge argument. That is a task someone else will have to perform.

10 Why gauge theories?

Since the presence of gauge freedom in a theory means that the theory employs quantities that lead to redundant descriptions in the form of a many–one correspondence between the state descriptions of the theory and the intrinsic physical state, why shouldn't the theory be purged of its 'surplus structure' (to use Michael Redhead's terminology) so as to achieve a one–one correspondence between state descriptions and physical states? I will now consider a series of possible responses.

10.1 Obstructions to getting an unconstrained Hamiltonian system

At least two kinds of obstructions can occur. (i) Quotienting out the gauge orbits may not produce a manifold. This situation occurs for GTR, but here the singularities are isolated and the reduced phase space is a disjoint union of manifolds (see Fischer and Moncrief, 1996). (ii) The reduced phase space is a manifold but this manifold is not the cotangent space of a reduced configuration space.

10.2 Ambiguities in quantization

Suppose now that no obstructions are encountered in passing to a reduced Hamiltonian phase space. But suppose that reduced phase space quantization gives a result that is physically inequivalent to Dirac constraint quantization, even allowing for operator ordering ambiguities. And suppose that the latter proves to be empirically correct. This is certainly a reason not to gauge out. But it also seems to be a reason to say that the 'gauge transformations' are not really gauge transformations, for it seems that relevant information is lost in passing to the reduced phase space. Such examples, however, may be fanciful. Dirac and reduced phase space quantization will coincide when the gauge group is unimodular;[21] and when the gauge group is not unimodular and the Dirac and reduced phase space quantizations are at odds, arguably the Dirac procedure is incorrect and a modified Dirac prescription is needed (see Duval *et al.*, 1991).

10.3 Future extensions and modifications of the theory

A gauge theory could be retained in order to preserve the possibility that a later development will provide a physically motivated way of breaking the original

[21] A group is unimodular if it carries a bi-metric volume.

gauge symmetry. This consideration has force to the extent that there are historical examples of such gauge breaking. However, such examples seem to be in short supply. The Aharonov–Bohm effect is sometimes touted as a relevant example, but it is better to take the moral of this example to be that gauge-invariant quantities must include non-local quantities such as the integral of the 4-potential around a closed loop.

10.4 Convenience

Here it is helpful to contrast various cases. (i) The use of the electromagnetic potentials in formulating Maxwell's theory seems to be purely a matter of convenience since Maxwell's equations can be stated and solved in any given application without use or mention of the potentials. (ii) More than mere convenience is involved in operating with gauge-dependent quantities in the toy example from section 4 of particle mechanics in Maxwellian spacetime. One can certainly work with gauge-independent inter-particle quantities; but for more than three particles, relative particle distances give an over-complete set of configuration variables for the reduced configuration space (see Belot, 2002), and writing unconstrained equations of motion requires an undemocratic choice of among these quantities. (iii) In our present state of knowledge, the use of gauge-dependent quantities in GTR – tensor fields on a manifold – cannot be ascribed to convenience or the desire to avoid an undemocratic choice. For at present we do not know how to do the mathematics of GTR purely in terms of the gauge-independent quantities; in particular, the usual way of working with differential equations is not an option since apparently the spacetime manifold disappears in the gauge-independent description, whatever exactly that may turn out to be (see section 5). Thus, at present it seems that treating GTR as a gauge theory is closer to *force majeure* than to convenience.

In sum, there is no one simple answer to 'Why gauge?' The answer will vary from case to case, and it can range from 'Because it makes the life of the physicist easier' to 'We don't seem have a choice in the matter.' But if the above is a fair summary, then the verdict must be that we don't presently have a satisfying answer to 'Why gauge?' since most of the reasons essayed, save for convenience, were of the in-principle but not-actually-in-practice form.

11 Conclusion

Nothing that I have said above is news to physicists. But it seems to me worth saying to a philosophical audience. Indeed, when I try to talk about constrained Hamiltonian systems to even my more knowledgeable colleagues in the philosophy of physics, they look at me as if I were speaking a Martian dialect. I certainly do not want to claim that the constraint formalism constitutes *the correct* approach to gauge; indeed, I doubt that there is an approach answering to that description, for gauge is such a broad and variegated concept that its explication requires the services of many different approaches and formalisms. But I do want to claim for the constraint formalism a number of virtues. In particular, among the extant approaches of which I am aware, it is the one with the broadest scope; for philosophers of science it has the advantage of immediately connecting to fundamental foundations issues, such as the nature of observables and the status of determinism; it explains how and under what conditions a fibre bundle structure emerges for theories which do not wear their bundle structure on their sleeves; and it calls attention to problems which arise in attempting to quantize gauge theories.

Independently of the merits of the constraint formalism, I want to urge that in getting a grip on the gauge concept, philosophers initially eschew the glitz of elementary particle physics, Yang–Mills theories, fibre bundles, etc., and concentrate instead on humbler examples. These examples often make it easier to see important conceptual connections, and they bring out the fact that the gauge concept is important for understanding not only Yang–Mills theories and theories of elementary particle physics, but Newtonian and classical relativistic theories as well.

Acknowledgements

I am grateful to Gordon Belot, Jeremy Butterfield, Katherine Brading, and Chris Martin for helpful comments and suggestions. Special thanks are due to Josep Pons for generously sharing his expertise on constrained Hamiltonian systems.

References

Ashtekar, A., and Bombelli, L. (1991). 'The covariant phase space of asymptotically flat gravitational fields'. In *Mechanics, Analysis and Geometry: 200 Years After Lagrange*, ed. M. Francaviglia, pp. 417–50. Amsterdam: Elsevier Science Publishers.

Batlle, C., Gomis, J., Pons, J. M., and Roman-Roy, N. (1986). 'Equivalence between the Lagrangian and Hamiltonian formalism for constrained systems'. *Journal of Mathematical Physics*, **27**, 2953–62.

Belot, G. (2002). 'Symmetry and gauge freedom'. *Studies in History and Philosophy of Modern Physics*, **34**, 189–225.

Bergmann, P. G. (1949). 'Non-linear field theories'. *Physical Review*, **75**, 680–5.

Bergmann, P. G., and Brunings, J. H. M. (1949). 'Non-linear field theories. II: Canonical equations and quantization'. *Reviews of Modern Physics*, **21**, 480–7.

Brading, K. A. (2002). 'Which symmetry? Noether, Weyl, and conservation of electric charge'. *Studies in History and Philosophy of Modern Physics*, **33**, 3–22.

Byers, N. (1999). 'E. Noether's discovery of the deep connection between symmetries and conservation laws'. *Israel Mathematical Society Proceedings*, **12**, 67–79.
Cabo, A., and Louis-Martinez, D. (1990). 'On Dirac's conjecture for Hamiltonian systems with first- and second-class constraints'. *Physical Review D*, **42**, 2726–35.
Dirac, M. (1950). 'Generalized Hamiltonian dynamics'. *Canadian Journal of Mathematics*, **2**, 129–48.
(1951). 'The Hamiltonian form of field dynamics'. *Canadian Journal of Mathematics*, **3**, 1–23.
(1964). *Lectures on Quantum Mechanics*. New York: Belfer Graduate School of Science Monographs Series.
Duval, C., Elhadad, J., Gotay, M. J., Śniatycki, J., and Tuynman, G. M. (1991). 'Quantization and bosonic BRST theory'. *Annals of Physics*, **206**, 1–26.
Earman, J. (1989). *World Enough and Spacetime: Absolute vs. Relational Theories of Space and Time*. Cambridge, MA: MIT Press.
(2002). 'Thoroughly modern McTaggart: or what McTaggart would have said if he had learned Einstein's general theory of relativity', *Philosophers' Imprint 2*. Available online at http://www.philosophersimprint.org.
Fischer, A. E., and Moncrief, V. (1996). 'A method of reduction of Einstein's equations of evolution and a natural symplectic structure on the space of gravitational degrees of freedom'. *General Relativity and Gravitation*, **28**, 207–19.
Gomis, J., Henneaux, M., and Pons, J. M. (1990). 'Existence theorem for gauge symmetries in Hamiltonian constrained systems'. *Classical and Quantum Gravity*, **7**, 1089–96.
Gotay, M. J., and Nester, J. M. (1984). 'Apartheid in the Dirac theory of constraints'. *Journal of Physics A*, **17**, 3063–6.
Gràcia, X., and Pons, J. M. (1988). 'Gauge generators, Dirac's conjecture, and degrees of freedom for constrained systems'. *Annals of Physics*, **187**, 355–68.
Henneaux, M., and Teitelboim, C. (1992). *Quantization of Gauge Systems*. Princeton, NJ: Princeton University Press.
Isham, C. J. (1992). 'Canonical quantum gravity and the problem of time'. In *Integrable Systems, Quantum Groups, and Quantum Field Theories*, ed. L. A. Ibot and M. A. Rodríguez, pp. 57–287. Boston, MA: Kluwer Academic.
Isham, C. J., and Kuchař, K. (1986a). 'Representations of spacetime diffeomorphisms. I: Canonical parametrized field theories'. *Annals of Physics*, **164**, 316–33.
(1986b). 'Representations of spacetime diffeomorphisms. II: Canonical geometrodynamics'. *Annals of Physics*, **164**, 288–315.
Kastrup, H. A. (1987). 'The contributions of Emmy Noether, Felix Klein and Sophus Lie to the modern concept of symmetries in physical systems'. In *Symmetries in Physics (1600–1980)*, ed. M. G. Doncel. Barcelona: Universitat Autonoma Barcelona.
Lee, J., and Wald, R. M. (1990). 'Local symmetries and constraints'. *Journal of Mathematical Physics*, **31**, 725–43.
Martin, C. (2002a). 'Gauge principles, gauge arguments and the logic of nature'. *Philosophy of Science*, **69**, S221–34.
(2002b). 'Gauging gauge: remarks on the conceptual foundations of gauge theory'. Ph.D. dissertation, University of Pittsburgh.
Maxwell, J. C. (1877). *Matter and Motion*. Reprinted (1951) New York: Dover.
Noether, E. (1918). 'Invariante Variationsprobleme'. In *Nachrichten von der Königlichen Gesellschaft der Wissenschaften zu Göttingen. Mathematisch-physikalische Klasse*, Book 2, 235–57.

Norton, J. D. (1993). 'General covariance and the foundations of general relativity: eight decades of dispute'. *Reports on Progress in Physics*, **56**, 791–858.

Olver, P. J. (1993). *Applications of Lie Groups to Differential Equations*, 2nd edn. New York: Springer-Verlag.

Sarlet, W. (1982). 'The Helmholtz conditions revisited: a new approach to the inverse problem of Lagrangian dynamics'. *Journal of Physics A*, **15**, 1503–17.

Wald, R. M. (1986). 'Spin-two fields and general covariance'. *Physical Review D*, **33**, 3613–25.

Weinberg, S. (1974). 'Recent progress in gauge theories of the weak, electromagnetic and strong interactions'. *Reviews of Modern Physics*, **46**, 255–77.

Wipf, A. (1994). 'Hamilton's formalism for systems with constraints'. In *Canonical Gravity: From Classical to Quantum*, ed. J. Ehlers and H. Friedrich, pp. 22–58. New York: Springer-Verlag.

9

Time-dependent symmetries: the link between gauge symmetries and indeterminism

DAVID WALLACE

1 Introduction

Mathematically, gauge theories are extraordinarily rich – so rich, in fact, that it can be all too easy to lose track of the connections between results, and become lost in a mass of beautiful theorems and properties: indeterminism, constraints, Noether identities, local and global symmetries, and so on.

One purpose of this short article is to provide some sort of a guide through the mathematics, to the conceptual core of what is actually going on. Its focus is on the Lagrangian, variational-problem description of classical mechanics, from which the link between gauge symmetry and the apparent violation of determinism is easy to understand; only towards the end will the Hamiltonian description be considered.

The other purpose is to warn against adopting too unified a perspective on gauge theories. It will be argued that the meaning of gauge freedom in a theory such as general relativity is (at least from the Lagrangian viewpoint) significantly different from its meaning in theories such as electromagnetism. The Hamiltonian framework blurs this distinction, and orthodox methods of quantization obliterate it; this may, in fact, be genuine progress, but it is dangerous to be guided by mathematics into conflating two conceptually distinct notions without appreciating the physical consequences.

The price paid by this article for abandoning the mathematics of gauge theory as far as possible is an inevitable loss of rigour. Virtually nothing will be 'proved' below; at most, the shape of proofs will be gestured at and strong plausibility-arguments advanced. For a more detailed understanding of the mathematics, the natural place to start is Earman's contribution to Part I of this volume (to which my own article can be seen as a commentary). Further details can be found in many standard texts on general relativity or quantum field theory (Peskin and Schroeder, 1995, is particularly clear; for a really in-depth mathematical analysis, consult Henneaux and Teitelboim, 1992).

A note on terminology: the word 'gauge', used extensively in this introduction, will not often appear again. Its meaning is now thoroughly ambiguous (as Earman notes) and I felt it simpler to resort to the marginally more cumbersome, but clearly definable, notion of a 'theory with a local symmetry group'. As will become clear below, there are genuine dynamical differences between general relativity and more 'conventional' gauge theories such as electromagnetism, but these differences are best appreciated on their own merits rather than being annexed to the essentially sterile debate as to whether or not general relativity is 'really' a gauge theory. As Humpty Dumpty has taught us, words mean just what we choose them to mean – neither more nor less.

2 Symmetries of the action

The basic setup of Lagrangian mechanics is the following. We are given some system, whose configuration is specified by a point in some *configuration space*, $\mathcal{Q}$. (The simplest example to keep in mind is ordinary N-particle mechanics, in which the configuration space is the space of all possible sets of positions for the N particles. However, the description is just as applicable to field theories provided we are prepared to abandon the demand for a manifestly covariant description: for a field theory, a point in $\mathcal{Q}$ is a specification of the field values at all spatial points for a given time.)[1]

A path through $\mathcal{Q}$ then specifies a possible history of the system: points on the path are labelled by time, so that the point labelled t gives the configuration of the system at time t. The only ontologically primary entities in this picture are the configurations and the paths through them: momentum, for instance, is only a derivative property of a path, and (unlike in Hamiltonian mechanics) cannot be regarded as on a par with configuration.

The task of Lagrangian mechanics is, then, to tell us which of the possible histories are allowed by the dynamics: that is, which histories are physically rather than merely logically possible. This is accomplished by means of the *action*, which is a rule assigning to each path γ a number $S[\gamma]$; the functional form of the action encodes everything there is to know about the dynamics of the system. The rule for specifying dynamically possible trajectories is then as follows:

(1) Pick any two points (say, q_1 and q_2), and two times (t_1 and t_2).
(2) Consider an arbitrary path (i.e. history) γ between q_1 and q_2 such that q_1 (q_2) is the state of the system at t_1 (t_2).

[1] See Goldstein (1980) or any classical mechanics textbook for a discussion of Lagrangian mechanics; Goldstein (1980) also gives an introduction to infinite-dimensional systems such as fields. For more mathematically rigorous treatments of finite-dimensional Lagrangian mechanics, consult Arnold (1989) or Abraham and Marsden (1978). Introductory treatments of the mathematically rigorous theory of infinite-dimensional systems are somewhat scarce, but chapter 3 of Marsden and Ratiu (1994) is one place to start.

(3) Evaluate the action $S[\gamma]$ for the path, as well as the actions $S[\gamma + \delta\gamma]$ for all small modifications $\delta\gamma$ of the path γ *that keep the end-points* q_1 *and* q_2 *fixed.*
(4) γ is a dynamically possible history only if the action is extremal under variations of the form above; that is, if it is extremal in the space of histories connecting $(q_1; t_1)$ with $(q_2; t_2)$.

One naive way of defining determinism might be as follows:

(A) A system is *naively deterministic* iff through any two time-labelled configurations $(q_1; t_1)$ and $(q_2; t_2)$ there exists only one dynamically possible history.

A moment's thought shows that this is too strong a restriction: consider, for instance, the dynamics of Earth's rotation around the Sun, where q_1 is the location of Earth on 1 January 2000 AD. Then through the time-labelled configurations $(q_1; 1/1/2000\,\text{AD})$ and $(q_1; 1/1/2001\,\text{AD})$ there exist (at least) two dynamically possible histories: the actual history, and one in which Earth orbited the Sun in the opposite direction.

Problems like this stem from a more general difficulty in the use of the least-action principle to describe dynamics: it treats dynamics as a boundary-value problem, where histories are specified by their initial and final configurations, rather than as an initial-value problem where sufficient information about the system at only one end of the history is enough to specify the rest of the history. This suggests, however, the following natural improvement to our definition of determinism.

(B) A system is *manifestly deterministic* iff an arbitrarily short segment of a dynamically possible history is sufficient to fix uniquely the rest of the history.

In the rest of this article we will investigate the link between symmetry and the violation of these two forms of determinism.

3 The definition of a symmetry

For our purposes, a (variational) symmetry can be defined as a transformation $\mathcal{T}$ on the space of possible histories, such that:

(1) The transformation takes each history to another history with the same action: $S[\mathcal{T}(\gamma)] = S[\gamma]$.
(2) The initial and final times (t_1, t_2) of a history are left unchanged.
(3) The transformation is locally defined, in the sense that the transformation $\mathcal{T}$ for any path may be found by breaking that path into arbitrarily small components and applying $\mathcal{T}$ separately to each component.

Obviously, the first of these is the conceptually central requirement, with the second and third being merely technical. The third condition does, however, play a crucial role in making the notion of a variational symmetry non-trivial; without it, a perfectly arbitrary permutation among all paths of a given action would count as a symmetry. (It has nothing to do with the distinction between 'global' and 'local' symmetries.)

In many cases the third rule can be implemented by requiring the symmetry to be defined in terms of some map f on the configuration space, so that $\mathcal{T}(\gamma)(t) \equiv f \cdot \gamma(t)$; most familiar symmetries (such as rotation or translation, or the electromagnetic gauge symmetry) can be specified in this way. However, not all can: the diffeomorphism symmetry of general relativity cannot, for instance, and in general nor can any symmetry which corresponds to some form of time translation.

For a consideration of determinism, the most important property of a symmetry is this: since it is action-preserving, in particular it preserves the extremality of a path. In other words, if γ is a path whose action is extremized under endpoint-preserving variations, so is $\mathcal{T}(\gamma)$. But since extremality is the necessary and sufficient condition for a path to count as a dynamically possible history, it follows that $\mathcal{T}(\gamma)$ is a dynamically possible history iff γ is. That is, symmetries map the space of dynamically possible histories onto itself.

4 Global and local symmetries

For our purposes, a global symmetry group is a group (in the technical sense) of symmetry transformations which can be specified in a time-independent way: that is, the form of the transformation on a given path does not depend on the initial and final times on that path. Rotation and translation are examples of global symmetries; so is time translation in non-relativistic particle mechanics, where the position of each particle is transformed to the position it would occupy Δt seconds later.

A local symmetry group, on the other hand, is a group of symmetry transformations whose action is time-dependent: that is, a group of transformations which can act independently on different segments of a path. The crucial difference between global and local symmetries is that if a global symmetry changes any segment of a history then (generically) it will change all of that history, whereas local symmetries may act non-trivially on only one segment of a history.

Our specification of a local symmetry as a time-dependent symmetry (i.e. one which is local in time) differs from the usual definition, in which the symmetry is taken as being dependent on both time and space. Bringing out the importance of *time* dependence is a major reason why this article uses the ordinary Lagrangian formalism, rather than the covariant Lagrangian formulation more normally used in field theory.[2] As we will see in the next section, it is time-dependence and not spacetime-dependence which is crucial to the breakdown of determinism in the presence of a local symmetry.

5 Symmetries and the breakdown of determinism

Either local or global symmetries can lead to a breakdown of naive determinism. Suppose that γ is some path from q_1 to q_2, and suppose that $\mathcal{T}$ is a symmetry which keeps the end-points of this particular γ unchanged: that is, $\mathcal{T}(\gamma)(t_1) = \gamma(t_1) = q_1$, and similarly for t_2. Then, since $\mathcal{T}(\gamma)$ is dynamically possible iff γ is, and since both have the end-points, naive determinism is violated.

This phenomenon occurs only for certain global symmetries, and then only for certain initial and final points: it does not occur at all for translational symmetry, for instance, and occurs for rotational symmetry only when the initial and final points are invariant under the rotation group or one of its subgroups. (The failure of naive determinism in the case of Earth's orbit, given above, occurs because the initial and final states are both invariant under rotation about the line between Earth and the Sun.)

It is, however, ubiquitous for local symmetries: given any two end-points, we can always find elements of the local symmetry group which leave those end-points fixed but change other parts of the paths between them.

Can a similar argument show the breakdown of manifest determinism? For global symmetries, no: the time-independence of a global symmetry means that if it is trivial on the initial part of arbitrary paths through q_1, it must be trivial on the whole path (see figure 1). But this is not the case for local symmetries, which can perfectly well leave one segment of a path fixed and change another (see figure 2). We can conclude, then, that whenever a theory has a local symmetry, that theory violates manifest determinism.

This is our reason for defining a local symmetry as one which is local in time rather than spacetime: local symmetries are interesting because of the breakdown of manifest determinism which they lead to, and nothing analogous occurs in spatially local symmetries. The interrelation between space and time which we have been accustomed to since Einstein and Minkowski should not blind us to the different dynamical roles which the two play.

(The purely spatial locality of a symmetry has no particularly interesting dynamical consequences: a symmetry which is spatially local but temporally global does

[2] By a 'covariant Lagrangian formulation' of a dynamical theory I mean one where the action S is expressed not as an integral over time of some Lagrangian function $L(t)$, but as an integral over *spacetime* of some Lagrangian density $\mathcal{L}(\mathbf{x}, t)$. Thus the Lagrangian density of the Klein–Gordon field is $\mathcal{L}(\mathbf{x}, t) = \frac{1}{2}(\partial^\mu \phi(\mathbf{x}, t)\partial_\mu \phi(\mathbf{x}, t) - \phi(\mathbf{x}, t)^2)$, whilst the Lagrangian function $L(t)$ is the integral of $\mathcal{L}(\mathbf{x}, t)$ over all $\mathbf{x}$.

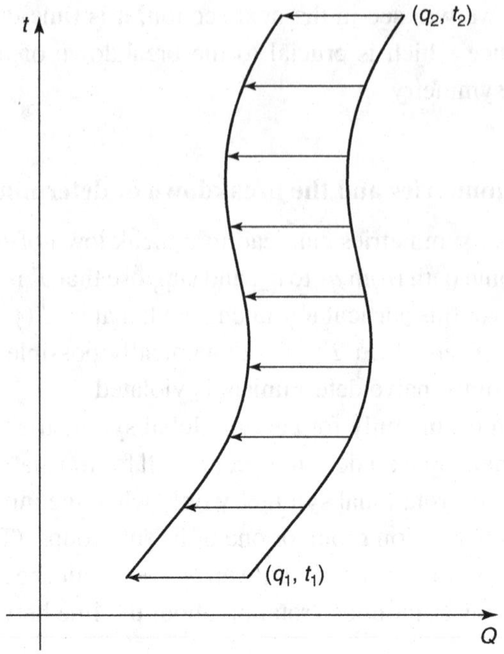

Figure 1. A global symmetry transformation.

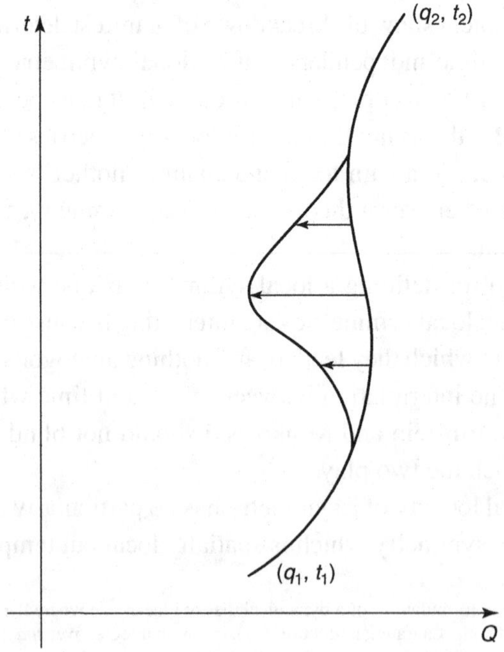

Figure 2. A local symmetry transformation.

not in any way threaten manifest determinism.[3] Inevitably, in a relatively covariant theory temporal locality probably implies spatial locality and vice versa, but as long as we are interested in the dynamics the distinction between space and time remains conceptually crucial.)

6 Two ways to repair determinism

Earman (this volume, Part I) regards it as an open question as to whether or not a given theory is deterministic; we will employ a complementary strategy, imposing determinism by fiat and determining the consequences for a theory with a local symmetry group.

Our strategy will be as follows: let $\mathcal{T}_1$ be any element of the local symmetry group which leaves the initial and final parts of paths fixed and changes other parts of them. The failure of manifest determinism occurs because any dynamically possible history will be taken by $\mathcal{T}_1$ to another such history with the same initial segment.

Our strategy for restoring determinism, then, is in essence simple:

> Whenever two histories are thus related by $\mathcal{T}_1$, they are in reality *the same history.*

This implies, of course, that there is not a one–one correspondence between the mathematics and the physics, a characteristic property of gauge theories which is discussed in Redhead (this volume).

One way of implementing this strategy – call it option A – is to insist on the following:

> **Option A.** Whenever two configurations are related by the action of $\mathcal{T}_1$ on paths through those configurations, they are in reality *the same configuration.*

This strategy is especially palatable when the symmetry itself is defined in terms of a map on the configuration space, instead of being given directly as a map on the space of paths. It is applied, for instance, in Earman's example of Maxwellian spacetime, where two configurations related by a translation are taken to be the same configuration. (It is also applicable to electromagnetism: two 4-potentials related by a gauge transformation are in general taken to be the same physical state.)

It is not, however, the only strategy: consider, instead, option B:

[3] I offer two concrete examples. In electromagnetism, the group of time-independent gauge transformations is a perfectly well-behaved global symmetry, leading to an infinite set of conserved quantities $\nabla \cdot \mathbf{E}(\mathbf{x})$, one for each $\mathbf{x}$, but not constraining these quantities to vanish or reducing the degrees of freedom of the theory. Similarly, in general relativity the group of time-independent spatial diffeomorphisms leads to the conservation of the momentum constraints but not to their vanishing.

Option B. Two histories related by $\mathcal{T}_1$ furnish descriptions of the same history in terms of two different sequences of configurations, without any claim that two different points in $\mathcal{Q}$ are the same configuration.

General relativity provides the example *par excellence* of option B: the configurations of general relativity are 3-geometries[4] and no two mathematically distinct 3-geometries are treated as physically the same, yet our freedom to foliate a manifold in many different ways lets us describe one and the same spacetime in terms of very different sequences of 3-geometries. For a more mundane example, consider Barbour's relational mechanics, in which time is defined only intrinsically to a path.[5] The time labels on paths in Barbour's configuration space, then, are arbitrary, and there exists a group of local symmetries which in effect change the time labels while keeping the points on the path fixed; thus, 'the configuration of the system at time t' will have changed, but this change is due only to a redescription of the history.

There is a very substantial conceptual difference between these two ways of reinterpreting a system in the light of a local symmetry. Option A essentially means supposing that configuration space is a redundantly large description of the actual set of configurations, suggesting that the 'real' theory lives on some sort of quotient of configuration space in which symmetry-connected configurations are identified; such a reformulation of the theory would no longer admit either indeterminism or local symmetry. No such reformulation is available for option B: in general relativity, for example, each 3-geometry is a perfectly legitimate configuration and if some of them are purged from the configuration space of general relativity then it will not be possible to formulate the dynamics.

As will be seen in the next section, this conceptual difference is essentially obliterated when we move from Lagrangian to Hamiltonian mechanics, and from classical to quantum theory.

7 Local symmetries from the Hamiltonian perspective

The transformation from the Lagrangian to the Hamiltonian viewpoint is reviewed in Earman (this volume, Part I), so I will not go into the details here. The important point about the transformation, for our purposes, is that a history of the system is now taken to be a path not in configuration space, but in phase space: that is, in the space not just of configurations but of the momenta conjugate to those configurations.

[4] I ignore, for simplicity, the purely spatial diffeomorphisms; if these are included then the configuration space becomes a space not of 3-geometries but of 3-metrics, and option A applies to 3-metrics related by a purely spatial diffeomorphism.

[5] For discussions of Barbour's approach, see Barbour (1994), Pooley and Brown (2002), and references therein.

As was mentioned above, this elevates momentum to the same ontological status as configuration; it is essentially equivalent to regarding the state of a system at a given time as being given not by its configuration at that time alone, but by the configuration and velocity jointly.

The local symmetry again leads to a breakdown of manifest determinism, and again we can recover determinism by insisting that mathematically different histories with the same initial and final conditions are in reality the same history. There are natural analogues of options A and B to implement this requirement:

Option A′. Whenever two states are related by a local symmetry transformation, they are in reality the same state. (This suggests that we should pass to the quotient space of phase space, dividing out by the action of the local symmetry.)

Option B′. Whenever two histories are related by a local symmetry transformation which keeps the initial and final conditions of the history unchanged, those two histories are in reality descriptions of the same history in terms of two different sequences of states.

However, option A′ is far more natural than option B′, for the following reason: specifying a history in terms of its initial and final conditions is not a natural strategy in Hamiltonian mechanics. This is because a single state, encoding as it does both position and velocity information, is in general (i.e. when manifest determinism holds) already enough information to specify a unique dynamical trajectory. To give *two* such states, one as the initial and one as the final state, is then to overdetermine the problem. If we wish to specify a history in this way we must give only the initial and final *configurations*, which breaks the symmetry between position and momentum which is so natural in Hamiltonian mechanics.

Furthermore, option A′ is less disastrous than option A for theories such as general relativity. Recall that applying option A to general relativity trivializes it, reducing its configuration space to a single point; however, if we are to identify only *states* and not configurations, related by the action of the symmetry, then the theory remains contentful.

However, applying option A′ to general relativity (or to Barbourian mechanics) remains conceptually unnatural. There is a perfectly natural interpretation of the local symmetry's action on states in general relativity: it represents time-evolution of those states to future states of the same history, and the freedom to choose which element of the symmetry group to apply corresponds to our freedom to define time in general relativity in many ways (i.e. to foliate spacetime in many ways). This is very different from the interpretation of the symmetry in, say, Maxwellian spacetime or electromagnetism, where it is interpreted as telling us simply that some apparently different states are really different descriptions of the very same state.

This would seem to suggest that in spite of its mathematical naturalness, we should avoid applying option A′ to theories such as general relativity. However, when we try to quantize a theory in the standard way, option A′ moves from being mathematically natural to mathematically compulsory: the Dirac quantization algorithm[6] requires the quantum state to be invariant under the action of the local symmetry. This forces the symmetry to be interpreted as a mere redescription of the physics, rather than as a physically meaningful transformation.

We can see, then, that the standard ('canonical') approach to the quantization of general relativity[7] is led by the mathematics of the quantization process to interpret the diffeomorphism symmetries in a way which is conceptually quite unnatural. It is resistance to this strategy which is a prime motivation in Barbour's alternative approach to quantum gravity.

8 Conclusion

The conceptually important difference between local and global symmetries is that the former, but not the latter, seem to imply a failure of determinism; this failure can in turn be traced to the fact that local symmetries allow us to transform only the mid-part of a system's history, keeping its initial and final states fixed. For this reason, it is conceptually helpful to *define* a local symmetry as one with this property – that is, as a temporally local symmetry. The spatial locality of the symmetry can be understood as conceptually uninteresting (at least from the dynamical point of view), a mere consequence of covariance.

Restoring determinism to a theory with a local symmetry requires us to drop the assumption that mathematics and physics are in one-to-one correspondence, treating mathematically distinct histories as physically the same. However, there are two very distinct ways of implementing this: for theories such as electromagnetism where the symmetry acts only on the configuration space, we regard the symmetry as telling us that certain configurations are really the same configuration, whilst for theories such as general relativity where time is not an external parameter, we regard it as telling us that the same history can be described by many different sequences of configurations.

This conceptual distinction is apparently lost when we apply standard methods of quantization to theories with local symmetries. We must await a working theory of quantum gravity before we can learn whether the loss of the distinction is an

[6] A thorough treatment of Dirac quantization may be found in Henneaux and Teitelboim (1992); see Matschull (1996) for a very clear account of the ideas involved.

[7] See Wallace (2000) for an elementary introduction to canonical quantum gravity.

important conceptual insight into gravity and time, or simply a case of following the mathematics one step too far.

References

Abraham, R., and Marsden, J. (1978). *Foundations of Mechanics*, 2nd edn. Reading, MA: Benjamin/Cummings.

Arnold, V. I. (1989). *Mathematical Methods of Classical Mechanics*, 2nd edn. Translated by K. Vogtmann and A. Weinstein. New York: Springer-Verlag.

Barbour, J. B. (1994). 'The timelessness of quantum gravity. I: The evidence from the classical theory'. *Classical and Quantum Gravity*, **11**, 2853–73.

Goldstein, H. (1980). *Classical Mechanics*, 2nd edn. Reading, MA: Addison-Wesley.

Henneaux, M., and Teitelboim, C. (1992). *Quantization of Gauge Systems*. Princeton, NJ: Princeton University Press.

Marsden, J. E., and Ratiu, T. S. (1994). *Introduction to Mechanics and Symmetry*. New York: Springer-Verlag.

Matschull, H.-J. (1996). Dirac's canonical quantization program. Available online from http://www.arXiv.org/abs/quant-ph/9606031.

Peskin, M. E., and Schroeder, D. V. (1995). *An Introduction to Quantum Field Theory*. Reading, MA: Addison-Wesley.

Pooley, O., and Brown, H. R. (2002). 'Relationism rehabilitated? I: Classical mechanics'. *British Journal for the Philosophy of Science*, **53**, 183–204.

Wallace, D. (2000). 'The quantization of gravity: an introduction'. Available online at http://www.arXiv.org/abs/gr-qc/00040005.

10

A fourth way to the Aharonov–Bohm effect

ANTIGONE M. NOUNOU

1 Three attempts for an explanation of the Aharonov–Bohm effect

The Aharonov–Bohm (A–B) effect is an effect one finds in every quantum field theory book and this is so for a very good reason. The prediction and subsequent experimental verification of the effect have been crucial cornerstones in the history of physics because they suggested that the gauge potential, also known as the A_μ field, might be interpreted as a real field, rather than just a mathematical artifact. Hence, ever since its discovery,[1] physicists have taken it for granted that A_μ does represent something at least as tangible as any matter field. However, when one examines these arguments more closely, one realizes that attributing the status of a really existent field to the A_μ field is not as straightforward as was originally thought. But first things first: we begin with an account of the effect itself, and then attempt to give some explanation for it.

1.1 The effect

The setting for the A–B effect is very similar to the two-slit experiment, with just one difference: immediately beyond the two slits and in between them is a very fine and long solenoid, ideally infinitely long, producing a magnetic field that is confined entirely within the tube of the solenoid. When the current in the solenoid is switched on, or in other words when there is a magnetic field **B** present inside the solenoid, the phase of the electrons changes and their interference pattern is shifted.

The configuration of the A–B effect is depicted in figure 1 and the shift is given by the relation

$$\begin{aligned}\delta = \Phi_1 - \Phi_2 &= \delta(\mathbf{B}=0) + \frac{q}{\hbar}\int_{(1)} \mathbf{A}\cdot d\ell - \frac{q}{\hbar}\int_{(2)} \mathbf{A}\cdot d\ell \\ &= \delta(\mathbf{B}=0) + \frac{q}{\hbar}\oint \mathbf{A}\cdot d\ell, \qquad (1)\end{aligned}$$

[1] Aharonov and Bohm (1959).

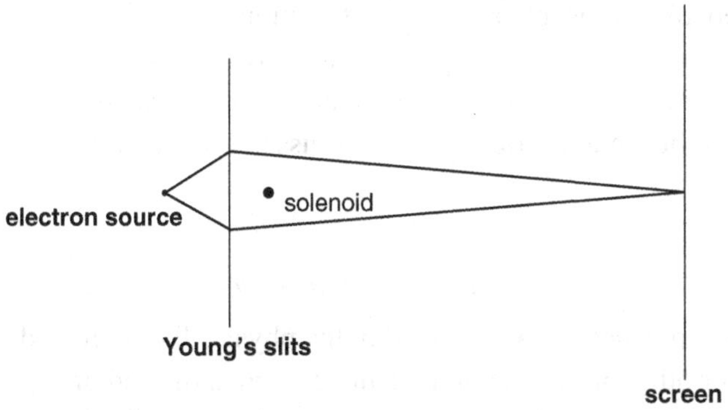

Figure 1. The Aharanov–Bohm experiment.

where Φ_1 and Φ_2 are the phases of the electrons passing through the slits 1 and 2 respectively. This equation tells us how the electron motion changes when some magnetic field is present. If we use Stokes' theorem at this point, the equation above becomes:

$$\delta = \Phi_1 - \Phi_2 = \delta(\mathbf{B}=0) + \frac{q}{\hbar}\oint_{\partial s} \mathbf{A}\cdot d\ell = \delta(\mathbf{B}=0) + \frac{q}{\hbar}\int_s \nabla \times \mathbf{A}\cdot d\mathbf{s} \quad (2)$$
$$= \delta(\mathbf{B}=0) + \frac{q}{\hbar}[\textit{flux of } \mathbf{B} \textit{ between paths (1) and (2)}].$$

Any choice of **A** which has the correct *curl* **A** = **B** gives the correct interference pattern; this point will be important later. Since the flux of **B** does not depend on which pair of paths we choose, provided that they surround the solenoid, for every arrival point there is the same phase change $x_0 = \frac{q}{\hbar}$[*flux of* **B** *between paths (1) and (2)*]. This means that the entire pattern is shifted by x_0.

These are the general ideas involved in the A–B effect, and the main consequence is that one has to make a choice. The first option would be that it is the magnetic field accounting for the effect; however, since it is confined inside the solenoid only, we have action at a distance. The second is that if we wanted to give a local causal account, the **A** field should be considered as real in the sense that the motion of the electron in some region is determined point by point by the presence of the **A** field in the same region.

1.2 The three approaches towards the A–B effect

In the philosophy literature there are three different explanations of the A–B effect that have been discussed in detail by Healey (1997; 1999; 2001), Leeds (1999), Lyre,[2] and Maudlin (1998), but they all face problems. There is, however, a fourth way to be found in the physics literature, which has not yet

[2] H. Lyre, 'A versus B! Topological non-separability and the Aharonov-Bohm effect', *Contribution for the International IQSA Conference: Quantum Structure V.* Cesena/Cesenatico, Italy, 2001.

been discussed by philosophers. My contribution to the debate will be to give a philosophical analysis of this option, after having discussed the fibre bundle formulation of field theories that gives rise to this explanation. The other three approaches and their main problems are discussed briefly in the remainder of this section.[3]

1.2.1 The first approach

The first two approaches are widespread in the physics literature and correspond to choosing one or other of the two options mentioned at the end of section 1.1. In the first attempt to explain how the phase shift of the electron comes about one considers that the origin of the shift is the magnetic field **B**. The appealing characteristic of the magnetic field is that it is gauge invariant and hence corresponds to a physically real entity; gauge invariance is considered to be a necessary condition for a mathematical entity to represent an observable because it is only gauge-invariant quantities that are single-valued and thus epistemically accessible.[4] However, the magnetic field is confined inside the solenoid only and is zero everywhere else. This means that the magnetic field is zero in the entire region from which the electrons are likely to pass and non-zero only in the region that is inaccessible to them, namely inside the solenoid.

Healey (1997) discerns two different notions of locality, both necessary for a process to be local. He calls them *local action* and *separability*, and he gives them the status of principles. So, for him, locality holds just in case both local action and separability hold. The principle of local action is expressed as follows: 'If S and P are spatially distant objects, then an external influence on S has no immediate effect on P' (Healey, 1997, p. 23). The principle of separability, on the other hand, may be stated as follows: 'any physical process occurring in spacetime region R is separable just in case it is supervenient upon an assignment of qualitative intrinsic physical properties at spacetime points in R' (*ibid.*, p. 24).

Note that although non-separability is a common feature of quantum mechanics, in this context the notion is somewhat different. For a detailed discussion see Healey (1997; 1999) and Maudlin (1998).

Clearly, this first explanation of the A–B effect entails that if we considered the magnetic field **B** to be the only existing interactive field, then we would have to succumb to action at a distance (which violates the principle of local action) and hence to non-locality.

[3] In what follows I will not engage in detailed criticisms of the other suggested explanations of the effect, due to lack of space. Criticisms will be the subject of another paper.

[4] For a discussion of the issue see, for example, Belot (1998), Martin (this volume), and Maudlin (1998).

1.2.2 The second approach

To avoid this problem of action at a distance, *one may take as a fact* that the physically interacting field is the vector potential **A**, and this constitutes the key assumption of the second approach. The main objection to the second approach is that the vector potential **A**, which is held responsible for the effect, is not gauge invariant and hence, since we require that physical objects are described only by gauge-invariant entities of the theory, this one does not qualify. Although at first sight the account involving this field seems to be local, the violation of separability is inevitable because the requirement that a physical process supervenes upon an assignment of qualitative intrinsic *physical* properties is not satisfied.[5]

Despite the fact that it seems impossible to regard the gauge potential as a real physical and, presumably, causal object there are two things about it that make its presence in an explanation of the effect very appealing: its occurrence is unavoidable in predictions of experimentally verifiable quantities in quantum field theory and it has probed gauge theories to an incredible degree over the last fifty years or so. These are very good reasons for not wanting to give up on the explanatory role the gauge potential may play, despite the fact that one would have to try a different way of involving it in the explanation.

1.2.3 The third approach

In the third approach, one considers as the real causal agent the so-called Dirac phase or holonomy. Despite the fact that the Dirac phase factor – or holonomy – is a good candidate because it is gauge invariant and, after all, it provides a measure of the phase shift, taking a closer look at these arguments we find that there is more that would need to be said in order to make this a good explanation of the effect and this has not been done.

The Dirac phase factor is expressed by the integral

$$S(C) = \exp\left(- (ie/\hbar) \oint_C \mathbf{A}(r) \cdot dr \right), \tag{3}$$

where the integral is taken over each closed loop C in spacetime. Healey (1997) and Lyre (*op. cit.*) argue that the holonomy is the quantity that expresses an intrinsic property of C, provided that C is a non-intersecting closed curve.[6] However, the holonomies do not supervene on assignments of qualitative intrinsic properties *at* spacetime points *in* the region through which the electrons pass, because by

[5] Healey (1997) has argued that even if one were prepared to bite the bullet and consider the gauge potential to be a physically real field, the A–B effect could not be rendered completely local. For a discussion that followed this argument see Healey (1999) and Maudlin (1998).

[6] Lyre (*op. cit.*) takes $I(C) = \oint \mathbf{A}(r) \cdot dr$, rather than $S(C)$, in order to get rid of the electronic charge e, and he chooses non-intersecting closed curves in order to avoid the difficulty arising from the fact that closed curves do not correspond uniquely to regions of space.

definition each $I(C) = \oint \mathbf{A}(r) \cdot dr$ supervenes upon the spacetime points of an arbitrary curve $C = \partial s$ which encircles the solenoid and not on the spacetime points through which a single electron passes. Therefore, if we choose the loop integral $I(C)$ to explain the A–B effect, we have, once again, violation of the separability principle.

1.3 Comments on the third approach and beyond

The main purpose in all these three attempts is to identify some mathematical structure that could be interpreted as a local – of some sort – causal agent that could be held responsible for the effect. However, a recurrent trait in all three approaches is that the A–B effect is non-local because the attempted explanation violates either the principle of local action or the principle of separability. What this discussion leaves us with is the realization that the A–B effect is a '*global*' effect and in my view any attempt to explain the effect should regard this 'global' characteristic as intrinsic. The word 'global' in this context means that the net effect on the phase of the electron is measured by the holonomy $I(C) = \oint_c \mathbf{A}(r) \cdot d\mathbf{r}$ along a closed curve that *surrounds* the solenoid. The curve along which we integrate is arbitrary and can get as close to or as far from the solenoid as we like, but in any case it has to be a curve that surrounds the solenoid. We cannot get the interference pattern by considering just open segments of the closed curves. It is in this sense that the phenomenon is 'global' but not universal.

So, the A–B effect is inherently non-local and this is a characteristic that any attempt to offer a good explanation of it needs to account for. This cannot be done by just restating its global nature, as in the third approach, nor by looking for a causal agent, as in the second. Despite the fact that it doesn't follow that *we have to* adopt a holistic explanation, attempting to give one is a good bet since we want to offer an explanation of a global effect.

2 The fibre bundle formulation of theories with gauge symmetries

It is a commonplace view in physics that physical objects interact and it is through their interactions that we observe them. Therefore we need a description that accounts for these interactions and explains our observations. One very fruitful[7] way of describing interactions is by using variational calculus and local symmetries but this way makes use of a mathematical structure which seems to be richer than the physical system it represents. Richer in this context means that not all the entities anticipated by the mathematical theory correspond to physical ones; the part of the

[7] Fruitful in this context means that the suggested way has given good descriptions/explanations and accurate predictions.

mathematical theory that contains the excess has been dubbed the *surplus structure*.[8] It is a matter of fact that in most – if not in all – theories of physics, for a given physical system there is a plethora of mathematical descriptions, which results in ambiguity of representation of physics by mathematics. However, the ambiguity of representation involved in gauge theories is of a distinct kind and it is associated with the presence or appearance of coupling terms that have been used successfully in representing interactions. There is more than one way of formulating such theories. A well-known way, favoured by many, makes use of *constrained Hamiltonian systems*.[9] But aside from this, or rather subsequent to it, there is another more elaborate formalism – the fibre bundle formalism – which, I believe, is more appropriate for describing interactive and interacting fields.

In their fibre bundle form, systems with non-Abelian symmetries first appeared in the Yang–Mills 1954 paper and they came to the forefront of research in physics from the late 1960s onwards. At the same time, since the 1930s, mathematicians who were studying relations between topology and geometry, and then from the 1950s onwards topologically non-trivial manifolds, developed the so-called fibre bundle formalism – a generic geometrical approach that encompasses the mathematical structures that describe systems with constraints imposed by gauge symmetries. But the utility of fibre bundles in gauge theories was introduced to the majority of the physics community by Wu and Yang (1975), who compiled a 'dictionary' translating between the physicist's terminology and the new mathematical terminology. The aim of this section is to aid our comprehension of how systems with gauge symmetries are described by this formalism and what its advantages are. Those readers familiar with this formalism may skip straight to section 3. What I hope to do is to describe and illustrate the concepts involved using the minimum of mathematical machinery. Inevitably, this entails a lack of rigour and precision, but I hope that the picture presented will be sufficient for what follows in the remainder of this paper.

2.1 What a fibre bundle is

Fibre bundles are a generalization of the Cartesian product in the following sense. A fibre bundle is a triplet $(\mathcal{M}, \pi, E)$ where $\mathcal{M}$ is what we call the base manifold, E is the total space, and π is a projection map $\pi : E \longmapsto \mathcal{M}$. The inverse image π^{-1} of the map π takes you from a point $x \in \mathcal{M}$ to E and it is called the fibre $F_x := \pi^{-1}(\{x\})$ over x. The total space E is $\mathcal{M}$ itself along with the bundle of all the fibres over all $x \in \mathcal{M}$, or $E := \cup_{x \in \mathcal{M}} F_x$. In certain cases the total space E is the product space $\mathcal{M} \times F$ which is a straightforward generalization of the Cartesian

[8] For a detailed discussion see Redhead (this volume).
[9] Eminent advocates of this formulation are Earman (2001; this volume, Part I) and Belot (1996).

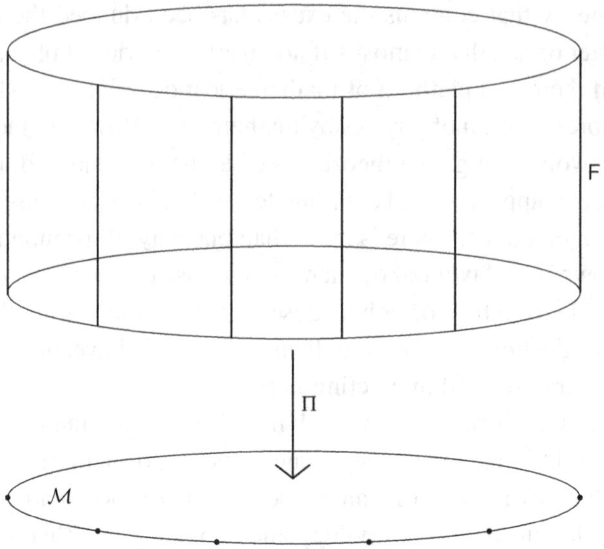

Figure 2. A product bundle.

product. As we know, if $\mathcal{M}_1$ and $\mathcal{M}_2$ are differentiable manifolds, then $\mathcal{M}_1 \times \mathcal{M}_2$ can be given a manifold structure where $\dim(\mathcal{M}_1 \times \mathcal{M}_2) = \dim(\mathcal{M}_1) + \dim(\mathcal{M}_2)$. But in fibre bundles the total space is not, in general, a product space and this will be made clear by two illustrative examples.

The first example is that of the product bundle.[10] The product bundle (figure 2) is one of the simplest examples of a fibre bundle and its three elements are: $\mathcal{M}$, $\pi = pr_1$ (the projection map taking you from any point of F_x, the fibre over x, to the point x on the manifold), and $E = \mathcal{M} \times F$.

Another example of a fibre bundle is the Mobius strip. Here, the base space $\mathcal{M}$ is the circle S_1 and the fibre could be taken to be the interval $[-1, 1]$. But the total space E is not the product space $\mathcal{M} \times [-1, 1]$, nor is it homeomorphic to it because the total space is twisted. It can be represented, instead, by a rectangle whose short edges identify as shown in figure 3.

2.2 Cross-sections

The notion of cross-section is very crucial in both the fibre bundle formalism and its application in physics, since all the matter fields, such as the electronic field, are defined as cross-sections of the tangent bundle, which is a special case of a fibre bundle as we will be discussing in the next section. In general, physical quantities are represented by real-valued differentiable functions on manifolds. The structure of the tangent space has a deep connection with the local differentiable properties

[10] For more details see Isham (1999, pp. 204–6).

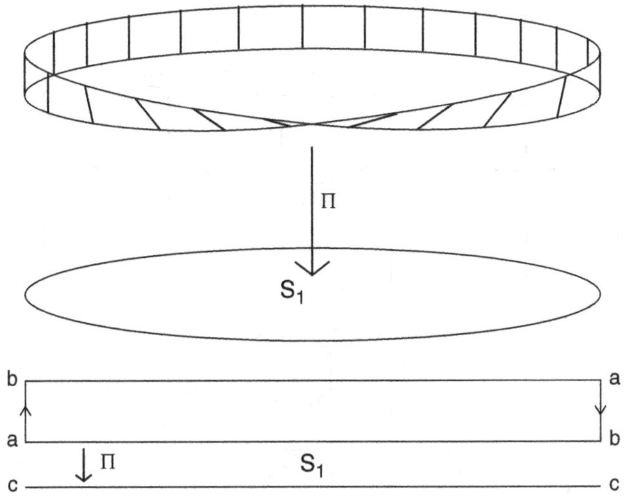

Figure 3. The Mobius strip.

of functions that are defined on a manifold. In the fibre bundle context, the generic way of describing fields that carry indices – indices which may express tensorial properties with respect to spacetime or with respect to internal structure groups – is by means of spacetime-valued fields that are defined as cross-sections on the tangent bundle over the manifold. The cross-section is a map $s : \mathcal{M} \longmapsto E$ such that the image of each point $x \in \mathcal{M}$ lies in $\pi^{-1}(\{x\})$. π and s are inverse to each other:

$$\pi \circ s = id_{\mathcal{M}} \tag{4}$$

So, here we are talking about some mathematical object (a field) which takes some *specific* values across the fibres as its location on the base manifold also changes. So far as the product bundle is concerned, the cross-section is defined uniquely and continuously everywhere (figure 4). However, in the case of the Mobius-strip bundle, which is a non-orientable surface, this is not the case. It becomes obvious from figure 5 that the cross-section is not continuous. Instead, it is equivalent to a function from S^1 to $[-1, 1]$ which is antiperiodic around the (circle) base manifold. The Mobius strip is just an example of a non-orientable fibre bundle, but from it we can see how the cross-section and its continuity depend upon the topology of the total space. At this point we have to make a leap. In general, in the cases of the so-called principal fibre bundles, where the bundles have the special structure of a vector space, the following theorem holds.

Theorem 1. *A principal fibre bundle has a continuous cross-section if and only if it is trivial.*[11]

[11] For a proof of this theorem see Isham (1999, p. 230).

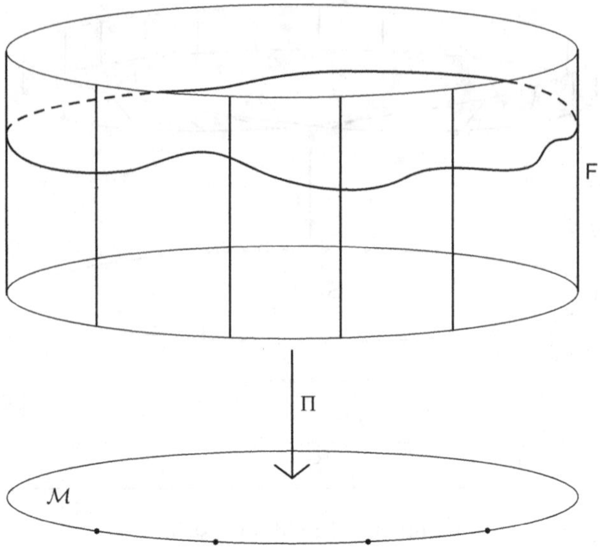

Figure 4. Cross-section of a product bundle.

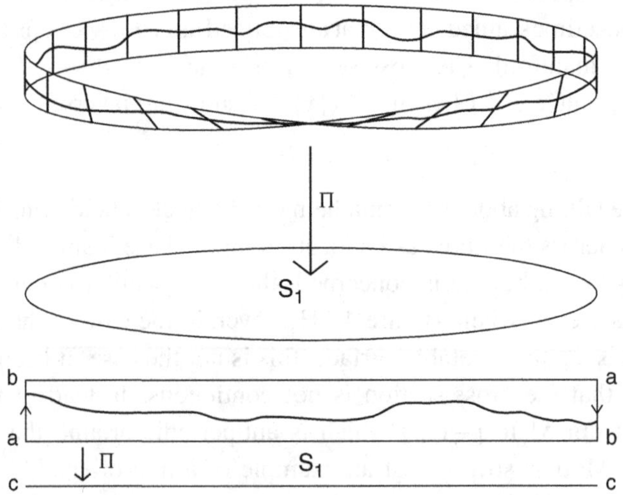

Figure 5. Cross-section of a Mobius-strip bundle.

One of the two things this theorem tells us is that when the topology of the base manifold is non-trivial, we will not find a continuous cross-section. So, if we take the base manifold to represent spacetime, then if the topology there is not trivial, we are not able to define vector fields continuously all over it, and this, as we shall see, is related to the well-known problem in gauge theories, the so-called

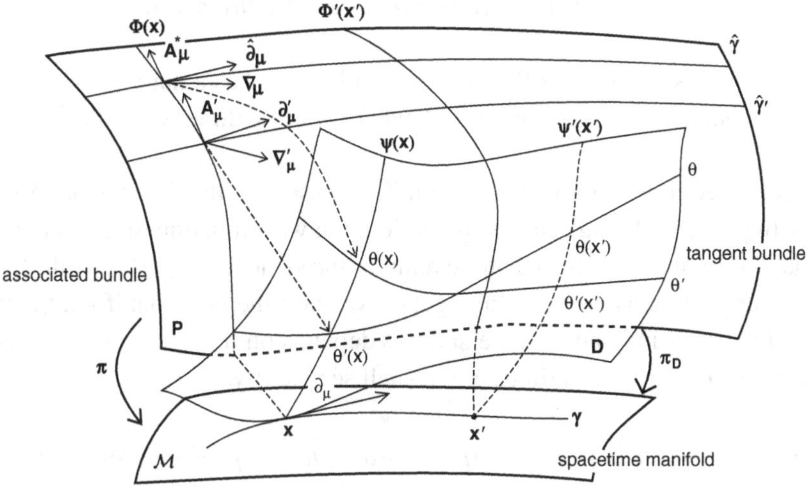

Figure 6. Associated and tangent bundles.

Gribov obstruction, which does not allow us to determine the gauge everywhere at once.[12]

2.3 Vector bundles, principal bundles, and connections

At this point we need to make another leap and try to visualize two more complicated examples of fibre bundles, having as a starting point the simple cases of the product and the Mobius bundle. The first case is that of the tangent bundle, which is the bundle of the tangent spaces at all points of a base manifold, while the second is the bundle of frames, which, as its name indicates, is the bundle of all frames at all points of the base manifold. In order to get a visual idea of what the various objects involved represent, we will use figure 6.[13]

2.3.1 The tangent bundle: a special example of a vector bundle

The base space $\mathcal{M}$ of the tangent bundle may be considered as the *4-dim* spacetime manifold. The fibre F_x over each point x of the manifold is the tangent space $T_x\mathcal{M}$ to $\mathcal{M}$ at the point x which is generated by the set of all the tangent vectors at this point; or, in other words, by the vectors of all the curves which pass through the point x and are tangent to x. The total space E, or the tangent bundle $T\mathcal{M}$, is defined as $T\mathcal{M} = \cup_{x\in\mathcal{M}} T_x\mathcal{M}$, the union of all tangent spaces at all points of the manifold $\mathcal{M}$. For each tangent space, the following theorem holds.

[12] See Gribov (1978) and Singer (1978).
[13] This illustration is based on Auyang (1995, p. 220, figure B3).

Theorem 2. *The tangent space $T_x\mathcal{M}$ carries a structure of a real vector space.*

It can also be shown[14] that the tangent bundle $T\mathcal{M}$ has a natural structure of a $2m$-dimensional differentiable manifold, where m is the dimension of the manifold $\mathcal{M}$ itself.

The cross-sections θ of this vector bundle are used for the description of matter fields with phase θ. Along each cross-section, the wavefunction of the matter field may take different values, but its phase remains the same, i.e. $\vartheta(x)$. The information encoded here is that as we move along a curve γ on the base manifold, the phase of the field may or may not change and this depends on the interactions which are accounted for by the connections, as we shall see shortly.

2.3.2 The bundle of frames: a special example of a principal fibre bundle

A more complicated case of a fibre bundle is the bundle of frames, which is a special case of what mathematicians call a principal bundle. A principal bundle is one whose fibres are Lie groups in a specific way. The principal fibre bundles 'have the important property that all non-principal bundles are *associated* with an underlying principal bundle. Furthermore, the twists in a bundle associated with a particular principal bundle are uniquely determined by the twists in the latter, and hence the topological implications of fibre bundle theory are essentially coded into the theory of principal fibre bundles' (Isham, 1999, p. 220). A typical example of a principal fibre bundle is the bundle of frames.

In the case of the bundle of frames, the base space $\mathcal{M}$ is, once again, an m-dimensional differentiable manifold which we may consider to be the *4-dim* spacetime manifold. A linear frame, or base, at the point $x \in \mathcal{M}$ is an ordered set $(b_1, b_2, \ldots, b_m)$ of basis vectors for the tangent space $T_x\mathcal{M}$. In this case, the projection map $\pi : \mathbf{B}(\mathcal{M}) \rightarrow \mathcal{M}$ is defined to be the function that takes a frame into the point x in $\mathcal{M}$ to which it is attached. The fibre over $x \in \mathcal{M}$ is, of course, the inverse image under the map π and it comprises the set of all the local frames that are associated with the point $x \in \mathcal{M}$. The total space of the bundle of frames, which we denote by $\mathbf{B}(\mathcal{M})$, is the *set* of *all frames* at *all points* of $\mathcal{M}$. $\mathbf{B}(\mathcal{M})$ is a space with the symmetry group $GL(m,\mathbb{R})$ acting on its vectors, as well as a differentiable manifold of dimension $m + m^2$.

In our graphic representation of the principal fibre bundle we can see the following. 'Over' each point x of the base space $\mathcal{M}$ there is the fibre of x, represented as a line with $\Phi(x)$ at the top. The cross-sections of this fibre bundle are depicted by the $\hat{\gamma}$-lines and they introduce a specific coordinate system along the curve γ so that as

[14] See Isham (1999, p. 89).

we are moving along the curve we have a fixed coordinate system or frame – this could be understood as an active transformation where the actual system is 'moving' but the frame remains the same. As we move along the fibre, the field Φ does not change but the frames do – this is what we could understand as a passive transformation, where the physical system remains fixed but its description changes.[15]

2.3.3 Connection on the bundles or moving around

Next, we need the notions of connection and pull-back. The connection tells us everything about how we move around in the bundle, while the pull-back is the operation we need in order to be able to 'move' from the total to the base space and the other way round.

The connection is a field defined on the bundle space and, as its name indicates, basically we need it so that we can connect or compare points in 'neighbouring' fibres in a way that is not dependent on any particular local bundle trivialization (i.e. choice of frame). This suggests that we should look for vector fields on the bundle space P that 'point' from one fibre to another.[16] What is needed, therefore, is some way of constructing vectors that point away from the fibre, i.e. elements of T_pP that complement the vertical vectors in V_pP.

What one can actually do with the connections is this: in the case of active transformations, and while still on the bundle of frames, the connections describe how the field of frames changes as we move along a spacetime path and therefore 'hop' from one fibre onto another. As a physical system moves along a spacetime curve γ, the tangent spaces change and so do the frame-fibres. In general, these tangent spaces are not in any natural relation to each other. The connection, represented by ∇_μ, allows us to compare these spaces, by expressing how $\hat{\partial}_\mu$ changes as we 'cross' different bundles. If the local representative of the connection was given the name A^*_μ,[17] this could be represented in a diagram as follows:

$$\begin{array}{ccc} A^*_\mu & \xrightarrow{G} & A'^*_\mu \\ \pi \uparrow & & \uparrow \pi \\ x & \longrightarrow & x' \end{array}$$

[15] If, instead of the bundle of frames, we had chosen a fibre bundle with the symmetry group $SU(2)$, we would have the $SU(2)$-bundle of the Yang–Mills theory. In this case, selecting a specific cross-section is also known as gauge choice or gauge fixing.

[16] See Isham (1999, p. 253) for a more detailed discussion.

[17] As a matter of fact, the connection is usually associated with a certain $L(G)$-valued 1-form ω on the bundle space P, while by Γ we denote the associated $L(GL(m, \mathbb{R}))$-valued 1-form on $U \subset \mathcal{M}$, and the symbol A^α_μ is used specifically for the Yang–Mills field, which can be regarded as a Lie-algebra-valued 1-form on $\mathcal{M}$, at least locally. In this paper, we chose to use the symbol A^*_μ for simplicity and to give some sort of unity. I would like to make it clear, though, that this 'unified' use of one symbol is not accurate and I would like to warn the reader that this may be confusing if they study, for example, Isham (1999).

All change is described by the connection but, as we should expect, this is done in a non-deterministic way; if there is no necessity to impose a choice of a specific cross-section, the evolved system may start from any point of the initial fibre and be found on any point of the final. However, as we can see from figure 6, when moving along γ and at the same time staying on the same cross-section, the initial coordination remains the same; which means that we know exactly where we will find our system when we are looking for it in the total space.

The passive view of the transformation is somewhat more difficult to describe correctly here, because the illustration we are using is inaccurate and incomplete;[18] but the intuitive idea is the following. For a description of the same spacetime point, we may use various different coordinations, which are related to each other by the action of the group $GL(m, \mathbb{R})$. Thus the 'location' on the bundle space, or the local trivialization, changes, while the physical system remains where it was on the spacetime manifold. In this case, the connections corresponding to the two different local trivializations are the transform of each other under the action of the group. In the form of a diagram, the situation could be illustrated as follows:

$$\begin{array}{ccc} A^*_\mu & \xrightarrow{G} & A'^*_\mu \\ \pi \nwarrow & & \swarrow \pi^{-1} \\ & x & \end{array}$$

In general relativity, the role of the connection is played by the well-known Christoffel symbols. In Yang–Mills theories, on the other hand, where the principal bundle is one with a Lie group acting on it, the role of the connection is played by the Yang–Mills field itself.

2.4 Gauge transformations

If we want to be more accurate, we have to say that the connection is an $L(G)$-valued 1-form on a principal bundle[19] and when the principal bundle is one with the $SU(2)$ group acting on it, it is such that locally the connection can be decomposed as the sum of a Yang–Mills field on $\mathcal{M}$ plus a fixed $L(G)$-valued 1-form on G. Since the latter $L(G)$-valued 1-form is fixed when we know the Yang–Mills field, we could 'identify', in this informal sense, the connection with a Yang–Mills field – as we have done in the previous sections. Here we need to take a closer look at how gauge transformations come up in fibre bundles.

In general, a gauge transformation is considered to be *any* automorphism of the bundle, and, depending on the interactions one wants to describe, the group that

[18] For more extended discussion see Isham (1999).
[19] For a detailed discussion see Isham (1999, pp. 254–62).

induces the automorphisms may be considered to be the $U(1)$ group of electromagnetism or the $SU(2)$ group of the weak interactions, to mention just two. In the case of passive transformations, the actual transformation map $\Phi : P \to P$ takes you from a coordinate chart to another – the two have domains U and U' that overlap. Then, it can be shown that the transformed connection is also a connection and that the transformation of the local representatives of the connection, i.e. of the Yang–Mills field, is our familiar gauge transformation. Along the same lines, when we consider active automorphisms of the bundle, the transformation on the manifold induces a transformation of the connection that locally is exactly like the familiar gauge transformation of the gauge field; the only difference here is that the automorphism is defined first on the manifold $\mathcal{M}$ as $h : \mathcal{M} \to \mathcal{M}$.

Mathematically speaking, the two different ways of viewing transformations, i.e. the active and the passive, are equivalent. Yet when we use this formalism to represent physical structures a problem arises. The active transformation is considered to correspond to an actual change of the physical system from one physical state to another. The passive transformation, on the other hand, changes only some part of the surplus structure, or the description of the system, one could say. In what sense, then, are the two equivalent when we talk physics? If we claimed that a transformation/change in the description of a structure corresponds to an altogether new 'reality' in a sense, similar to that of a physical system that has been transported to a new spacetime region, would we do justice to the mathematical equivalence? Or is this a far-fetched assumption? To answer the first question in the positive, one has to take into account that in the active case, there is some actual change of the physical system we study, but in the passive case there is not. In our case, however, there does not seem to be an actual change happening, unless we consider that some of the mathematical objects that live in the bundle space – which undergo the change, the connections for example – do correspond to physical systems. If that were the case, we could comprehend how the two types of transformation are equivalent in a physical sense. But the question of whether the connections have physical status is one to which we cannot give a straightforward answer, at least not right now, because although we make use of the connections to represent the interactive fields, we cannot say before we give it some further thought that these are indeed 'tangible' physical objects. Note in passing that the same sort of question is addressed by Redhead (2001), who claims that when the automorphisms of the physical system and the mathematical structure are in one-to-one correspondence and if the symmetries of the physical system express important structural properties of it, then so would the symmetries of the mathematical structure. Things are somewhat different, though, when the symmetries are present not in the parts of the mathematical structure that directly correspond to physical entities but in the 'excessive' parts of the theory, the so-called surplus structure, in which case the mathematical symmetry gives coupling terms that may be interpreted as interactions of the physical structure.

2.5 Associations

The tangent bundle and the principal fibre bundle that can be seen in figure 6 are associated bundles. In general, the basic intuition that underlies their association is that 'given a particular principal bundle $(P, \pi, \mathcal{M})$ with structure group G, we can form a fibre bundle with fibre F for each space F on which G acts as a group of transformations' (Isham, 1999, p. 232). In our specific example, the group of the bundle of frames $GL(m, \mathbb{R})$ acts on the tangent space on each point of $\mathcal{M}$ and the result of the action is the change of the mathematical expression of the local coordinate chart either in a passive way (if x does not change and therefore we are still on the same fibre) or in an active way (when x changes as well). So, for the same x, a passive symmetry transformation could take the connection field $A^*_\mu(x)$ to $A'^*_\mu(x)$, while an active one could take it to $A^*_\mu(x')$ or $A'^*_\mu(x')$ depending on whether we stayed on the same cross-section or not. These changes on the principal bundle are linked with changes on the associated tangent bundle in the following way. When we are considering passive transformations, the action of $GL(m, \mathbb{R})$ on the vector space of the tangent bundle can be understood as changing the direction of the tangent vector on x, while still remaining on the same tangent 'plane' or fibre; so it takes you from $\psi(x)$ to $\psi'(x)$. When the transformations are active, there is a total change of the ψ-field – i.e. a change which affects both the spacetime point and the fibre. So, if the transformation leaves the field on the same cross-section, the transformed field will be $\psi(x')$ while if not on the same cross-section, the transformed field will be $\psi'(x')$.

When the group acting on the principal bundle is a gauge group, the action of the group on the associated bundle will be expressed as a change of the phase of the matter field – with or without simultaneous change of its spacetime location, depending on whether the transformation is considered as active or passive respectively. In the passive case, starting with phase $\vartheta(x)$, we end up with phase $\vartheta'(x)$. On the other hand, an active action of the group projects the original point of the total space to some other point which lies on a different fibre altogether. In this case, if we are still on the same cross-section, the transformed phase will be $\vartheta(x')$, while if we are not, the new phase will be $\vartheta'(x')$.

This association between principal and vector bundles is what allows for coupling terms to appear, terms that can be interpreted as interaction terms between gauge and matter fields. Moreover, the information about the topology of the base manifold that is not encoded by the physical objects living on the base manifold is encoded by the entities on the bundle. Hence, in the case of the A–B effect, the physical objects such as the magnetic field inside the solenoid may tell us that the 'curvature' is non-zero in the region of spacetime they occupy but they do not inform us about the effect of this curvature on matter fields that are transported around them. This information is provided by the gauge potentials, or connections.

In concluding this section we need to address an important question. If we had to take one of the two spaces as physically real, namely E or $\mathcal{M}$, what should we consider, the spacetime manifold or the total bundle? Considering the base manifold as physically real, and along with it all and only the objects that live in it, we would be left with gauge-invariant quantities only, which, as we have seen in the case of the A–B effect, do not explain much. On the other hand, if we regard as physically real the total space, we would have to justify in what sense the mathematical structures of the fibre space could be given the status of real physical objects. In this paper as well as in the literature, though, we have seen that this latter option is also problematic because no matter how hard we tried, it was not possible to attribute to the gauge potentials the status of physical, experimentally detectable, and locally interacting objects. This realization points towards a different interpretation of the gauge fields and their role in providing explanations, an altogether different kind of explanation, which I call topological.[20]

3 Topological explanation: a fourth way to the A–B effect

The fourth way to the A–B effect provides a holistic explanation of the phenomenon. This kind of explanation takes into consideration the entire system rather than small parts of it considered to be causally related to each other, and crucially relies on the topology of the entire system. For this reason it is called *topological explanation.* The fibre bundle formulation of gauge theories was developed so that topological considerations can be described lucidly and, therefore, it constitutes their most favourable mathematical expression. Hence in what follows we use the fibre bundle formalism to give a topological explanation of the A–B effect.

3.1 The holistic character of the topological explanation

The topological explanation of the A–B effect is the approach favoured by many mathematical physicists[21] and we were directed towards it for several reasons. The fact that there does not seem to be a satisfactory bit-by-bit causal account of the phenomenon, which is the result of the non-separability present in the last two

[20] Note that Leeds (1999) attempts an explanation of the A–B effect that arises from the fibre bundle formalism, but he falls prey to the same assumption as others, namely that the gauge field should be interpreted as a causal agent.

[21] For topological accounts and explanations of the effect see, for example, Nakahara (1990), Nash and Sen (1983), and Ryder (1985).

attempts to explain the effect, indicates that one should take more into account than just the (speculated) events and physical processes along the path of the electron. Knowing what is going on in each part of our physical system is not enough since this knowledge leaves out information about the whole that cannot be retrieved. Therefore we require a formalism that contains all the necessary information for a good comprehension and explanation of the events. This formalism, we suggest, is the fibre bundle formalism, in which the mathematical entities of the surplus structure *register* all the information – not just bits and pieces of it – about the topology of the base manifold. Consequently, the mathematical objects involved do not dictate the behaviour of physical objects as though they were the *causal agents* acting on those physical objects, nor are they held responsible for a signalling process that takes place – allegedly – between solenoid and electrons. Instead, we claim, they are descriptive tools that encode all the information about the modification of spacetime, and for this reason they account for the effect in terms of the relations between the spacetime points and the physical objects, i.e. the electrons, involved. From this perspective one could say that the solenoid has modified not just the spacetime points that it occupies but also the region around it and therefore the shift in the phase of the electrons happens as a result of this modification. The gauge field does not participate in this modification, it just encodes it; it gives us a mathematical tool for calculating the measurable results this change brings about. A measure of the results is provided by the holonomies, which do not generate them. In this way, we gain full *awareness* of all the elements involved and the factors affecting the electron, and a good understanding of its behaviour. But let us examine how this is done.

3.1.1 Holonomies, homotopy, and the $U(1)$ group of electromagnetism

As we have already seen, fibre bundles involve mappings between a base and some other manifold and these mappings describe completely the structural characteristics, or the topology, of these spaces. The discussion in this context is related closely to the discussion of the topological non-separability of the A–B effect and the holonomies that are involved; let us explore here how the discussion of loop integrals in topologically non-trivial manifolds fits into the more general picture of fibre bundles and how this relates to the A–B effect and its explanation.

Assume that the base space is the spacetime manifold with an infinitely long cylindrical hole in it. For the sake of simplicity, we can consider a slice of it, which, for a given time, is described mathematically as a plane with a hole. The hole could represent the area occupied by the solenoid that is inaccessible to the electron.[22]

[22] One might object that the wavefunction of a quantum particle, such as the electron, is extended over the entire space between the slits and the screen, but the truth is that the energies involved are so small that the electron cannot penetrate the solenoid, whether this is switched on or not. See also section 3.1.2.

The presence of the hole renders the configuration space topologically non-trivial or, in other words, not simply connected. The infinitely many curves surrounding the hole are equivalent in the sense that they can be deformed into each other continuously, but they cannot become zero. We say that the functions representing these curves are *homotopic* – i.e. map preserving – and they belong to a group called the *fundamental group* or *first homotopy group*. The functions describing the curves have parameters that take values from the interval $[0, 1]$. Hence this space, call it X, topologically corresponds to the direct product of the line $\mathbb{R}^1$ and the circle S^1, namely $\mathbb{R}^1 \times S^1$.

The electromagnetic field that is involved in the origination of the A–B effect is a physical entity that is described using the $U(1)$ group G, and the topology associated with this group is also that of the circle S^1. A fibre bundle is generated by the base manifold and the group, and its structure is as has been described above. The connection in this fibre bundle is the field A_μ[23] and the electromagnetic field is represented by the four-dimensional curl of A_μ which is also known as the curvature. Given that the actual magnetic field, or curvature, is zero everywhere on the manifold, we are talking about the vacuum here, where the curvature is zero, but the connection is not necessarily so.

The gauge function χ may be considered as a mapping from the group space G onto the configuration space X, that is $\chi : G \rightarrow X$, whose non-trivial part is given by $\chi : S^1 \rightarrow S^1$. In the terminology we have introduced above, this means that the connection 1-form is pulled back to our base space. We have already said that in the fibre bundle formalism all the information about the topology of the base space is included in the structure of the bundle space and vice versa. Here we can see how this is realized in the A–B case, where the non-trivial topology of the base space is reflected by the non-trivial topology of the group that one uses in order to define the principal bundle.

3.1.2 Topological explanation (1): the manifold first

There are two variations of what we might consider as a topological explanation of the A–B effect. In the first, one notices that the difference in magnitude between the electron and the solenoid is of the order of 10^{10}. Given that the energies we are talking about are very low, this means that a very big chunk of space, 10 000 000 000 bigger than the electron itself, cannot be accessed by it. So, from the perspective of the electron, *it is as though* spacetime is topologically non-trivial where the solenoid is, and that might be considered as a very good approximation. Moreover, even from the point of view of the human observer, treating the space outside the solenoid as topologically non-trivial is not a far-fetched idea if one considers the

[23] The connection follows the general transformation rule $A^\mu \rightarrow A^\mu + \partial^\mu \chi$. Because in our case we are in the vacuum, we can write that $A^\mu = \partial^\mu \chi$.

limiting case where the solenoid is shrunk to a point. Such a point solenoid cannot be made to disappear completely and hence one has to accept that the spacetime manifold is not simply connected. Therefore, an explanation of the phenomenon can be provided by the non-trivial topology which, as we have seen, results in non-vanishing holonomies, and it might be appropriate since anything else – that is, the magnetic field and/or the gauge potentials – cannot adequately explain what is happening there. The fact that the phase of the matter field – defined on the associated bundle – changes as the field is parallel transported along the base manifold is a consequence of its non-trivial topology, and this explains it, while at the same time the holonomy allows for its measurement.

One important point to clarify here is that what is really important for the effect to happen is not just the material presence of the solenoid in the setup, for one then might claim that even when the solenoid is switched off the region inside it is still inaccessible to the electron and yet there is no A–B effect. What is crucial for the effect to happen is that the flux of the electromagnetic field inside modifies the spacetime around it, and this modification is encoded by the connection.

3.1.3 Topological explanation (2): the bundle first

The topological explanation of the effect may be given a different gloss. One may assert that it is not the presence of the solenoid that makes the topology of $\mathcal{M}$ non-trivial; rather, it is the topology of the bundle vacuum itself – and hence of the configuration vacuum – that is non-trivial, and as a consequence the phase of the electron field is shifted as it passes through (the vacuum in this context is defined as a region where the energy of the electromagnetic field is zero). The connection of the principal bundle – that is to say, the gauge field A_μ – describes how the phase shift occurs and it is not the causal agent responsible for the shift but an information bearer instead: it just contains all the information about how the matter fields should behave as they move along the spacetime manifold. The curvature of the total space is nothing other than the familiar electromagnetic field, which cannot be considered to be the causal agent responsible for the effect either. Instead, it may be regarded as a property of the spacetime points, conferred to them by the modified topology of the base manifold.

At this point it is worth making some brief remarks on the second and the third approach mentioned in section 1, above. The fourth way differs from what I have named the second approach to the A–B effect because here we do not need to rely on the reality or the locality of the gauge field. What matters in the first version of the topological explanation is the non-triviality of the base manifold which affects the bundle space in such a way that the phase of the matter field is shifted, while in the second version it is the non-triviality of the total space that is explanatory via the non-trivial topology of the vacuum (which results in a phase shift, accounting for the effect). Moreover, since we do not need to rely on the reality of the holonomy either, it differs from the third approach as well: we do not need to consider the holonomies as the fundamental causal entities because it is the topology, rather than the holonomy, which constrains and controls the effects on the physical objects. Hence in this approach we obtain a holistic causal picture where the ultimate 'cause' of the shift is the topology of the base manifold and therefore of the total space, as in the first variation, or the total space and therefore of the base manifold, as in the second. The modified topology endows the spacetime with some properties, which in turn affect the physical objects that move around in it. The importance of the fibre bundle formalism is that it provides a complete tool for the precise description of the phenomenon and for the calculation of quantities that are measurable.

There are several reasons why one may or may not like the approaches I have just presented. First of all, and before we actually assess the topological explanation, I would like to mention two possible objections to – or reasons for not liking – it that would persist even if the topological explanation turned out to be a *bona fide* explanation. The first one is that we give up completely the idea of ever getting a local causal account by making use of the mathematical entities – at least within this formalism. The second is that we also part with determinism in the sense that since up there, in the bundle space, we have more entities than down here, such that there are infinitely many gauge fields corresponding to one electromagnetic field; hence starting from well-defined initial conditions, we may end up in one out of infinitely many possible final states of the total space. But then it seems that these are characteristics inherent to the way physical objects are represented by gauge theories and the account I offer does justice to this; moreover, this approach gives us a rich way of understanding what is going on.

4 Assessing the topological explanation

The topological explanation offered in the previous section of this paper belongs to a category of its own kind because, despite the fact that one could claim that topological explanations are deductive-nomological, or are merely holistic or non-local explanations that follow from a unification model, in fact there are some different and some additional elements to them.

The topological explanation is not a deductive-nomological explanation because although in it one uses a mathematical structure derivable from a unification model, the entities of the structure represent the physical entities involved along with a whole lot of entities that belong to the surplus structure. This surplus structure is indispensable because without it one does not get a satisfactory explanation and, as it stands, it minimally encodes all the information of the entire system, albeit using those entities that may not correspond directly to physical ones. This makes the assignment of truth value to statements that include these entities problematic; hence it poses difficulties for the deductive-nomological approach.

Given that in it one takes into account the entire physical system rather than what one might consider to be the 'causally relevant' constituent parts of it, the topological explanation is holistic indeed. However, there is the additional element that the notion of topology plays a crucial explanatory role. It is because of the non-trivial topology, not just because we study the physical system in its entirety, that we get an explanation of the effect at all, hence the preference for the term topological.

At the same time, although the topological explanation of the A–B effect encompasses its non-separable nature, it is not non-local because in it the actual topology is described locally and there is no kind of action at a distance involved. Finally, the fields that could be held responsible for non-locality do not play an actively causal role in the explanation, as I have argued.

Distinct as it is, the topological explanation offers more than all the other attempts for an explanation of the effect. The first approach to the effect gives a very unsatisfactory and insufficient account of it, an account that yields to action at a distance, as we have seen. Compared to this first attempt, all other approaches are definitely more valuable as explanations. A comparison of the topological explanation with the other two approaches presents more challenging tests. The flaw that turned the second approach into an unsatisfactory explanation of the effect was the fact that the gauge potential could not be interpreted as a locally acting and physically real causal agent. The topological explanation manages to overcome this flaw by considering the potential to be the bearer of all the information needed in order to describe the global topology and hence the results of a change in it. As the A_μ field does not have the status of a causal agent any more, there is no non-separability as in the second approach and therefore no need to remedy it. In the third approach it was considered that it was the holonomy that brings about the shift in the phase of the electron. But it is hard to imagine what kind of a physical, causally interacting with the electrons, entity that integral might represent. After all, it can be seen from equation (2) that when one uses Stokes' theorem the integral *I(C)* yields

$$I(C) = \oint_{\partial s} \mathbf{A} \cdot d\ell = \int_s \nabla \times \mathbf{A} \cdot ds = \textit{flux of magnetic field}. \tag{5}$$

In other words, the holonomies are just a measure of the effect that provides a good indication of its holistic character but which offers no explanation of it. On the other hand, the very fact of a non-vanishing holonomy is explained by the topological explanation: non-trivial topology entails non-vanishing holonomies.

The fibre bundle formalism of the theory, with the mappings and the pull-back operations defined in it, makes it easy, at least mathematically, to go either way: one can calculate the value of the holonomy if one knows the value of the magnetic field and then work out the connection or, conversely, one can begin with the gauge group and the connection and then figure out what the non-vanishing holonomy is. But what truly does the explaining here is the non-trivial topology, although in physics we started from the measurable, non-zero flux of the magnetic field. Furthermore, not only does it explain the effect itself, but it also sheds light on the role played by the entities involved in it, whether they are physical or not.

As a result of all this discussion, what we are left with is an explanation that is superior to both the second attempt, since it does not try to turn the connection into a real interacting field, and the third attempt, since it goes beyond the evidence and explains where the evidence stems from. However, the fourth way needs to address the following question. Notwithstanding the hat trick of gauge theories in their fibre bundle formulation, namely that they provide all the information which may reveal possible causal links between the physical objects involved, that they predict behaviours as well as measurable quantities, and that they effectively probed theoretical and experimental physics, one is more than justified to ask: does the claim that the topology is non-trivial provide a *deep* explanation? The answer is *no*, at least so far as the A–B effect is concerned, for, to begin with, the topological claims in the attempted explanations of the A–B effect are not true, strictly speaking. Here are the reasons why.

4.1 Assessment of topological explanation (1)

So far as our first attempt is concerned, there is a crucial disparity between the alleged approximate explanation of the A–B effect and what one would consider as legitimate approximate explanations in physics. We make use of a model that clearly involves what Hesse (1963) would call a negative analogy: namely, we consider that a spacetime manifold with a solenoid in it is non-trivial. Moreover, we require that this very analogy causally explains the physical events, hence its non-inclusion would undermine even the positive analogies between the model and the physical system. From this perspective, then, tempting though the approximation may be, it does not constitute a legitimate explanation. However, the fact that the holonomies are non-vanishing, along with the fact that the fibre bundle formalism explains the origin of these holonomies, provide a good indication that there may be something about gauge fields that points towards topological explanations. Let us try and see whether the second version of the topological explanation does any better.

4.2 Assessment of topological explanation (2)

The vacuum state that this interpretation of the topological explanation requires is a state where the electromagnetic field is zero. The fact that there is a solenoid with electromagnetic flux inside in some finite region of the spacetime manifold means that one could consider that the vacuum extends over the rest of spacetime except for the region occupied by the solenoid itself. But surely in this second attempt to provide a topological explanation, the alleged vacuum state is not really a vacuum because of the presence of the solenoid and the magnetic field inside it. Therefore things seem to be at least as bad as in the previous attempt: although the claim that there is a vacuum outside the solenoid is true, the problem now is that the vacuum in classical field theories is a global state of the field. This fact does not allow for any concessions because if the state were really a vacuum state, then the global vacuum would imply local vacua and this is certainly not the case in the region inside the solenoid. It seems, therefore, that once again our attempts to salvage the topological explanation of the A–B effect using approximations have failed.

The situation we encounter in the explanation of the A–B effect could be compared to the classical case of projectile motion.[24] In projectile motion, in order to explain the parabolic trajectories, one has to assume that the gravitational field strength g is constant throughout the path of the projectile and with direction perpendicular to the surface of the flat Earth. So, one considers the curvature of the Earth to be zero, locally, and hence one changes its global topology from that of a sphere to that of a plane. In both the A–B and the projectile cases, we have exchanged the actual topology of the physical system with a different one and we therefore use a negative analogy for explanatory purposes. At the same time, in the A–B case, as well as in the gravitational, it is not the change in the topology that provides the deep (that is to say the true causal) explanation for the phenomena, rather it the presence of the solenoid in the former and that of the gravitational field in the latter.

One may claim that there is a major difference between the two approaches: in the A–B case either there is or there is not a vacuum, while in the projectile motion case the change of topology may be thought of as just an approximation where the gravitational field lines are approximately parallel lines and the surface of the Earth is approximately a plane, therefore the trajectory is approximately part of a parabola. The argument goes, then, that in the case of projectile motion we just approximate the actual physical situation with some mathematical structure that does not essentially misrepresent it and this is because the negative analogy in this case does not causally affect essential properties of the system. The truth of the matter, though, is that the negative analogy does affect the essential property that the gravitational field strength is inversely proportional to r^2; the conclusion that

[24] This analogy was an idea of Professor M. Redhead, to whom I am grateful.

follows is that although we might consider a gravitational field with parallel lines near the surface of the Earth as a good approximation, the alleged change in topology fails to serve any explanatory purposes. In both cases, then, by using topological considerations one exceeds by far what one might consider as reasonable limits of approximation and idealization. Yet in both cases we get useful and fruitful – in an explanatory sense – insights about the relations between the physical objects involved in the processes, while from the formalism as a whole we get very good predictions about their future behaviour and certain measurable quantities.

Taking all the above into account, we are justified in saying that a topological explanation like the one we employed for the A–B effect misrepresents reality. For one reason, the base space manifold is trivial despite the presence of the solenoid in it and, for another, the vacuum is not really a vacuum because of the presence of the solenoid. However, this 'failure' of the non-trivial topology of the mathematical structure to 'explain' the physical events is not a sufficient reason to reject the theory or to undermine its *heuristic* power.

Moreover, the good news is that things take a different turning in relativistic quantum field theories because, as Redhead (1995a; 1995b) showed, the straightforward relation between the global and the local vacuum state that we mentioned above breaks down in there since a global vacuum state allows for non-vacuum local states due to quantum fluctuations. Of course here one has to make a leap, and starting from a classical discussion one draws conclusions about relativistic quantum objects, but one is justified in doing so because whatever we have discussed so far applies in the relativistic quantum case as well and because we are not really interested in what is going on in the classical cases only – these just provide a stepping stone. What we could say then about the topological explanation (2) of the A–B effect in the case of a relativistic vacuum, where a global vacuum state does not prevent observables that result from quantum fluctuations, is that since 'these vacuum fluctuations of local observables are a characteristic feature of the *relativistic* vacuum' (Redhead, 1995b, p. 81), one is justified in claiming that in the A–B case the state of the field is indeed a vacuum state despite the fact that locally it takes non-zero values. To take the old Aristotelian line of argument, one could claim here that the vacuum state of the relativistic quantum fields is not space(time) empty of objects. Rather, it is a field defined over spacetime that allows for either manifestation or not of observables, locally, due to its quantum fluctuations. Hence a vacuum state that is compatible with the presence of objects in it is reminiscent of Aristotle's wooden cube immersed in water, only in this case the water-field penetrates the cube-solenoid throughout its extent and so interpenetration and therefore coexistence become possible.[25]

[25] For detailed discussions about the Aristotelian notions of space and the vacuum and its relations to the modern notion of the vacuum, see Aristotle (1991), Jammer (1954), and Grant (1981).

4.3 Topological solutions

I feel compelled at this point to stress that the main aim of gauge theories is to describe elementary particles and fundamental interactions, both of which are quantum and relativistic physical entities, in a unified way, if possible, and to a great extent they have done so. In these attempts, topological considerations and non-trivial topologies are used as positive or neutral analogies and play a fundamental role in explaining as well as in probing the theories. Besides, the above reflections about the vacuum state of fields and the possibility of a base space with a non-trivial topology become unquestionably legitimate and worthwhile when one considers stable extended solutions to the Euler–Lagrange equations of motion of non-linear field theories. The Yang–Mills theories are non-linear and the topological solutions offered are well-defined topological objects with finite energy, which have the general name *solitons*; *monopoles* and *instantons* – or *pseudo-particles* – are soliton solutions too. Soliton solutions have been given serious thought by theoretical physicists over the past twenty-five years or so because they sidestep the problems of infinities and renormalization. The problem of infinities impairs quantum field theories that describe basic matter fields of nature as though they were point objects. However successful the point-particle theories may be, the quest for something more satisfactory, a theory that would not even have to be renormalizable, continues. The stability and finitarity of the topological solutions has been very promising, in terms of the explanations it provides, and alluring, so far as its heuristic powers are concerned. The only missing link in the dialectic relation between physics, mathematics, and nature is experimental verification – or falsification – that is eagerly awaited and related to the discovery – or not – of the Yang–Mills–Higgs monopole. Then again, research in theoretical physics seems to be sustained by intuition and an eagerness of the researchers to see how far the existing theories could reach, rather than by how much evidence there is for them. With this in mind, one has to admit that the combination of gauge theories, fibre bundles, and topological considerations have taken twentieth-century physics very far, despite the fact that there may be some explanatory and evidence-dependent loose ends.

Acknowledgements

The first two versions of this paper were presented in the research students' seminar at the London School of Economics and in the Symmetry Workshop 2000 in Oxford. I would like to thank the participants in both events for valuable comments and criticisms. I am indebted to Professor Isham, who introduced me to fibre bundles and patiently explained to me again and again the most difficult parts of this amazing formalism; to Professor Redhead who read the manuscripts and commented in detail on them, sharing with me his immense understanding and knowledge; and to Dr Hoefer who encouraged me to write a paper about fibre bundles without mathematics. Finally, I am grateful to Dr Brading and Dr Castellani for both organizing the Symmetry Workshop and carefully reading the final drafts of this paper and suggesting detailed corrections that motivated me to improve it.

References

Aharonov, Y., and Bohm, D. (1959). 'Significance of electromagnetic potentials in quantum theory'. *Physical Review*, **115**, 485–91.

Aristotle (1991). *Physics*, ed. L. Judson. Oxford: Clarendon Press.

Auyang, S. Y. (1995). *How is Quantum Field Theory Possible?* Oxford: Oxford University Press.

Belot, G. (1996). 'Whatever is never and nowhere is not'. Ph.D. dissertation, University of Pittsburgh.

(1998). 'Understanding electromagnetism'. *British Journal for the Philosophy of Science*, **49**, 532–55.

Grant, E. (1981). *Much Ado About Nothing.* Cambridge: Cambridge University Press.

Gribov, V. N. (1978). 'Quantization of non-Abelian theories'. *Nuclear Physics B*, **139**, 1–19.

Healey, R. (1997). 'Non-locality and the Aharonov–Bohm effect'. *Philosophy of Science*, **64**, 18–41.

(1999). 'Quantum analogies: a reply to Maudlin'. *Philosophy of Science*, **25**, 440–7.

(2001). 'On the reality of gauge potentials'. *Philosophy of Science*, **84**, 432.

Hesse, M. (1963). *Models and Analogies in Physics*. London: Sheed and Ward.

Isham, C. J. (1999). *Modern Differential Geometry for Physicists*. Singapore: World Scientific.

Jammer, M. (1954). *Concepts of Space*. Cambridge, MA: Harvard University Press.

Leeds, S. (1999). 'Gauges: Aharonov, Bohm, Yang, Healey'. *Philosophy of Science*, **66**, 607–27.

Maudlin, T. (1998). 'Discussion: Healey and Aharonov–Bohm'. *Philosophy of Science*, **65**, 361–8.

Nakahara, M. (1990). *Geometry, Topology and Physics*. Bristol: Institute of Physics Publishing.

Nash, C., and Sen, S. (1983). *Topology and Geometry for Physicists*. New York: Academic.

Redhead, M. (1995a). 'More ado about nothing'. *Foundations of Physics*, **4**, 1443–7.

(1995b). 'The vacuum in relativistic quantum field theory'. In *Philosophy of Science Association 1994*, vol. 2, ed. D. Hull, M. Forbes, and R. Burian, pp. 77–87. East Lansing: Philosophy of Science Association.

(2001). 'The intelligibility of the universe'. In *Philosophy at the New Millennium*, ed. A. O'Hear. Cambridge: Cambridge University Press.

Ryder, L. H. (1985). *Quantum Field Theory*. Cambridge: Cambridge University Press.

Singer, I. M. (1978). 'Some remarks on the Gribov ambiguity'. *Communications in Mathematical Physics*, **60**, 7–12.

Wu, T. T., and Yang, C. N. (1975). 'Concept of non-integrable phase factors and global formulation of gauge fields'. *Physical Review D*, **12**, 3845–57.

Part II

Discrete symmetries

11

Classic texts: extracts from Leibniz, Kant, and Black

The Leibniz–Clarke correspondence

GOTTFRIED W. LEIBNIZ

Leibniz's third letter to Clarke, paragraph 5

I have many demonstrations, to confute the fancy of those who take space to be a substance, or at least an absolute being. But I shall only use, at the present, one demonstration, which the author here gives me occasion to insist upon. I say then, that if space was an absolute being, there would something happen for which it would be impossible there should be a sufficient reason. Which is against my axiom. And I prove it thus. Space is something absolutely uniform; and, without the things placed in it, one point of space does not absolutely differ in any respect whatsoever from another point of space. Now from hence it follows, (supposing space to be something in itself, besides the order of bodies among themselves,) that 'tis impossible there should be a reason, why God, preserving the same situations of bodies among themselves, should have placed them in space after one certain particular manner, and not otherwise; why every thing was not placed the quite contrary way, for instance, by changing East into West. But if space is nothing else, but that order or relation; and is nothing at all without bodies, but the possibility of placing them; then those two states, the one such as it now is, the other supposed to be the quite contrary way, would not at all differ from one another. Their difference therefore is only to be found in our chimerical supposition of the reality of space in itself. But in truth the one would exactly be the same thing as the other, they being absolutely indiscernible; and consequently there is no room to enquire after a reason of the preference of the one to the other.

Leibniz's third letter to Clarke, paragraph 6

The case is the same with respect to time. Supposing anyone should ask, why God did not create every thing a year sooner; and the same person should infer from

Note. Extract from: 1956, Manchester: Manchester University Press.

thence, that God has done something, concerning which 'tis not possible there should be a reason, why he did it so, and not otherwise: the answer is, that his inference would be right, if time was any thing distinct from things existing in time. For it would be impossible there should be any reason, why things should be applied to such particular instants, rather than to others, their succession continuing the same. But then the same argument proves, that instants, consider'd without the things, are nothing at all; and that they consist only in the successive order of things: which order remaining the same, one of the two states, viz. that of a supposed anticipation, would not at all differ, nor could be discerned from, the other which now is.

Concerning the ultimate ground of the differentiation of directions in space

IMMANUEL KANT

What we are trying to demonstrate, then, is the following claim. The ground of the complete determination of a corporeal form does not depend simply on the relation and position of its parts to each other; it also depends on the reference of that physical form to universal absolute space, as it is conceived by the geometers. This relation to absolute space, however, cannot itself be immediately perceived, though the differences, which exist between bodies and which depend exclusively on this ground alone, can be immediately perceived. If two figures drawn on a plane surface are equal and similar, then they will coincide with each other. But the situation is often entirely different when one is dealing with corporeal extension, or even with lines and surfaces, not lying on a plane surface. They can be exactly equal and similar, and yet still be so different in themselves that the limits of the one cannot also be the limits of the other. The thread of a screw which winds round its pin from left to right will never fit a nut of which the thread runs from right to left. Even if the size of the screw is the same as the size of the nut, and even if the number of times which the thread winds round the pin of the screw is the same as the number of times which the thread winds round the inside of the nut, the nut and the screw will never match each other. A spherical triangle can be exactly equal and similar to another such triangle, and yet still not coincide with it. But the most common and clearest example is furnished by the limbs of the human body, which are symmetrically arranged relative to the vertical plane of the body. The right hand

Note. Extract from: *The Cambridge Edition of the Works of Immanuel Kant: Theoretical Philosophy, 1755–1770*, trans. and ed. D. Walford and R. Meerbrote, 1992, Cambridge: Cambridge University Press, pp. 365–72.

is similar and equal to the left hand. And if one looks at one of them on its own, examining the proportion and the position of its parts to each other, and scrutinising the magnitude of the whole, then a complete description of the one must apply in all respects to the other, as well.

I shall call a body which is exactly equal and similar to another, but which cannot be enclosed in the same limits as that other, its *incongruent counterpart.* Now, in order to demonstrate the possibility of such a thing, let a body be taken consisting, not of two halves which are symmetrically arranged relatively to a single intersecting plane, but rather, say, a *human hand.* From all the points on its surface let perpendicular lines be extended to a plane surface set up opposite to it; and let these lines be extended the same distance behind the plane surface, as the points on the surface of the hand are in front of it; the ends of the lines, thus extended, constitute, when connected together, the surface of a corporeal form. That form is the incongruent counterpart of the first. In other words, if the hand in question is a right hand, then its counterpart is a left hand. The reflection of an object in a mirror rests upon exactly the same principles. For the object always appears as far behind the mirror as it is in front of it. Hence, the image of a right hand in a mirror is always a left hand. If the object itself consists of two incongruent counterparts, as the human body does if it is divided by means of a vertical intersection running from front to back, then its image is congruent with that object. That this is the case can easily be recognised if one imagines the body making half a rotation; for the counterpart of the counterpart of an object is necessarily congruent with that object.

Let that suffice to explain the possibility of spaces which are perfectly similar and equal and yet incongruent. Let us now proceed to the philosophical application of these concepts. It is apparent from the ordinary example of the two hands that the shape of the one body may be perfectly similar to the shape of the other, and the magnitudes of their extensions may be exactly equal, and yet there may remain an inner difference between the two, this difference consisting in the fact, namely, that the surface which encloses the one cannot possibly enclose the other. Since the surface which limits the physical space of the one body cannot serve as a boundary to limit the other, no matter how that surface be twisted and turned, it follows that the difference must be one which rests upon an inner ground! This inner ground cannot, however, depend on the difference of the manner in which the parts of the body are combined with each other. For, as we have seen from our example, everything may in this respect be exactly the same. Nonetheless, imagine that the first created thing was a human hand. That human hand would have to be either a right hand or a left hand. The action of the creative cause in producing the one would have of necessity to be different from the action of the creative cause producing the counterpart.

Suppose that one were to adopt the concept entertained by many modern philosophers, especially German philosophers, according to which space simply consists in the external relation of the parts of matter which exist alongside each other. It would follow, in the example we have adduced, that all actual space would simply be *the space occupied by this hand.* However, there is no difference in the relation of the parts of the hand to each other, and that is so whether it be a right hand or a left hand; it would therefore follow that the hand would be completely indeterminate in respect of such a property. In other words, the hand would fit equally well on either side of the human body; but that is impossible.

Our considerations make it plain that the determinations of space are not consequences of the positions of the parts of matter relative to each other. On the contrary, the latter are the consequences of the former. Our considerations, therefore, make it clear that differences, and true differences at that, can be found in the constitution of bodies; these differences relate exclusively to *absolute* and *original space,* for it is only in virtue of absolute and original space that the relation of physical things to each other is possible. Finally, our considerations make the following point clear: absolute space is not an object of outer sensation; it is rather a fundamental concept which first of all makes possible all such outer sensation. For this reason, there is only one way in which we can perceive that which, in the form of a body, exclusively involves reference to pure space, and that is by holding one body against other bodies.

The identity of indiscernibles

MAX BLACK

A. Here is a different argument that seems to me quite conclusive. The only way we can discover that two different things exist is by finding out that one has a quality not possessed by the other or else that one has a relational characteristic that the other hasn't.

If *both* are blue and hard and sweet and so on, and have the same shape and dimensions and are in the same relations to everything in the universe, it is logically impossible to tell them apart. The supposition that in such a case there might really be two things would be unverifiable *in principle.* Hence it would be meaningless.

Note. Extract from: 1952, *Mind*, **LXI** (242), 153–64.

B. You are going too fast for me.

A. Think of it this way. If the principle were false, the fact that I can see only two of your hands would be no proof that you had just two. And even if every conceivable test agreed with the supposition that you had two hands, you might all the time have three, four, or any number. You might have nine hands, different from one another and all indistinguishable from your left hand, and nine more all different from each other but indistinguishable from your right hand. And even if you really did have just two hands, and no more, neither you nor I nor anybody else could ever know that fact. This is too much for me to swallow. This is the kind of absurdity you get into, as soon as you abandon verifiability as a test of meaning.

B. Far be it from me to abandon your sacred cow. Before I give you a direct answer, let me try to describe a counterexample.

Isn't it logically possible that the universe should have contained nothing but two exactly similar spheres? We might suppose that each was made of chemically pure iron, had a diameter of one mile, that they had the same temperature, colour, and so on, and that nothing else existed. Then every quality and relational characteristic of the one would also be a property of the other. Now if what I am describing is logically possible, it is not impossible for two things to have all their properties in common. This seems to me to *refute* the Principle.

A. Your supposition, I repeat, isn't verifiable and therefore can't be regarded as meaningful. But supposing you *have* described a possible world, I still don't see that you have refuted the principle. Consider one of the spheres, *a*,...

B. How can I, since there is no way of telling them apart ?
Which one do you want me to consider?

A. This is very foolish. I mean either of the two spheres, leaving you to decide which one you wished to consider. If I were to say to you 'Take any book off the shelf' it would be foolish on your part to reply 'Which?'

B. It's a poor analogy. I know how to take a book off a shelf, but I don't know how to identify one of two spheres supposed to be alone in space and so symmetrically placed with respect to each other that neither has any quality or character the other does not also have.

A. All of which goes to show as I said before, the unverifiability of your supposition. Can't you imagine that one sphere has been designated as '*a*'?

B. I can imagine only what is logically possible. Now it is logically possible that somebody should enter the universe I have described, see one of the spheres on his left hand and proceed to call it '*a*'. I can imagine that all right, if that's enough to satisfy you.

A. Very well, now let me try to finish what I began to say about *a* . . .

B. I still can't let you, because you, in your present situation, have no right to talk about *a.* All I have conceded is that if something were to happen to introduce a change into my universe, so that an observer entered and could see the two spheres, one of them could then have a name. But this would be a different supposition from the one I wanted to consider. My spheres don't yet have names. If an observer were to enter the scene, he could perhaps put a red mark on one of the spheres. You might just as well say 'By "*a*" I mean the sphere which would be the first to be marked by a red mark if anyone were to arrive and were to proceed to make a red mark!' You might just as well ask me to consider the first daisy in my lawn that would be picked by a child, if a child were to come along and do the picking. This doesn't now distinguish any daisy from the others. You are just pretending to use a name.

A. And I think you are just pretending not to understand me. All I am asking you to do is to think of one of your spheres, no matter which, so that I may go on to say something about it when you give me a chance.

B. You talk as if naming an object and then thinking about it were the easiest thing in the world. But it isn't so easy. Suppose I tell you to name any spider in my garden: if you can catch one first or describe one uniquely you can name it easily enough. But you can't pick one out, let alone 'name' it, by just thinking. You remind me of the mathematicians who thought that talking about an Axiom of Choice would really allow them to choose a single member of a collection when they had no criterion of choice.

A. At this rate you will never give me a chance to say anything. Let me try to make my point without using names. Each of the spheres will surely differ from the other in being at some distance from that other one, but at no distance from itself – that is to say, it will bear at least one relation to itself – *being at no distance from,* or *being in the same place as* – that it does not bear to the other. And this will serve to distinguish it from the other.

B. Not at all. *Each* will have the relational characteristic *being at a distance of two miles*, say, *from the centre of a sphere one mile in diameter*, etc. And each will have the relational characteristic (if you want to call it that) of *being in the same place as itself.* The two are alike in this respect as in all others.

A. But look here. Each sphere occupies a different place; and this at least will distinguish them from one another.

B. This sounds as if you thought the places had some independent existence, though I don't suppose you really think so. To say the spheres are in 'different places' is just to say that there is a distance between the two spheres; and we have already seen that will not serve to distinguish them. Each is at a distance – indeed the same distance – from the other.

A. When I said they were at different places I didn't mean simply that they were at a distance from one another. That one sphere is in a certain place does not entail the existence of any *other* sphere. So to say that one sphere is in its place, and the other in its place, and then to add that these places are different seems to me different from saying the spheres are at a distance from one another.

B. What does it mean to say 'a sphere is in its place'? Nothing at all, so far as I can see. Where else could it be? *All* you are saying is that the spheres are in different places.

A. Then my retort is, What does it mean to say 'Two spheres are in different places'? Or, as you so neatly put it, 'Where else could they be?'

B. You have a point. What I should have said was that your assertion that the spheres occupied different places said nothing at all, unless you were drawing attention to the necessary truth that different physical objects must be in different places. Now if two spheres must be in different places, as indeed they must, to say that the spheres occupy different places is to say no more than they are two spheres.

A. This is like a point you made before. You won't allow me to deduce anything from the supposition that there are two spheres.

B. Let me put it another way. In the two-sphere universe, the only reason for saying that the places occupied were different would be that different things occupied them. So in order to show the places were different you would first have to show, in some other way, that the spheres were different. You will never be able to distinguish the spheres by means of the places they occupy....

You are assuming that in order to verify that there are two things of a certain kind, it must be possible to show that one has a property not possessed by the other. But this is not so. A pair of very close but similar magnetic poles produce a characteristic field of force which assures me that there are two poles, even if I have no way of examining them separately. The presence of two exactly similar stars at a great distance might be detected by some resultant gravitational effect or by optical interference –

or in some such similar way – even though we had no way of inspecting one in isolation from the other. Don't physicists say something like this about the electrons inside an atom? We can verify *that* there are two, that is to say a certain property of the whole configuration, even though there is no way of detecting any character that uniquely characterises any element of the configuration.

A. But if you were to approach your two stars one would have to be on your left and one on the right. And this would distinguish them.

B. I agree. Why shouldn't we say that the two stars are distinguishable – meaning it would be possible for an observer to see one on his left and the other on his right, or more generally, that it would be *possible* for one star to come to have a relation to a third object that the second star would not have to that third object.

A. So you agree with me after all.

B. Not if you mean that the two stars do not have all their properties in common. All I said was that it was logically possible for them to enter into different relationships with a third object. But this would be a change in the universe.

A. If you are right, nothing unobserved would be observable. For the presence of an observer would always change it, and the observation would always be an observation of something else.

B. I don't say that every observation changes what is observed. My point is that there isn't any *being to the right* or *being to the left* in the two-sphere universe until an observer is introduced, that is to say until a real change is made.

A. But the spheres themselves wouldn't have changed.

B. Indeed they would: they would have acquired new relational characteristics. In the absence of any asymmetric observer, I repeat, the spheres would have all their properties in common (including, if you like, the power to enter into different relations with other objects). Hence the principle of Identity of Indiscernibles is false.

A. So perhaps you really do have twenty hands after all?

B. Not a bit of it. Nothing that I have said prevents me from holding that we can verify *that* there are exactly two. But we could know *that* two things existed without there being any way to distinguish one from the other. The Principle is false.

A. I am not surprised that you ended in this way, since you assumed it in the description of your fantastic 'universe'. Of course, if you began by assuming that the spheres were numerically different though qualitatively alike, you could end by 'proving' what you first assumed.

B. But I wasn't 'proving' anything. I tried to support my contention that it is logically possible for two things to have all their properties in common by giving an illustrative description. (Similarly, if I had to show it is logically possible for nothing at all to be seen I would ask you to imagine a universe in which everybody was blind.) It was for you to show that my description concealed some hidden contradiction. And you haven't done so.

A. All the same I am not convinced.

B. Well, then, you ought to be.

12

Understanding permutation symmetry

STEVEN FRENCH AND DEAN RICKLES

> If a system in atomic physics contains a number of particles of the same kind, e.g. a number of electrons, the particles are absolutely indistinguishable one from another. No observable change is made when two of them are interchanged . . . A satisfactory theory ought, of course, to count any two observationally indistinguishable states as the same state and to deny that any transition does occur when two similar particles exchange places.
>
> (*Dirac, 1958, p. 207*)

1 Introduction

In our contribution to this volume we deal with *discrete* symmetries: these are symmetries based upon groups with a discrete set of elements (generally a set of elements that can be enumerated by the positive integers). In physics we find that discrete symmetries frequently arise as 'internal', non-spacetime symmetries. Permutation symmetry is such a discrete symmetry, arising as the mathematical basis underlying the statistical behaviour of ensembles of certain types of indistinguishable quantum particle (e.g. fermions and bosons). Roughly speaking, if such an ensemble is invariant under a permutation of its constituent particles (i.e. permutation symmetric) then one doesn't 'count' those permutations which merely 'exchange' indistinguishable particles; rather, the exchanged state is identified with the original state.

This principle of invariance is generally called the 'indistinguishability postulate' (IP), but we prefer to use the term 'permutation invariance' (PI). It is this symmetry principle that is typically taken to underpin and explain the nature of (fermionic and bosonic) quantum statistics (although, as we shall see, this characterization is not uncontentious), and it is this principle that has important consequences regarding the metaphysics of identity and individuality for particles exhibiting such statistical behaviour.

In this paper we will largely be dealing with the following two types of problem.

How are we to understand the metaphysics of PI?

For instance, do we follow the 'Received View' and say that permutation invariance shows us that quantum particles are not individuals? Do we maintain that they are individuated by their spatiotemporal location, or perhaps by some extra-theoretical property (e.g. the 'primitive thisness' of the object)? Given this individuation how are we to understand PI? Maybe we can resolve the issue in some completely different way, with 'structures' replacing 'objects' perhaps? It is clear that such questions readily relate to 'traditional' metaphysical issues connected to identity and individuality.

How are we to understand the status of PI, theoretically and empirically?

For example, should PI be considered as an axiom of quantum mechanics? Or should it be taken as justified empirically? Why do there appear to be only bosons and fermions in the world when PI allows the possibility of many more types? This is usually resolved by postulating, *ad hoc*, some 'superselection rule', called the 'symmetrization postulate' (SP), restricting the state vector to the fermionic and bosonic subspaces of the systems' Hilbert space. However, rather than resolving the difficulty, this simply moves the explanatory task one step backwards (i.e. how are we then to understand SP?). Alternatively, the extra, possibly redundant, mathematical structure responsible for the extra possibilities regarding symmetry types of particles might be understood as 'surplus structure' (in the sense of Redhead, 1975). One often finds such surplus structure in theories possessing lots of symmetry, and it frequently points to the existence of 'gauge freedom' in a theory (e.g. in general relativity, Yang–Mills theory, and electromagnetism). It is here that a possible relation of permutation invariance to diffeomorphism invariance (the symmetry underlying the general covariance of general relativity) becomes apparent.

In this paper we survey a number of these issues and their consequences, introducing the reader to the various schools of thought regarding the status and interpretation of PI (and, likewise, though to a lesser extent, SP). Let us begin with a brief introduction to the formal aspects of PI and relevant related topics in group theory and classical/quantum statistical mechanics.

2 The mathematics and physics of permutation symmetry

Permutation symmetry is a discrete symmetry supported by the permutation group $Perm(\mathcal{X})$ of bijective maps (the permutation operators, $\hat{P}$) of a set $\mathcal{X}$ onto itself.[1] When $\mathcal{X}$ is of finite dimension $Perm(\mathcal{X})$ is known as the symmetric group

[1] The fact that the set $Perm(\mathcal{X})$ has the structure of a group simply means that: (i) we can combine any two elements ($\hat{P}_1, \hat{P}_2 \in Perm(\mathcal{X})$) in the set to produce another element ($\hat{P}_3 = \hat{P}_1 \cdot \hat{P}_2$) that is also contained within that set ($\hat{P}_3 \in Perm(\mathcal{X})$); and (ii) each element $\hat{P} \in Perm(\mathcal{X})$ also has an inverse, $\hat{P}^{-1} \in Perm(\mathcal{X})$.

S_n (where n refers to the dimension of the group). For instance, $\mathcal{X}$ might be the set consisting of the labels of the two sides of a coin: heads 'H' and tails 'T'; or perhaps the 'names' of n particles making up some quantum mechanical system, an He^4 atom for example. If we take the coin as our example, then $\mathcal{X} = \{H, T\}$ and $Perm(\mathcal{X})$ is an order two group, S_2, consisting of two elements (computed as having 2! elements via the dimension, $n = 2$, of the group): (i) the identity map, $id_{\mathcal{X}}$, which maps H to H and T to T; and (ii) the 'flip' map (or 'exchange' operator), $\hat{P}_{HT}$, which maps H to T and T to H.

Now, to say that some object (i.e. a set or the total state vector of a system of particles) is 'permutation symmetric' means that it is invariant under the action of $Perm(\mathcal{X})$: it remains unchanged (in some relevant sense) when it is operated upon by the elements (i.e. the permutation operators) of $Perm(\mathcal{X})$, including (for $n \geq 2$) the elements that 'exchange' the components of the object (in this case the labels of the sides of the coin or the labels of the particles in a quantum system).

The coin clearly is not permutation symmetric (i.e. does not satisfy PI), since we must distinguish 'heads' from 'tails'; that is, there is an *observable* difference between these two states of the coin. However, when we consider systems containing several indistinguishable particles,[2] each with several possible states (particles such as electrons, neutrons, and photons), we find that they are indeed permutation symmetric, and that this symmetry 'shrinks' the number of possible states of the total system, thus altering the statistical behaviour of the ensemble. In this way PI is generally taken to *explain* the divergence of quantum statistics from classical statistics.[3]

To see how these 'altered statistics' follow from PI, and what they look like, let us compare classical and (bosonic/fermionic) quantum statistics using a simple example.

Consider the distribution of a system of n indistinguishable objects (e.g. free particles) over m microstates. It is helpful at this stage to view the objects as balls and the microstates as the two halves of a box (making each side big enough to accommodate all n balls).[4] Statistical mechanics is, very loosely, the study of the number of ways one can redistribute the objects over the microstates without altering the macrostate. Let us consider the simple case where we have two objects

[2] Particles are said to be indistinguishable in that they possess the same state-independent (intrinsic) properties, such as rest mass, charge, and spin. Since these quantities have a continuous spectrum in classical mechanics we can still individuate particles by their variations with respect to these properties. If it were the case that we had a classical system containing particles that exactly matched in these properties, then we could still distinguish the subsystems by their spatiotemporal location. Such luxuries are not available in quantum theory because of discrete spectra and the absence of definite trajectories.

[3] Note, however, that this explanatory link has been contested by Huggett (1999a).

[4] The separators in the diagrams are there as an aid to visualization rather than as part of the system we are considering.

(balls) and two microstates (boxes). Let us label the balls ‘a’ and ‘b’, and the sides of the box by ‘L’ (left) and ‘R’ (right). Let ‘L(a)’ be the state where ball a is in the left-hand side (LHS) of the box; let ‘L(ab)’ be the state where both balls are in the LHS; and let ‘L(0)’ mean that the LHS is empty (similarly, *mutatis mutandis*, for the right-hand side (RHS) and ball b). Classically, we have four possible distributions:

L(a)	R(b)
L(b)	R(a)
L(ab)	R(0)
L(0)	R(ab)

Each possible permutation of the balls is counted in the statistics and, if we assume equiprobability, each configuration has a probability of $\frac{1}{4}$ of being realized. Such a distribution is known as a Maxwell–Boltzmann distribution, and it follows the corresponding statistics for such distributions.[5]

The situation is different when we consider quantum particles because, in addition to being indistinguishable, they are subject to PI. There are two types of statistical behaviour for particles in quantum mechanics having to do with the ways in which they can combine in ensembles.[6] Firstly, we have bosons (particles with integer spins, e.g. photons) behaving according to Bose–Einstein statistics: meaning, *inter alia*, that these particles can occupy the same state in a quantum system (the balls can reside in the same side of the box). Secondly, we have fermions (particles with half-integer spins, e.g. electrons) behaving according to Fermi–Dirac statistics: meaning, *inter alia*, that these particles *cannot* occupy the same state in a quantum system (the balls cannot reside in the same side of the box). This latter principle – not directly connected to PI – is generally known as Pauli’s Exclusion Principle.

These two points have an impact on the possible configurations we can count in the statistics. For instance, in the case of bosons we identify those configurations which differ only by an exchange of identical particles (i.e. the first and second configurations from the classical statistics above), but we can allow those configurations in which two objects occupy the same state. So if we consider the balls as bosons we get the following three distinct possibilities (where ‘L(1)’ means

[5] Huggett (1999a) has argued that Maxwell–Boltzmann statistics do not necessarily imply that we must count permutations as distinct: when there are many states available to each particle the rule breaks down. In the case of the present example we are dealing with many particles per state, and so the relation between Maxwell–Boltzmann statistics and counting permutations as distinct still holds.

[6] Of course, we are, for the moment, ignoring the case of para-statistics; namely, types of quantum statistics that violate SP, on which see sections 3 and 4.

that 'some' particle is in the left-hand side – similarly, *mutatis mutandis* for the right-hand side):

L(1)	R(1)
L(ab)	R(0)
L(0)	R(ab)

So we have removed a classically possible state by identifying 'exchanged states'.[7] This has the consequence that the probabilities for finding a system in a certain state (still assuming equiprobability) each go from 1/4 to 1/3. Following a similar procedure with fermions, and then applying the Exclusion Principle, we get just one possible state:

L(1)	R(1)

– which, of course, has a probability of 1 of being realized. All we have done here is to identify those configurations which differ only in which ball occupies which side of the box (following Dirac's intuition expressed in the opening quote) and then we have forbidden two balls to occupy the same side. In both the quantum cases the systems (or, more formally, their state vectors) are invariant under the action of the permutation group: when we apply the permutation operators to the state vectors they continue to describe the same physical state; following Dirac's intuition we identify the states. Hence, the quantum systems satisfy PI, unlike the classical system. Let us now make some of these ideas more exact by introducing some elementary quantum theory.

States of quantum systems (single- or many-particle) are represented by rays Ψ in a Hilbert space $\mathcal{H}$. For many-particle systems the Hilbert space is the joint space constructed by tensoring together the component particles' Hilbert spaces. The observables $\hat{O}$ of a quantum system are represented by Hermitian operators acting upon that system's Hilbert space.

Now consider a system consisting of two indistinguishable particles. The Hilbert space for this system is: $\mathcal{H}_{total} = \mathcal{H}_1 \otimes \mathcal{H}_2$, where the subscripts '1' and '2' label the composite particles, and $\mathcal{H}_1 = \mathcal{H}_2 = \mathcal{H}$. If the particles are in the pure states ϕ and ψ respectively, then the composite system is in the (pure) state $\Psi = \phi \otimes \psi$.

[7] Note that we have simplified the first configuration here, since what we actually have, formally, is the state: [(L(a) and R(b)) + (L(b) and R(a))]. In the fermionic case we find a similar superposition only with a change in sign (when the permutations are odd): [(L(a) and R(b)) − (L(b) and R(a))]. Note that the change of sign has no effect on the observable properties (expectation values) of the system.

The permutation operators act upon Ψ as follows: (i) $\hat{P}_{id}(\Psi) = (\phi \otimes \psi)$ and (ii) $\hat{P}_{\phi\psi}(\Psi) = (\psi \otimes \phi)$.

The Hamiltonian, $\hat{H}_\Psi = \hat{H}(\phi \otimes \psi)$, of the composite system is symmetric with respect to ϕ and ψ. Hence, $\hat{H}_\Psi$ is invariant under the action of the permutation group of permutations of the composite particles' labels: $[\hat{H}, \hat{P}] = 0, \forall \hat{P}$. By an invariance of a quantum state under the action of the permutation group (i.e. PI) we then mean that every physical observable $\hat{O}$ commutes with every permutation operator $\hat{P}$: $[\hat{O}, \hat{P}] = 0, \forall \hat{O} \forall \hat{P}$ – the physical interpretation of this being that there is no measurement that we could perform which would result in a discernible difference between permuted (final) and unpermuted (initial) states. This has the consequence that expectation values for unpermuted states are equal to expectation values for permutations of that state. Or, more formally, for any arbitrary state ψ, Hermitian operator $\hat{O}$, and permutation operator $\hat{P}$:

$$\langle \psi \mid \hat{O} \mid \psi \rangle = \langle \hat{P}\psi \mid \hat{O} \mid \hat{P}\psi \rangle = \langle \psi \mid \hat{P}^{-1}\hat{O}\hat{P} \mid \psi \rangle \tag{1}$$

It is this result – basically a formal expression of PI – which motivates the claim that PI can be understood as a restriction on the possible observables of a system given its state and, as such, it can be viewed as a superselection rule determining which observables are physically relevant. We shall return to this claim and, more generally, the status of PI in later sections.

Finally, let us turn to the mathematical representation of particle types. For this we need the concept of an 'irreducible representation'. Firstly, a representation ρ of a group G on a linear space V is simply a map that assigns to each element of the group $g \in G$ a linear operator $\hat{O}(V)$ on the space. When the linear space is the (joint) Hilbert space $\mathcal{H}$ spanned by the states, $\{\phi \otimes \phi, \phi \otimes \psi, \psi \otimes \phi, \psi \otimes \psi\}$, of two indistinguishable particles, and the group is the permutation group, the representation will associate a unitary operator acting on $\mathcal{H}$ (i.e. on the state vector $\Psi \in \mathcal{H}$) to each permutation operator $\hat{P} \in Perm(\mathcal{X})$. We can represent this schematically as follows (beginning with the group element, then the representation of that element, and finally the physical operation):[8]

(i) $\hat{P}_{\phi\psi}\Psi \Rightarrow \begin{pmatrix} 0 & 1 \\ 1 & 0 \end{pmatrix} \Psi \Rightarrow$ 'exchanging the particles';

(ii) $\hat{P}_{id}\Psi \Rightarrow \begin{pmatrix} 1 & 0 \\ 0 & 1 \end{pmatrix} \Psi \Rightarrow$ 'leaving them alone'.

[8] We should point out that this way of doing things is an oversimplification in the following respects: firstly, since the joint states lie in a four-dimensional Hilbert space they are represented by 4-vectors, but here we are assuming that they are 2-vectors. Also, the permutation operators should properly be 4×4 matrices; here we write them as 2×2 matrices. However, since nothing of import depends on this, we prefer to keep things simple in this way to facilitate understanding.

This pair of matrices gives a unitary representation of $Perm(\mathcal{X})$ on $\mathcal{H}$. The only matrix that commutes with both of them is the unit matrix or some scalar multiple of the unit matrix. Representations of this kind are said to be 'irreducible'. Alternatively, a representation is said to be irreducible if the only invariant subspaces it possesses are $\{0\}$ and $\mathcal{H}$ (i.e. the zero vector and the whole space) – where a subspace $\mathcal{H}'$ of $\mathcal{H}$ is invariant if $\Psi \in \mathcal{H}'$ implies $\rho(\hat{P})\Psi \in \mathcal{H}', \forall \hat{P}$.

We are interested in the irreducible representations of the permutation group because each such representation is 'carried' by an irreducible subspace of the Hilbert space, where each such subspace is invariant under the action of the permutation operators. Thus, the subspaces represent symmetry sectors corresponding to the possible types of permutation symmetry possessed by the particles whose state vectors lie in that subspace. In the case we are considering we find that the total Hilbert space is partitioned into two subspaces invariant under the permutation group: (i) a (three-dimensional) symmetric subspace (spanned by three vectors: $\{\phi \otimes \phi, \psi \otimes \phi, \phi \otimes \phi + \psi \otimes \phi\}$) corresponding to bosons; and (ii) a (one-dimensional) antisymmetric subspace (spanned by one vector: $\{\phi \otimes \phi - \psi \otimes \phi\}$) corresponding to fermions.[9] The symmetric subspace is quite clearly reducible, but the three subspaces spanning it are one-dimensional and, therefore, irreducible: they contain no permutation invariant proper subspaces. Hence, the irreducible representations correspond to *types* of particle.[10]

However, when we consider more than two particles (giving a non-Abelian permutation group) we find that we get more than the two symmetry types that we observe in nature. For instance, for three indistinguishable particles we have 3! irreducible representations of the permutation group in the joint Hilbert space. In addition to the standard symmetry types (bosons and fermions), we also obtain 'parabosons' and 'parafermions', transforming differently under the action of the permutation group, and leading to alternative kinds of statistical behaviour known as 'parastatistics'.[11]

Thus, with the framework we have built up so far we can see that there is more 'mathematical structure' than there is 'physical structure': nature has shown us (so far) that there are bosons and fermions, whilst the theory allows for particles with different symmetry types (with potentially observable differences). In order to overcome this problem Messiah (1962, p. 595) introduced a postulate, the 'symmetrization postulate' (SP), which served to restrict the possible particles to those

[9] These 'spanning' vectors correspond, of course, to the possible outcomes in the quantum statistics.

[10] The symmetry properties mentioned earlier mean also that symmetry type is conserved; that is, state vectors remain in one or the other subspace over time – once a boson (fermion) always a boson (fermion)! However, this rule breaks down in supersymmetric theories, since such theories possess a symmetry relating Fermi (matter) to Bose (force); however, we shall ignore this complication here.

[11] The statistical behaviour of these 'higher-dimensional' irreducible representations is best modelled by the 'braid group' rather than the 'permutation group'.

two classes that we have so far found the world to be grouped into. This postulate can be stated simply: states of identical particle systems must be either symmetrical or antisymmetrical under the action of permutation operators. We can now turn to the philosophical implications of these ideas.

3 The relationship between permutation invariance and the symmetrization postulate

Let us consider in a little more detail the relationship between SP and PI. The former is obviously a restriction on the states of the assembly. If the latter is likewise understood, then it is easy to see that SP is sufficient but not necessary for PI. Understood in this way, the fact that PI is implied by SP means it picks up indirect empirical support from the latter (assuming that all known particles are either bosons or fermions – an assumption which has been questioned (see the papers in Hilborn and Tino, 2000)).

However, as Greenberg and Messiah (1964) argued, PI should be interpreted, not as a restriction on the states, but rather as a restriction on the possible observables for the assembly. On this view, PI dictates that any permitted observable must commute with any permutation operator and this in turn implies that the observable must be a symmetric function of the particle labels. The difference between SP and PI can thus be expressed as follows: SP expresses a restriction on the states for all observables, Q; whereas PI expresses a restriction on the observables, Q, for all states. From this perspective what PI does is restrict the accessibility of certain states, such that once in a certain set of states, whether Bose–Einstein, Fermi–Dirac or parastatistical of a given order, the particles cannot move into a different set. However, the question as to its status now becomes acute. Before we discuss this question in more detail, we shall consider one further aspect of the formal representation of quantum statistics and PI.

4 Permutation invariance and the topological approach to particle identity

As is well known, Schrödinger's early attempt to give a broadly classical interpretation of the new quantum mechanics foundered on the point that the appropriate space for a many-particle wave function had to be multi-dimensional. Even then, as Einstein pointed out, use of the full configuration space formed by the N-fold Cartesian product of three-dimensional Euclidean space appeared to conflict with the new quantum statistics insofar as within this full space, configurations related by a particle permutation are regarded as distinct. The standard resolution of this problem, of course, is to move to the reduced quotient space formed by the action of the permutation group on the full configuration space, in which points corresponding to a permutation of the particles are identified, and then apply appropriate quantum conditions (see Leinaas and Myrheim, 1977). In this context PI is effectively coded into the topology of configuration space itself, and the different statistical types then correspond to different choices of boundary conditions on the wave function (see, for example, Bourdeau and Sorkin, 1992).

This reduced configuration space is not in general a smooth manifold since it possesses singular points where two or more particles coincide. This leads to two technical difficulties: firstly, it is not clear how one might define the relevant Hamiltonian at such singularities (Bourdeau and Sorkin, 1992, p. 687) and secondly, the existence of these singular points is not compatible with Fermi–Dirac statistics. The obvious, and now standard, solution is to simply remove from the configuration space the subcomplex consisting of all such coincidence points, yielding a smooth manifold. The relevant group for n particles is then the n-string braid group as we noted above, and the irreducible unitary representations of this group can be used to label the different statistics that are possible. Imbo, Shah Imbo, and Sudarshan (1990) have provided a definition of the 'statistical equivalence' of two such representations in terms of which they obtain not merely ordinary statistics, parastatistics and fractional or anyon statistics but more exotic forms which they call 'ambistatistics' and 'fractional ambistatistics'. The deployment of the braid group in this manner may appear to conflict with the suggestion that one of the advantages of the configuration space approach is that it actually excludes the possibility of non-standard statistics. This claim is based on the work of Leinaas and Myrheim (1977) which apparently demonstrates that for a space of dimension 3 or greater only the standard statistics are possible. The conclusion drawn is that 'the (anti-)symmetrization condition on the wave function is now seen to be related to the dimensionality of space, in contrast to the Messiah and Greenberg analysis wherein the (anti-)symmetrization condition receives the status of a postulate' (Brown, Sjöqvist, and Bacciagaluppi, 1999, p. 230).

It turns out, however, that this treatment assumes the standard quantization procedure which incorporates one-dimensional Hilbert spaces only. From the group-theoretic perspective this amounts to allowing only one-dimensional representations and so it should come as no surprise that para- and ambi-statistics cannot arise. Effectively what Leinaas and Myrhiem have done is to ignore the 'kinematical ambiguity' inherent in the quantization procedure which derives from the (mathematical) fact that the set of irreducible representations of the permutation group contains not just the trivial representation manifested above but also others corresponding to exotic statistics (Imbo *et al.*, 1990, pp. 103–4). In what follows we shall occasionally return to the topological approach to see if it can shed any light on the issue of the status of PI.

5 Permutation invariance and the metaphysics of individuality

5.1 The Received View: quantum particles as non-individuals

One well-known approach to this issue takes PI to be profoundly related to the peculiar metaphysical character of quantum particles, namely that they are 'identical' or 'non-individual', in some sense. Referring back to our illustration of the difference between classical and quantum statistics above, the argument for such a view goes like this: in classical Maxwell–Boltzmann statistics, a permutation of the particles is taken to give rise to a new, countable arrangement. Since the particles are indistinguishable, in the sense of possessing all intrinsic or state-independent properties in common (that is, properties such as rest mass, charge, spin, etc.), this generation of new arrangements must reflect something about the particles which goes beyond their intrinsic properties, something which allows us to treat them as distinct individuals. In the quantum case, whether Fermi–Dirac or Bose–Einstein, the distribution is permutation invariant and a permutation of the particles does not yield a new arrangement. Hence the statistical weight in quantum statistics – of either form – is appropriately reduced. Since the particles are regarded as indistinguishable in the same sense as their classical counterparts, this reduction in the count, due to PI, must reflect the fact that the particles can no longer be regarded as individuals – they are, in some sense, 'non-individuals'. In other words, according to this argument, PI implies non-individuality.

We shall call this view – that quantum particles are, in some sense, not individuals – the Received View. It became fixed in place almost immediately after the development of quantum statistics itself (and in its modern incarnation it can be found in Dieks, 1990, for example). Thus at the famous Solvay Conference of 1927, Langevin noted that quantum particles could apparently no longer be identified as individuals, and in that same year both Born and Heisenberg insisted that quantum statistics implied that the 'individuality of the corpuscle is lost' (Born, 1926; see Miller, 1987, p. 310). Some years later, in 1936, Pauli wrote to Heisenberg that he considered this loss of individuality to be 'something much more fundamental than the space-time concept' (see von Meyenn, 1987, p. 339).

5.2 Challenges to the Received View

The Received View has been challenged on a variety of grounds over the past fifteen years or so. These challenges come at the issue from two directions, and in both cases it is denied that there exists some fundamental metaphysical difference between quantum and classical particles. The first challenge insists that classical particles, like their quantum counterparts, are not only 'indistinguishable', in the above sense, but should also be understood as subject to PI. The grounds for this claim rest on a positivistic understanding of the meaning of 'non-individuality' which takes the latter notion to be determined experimentally (Hestenes, 1970). The idea is that non-individuality follows from the requirement that in order for the entropy to be extensive, the relevant expression must be divided by $N!$, where N is the number of particles in the assembly. It is this extensivity of entropy which, it is claimed, resolves the infamous Gibbs' paradox: if like gases at the same pressure and temperature are mixed, then there is no change in the experimental entropy. This is in disagreement with the result obtained from Maxwell–Boltzmann statistics, incorporating the considerations of particle permutations sketched above. Excluding such permutations from the calculation of the statistical entropy by dividing by $N!$ is then understood as resolving the 'paradox'. If this were correct, then – it is claimed – classical statistical mechanics would have to be regarded as permutation invariant also (see, for example, Saunders, this volume, note 13) and the contrast with quantum physics would have to be sought elsewhere.

Historically, however, the failure of extensivity and Gibbs' paradox were seen as revealing a fundamental flaw in the Maxwell–Boltzmann definition of entropy and one which is corrected by shifting to an understanding of the particles as quantum in nature.[12] In other words, the force of the argument can be turned around: what it shows is that the world is actually quantum in nature, as one would expect. What the exclusion of the permutations (by dividing by $N!$) is a manifestation of is precisely that the particles are not just indistinguishable in the classical sense. If this aspect is incorporated into the analysis from the word go, the so-called 'paradox' simply does not arise.[13]

The second challenge suggests that, contrary to the Received View, quantum particles, just like their classical counterparts, can also be regarded as individuals. The question immediately arises: if this were the case, how would one account for the different treatment of permutations; that is, how would one account for the difference between classical and quantum statistics? We recall that this difference lies in the drop in statistical weight assigned to the relevant arrangements. This can be accounted for, without appealing to the supposed 'non-individuality' of the particles, by focusing explicitly on the role of PI, understood as a kind of initial 'accessibility' condition (French, 1989a).

To see this, let us recall that, understood as a restriction on the observables, PI acts as a superselection rule which divides up the relevant Hilbert space into a number of irreducible subspaces, corresponding to irreducible representations of

[12] Here we are following Post who writes: 'the flaw in classical statistical mechanics represented by Gibbs' paradox points to a radical theory of non-individuality such as Bose's' (1971, p. 23, note 50). By 'flaw' Post means a 'neuralgic point' which can act as the heuristic stimulus for new theoretical development.

[13] Historically, the $N!$ division was the subject of a vigorous dispute between Planck, who defended the move, and Ehrenfest, who argued that it was *ad hoc* as it stood and hence required further justification.

the symmetric group. It can be shown that transitions between such subspaces are (generally) forbidden (at least in non-supersymmetric theories). Hence PI imposes a restriction on the states of the assembly such that once a particle is in a given subspace, the others – corresponding to other symmetry types – are inaccessible to it. Returning to the argument above, the reduction in statistical weight is now explained by the inaccessibility of certain states, rather than by a change in the metaphysical nature of the particles. In the simple case of two particles distributed over two one-particle states, the only subspaces available are the symmetric and antisymmetric and so only one of the two possible states formed by a permutation is ever available to the system. Thus the statistical weight corresponding to the distribution of one particle in each such state is half the classical value.

We shall examine the two components of this alternative to the Received View in a little more detail. First of all, with regard to the role of PI, the idea of restrictions on the set of states accessible to a system can also be found in classical statistical mechanics, of course. There it is the energy integral which imposes the most important restriction as it determines which regions of the relevant phase space are accessible. Other uniform integrals of the motion may also exist for a particular assembly but these are not generally thermodynamically significant. What PI represents is an additional constraint or initial condition, imposed on the situation. In particular, the symmetry type of any suitably specified set of states is an absolute constant of motion equivalent to an exact uniform integral in classical terms (see Dirac, 1958, pp. 213–16). Of course, some may wonder whether this actually sheds much light on the status of PI, since it appears to leave it standing as a kind of 'brute fact' but at least it is no more brutish than the other, classical, constraints.

Secondly, there is the issue of how we are to understand the individuality of the particles. As is well known, we have a range of options to choose from, some more attractive than the others:

(a) haecceity or primitive thisness (Adams, 1979);
(b) some form of Lockean substance;
(c) spatiotemporal location;
(d) some subset of properties.

The first two are perhaps the least attractive as far as broadly 'empiricist' philosophers are concerned, since they appeal to factors which are utterly empirically superfluous (see French and Redhead, 1988; Redhead and Teller, 1991 and 1992; van Fraassen, 1991). However, they share what some would see as the advantage of making manifest the conceptual distinction between individuality and distinguishability, where the latter is to be understood in terms of some difference in properties. The third and fourth options collapse this distinction by taking that which renders the entity distinguishable as that which also 'confers' individuality. Each requires some extra postulate, however, in order to effect this collapse. In the case of option (c), this extra 'something' is the postulate that the particles are impenetrable, since if they were not, their spatiotemporal locations could not serve to distinguish and hence individuate them. As is well known, however, this option is problematic in the quantum context as standardly understood, since it can be shown that the family of observables corresponding to the positions of single particles cannot provide distinguishing spatiotemporal trajectories (as Huggett and Imbo, 2000, emphasize, one can prove the more general result that no family of observables can provide such trajectories).[14]

Nevertheless, considerations of impenetrability do feature within the topological approach. We recall that, according to this perspective, we must move to the reduced configuration space formed by removing the points where the particles would coincide. An obvious justification for this adverts to the impenetrability of the particles, understood, in turn, as due to certain repulsive forces holding between them. Such a conjecture has been made in the case of anyons (see Aitchison and Mavromatos, 1991) and more generally this has been taken to confer a further advantage on the configuration space approach, in the sense that

> it allows particle statistics to be understood as a kind of 'force' in essence similar to other interactions with a topological character, like the interaction between an electric and magnetic charge in three spatial dimensions, or the type of interaction in two dimensions which is responsible for the Bohm–Aharonov effect and fractional statistics.[15]

Some have regarded such an understanding as suspiciously *ad hoc* (see Brown *et al.*, 1999). One way of eliminating this 'ad hocness' is to shift to the framework of de Broglie–Bohm pilot wave theory.[16] Here, as is well known, there is a dual ontology of point particles plus pilot wave, where the role of the latter is to determine the instantaneous velocities of the former through the so-called 'guidance equations' (*ibid.*). Since these equations are first order, the trajectories of two particles which are non-coincident to begin with will never coincide. In effect the impenetrability of the particles is built into the guidance equations and the singularity points

[14] They also stress that this lack of trajectories does not imply anything about the status of PI.

[15] Bourdeau and Sorkin (1992), p. 687. Again, we can't help but recall some relevant history here. The suggestion that the non-classical aspects of quantum statistics reflect a lack of statistical independence and hence a kind of correlation between the particles can be traced as far back as Ehrenfest's early reflections on Planck's work and crops up again and again in the literature. On the philosophical side, Reichenbach (1956, pp. 234–5) argued that such correlations – taken realistically in this sense – represent causal anomalies in the behaviour of the particles: for bosons these anomalies consist in a mutual dependence in the motions of the particles which could be characterized as a form of action at a distance; for fermions, the anomaly is expressed in the Exclusion Principle if this is interpreted in terms of an interparticle force. As far as Reichenbach himself was concerned, such acausal interactions should be rejected and thus he preferred the account of quantum statistics which emphasizes the metaphysical lack of individuality of the particles and in which the correlations are not regarded in force-like terms.

[16] Of course, adopting such a framework means abandoning the standard eigenvalue–eigenstate link and the latter is precisely what is assumed in the above proofs that spatiotemporal trajectories cannot serve to distinguish between particles.

remain inaccessible. Hence the conclusion that 'within the topological approach to identical particles the removal of the set ... of coincidence points from the reduced configuration space ... follows naturally from de Broglie–Bohm dynamics as it is defined in the full space' (*ibid.*, p. 233).[17]

It is part of the attraction of this framework that it retains, or appears to retain, a form of classical ontology which meshes well with the metaphysical view of particles as individuals.[18] However, it is also important to recognize that the topological approach can also accommodate the Received View of particles as non-individuals. Indeed, one of the motivations given by Leinaas and Myrheim (1997, p. 2) is that it allows one to dispense with the whole business of introducing particle labels and then effectively emasculating their ontological force by imposing appropriate symmetry constraints (cf. also Bourdeau and Sorkin, 1992, p. 687). Of course, if one is going to insist that the particles are non-individuals, then some alternative justification for the removal of the coincidence points must be sought for. One possibility is to tackle the problem of collisions directly. Bourdeau and Sorkin (1992), for example, show that for fermions, the self-adjoint extension of the Hamiltonian to cover the singularities is unique, at least in the two-dimensional case, so that collisions are strictly forbidden, whereas in the case of both Bose–Einstein and fractional statistics there is a range of alternative extensions, some of which allow collisions but some do not.[19] By requiring that the wave function remains finite at the coincident point they argue that a unique choice of Hamiltonian can then be made and it turns out that collisions are allowed only in the case of Bose–Einstein statistics. Thus, whereas for fermions it doesn't really matter whether the singular points are retained or not, for bosons and anyons, on this account, it does, since these points are either the locations of collisions in the boson case or the locations of vanishing Ψ for anyons.

6 Individuality and the identity of indiscernibles

Let us return to our list of options for understanding quantum individuality. Option (d) attempts to ground it in some subset of properties of the particles. This also requires a supplementary principle in order to block the possibility of two individuals sharing the same subset of properties, and this is provided, of course, by the

[17] There is the worry that this might exclude the possibility of Bose–Einstein statistics, if it is the case that the de Broglie–Bohm trajectories never in fact cross. Brown *et al.* (1999, p. 233) argue that this latter claim is simply not correct since symmetry considerations demonstrate that if the particles coincide at all, they coincide for ever. If the bosons are initially separated, then the relative velocity vanishes at the coincident point which together with the first-order nature of the guidance equations means that they can never cross.

[18] Brown *et al.* (1999) explicitly note this point. Nevertheless it is not clear that the particles can in fact be regarded straightforwardly as individuals in the classical sense (see French, 2000). Further criticisms of the view that de Broglie–Bohm theory is philosophically classical can be found in Bedard (1999). For a response see Dickson (2000).

[19] Thus they argue that simply cutting out the singular points results in a loss of information.

Principle of the Identity of Indiscernibles (PII). In terms of second-order logic with equality, PII can be written as

$$\forall \Gamma\{\Gamma(a) \equiv \Gamma(b)\} \rightarrow a = b \tag{2}$$

where a and b are individual constants designating the entities concerned and Γ is a variable ranging over the possible attributes of these entities. Different forms of PII then arise depending on what sort of attributes feature in the range of Γ.[20] The logically weakest form, PII(1), states that it is not possible for two individuals to possess all properties and relations in common; PII(2) excludes properties and relations which can be described as spatiotemporal; while the strongest form PII(3), includes only monadic, non-relational properties. Before we consider the status of these forms of PII in quantum physics, it is worth noting, first of all, that PII(1) has often been taken as necessarily true on the grounds that no two individuals can possess exactly the same spatiotemporal properties or enter into exactly the same spatiotemporal relations (see for example Quinton, 1973, p. 25). This obviously assumes that the individuals concerned are impenetrable and amounts to a form of option (c) above. Both PII(1) and PII(2) allow for the possibility that relations might be capable of distinguishing entities and hence confer individuality (see for example Casullo, 1984, who argues that the view of entities as nothing more than bundles of properties and relations is plausible only if based on PII(1) with relations given the capacity to individuate). However, such a possibility has been vigorously disputed on the grounds that since relations presuppose numerical diversity, they cannot account for it (see Russell, 1956, and Armstrong, 1978, pp. 94–5). We shall return to this possibility shortly.

When it comes to the status of PII in quantum physics, if the non-intrinsic, state-dependent properties are identified with all the monadic or relational properties which can be expressed in terms of physical magnitudes associated with self-adjoint operators that can be defined for the particles, then it can be shown that two bosons or two fermions in a joint symmetric or antisymmetric state respectively have the same monadic properties and the same relational properties one to another (French and Redhead, 1988; see also Butterfield, 1993). On the basis of such an identification, even the weakest form of the principle, PII(1), fails for both bosons and fermions (French and Redhead, 1988; French, 1989b).[21] Hence the Principle of the Identity of Indiscernibles cannot be used to effectively guarantee individuation via the

[20] We exclude the attribute of 'being identical with a', since PII would then simply be a theorem of second-order logic. Furthermore, Adams (1979) identifies haecceity or 'primitive thisness' with precisely this attribute and hence admitting it here would be tantamount to adopting option (a).

[21] Margenau (1944) had earlier concluded that PII(3) fails, since the same reduced state can be assigned to the fermions in an antisymmetric state and hence they possess the same monadic properties. This conclusion has been criticized on the grounds that such reduced states cannot be regarded either as ontologically separate or as encoding genuinely monadic properties; see Castellani and Mittelstaedt (2000); Massimi (2001).

state-dependent properties and option (d) fails,[22] leaving Lockean substance or primitive thisness as the only alternatives.

However, there may still be hope for this option. Saunders (this volume, section 1) has recently revived Quine's proposal for the analysis of identity, which he understands as yielding a version of PII. Roughly speaking this is the condition that $x = y$ (where $x, y, u_1, u_2, \ldots$ are variables) if and only if, for all unary predicates A, binary predicates $B, \ldots$, n-ary predicates P, we have

- $A(x) \equiv A(y)$
- $B(x, u_1) \equiv B(y, u_1)$; $B(u_1, x) \equiv B(u_1, y)$
- $P(x, u_1, \ldots, u_n) \equiv B(y, u_1, \ldots, u_n)$ and permutations

together with all universal quantifications over the free variables $u_1, \ldots, u_n$ other than x and y (*ibid.*). If the relevant language contains monadic predicates only, then this principle amounts to the claim that two entities are identical if and only if they have all properties in common. Two entities are said to be absolutely discernible if there is a formula with only one free variable which applies to one entity but not the other. With only monadic predicates allowed, the principle states that numerically distinct entities are absolutely discernible. If relations are admitted, however, one can have entities which are not identified by the principle yet are not absolutely discernible. Two entities are said to be relatively discernible if there is a formula in two free variables which applies to them in any order. But there is a further category: if the admitted relations include some which are irreflexive, then one can have entities which are counted as distinct according to the principle but are not even relatively discernible. These are said to be weakly discernible, and the principle excludes the possibility of entities neither absolutely, relatively, or weakly discernible.

This, it is claimed, is more natural from a logical point of view (being immune to the standard counter-examples – such as the infamous two globes – which beset PII; see Saunders, *ibid.*), and is also better suited to quantum mechanics in that, unlike traditional versions of PII, it is not violated by fermions at least, since an irreflexive relation always exists between them. Consider, for example, two fermions in a spherically-symmetric singlet state. The fermions are not only indistinguishable but also have exactly the same spatiotemporal properties and relations in themselves and everything else. However, each satisfies the symmetric but irreflexive relation of 'having opposite direction of spin to' and so are weakly discernible. Thus for fermions, at least, we have the possibility of grounding their individuality via a version of PII, without having to appeal to anything like primitive

[22] For alternative discussions see Cortes (1976); Barnette (1978); Ginsberg (1981); Teller (1983); and van Fraassen (1984; 1991).

thisness.[23] However, there is an obvious concern one might have here, which reprises the worry hinted at above, regarding the individuating power of relations: doesn't the appeal to irreflexive relations in order to ground the individuality of the objects which bear such relations involve a circularity? Such concerns are rooted in the – apparently plausible – view that relata have ontological priority over relations, such that the former can be said to 'bear' the latter. Suppose we were to drop such a view. Of course, in order to describe the relation – either informally as above or set-theoretically in terms of an ordered tuple $\langle x, y\rangle$ – we have to introduce some form of label, as in the example above, but description should not dictate conceptualization. The label can be understood as a kind of place-holder and instead of talking of relata 'bearing' relations, one can talk of the intersection of relations as constituting relata, as Cassirer did, or of relata as unifying relations, as Eddington did. The names here give the game away – what this amounts to is some kind of structuralist ontology which allows for individuation via relations.

7 PI is neither sufficient nor necessary for non-individuality

Returning now to the issue of the status of PI, the point we want to emphasize is that something further needs to be added to get from it to the Received View of particles as non-individuals. In other words, PI is not sufficient for non-individuality. The question remains, is it necessary?

Van Fraassen (1991, p. 375), for example, has answered that it is not, while acknowledging that the claim – that 'when identity is properly understood, it entails Permutation Invariance tautologically' – nevertheless contains a 'core' of truth. Butterfield (1993, p. 457) has disagreed, however, insisting that this claim merely summarizes the motivation for PI, namely that expectation values for the composite system cannot be sensitive to the differences between ϕ and $P\phi$. At issue here, of course, is what is meant by 'identity', or non-individuality, being 'properly understood'. Butterfield's understanding appears to correspond to Dirac's above, but if one were to reject this as broadly positivistic, how do the alternatives stand up?

In order to examine this question, we need to consider what we mean by non-individuality in this context. All of the options considered here are constructed through a combination of indistinguishability – of course – together with the denial of some 'principle' of individuality, whether that be substance, primitive thisness, spatiotemporal trajectories, or PII. The question then is whether any of these

[23] No such possibility exists for bosons; however, Saunders adopts the Redhead and Teller option of regarding them as non-individual field quanta. He takes this metaphysical difference as tracking the physical one between the 'stable constituents of ordinary matter' (fermions) and gauge quanta (bosons), although it is not clear why the metaphysics should follow the physics in this particular way, or at all.

alternatives can provide the restriction on observables that PI demands. Clearly neither substance nor primitive thisness can, since by their very nature neither can be expressed in terms of observables! What about indistinguishability plus the denial of spatiotemporal trajectories? Huggett and Imbo (2000) have recently considered this case and have argued that here too PI does not follow. Their argument considers each conjunct separately: first of all, the absence of spatiotemporal trajectories follows from the dynamics of quantum mechanics together with the general way observables are treated and again this imposes no restrictions on these observables. Hence, the lack of spatiotemporal trajectories is simply irrelevant in this case. Furthermore, there exists the possibility of particles, such as first quantized versions of Greenberg's quons (Greenberg, 1991)[24] which are indistinguishable but for which every Hermitian operator is an observable – hence some of the observables are non-symmetric and PI is violated.

Of course, if PII were to hold for quons, then the implication would be restored for that particular understanding of non-individuality; that is, if quons could not be considered as non-individuals in this sense, then the violation of PI would not be indicative of the failure of the entailment. However, what would have to be shown is that there are no possible quon states for which PII is violated in the same ways as for bosons and fermions. In the case of paraparticles, French and Redhead (1988) showed that although states do exist for which the monadic properties of all the separate particles are the same, there also exist possible paraparticle states for which PII is violated. So far as we know, nothing equivalent has been demonstrated for quons, although in what little philosophical discussion there has been about them, it appears to be assumed that such PII violating states exist (Hilborn and Yuca, 2002).

The choice, then, is stark: either adopt the Dirac/Butterfield understanding of non-individuality, in which case PI is indeed an expression of it but not in a metaphysically interesting way, or accept that non-individuality does not imply permutation invariance. The latter option leaves PI metaphysically ungrounded.

8 Underdetermination and the structuralist view of particles

This metaphysical 'detachment' of PI has been expressed in terms of a kind of underdetermination which holds between the Received View, in which PI is tied up with the non-individuality of the particles and the alternative account of particles as individuals, with PI taken as some sort of initial condition (French and Redhead, 1988). As we have indicated, PI does not discriminate between these

[24] This possibility is not unproblematic, of course. The q-mutator formalism apparently requires some observables to be non-local and there is the further issue as to whether this possibility is ruled out on experimental grounds (again see the papers in Hilborn and Tino, 2000).

conceptual possibilities (French, 1989a, 1998; van Fraassen, 1991; Huggett, 1997; Balousek, 2000) and hence any argument for one or the other is going to have to proceed on different grounds. Most famously, perhaps, Redhead and Teller have elaborated a methodological argument to the effect that the Received View meshes better with the metaphysics of Quantum Field Theory where – it is claimed – individuality is not assumed from the word go (this argument goes back to Post, 1963). This last claim has been disputed (by de Muynck, 1975, and van Fraassen, 1991; for criticisms of the latter, see Butterfield, 1993), and Balousek (2000) has insisted that to appeal to methodology to break the underdetermination is to accede to a form of conventionalism (for a response see Teller and Redhead, 2001).

We shall not pursue the ins and outs of this debate here. The alternative is to accept the underdetermination and explore its implications. As far as the particles themselves are concerned, it motivates a shift – some might say retreat – to a structuralist view of entities which eschews talk of individuality or non-individuality entirely (see Ladyman, 1998). According to such a view, the particles are nothing but 'nodes' or 'intersections' in some kind of physical structure, which now bears all the ontological weight. We shall consider how PI looks from such a position shortly.

9 The experimental and theoretical status of PI

Returning to the issue of the status of PI, let us consider whether there could be direct evidence for the principle. We can examine this question in a broader context: what is required for any symmetry principle to be observable? Kosso (2000) has argued that the distinction between observable and unobservable symmetries matches that between global and local. Thus Lorentz invariance, for example, is directly observable whereas general covariance is not, since a specific dynamical principle such as the principle of equivalence must be assumed in order to infer the symmetry from the observation. Now, a problem arises when it comes to symmetries such as PI: in order to observe whether a symmetry holds or not, one must be able to observe that (i) the specified transformation has taken place; and (ii) the specified invariant property remains the same under the transformation (*ibid.*, p. 86). The first condition requires there to be a fixed point of reference with respect to which the transformation can be measured. In the case of the permutation of protons and neutrons underlying isospin symmetry, for example, there exists a 'fixed standard' of what it is to be a neutron and what it is to be a proton with respect to which the permutation makes sense (*ibid.*). If, however, the symmetry is 'observationally complete', in the sense that all of the observable properties of the system are invariant under the transformation, then the symmetry will be unobservable in principle (*ibid.*, p. 88). Kosso does not consider PI in his discussion but it is observationally complete in precisely this sense: in order to test whether it holds or not, one would have to be able to distinguish experimentally the states represented by $|\Psi\rangle$ and $\hat{P}|\Psi\rangle$ to begin with. However, that is precisely what the principle itself denies (Balousek, 1999, p. 20).

If there can be no direct evidence for PI, is it demanded by the theory of quantum mechanics itself? It is often claimed that PI is not logically required by the axioms of the theory (Balousek, 1999), but of course, this depends on what the latter are taken to include. Adopting a crude historical perspective, the work of Weyl, Dirac, and von Neumann can be seen as an attempt to impose order upon what was, in the late 1920s, a bit of a hodgepodge of laws and principles, some of them 'phenomenological' in nature, such as the Exclusion Principle. Weyl's framework was explicitly group-theoretical and here, as we have indicated, PI can be incorporated within this framework as an expression of the metaphysical nature of quantum objects. Dirac, on the other hand, eschewed both group theory and metaphysics, preferring his own 'bra' and 'ket' framework in which PI is seen as nothing more than an expression of the observational indistinguishability of $|\Psi\rangle$ and $\hat{P}|\Psi\rangle$, from a rather crude verificationist perspective. In the context of von Neumann's Hilbert space formalism, PI does appear to be an extra postulate reflecting either the metaphysical nature of quantum particles or some kind of 'initial condition' as we noted above but of course one could make the case that it should be added to the standard axioms of quantum mechanics whatever they are in order to extend the theory to give a quantum statistics. Whatever framework one chooses, claims that PI is, in some sense, '*ad hoc*' must be treated with caution.

Where does all this leave the status of PI? It appears to be a kind of 'free-floating' principle, one that is required neither experimentally nor metaphysically. Huggett (1999b) has proposed that it be regarded straightforwardly as a symmetry on a par with rotational symmetry, for example. Now, of course, as Huggett acknowledges, the two symmetries are very different,[25] but, nevertheless, PI is implied by the conjunction of a further symmetry principle which spacetime symmetries also obey together with the formal structure of the permutation group. This further principle is what he calls 'global Hamiltonian symmetry' which implies that the relevant symmetry operator commutes with the relevant Hamiltonian.[26] With regard to the permutation group, of course, permutations of a subsystem are permutations of

[25] A quantum system of the kind we have been considering is not just *covariant* with respect to permutations but *invariant*: permutations are not just indistinguishable to appropriately transformed observers but to *all* observers.

[26] What we take the relevant Hamiltonian to cover is crucial here because, again as Huggett acknowledges, the principle would appear to be violated in the case where, for example, we have a non-central potential term in the Hamiltonian of an atomic system, but, he insists, the symmetry is restored if we consider the 'full' Hamiltonian of system plus field, which does commute with the operators of the rotation group. As he points out (1999b, p. 345), if observers are taken to be systems too, this symmetry principle is equivalent to covariance for spacetime symmetries.

the whole system and this 'global Hamiltonian symmetry' very straightforwardly implies PI, without any additional assumptions concerning the structure of state space (Huggett, 1999b, pp. 344–5).[27] Hence, Huggett concludes (p. 346),

> we should view permutations in a similar light to rotations: we should not take [permutation invariance] as a fundamental symmetry principle in order to explain quantum statistics. Instead we should recognize that it is a particular consequence of global Hamiltonian symmetry given the group structure of the permutations. Further, if we accept the similarity of permutation and rotation symmetry, it becomes natural to see quantum statistics as a natural result of the role symmetries play in nature.

However, as Huggett acknowledges, permutation invariance only follows from his general symmetry principle given the particular structure of the permutation group. So the issue of the status of PI is pushed back a step: what is the status of the structure of the permutation group? Or, to put it another way, why should that particular group structure be applicable? At one extreme we have the view that it is *a priori*. As is well known, Weyl (1952, p. 126) insisted that 'all *a priori* statements in physics have their origin in symmetry'. Not surprisingly, empiricists such as van Fraassen have tended to resist this line (see van Fraassen, 1989) and move to the other end of the spectrum, offering a broadly pragmatic answer to our question. From this perspective, PI comes to be seen as nothing more than a problem-solving device (see Bueno, 2001). Occupying the middle ground we have the following alternative answers to our question:

(a) It is just a brute fact. We have already encountered this option in our discussion of PI as an 'initial condition' imposed on the situation.
(b) It is to be understood as reflecting the metaphysically peculiar nature of the particles themselves. However, given that the particles can also be described in a metaphysically straightforward way – as individuals – this option is always going to require some further principle whose status may be less well grounded than that of PI itself.[28]
(c) It is to be understood as reflecting a structural aspect of the world. From this perspective, that the permutation group is applicable is neither a simple 'brute fact' nor metaphysically derivative, in the sense of mathematically representing the nature of the particles, but rather it represents something profoundly structural about the world.

Huggett rejects the first two options but then leaves the metaphysical status of PI hanging. We shall pursue the last option a little further in another context, namely the connection between permutations of particles and diffeomorphisms of spacetime points.

[27] It does, however, assume that the system being measured and the measurement apparatus are composed of the same indistinguishable particles, otherwise the Hamiltonian will not remain unchanged. Thanks to Nick Huggett for pointing this out.

[28] This is, in essence, the heart of the dispute between Balousek, and Redhead and Teller.

10 Permutation symmetry, structuralism, and diffeomorphism invariance

Structuralism has a long and interesting history which is intimately bound up with developments in physics. Both general relativity and quantum mechanics had a profound impact on the work of early structuralists such as Cassirer and Eddington (for discussion see French, 2001). Putting it crudely, the central idea of this programme is to effectively deconstruct the 'object of knowledge' – whether spacetime or quantum particles – into a web of relations bound together by symmetry principles represented group theoretically. If we focus on the group-theoretic representation of invariant properties such as mass and spin, what this 'deconstruction' yields are kinds of particles (see Castellani, 1993, 1998). Similarly, PI can be seen as embodying a form of structuralist representation of 'broader' kinds, namely bosons, fermions, parabosons, parafermions, and so on. In other words, the status of PI, from this perspective, is that of one of the fundamental symmetry principles which effectively binds the 'web of relations' constituting the structure of the world into these broad kinds.

10.1 Permutation symmetry and structural realism

It is this kind of structuralist deconstruction of objects which is incorporated into Ladyman's 'ontic' form of structural realism, alluded to above. As we indicated, this attempts to avoid the metaphysical underdetermination that PI yields by reconceptualizing quantum objects entirely in structural terms (see French and Ladyman, 2002; Saunders, 2002). However, the following objection has been raised to such a move: if structure is understood in 'relational' terms – as it typically is – then there need to be 'relata', and these cannot be relational themselves. The force of the objection is clearly seen in the case of PI: we began by considering the distribution of objects over states and the effects of permutations on such objects. How can PI play a part in the 'deconstruction' of objects into structures when its very articulation is based on the assumption that there are objects to begin with? In responding to this objection, structuralists typically appeal to the following manoeuvre (it can be found in Eddington and, before him, Poincaré, for example): we recall that we begin by introducing particle labels and it is upon these that the particle permutation operator acts. We then assume that these labels denote objects – an assumption that may be supported by the observation of the individual flashes on a scintillation screen, for example – and this allows us to apply the mathematics of group theory (with its underlying standard set theory). However, once we have obtained the relevant structure, we can dispense with our original assumption, regarding it as no more than a heuristic crutch and the labels as simply convenient place-holders which serve only to help us focus attention on what is metaphysically fundamental – PI in this case. To use a famous metaphor, the object is a kind of 'ladder' which we use to reach the structure but which we can then 'throw away', or 'deconstruct'. Of course, there are other objections to structural realism which must be addressed (see, for example, Bueno, 2001, and Chakravartty, 2001) but our intention here is just to indicate what may be a natural home for the structuralist understanding of PI. Furthermore, this sort of picture can accommodate a structuralist conception of spacetime as well.

10.2 Diffeomorphisms, permutations, and the structuralist conception of spacetime

Stachel (2002) has recently explored the connections between the interpretation of general covariance and permutation invariance on the one hand and the metaphysical analyses of spacetime points and quantum particles on the other. He begins by abstracting the differentiability and continuity properties of manifolds leaving a bare set of points. The (continuous) principle of diffeomorphism invariance then becomes (discrete) permutation invariance. A version of the hole argument can then be seen to apply to this set (see Norton, 1988, for an elementary account of this argument). We have already seen how such a set, along with PI, models the statistical behaviour of ensembles of indistinguishable quantum particles. Thus the analogy is complete and extends, *mutatis mutandis*, to any theory which 'demands the complete indistinguishability of its fundamental objects' (Stachel, 2002, p. 248).

On this basis the choice between 'substantivalist' and 'relationist' conceptions of spacetime is rendered as that between the 'individual' and 'non-individual' metaphysics of points. Stachel himself opts for the latter package (applying the result to both the spacetime and particle cases). The 'reduced phase space' method of solving the hole argument[29] is then understood as applying to the permutation case (where the gauge orbits are equivalence classes of permuted states). Since that solution is seen by Stachel as corresponding to a relational solution, the particle case is understood similarly. The idea is that the objects in a set (be they the points of a manifold or the members of a quantum ensemble) are individuated only by the relations of that set: indistinguishable objects are not individuals intrinsically but derive that property from relations. We have been here before, of course, with Saunders' Quinean approach to PII above and, indeed, Saunders also applies this approach to the spacetime case.

The central idea here, then, is that any theory that demands the complete indistinguishability of its fundamental objects *requires* invariance under the full permutation group for discrete symmetries or the diffeomorphism group for continuous symmetries. Stachel explicitly draws the analogy between substantivalism in the spacetime case and assuming individuality for quantum particles, and relationism

[29] This solution takes the equivalence classes of diffeomorphic states (i.e. the gauge orbits) as the points of a new 'reduced' or 'physical' phase space.

and non-individuality, respectively. Moving in the other direction both Maidens (1993) and Hoefer (1996) have explored the idea of regarding spacetime points as 'non-individuals' in some sense (Saunders, 2002, endorses Hoefer's conclusions). What then becomes crucial is the *sense* in which the points are regarded as non-individuals, just as we have discussed for particles. Stachel, in particular, understands the non-individuality of particles as their being individuated 'entirely in terms of the relational structures in which they are embedded' (hence the analogy with relationism on the spacetime side). But then it is not clear what metaphysical work the notion of 'non-individuality' is doing, when we still have 'objects' which are represented by standard set theory (and this is precisely the criticism that can be levelled against attempts to import non-individuality into the spacetime context). What one needs to do to flesh out such an account is to apply to spacetime points the kind of 'quasi-set' theory that has been applied to non-individual quantum particles (Krause, 1992). Of course, Stachel could still maintain individuality in both cases but at the price of introducing inaccessible states in quantum mechanics and indeterminism in spacetime theory. In both cases we seem to have a kind of metaphysical underdetermination.

Again the alternative, 'middle way' is to drop objects out of the ontology entirely, regarding both spacetime and particles in structural terms. Indeed, this appears to be the more appropriate way of understanding both Stachel's talk of individuating objects 'entirely in terms of the relational structures in which they are embedded' (2002, p. 232) and, as we have seen, Saunders' account of 'weakly discernible' entities. However, rather than thinking of the objects being individuated, we suggest they should be thought of as being structurally constituted in the first place. In other words, it is the relational structures which are regarded as metaphysically primary and the objects as secondary or 'emergent'. The labels that appear in both cases, assigned to spacetime points and particles respectively, are just mathematical devices which allow us to apply our set-theoretical resources. And in both cases, the relevant symmetries, encoded in diffeomorphism invariance and PI respectively, will be seen as essential components of this structural metaphysics.

There is, for sure, plenty of work to be done here; but this structural perspective places symmetry at the heart of our metaphysics, and that surely makes the task a tantalising one!

References

Adams, J. (1979). 'Primitive thisness and primitive identity'. *Journal of Philosophy*, **76**, 5–26.

Aitchison, I. J. R., and Mavromatos, N. E. (1991). 'Anyons'. *Contemporary Physics*, **32**, 219.

Armstrong, D. (1978). *Nominalism and Realism*, Vols. 1 and 2. Cambridge: Cambridge University Press.

Balousek, D. (1999). 'Indistinguishability, individuality and the identity of indiscernibles in quantum mechanics'. Preprint.

(2000). 'Statistics, symmetry and the conventionality of indistinguishability in quantum mechanics'. *Foundations of Physics*, **30**, 1–34.

Barnette, R. L. (1978). 'Does quantum mechanics disprove the principle of the identity of indiscernibles?' *Philosophy of Science*, **45**, 466–70.

Bedard, K. (1999). 'Material objects in Bohm's interpretation'. *Philosophy of Science*, **66**, 221–42.

Born, M. (1926). 'Quantenmechanik der Stobvurgnge'. *Zeitschrift für Physik*, **38**, 803–27.

Bourdeau, M., and Sorkin, R. D. (1992). 'When can identical particles collide?' *Physical Review D*, **45**, 687–96.

Brown, H. R., Sjöqvist, E., and Bacciagaluppi, G. (1999). 'Remarks on identical particles in de Broglie–Bohm theory'. *Physics Letters A*, **251**, 229–35.

Bueno, O. (2001). 'Weyl and von Neumann: symmetry, group theory, and quantum mechanics'. Available online at http://philsci-archive.pitt.edu/documents/disk0/00/00/04/09/index.html.

Butterfield, J. (1993). 'Interpretation and identity in quantum theory'. *Studies in History and Philosophy of Science*, **24**, 443–76.

Casullo, A. (1984). 'The contingent identity of particulars and universals'. *Mind*, **123**, 527–54.

Castellani, E. (1993). 'Quantum mechanics, objects and objectivity'. In *The Foundations of Quantum Mechanics – Historical Analysis and Open Questions*, ed. C. Garola and A. Rossi, pp. 105–14. Dordrecht: Kluwer.

(1998). 'Galilean particles: an example of constitution of objects'. In *Interpreting Bodies: Classical and Quantum Objects in Modern Physics*, ed. E. Castellani, pp. 181–94. Princeton, NJ: Princeton University Press.

Castellani, E., and Mittelstaedt, P. (2000). 'Leibniz's principle, physics and the language of physics'. *Foundations of Physics*, **30**, 1587–604.

Chakravartty, A. (2001). 'The structuralist conception of objects', forthcoming in *Philosophy of Science*.

Cortes, A. (1976). 'Leibniz's Principle of the Identity of Indiscernibles: a false principle'. *Philosophy of Science*, **43**, 491–505.

de Muynck, W. (1975). 'Distinguishable and indistinguishable-particle descriptions of systems of identical particles'. *International Journal of Theoretical Physics*, **14**, 327–46.

Dickson, M. (2000). 'Discussion: are there material objects in Bohm's theory?' *Philosophy of Science*, **67**, 704–10.

Dieks, D. (1990). 'Quantum statistics, identical particles and correlations'. *Synthese*, **82**, 127–55.

Dirac, P. A. M. (1958). *The Principles of Quantum Mechanics*. Oxford: Clarendon Press.

French, S. (1989a). 'Identity and individuality in classical and quantum physics'. *Australasian Journal of Philosophy*, **67**, 432–46.

(1989b). 'Why the Principle of Identity of Indiscernibles is not contingently true either'. *Synthese*, **78**, 141–66.

(1998). 'On the withering away of physical objects'. In *Interpreting Bodies*, ed. E. Castellani, pp. 93–113. Princeton, NJ: Princeton University Press.

(2000). 'Putting a new spin on particle identity'. In *Spin-Statistics Connection and Commutation Relations: Experimental Tests and Theoretical Implications*, ed. R. Hilborn and G. M. Tino, pp. 305–17. Melville, NY: American Institute of Physics.

(2001). 'Symmetry, structure and the constitution of objects'. Available online at http://philsci-archive.pitt.edu/documents/disk0/00/00/03/27/index.html.

French, S., and Ladyman, J. (2002). 'Remodelling structural realism: quantum physics and the metaphysics of structure', forthcoming in *Synthese*.

French, S., and Redhead, M. (1988). 'Quantum physics and the identity of indiscernibles'. *British Journal for the Philosophy of Science*, **39**, 233–46.

Ginsberg, A. (1981). 'Quantum theory and the identity of indiscernibles revisited'. *Philosophy of Science*, **48**, 487–91.

Greenberg, O. W. (1991). 'Interactions of particles having small violations of statistics'. *Physica*, **180**, 419–27.

Greenberg, O. W., and Messiah, A. M. L. (1964). 'Symmetrization postulate and its experimental foundation'. *Physical Review B*, **136**, 248–67.

Hestenes, D. (1970). 'Entropy and indistinguishability'. *American Journal of Physics*, **38**, 840–5.

Hilborn, R. C., and Tino, G. M., eds. (2000). *Spin-Statistics Connection and Commutation Relations: Experimental Tests and Theoretical Implications*. Melville, NY: American Institute of Physics.

Hilborn, R. C., and Yuca, C. L. (2002). 'Identical particles in quantum mechanics revisited'. *British Journal for the Philosophy of Science*, **53**, 355–89.

Hoefer, C. (1996). 'The metaphysics of spacetime substantivalism'. *Journal of Philosophy*, **93**, 5–27.

Huggett, N. (1997). 'Identity, quantum mechanics and common sense'. *The Monist*, **80**, 118–30.

(1999a). 'Atomic metaphysics'. *Journal of Philosophy*, **96**, 5–24.

(1999b). 'On the significance of the permutation symmetry'. *British Journal for the Philosophy of Science*, **50**, 325–47.

Huggett, N., and Imbo, T. D. (2000). 'What is an elementary quarticle?' Chicago, IL: University of Chicago Preprint.

Imbo, T. D., Shah Imbo, C., and Sudarshan, E. C. G. (1990) 'Identical particles, exotic statistics and braid groups'. *Physics Letters B*, **234**, 103–7.

Kosso, P. (2000). 'The empirical status of symmetries in physics'. *British Journal for the Philosophy of Science*, **51**, 81–98.

Krause, D. (1992). 'On a quasi-set theory'. *Notre Dame Journal of Formal Logic*, **33**, 402–11.

Ladyman, J. (1998). 'What is structural realism?' *Studies in History and Philosophy of Science*, **29**, 409–24.

Leinaas, J. M., and Myrheim, J. (1997). 'On the theory of identical particles'. *Nuovo Cimento B*, **37**, 1–23.

Maidens, A. V. (1993). 'The hole argument: substantivalism and determinism in general relativity.' Ph.D. thesis, University of Cambridge.

Margenau, H. (1944). 'The exclusion principle and its philosophical importance'. *Philosophy of Science*, **11**, 187–208.

Massimi, M. (2001). 'Exclusion principle and the identity of indiscernibles: a response to Margenau's argument'. *British Journal for the Philosophy of Science*, **52**, 303–30.

Messiah, A. M. L. (1962). *Quantum Mechanics*, Vol. 2. Amsterdam: North-Holland.

Miller, A. (1987). 'Symmetry and imagery in the physics of Bohr, Einstein and Heisenberg'. In *Symmetries in Physics (1600–1980)*, ed. M. G. Doncel *et. al.*, pp. 300–25. Barcelona: Servei de Publicacions.

Norton, J. (1988). 'The hole argument'. In *PSA 1988*, ed. A. Fine and J. Leplin, Vol. 2, pp. 56–64. East Lansing, MI: Philosophy of Science Association.

Post, H. (1963). 'Individuality and physics'. *The Listener*, **70**, 534–7.
(1971). 'Correspondence, invariance and heuristics'. *Studies in History and Philosophy of Science*, **2**, 213–55.
Quinton, A. (1973). *The Nature of Things*. London: Routledge and Kegan Paul.
Redhead, M. L. G. (1975). 'Symmetry in intertheory relations'. *Synthese*, **32**, 77–112.
Redhead, M., and Teller, P. (1991). 'Particles, particle labels, and quanta: the toll of unacknowledged metaphysics'. *Foundations of Physics*, **21**, 43–62.
(1992). 'Particle labels and the theory of indistinguishable particles in quantum mechanics'. *British Journal for the Philosophy of Science*, **43**, 201–18.
Reichenbach, H. (1956). *The Direction of Time*. Berkeley, CA: University of California Press.
Russell, B. (1956). 'On the relations of universals and particulars'. In *Logic and Knowledge*, ed. R. C. Marsh, pp. 105–24. New York: Allen and Unwin.
Saunders, S. (2002). 'Scientific realism, again', forthcoming in *Synthese*.
Stachel, J. (2002). '"The relations between things" versus "the things between relations": the deeper meaning of the hole argument'. In *Reading Natural Philosophy: Essays in the History and Philosophy of Science and Mathematics*, ed. D. Malament, pp. 231–66. Chicago and LaSalle, IL: Open Court.
Teller, P. (1983). 'Quantum physics, the identity of indiscernibles and some unanswered questions'. *Philosophy of Science*, **50**, 309–19.
Teller, P., and Redhead, M. (2001). 'Is indistinguishability in quantum mechanics conventional?' *Foundations of Physics*, **30**, 951–7.
van Fraassen, B. (1984). 'The problem of indistinguishable particles'. In *Science and Reality: Recent Work in the Philosophy of Science*, ed. J. T. Cushing, C. F. Delaney, and G. M. Gutting, pp. 153–72. Notre Dame, IL: University of Notre Dame Press.
(1989). *Laws and Symmetry*. Oxford: Oxford University Press.
(1991). *Quantum Mechanics: An Empiricist View*. Oxford: Oxford University Press.
von Meyenn, K. (1987). 'Pauli's belief in exact symmetries'. In *Symmetries in physics (1600–1980)*, ed. M.G. Doncel *et al.* Barcelona: Servei de Publicacions.
Weyl, H. (1952). *Symmetry*. Princeton, NJ: Princeton University Press.

13

Quarticles and the identity of indiscernibles

NICK HUGGETT

1 Introduction

In sections 6 and 7 of their paper in this volume, French and Rickles raise the question of the logical relations between the indistinguishability postulate (IP) and the various senses in which particles might fail to be individuals. In section 6 they refer to the convincing arguments of French and Redhead (1988) and of Butterfield (1993) that IP does not logically entail non-individuality, understood several ways – even though, as all seem to concede, there *is* something perverse about taking bosons and fermions to be individuals. Going the other way, the possibility of IP violating 'quons' (Greenberg, 1991) shows that if non-individuality is taken to mean the absence of continuous distinguishing trajectories, characteristic of standard quantum mechanics (QM), then non-individuality does not entail IP. Nor, as French and Rickles point out, do substance or haecceity views of individuality.

But what if we conceive of individuality in terms of the Principle of the Identity of Indiscernibles (PII)? First, French and Redhead (1988) and Butterfield (1993) have given theorems showing that bosons and fermions violate PII, while the former have also demonstrated violations of PII in the case of a certain paraparticle state. But these cases, as I will explain (and as French and Rickles point out), cover just a very few of the possible kinds of quantum particles, and so for each kind the question arises as to whether it violates PII. I will give an answer to this question here, rather more general than – though based on – those previously offered.

So consider the theorems proven about PII. Suppose one has an n-particle system. The states of such a system lie in the Hilbert space that is the tensor product of n Hilbert spaces, one for each of the particles. Suppose further that the Hilbert space for particle i or j – appearing as the ith and the jth factors of the tensor product space – is H. And suppose finally that Q is an observable on H with eigenvalues p and q (and let I represent the identity on any of the factors). Then,

$Q_i = I \otimes I \otimes \ldots \otimes Q \otimes \ldots \otimes I$ (with Q as the ith factor) is intuitively the observable corresponding to a measurement of the value of Q for the ith particle, and *mutatis mutandis* Q_j.

Then, French and Redhead (1988) showed the following: for any state Ψ such that $P_{ij}\Psi = \pm\Psi$ (where P_{ij} transposes the ith and the jth factors of the tensor product space) so that i and j are either symmetrized (if $+$) or antisymmetrized (if $-$),[1]

$$\mathrm{pr}^{\Psi}(Q_i = q) = \mathrm{pr}^{\Psi}(Q_j = q) \tag{1}$$

and

$$\mathrm{pr}^{\Psi}(Q_i = q | Q_j = p) = \mathrm{pr}^{\Psi}(Q_j = q | Q_i = p). \tag{2}$$

That is, for example, the probability that the system, while in state Ψ, would possess value q for observable Q_i if measured is equal to the probability that it would possess that value for Q_j if measured – roughly, the chance that particle i has value q is the same as the chance that particle j does. Butterfield (1993) extends these results by considering a second observable on H, Q', with eigenvalues q' and p' and the corresponding observables Q'_i and Q'_j on the tensor product space. He also considers a third particle whose Hilbert space is the kth factor and for which Q' is also an observable.[2] Then he shows – again for bosons and fermions – that

$$\mathrm{pr}^{\Psi}(Q_i = q | Q'_j = p') = \mathrm{pr}^{\Psi}(Q_j = q | Q'_i = p') \tag{3}$$

and

$$\mathrm{pr}^{\Psi}(Q'_k = q' | Q_i = p) = \mathrm{pr}^{\Psi}(Q'_k = q' | Q_j = p). \tag{4}$$

Since the probabilities for the possession of eigenvalues of observables capture the dynamical properties of a quantum system exhaustively, the authors conclude that these four results show that any pair of (anti)symmetrized particles in a system are indistinguishable by monadic (the first result) or relational (the other results) dynamical properties. The possibility remains that the particles are distinguished by their intrinsic properties, those that are state-independent, and can be treated as 'c-number' quantities, such as spin magnitude, mass, charge, or colour. So let's suppose that the particles are 'identical', sharing all their intrinsic properties. Then, as Butterfield (1993) and French and Redhead (1988) conclude, i and j in state Ψ violate both logically strong – that individuals sharing all monadic properties are

[1] Note that we have assumed nothing at this stage about whether the particles are identical, or (anti)symmetrized in every state, or what the effect of other permutations of Ψ might be. We will consider these factors below.

[2] Now k need not be identical to the other particles – the proof certainly does not assume that it is (anti)symmetrized with respect to i or j. However, the proof does assume that it has the same Hilbert space H, or at the very least that there is a correspondence between the observable Q' on H and an observable on k's Hilbert space.

identical – and weak – that individuals sharing all monadic *and relational* properties are identical – forms of the Principle of the Identity of Indiscernibles.

But of course there is more. For if i and j are identical bosons or fermions then $P_{ij}\Psi = \pm\Psi$ is always true: they satisfy the 'symmetrization postulate'. And so, conclude Butterfield and Co., a pair of identical bosons or fermions always violate PII.

A word of clarification is in order. One might well think that if we are, say, talking about bosons, then since the system is only ever in symmetrized states, the appropriate Hilbert space is just the *symmetrized sector* of the tensor product space – just those states for which $P_{ij}\Psi = \Psi$. But strictly Q_i and Q_j are not observables on the symmetric sector, but only on the full tensor product space, so clearly these results refer to the full space. Now, it's easy to see that the *restrictions* of Q_i and Q_j to the symmetric sector are identical, and so one might say that they are not really different observables at all for bosons. But French and Redhead (1988) are sensitive to this kind of worry, for they argue that one can indeed take the full tensor product space as the state space for the bosons, taking all states, symmetrized or not, to be possible but – in a terminology I propose – taking only the symmetric ones to be 'preparable'. After all, they point out, we can understand the unpreparability of the unsymmetrized states in terms of a boundary condition (that the initial state was symmetrized) and the symmetrization of the Hamiltonian (which guarantees that the symmetry type is preserved).[3] So we can – and will for the purposes of this discussion – take Q_i and Q_j to be distinct operators, even though their restrictions to the symmetric sector are not distinct. Finally, however, one might object that Q_i and Q_j so construed are no longer observables, since they violate the 'indistinguishability postulate' (IP), according to which every observable must commute with every permutation of identical particles: $[P_{ij}, O] = 0$ in this case. French and Redhead's response is analogous to their reply to the first objection: they allow 'observables' that cannot be observed, so that, in principle, any Hermitian operator, whether commuting or not, might be used to discern particles. I will follow this suggestion, but (until section 3) I will use 'observable' to refer *only* to those Hermitian operators that satisfy IP. Hence we want to know whether PII is violated not for the smaller set of properties (those which correspond to observables), but for the larger set (those corresponding to all Hermitian operators) which contains the smaller as a subset. (Of course if the PII is violated for the latter it is automatically violated for the former, since it is a subset, and hence contains no more properties that might discern particles.) What the results show is that two identical bosons are indeed indiscernible even in this broader sense (and so are identical fermions).[4]

[3] This view has its problems, especially those raised by Redhead and Teller (1991).

[4] Two remarks: first, when we look at other kinds of particles we will find that it is quite possible to consider observables that violate IP, and so we can avoid the debate about whether French and Redhead's scheme is legitimate. Second, Saunders (this volume) argues for a rather different and highly plausible understanding of PII in quantum mechanics (see also French and Rickles, this volume).

Consider how the proofs of these results go, following the style adopted by Butterfield (1993). What we need in order to calculate these probabilities is the probabilities – for state Ψ – of such propositions as the atomic $Q_i = q$ and conjunctive $Q_i = q$ & $Q'_j = p$, since the conditional probabilities can be found in terms of these unconditional probabilities. But these quantities can be expressed in terms of the expectation values for the projection operators onto the eigenstates with the appropriate eigenvalues (and products of such operators). For example, let the projection operators onto the subspace of states that are eigenstates of Q_i and Q'_j with eigenvalues q and p respectively be denoted ${}^qQ_i$ and ${}^pQ'_j$ respectively. And so on in the obvious way for the other projection operators. Then according to the standard QM algorithm, for instance:

$$\mathrm{pr}^{\Psi}(Q_i = q) = \langle\Psi|{}^qQ_i|\Psi\rangle \tag{5}$$

and

$$\mathrm{pr}^{\Psi}(Q_i = q \,\&\, Q'_j = p) = \langle\Psi|{}^qQ_i\,{}^pQ'_j|\Psi\rangle \tag{6}$$

Finally, the purely algebraic relation $(P_{ij})^2 = I$ plus the fact that $Q_j = P_{ij}Q_iP_{ij}$ – let's call this the 'conjugacy condition' (CC) – which implies that ${}^qQ_j = P_{ij}\,{}^qQ_iP_{ij}$, will deliver our proofs. Given $P_{ij}\Psi = \pm\Psi$,

$$\begin{aligned}\mathrm{pr}^{\Psi}(Q_i = q) &= \langle\Psi|{}^qQ_i|\Psi\rangle\\ &= \langle\Psi|P_{ij}\,{}^qQ_jP_{ij}|\Psi\rangle\\ &= \langle\Psi|{}^qQ_j|\Psi\rangle\\ &= \mathrm{pr}^{\Psi}(Q_j = q)\end{aligned} \tag{7}$$

and

$$\begin{aligned}\mathrm{pr}^{\Psi}(Q_i = q \,\&\, Q'_j = p) &= \langle\Psi|{}^qQ_i\,{}^pQ'_j|\Psi\rangle\\ &= \langle\Psi|P_{ij}\,{}^qQ_jP_{ij}P_{ij}\,{}^pQ'_iP_{ij}|\Psi\rangle\\ &= \langle\Psi|{}^qQ_j\,{}^pQ'_i|\Psi\rangle\\ &= \mathrm{pr}^{\Psi}(Q_j = q \,\&\, Q_i = p)\end{aligned} \tag{8}$$

and so on for any of the atomic or conjunctive propositions that we need in order to derive the conditional probabilities, and hence they are equal too, proving the theorems.

2 Generalizations

I want to generalize these results still further, in two ways. First, why stop with conditionalizing on only one or two atomic propositions? Why not take the most general relational probability to be

$$\begin{aligned}\mathrm{pr}^{\Psi}(\Pi_{ij}) \equiv \mathrm{pr}^{\Psi}(Q_i = q_i \;\&\; Q'_i = p_i \;\&\; \cdots \;\&\; R_j = r_j \;\&\; S_j = s_j \;\&\; \cdots \\ \&\; T_k = t_k \;\&\; U_k = u_k \;\&\; \cdots \;|V_i = v_i \;\&\; W_i = w_i \;\&\; \cdots \\ X_j = x_j \;\&\; Y_j = y_j \;\&\; \cdots \;\&\; Z_k = z_k \;\&\; A_k = a_k \;\&\; \cdots) \quad (9)\end{aligned}$$

where all the operators are of the same form as Q_i, with a suitable Hermitian operator on H replacing Q in the appropriate 'slot' in the tensor product.[5] (And why not let the ellipses be filled in with propositions concerning fourth and fifth and so on particles?) Then we need to show that if $P_{ij}\Psi = \pm\Psi$ then $\mathrm{pr}^{\Psi}(\Pi_{ij}) = \mathrm{pr}^{\Psi}(\Pi_{ji})$, where the latter argument is obtained from Π_{ij} by transposing the i and j labels on the observables throughout. And to show this we need – much as in the cases considered so far – to show that the probability of an arbitrary conjunction of atomic propositions is equal to that of its transposition. And to do this we need one more fact about these operators, namely that for $k \neq i, j$, $P_{ij}Q_kP_{ij} = Q_k$, from which it follows that $P_{ij}\,{}^{q}Q_kP_{ij} = {}^{q}Q_k$ or $P_{ij}\,{}^{q}Q_k = {}^{q}Q_kP_{ij}$ (and similarly for any of the other operators in question) – let's call this the 'independence condition' (IC).[6] Then we see for example that

$$\begin{aligned}&\mathrm{pr}^{\Psi}(Q_i = q_i \;\&\; Q'_i = p_i \;\&\; \cdots \;\&\; R_j = r_j \;\&\; S_j = s_j \;\&\; \cdots \\ &\qquad\qquad \&\; T_k = t_k \;\&\; U_k = u_k \;\&\; \cdots) \\ &\quad = \langle\Psi|{}^{q_i}Q_i\,{}^{p_i}Q'_i \ldots {}^{r_j}R_j\,{}^{s_j}S_j \ldots {}^{t_k}T_k\,{}^{u_k}U_k \ldots |\Psi\rangle \quad (10)\\ &\quad = \langle\Psi|P_{ij}\,{}^{q_i}Q_jP_{ij}P_{ij}\,{}^{p_i}Q_jP_{ij} \ldots \\ &\qquad P_{ij}\,{}^{r_j}R_iP_{ij}P_{ij}\,{}^{s_j}S_iP_{ij} \ldots {}^{t_k}T_k\,{}^{u_k}U_k \ldots |\Psi\rangle \quad (11)\end{aligned}$$

by CC

$$= \langle\Psi|P_{ij}\,{}^{q_i}Q_j\,{}^{q'_i}Q'_j \ldots {}^{r_j}R_i\,{}^{s_j}S_i \ldots {}^{t_k}T_k\,{}^{u_k}U_k \ldots P_{ij}|\Psi\rangle, \quad (12)$$

using $P_{ij}^2 = I$ and applying IC to shift the remaining P_{ij} all the way to the right, to obtain for (anti)symmetrized Ψ

$$\begin{aligned}&= \langle\Psi|{}^{q_i}Q_j\,{}^{p_i}Q'_j \ldots {}^{r_j}R_i\,{}^{s_j}S_i \ldots {}^{t_k}T_k\,{}^{u_k}U_k \ldots |\Psi\rangle \quad (13)\\ &= \mathrm{pr}^{\Psi}(Q_j = q_i \;\&\; Q'_j = p_i \;\&\; \cdots \;\&\; R_i = r_j \;\&\; S_i = s_j \;\&\; \cdots \\ &\qquad\qquad \&\; T_k = t_k \;\&\; U_k = u_k \;\&\; \cdots), \quad (14)\end{aligned}$$

[5] Note that in this case, and as we go on, the observables in question – and hence their projection operators – need not all commute. Thus the conditional probability, found as before in terms of the expectation values of the corresponding products of projection operators is dependent on the order of the operators. We will adopt the convention that the order of the projection operators is the same as the order of the propositions in the expression for the probability.

[6] In fact, as you will have found if you've worked through the proofs, this condition is also required to prove the fourth result of Butterfield and Co.

as we required. Then, since the probabilities for an arbitrary conjunction and its $i \leftrightarrow j$ transposition agree, no conditional probabilities for the system taking on values for the Hermitian operators in question will serve to discern identical bosons or fermions in a system.

There is, however, a further generalization that I think should be made, one which also preserves the result regarding PII. Why assume that only operators of the form $I \otimes \ldots \otimes Q \otimes \ldots \otimes I$ can represent single-particle properties? Of course they do, but, I would suggest, as a special case. Here's the more general notion that I have in mind. Suppose that for an n-particle system we find a 'family' of n Hermitian operators, $\{O_1, \ldots O_i, \ldots O_j, \ldots O_n\}$, that satisfy CC pair-wise: $O_j = P_{ij} O_i P_{ij}$ for all i, j. Then it seems to me that they are candidates for representing single-particle properties, since the natural interpretation is that whatever quantity O_i represents of the ith particle, O_j represents of the jth particle, and so on. Clearly Q_i, Q_j and so on satisfy this condition, but it allows other possibilities too: for example, $A \otimes A \otimes \ldots \otimes B \otimes \ldots \otimes A$ (A, B observables on H) and certain sums of such operators. Now one might argue that being a *single-particle* Hermitian operator requires more than membership in such a family, but no matter for our purposes: the 'minimal condition' of *membership in such a family of single-particle operators* is enough to prove all our results. For these results only depend on two properties of the single-particle operators: CC and IC. CC for Q_i and Q_j etc. follows immediately (as a special case) of the minimal condition, and IC can be easily shown. Suppose that the minimal condition is satisfied by Hermitian operators $\{O_1, \ldots O_n\}$. Then

$$O_k = P_{jk} O_j P_{jk} = P_{ij} P_{ik} P_{ij} O_j P_{ij} P_{ik} P_{ij} = P_{ij} P_{ik} O_i P_{ik} P_{ij} = P_{ij} O_k P_{ij}, \quad (15)$$

where the second step uses the algebraic identity $P_{jk} = P_{ij} P_{ik} P_{ij}$, which holds if $k \neq i, j$, and the other steps by the minimal condition. And so the general proof goes through – for bosonic and fermionic i, j – as long as the minimal condition is satisfied; so identical bosons and fermions are indistinguishable by anything we might consider to be a single-particle property.

This way of constructing the proofs is nice, I think, because it shows very clearly just what assumptions are doing the work, and because it points the way to extending these considerations to kinds of particles other than bosons and fermions, as French and Rickles suggest.

3 Quarticles

First though, what are these other kinds of particles? As French and Rickles describe (and see for example Greiner and Muller, 1989, chapter 9, for more details), the bosonic and fermionic representations of the permutation group S_n are only two possible representations – the one-dimensional ones. As the number of particles increases, one finds representations of higher and higher dimensions. This raises the question of whether there could be a species of quantum particle for every representation – as bosons and fermions correspond to the symmetric and antisymmetric representations – and indeed whether there could be a species of particle for every direct sum of representations. This is the question explored and answered by Hartle, Stolt, and Taylor (e.g. 1970). More precisely, we want to characterize each (statistical) kind of particle by a rule that specifies, for each n, what the space of allowed states is – after all, a species of particle can come in any number. In this case it is natural to impose a condition of 'cluster decomposition'. First, suppose that one looks at m of the particles in a system of n: the rule for n such particles determines their state space, from which we can extract the state space of the first m, which must carry exactly the same representations as the rule for m particles demands. Second, in the other direction, if we take two systems of m and n particles respectively, then the rule determines their state spaces, and then the joint state space obtained from them must satisfy the rule for $n + m$ particles. That is, cluster decomposition demands that parts and wholes must be related consistently by the state space rule for any species of particle.

Hartle, Stolt, and Taylor showed that this condition entails a very simple classification of the species of possible particles. They found a correspondence between pairs of natural numbers (p, q) satisfying $p + q > 0$ *together with* ∞ and the allowed species of particles. While these numbers determine which representations are allowed to any species of particle in a fairly direct way, space does not permit that I explain how here.[7] There are, however, a few points and illustrations that I can usefully make.

First, for bosons (or fermions) all Hermitian operators on the space of symmetric (or antisymmetric) states satisfy IP, so all are observables as we defined them earlier. In the higher dimensional representations this is not so: there are Hermitian operators on a (p, q) state space that do not commute with all permutations. Hence we should no longer define 'observables' to be only those operators that satisfy IP – any Hermitian operator on the allowed space could in principle represent a distinct observable quantity. And so corresponding to (almost) every (p, q) and ∞ there is both a possible 'distinguishable' kind of particle for which all Hermitian operators on the space are observables and a distinct possible 'indistinguishable' kind of particle for which IP is imposed on observables.[8] That is, in choosing the properties of a type of particle, nature needs to decide not just p and q, but also whether any physical quantities correspond to observables violating IP.

Then, for ∞ particles the whole tensor product space is the state space, and if they are distinguishable then they are called 'quantum Maxwell–Boltzmann particles',

7 See for example N. Huggett, in preparation, 'What is an elementary quarticle?', appendix, for an account.

8 R. Espinoza, T. D. Imbo, and M. Satriawan, 'Identicality, (in)distinguishability and quantum statistics', in preparation, ask what other sets of Hermitian operators – fewer than all those on the state space but more than just those that satisfy IP – can be taken as the observables, and show how to classify all possible answers.

since the number of independent states available to them is always the same as the number of corresponding classical states. Bosons and fermions are the only cases – because all operators on the (anti)symmetric sector satisfy IP – for which there is no distinguishable kind. 'Parabose' particles of order p (those for which up to p particles may be mutually antisymmetrized) are $(p, 0)$ particles, while 'parafermi' particles of order q are $(0, q)$ particles. Now it is usual to speak of these last two kinds of particles as 'paraparticles' but we are without a convenient terminology to refer to all particles in the (p, q) classification (other than the bland '(p, q)-particles'). To show that these kinds of particles interpolate between quanta (bosons and fermions) and classical(-like) particles (quantum Maxwell–Boltzmann particles) let's call them 'quarticles', of which paraparticles are a special class. And so, French and Rickles' question concerns PII for quarticles: When (if ever) does it hold? When (if ever) does it fail?[9] Here's a (fairly) comprehensive answer to this question, in three parts:

(i) For any number of any kind of identical quarticle (even as a subsystem of a system including other non-identical particles) there are states in which no particles are discernible. This follows because for any quarticles either totally symmetrized or totally antisymmetrized states (for which $P_{ij}\Psi = \pm\Psi$, where i, j label *any* pair of the identical quarticles) are allowed, and we have already seen that they are states of indiscernible particles.

(ii) For any number greater than two of any kind of identical quarticle but quanta there are states in which some particles are discernible and some are indiscernible.

Suppose there are m identical quarticles. Again let Q be a Hermitian operator on the single particle Hilbert space H, and define Q_1 etc. as before (i.e. $Q_1 = Q \otimes I \otimes \ldots \otimes I$ etc.), but now supposing that Q is non-degenerate with eigenstates $\phi_1, \phi_2, \ldots, \phi_m, \ldots$ with corresponding eigenvalues $q_1, q_2, \ldots, q_m, \ldots$ (so that H is at least m-dimensional). Let $S_{ab\ldots c}$ be the operator on the tensor product space that totally symmetrizes states with respect to transpositions of the ath, bth, $\ldots c$th factors, and similarly let $A_{ab\ldots c}$ be the operator on the tensor product space that totally *anti*symmetrizes states with respect to transpositions of the ath, bth, $\ldots c$th factors.[10]

Then for any system of $m > 2$ quarticles of any kind but bosons and fermions either

$$\Psi_s = S_{23\ldots m} A_{12} \phi_1 \phi_2 \ldots \phi_m \tag{16}$$

[9] French and Redhead (1988) consider a specific three-particle state in a two-dimensional representation (which is allowed to infinitely many kinds of quarticle) and show that two of the particles are indistinguishable while either can be distinguished from the third.

[10] Thus, for example, $S_{12\ldots m}\phi_1\phi_2 \ldots \phi_m \ldots \phi_n$ is the normalized sum of every permutation of the first m factors of $\phi_1\phi_2 \ldots \phi_m \ldots \phi_n$, and $A_{12\ldots m}\phi_1\phi_2 \ldots \phi_m \ldots \phi_n$ the normalized sum of the even permutations minus the odd permutations of the first m factors. For further discussion and the explicit form of these operators see for example Greiner and Muller (1989), chapter 9.

or

$$\Psi_a = A_{23\ldots m} S_{12} \phi_1 \phi_2 \ldots \phi_m \tag{17}$$

is an allowed state (of course *not* the only kind of allowed state). In either case the total (anti)symmetrization of quarticles $2, 3, \ldots, m$ entails that for $2 \leq i, j \leq m$, $P_{ij}\Psi = \pm\Psi$; so from our earlier proofs quarticles $2, 3, \ldots, m$ are indiscernible. However, we can easily calculate $\mathrm{pr}^{\Psi}(Q_1 = q_1)$ and $\mathrm{pr}^{\Psi}(Q_i = q_1)$ to see that these probabilities discern the quarticle 1 from every other quarticle. For example,

$$\Psi_s = S_{2\ldots m}(\phi_1 \phi_2 \ldots \phi_m - \phi_2 \phi_1 \ldots \phi_m), \tag{18}$$

a sum of $2 \cdot (m-1)!$ terms with equal coefficients (one for each permutation of the factors $2, 3, \ldots m$). $(m-1)!$ of the terms have ϕ_1 as the first factor – all those with positive coefficients. But only $(m-2)!$ of the terms have ϕ_1 as the ith factor – the number of permutations of the term with the negative coefficient which have ϕ_1 in the ith place. Therefore,

$$\mathrm{pr}^{\Psi_s}(Q_1 = q_1) = \langle \Psi |^{q_1} Q_1 | \Psi \rangle = (m-1)!/(2 \cdot (m-1)!) = 1/2 \tag{19}$$

while (for $2 \leq i \leq m$)

$$\mathrm{pr}^{\Psi_s}(Q_i = q_1) = \langle \Psi |^{q_1} Q_1 | \Psi \rangle = (m-2)!/(2 \cdot (m-1)!) = 1/(2 \cdot (m-1)), \tag{20}$$

which are never equal for $m > 2$. Hence quarticles 1 and 2 are indeed discernible (as they are in state Ψ_a).

(iii) For any number greater than two of any kind of identical quarticle but quanta there are states in which all the particles are discernible.

The following is an allowed state for $m > 2$ of any kind of quarticle but bosons and fermions:

$$\Psi_d = \sum_{i=1}^{m} S_{12\ldots(i-1)(i+1)\ldots m} A_{i(i+1)} \phi_1 \phi_2 \ldots \phi_m, \tag{21}$$

where $i \pm 1$ is taken modulo m, and Ψ_d is a sum of states similar to those considered in equation (16); for each of the first m factors of the tensor product, there is contribution to Ψ_d in which the factor and its successor are antisymmetrized before the remaining factors are symmetrized. Then – as in the proof of (ii) – we count the total number of terms in the sum in which ϕ_i appears as the ith (or jth) factor to find the numerator for $\mathrm{pr}^{\Psi_d}(Q_i = q_i)$ (or $\mathrm{pr}^{\Psi_d}(Q_j = q_i)$) and – since all the coefficients are equal – the denominator is just the total number of terms in the sum. Omitting the details we find that:

$$\mathrm{pr}^{\Psi_d}(Q_i = q_i) = \{(m-1)! + (2m-3) \cdot (m-2)!\}/(2 \cdot m!) \tag{22}$$

while for $i \neq j$

$$\mathrm{pr}^{\Psi_d}(Q_j = q_i) = \{(2m-4) \cdot (m-2)!\}/(2 \cdot m!), \tag{23}$$

which are unequal for all m. Hence every quarticle is discernible from every other by the probability for possessing an appropriate quantity.

4 Concluding remarks

In conclusion there are a couple of remarks I'd like to add. First, I have addressed the question of whether quarticles are discernible by any Hermitian operators, since this is how the issue was originally raised. But what if we consider instead just those Hermitian operators that satisfy IP, as the Q_i of the proofs do not? As we noted, if the quarticles are bosons (or fermions) then any Hermitian operator on the possible (preparable) states – the (anti)symmetrized sector – satisfies IP automatically, and so IP is not a *further* limitation on what quantities are observable. However, things are very different for other kinds of quarticle, since then the space of states allows IP violating Hermitian operators, in which case IP actually does limit which quantities are observable. Hence if we ask again the question about PII for quarticles, but now consider only those Hermitian operators that are observables, we should expect different answers depending on whether we impose IP or not.

To indicate that we are now asking a different question we should have a new name for the principle at stake. The principle is that for every pair of (identical) quarticles there is some conditional probability for the possession of single particle *observable* quantities on which they disagree: let's call this the 'Principle of the Identity of *Indistinguishables*'. We will investigate whether this principle holds exactly as before, except we demand that the probabilities we consider are only for observable quantities.

Clearly if IP is maximally violated, in the sense that every Hermitian operator is an observable, then the conclusions concerning PII will follow here too, since the Hermitian operators in those arguments (and their restrictions to the quarticle state spaces) will be observables. If, on the other hand, the IP is imposed then we will find that the identity of indistinguishables is maximally violated – any pair of identical quarticles in a system will have all probabilities for the possession of observable quantities in common (just like bosons and fermions).[11]

The second remark is a question. Our calculations made clear that the (anti)symmetrization of a pair of quarticles is sufficient for a violation of PII.

[11] This is easy to see intuitively: if the minimal condition for being a single particle observable – inclusion in a family of observables related by CC pair-wise – is satisfied, then we have for all i, j, $O_j = P_{ij} O_i P_{ij}$. But by IP $P_{ij} O_i P_{ij} = O_i$ so $O_i = O_j$. Thus the probability for the ith quarticle to possess some given value of whatever quantity O_i represents must be the same as the probability for the system to possess the same value for O_j, which represents the same property but for the jth particle. For the conjunctive probabilities required to find the more general probabilities note that if a Hermitian operator satisfies the IP so do the corresponding projection operators and products of projection operators. Thus a calculation like that of equations 10–14 shows that IP is sufficient to show that for any Ψ, $\mathrm{pr}^{\Psi}(\Pi_{ij}) = \mathrm{pr}^{\Psi}(\Pi_{ji})$ as required.

But what about the converse? If none of the conditional probabilities for the possession of values of any Hermitian operators will discern particles i and j in some state, Ψ, can we infer that $P_{ij}\Psi = \pm\psi$? Consider, for example, a two-particle case in which $Q_1 = Q \otimes I$ and $Q_2 = I \otimes Q$, with ϕ_i the eigenstate of Q with eigenvalue q_i. Then if $\Psi = \phi_1\phi_2 + e^{i\theta}\phi_2\phi_1$ it is easy to check that, for instance, $\mathrm{pr}^{\Psi}(Q_1 = q_1) = \mathrm{pr}^{\Psi}(Q_2 = q_1)$. And so even though $P_{ij}\Psi \neq \pm\Psi$ for $\theta \neq n\pi/2$, the Hermitian operators we used to discern quarticles will not discern these two particles. So the question is, even though these probabilities do not discern the particles, does $P_{ij}\Psi \neq \pm\Psi$ entail that *some* conditional probability for the possession of values for some single-particle Hermitian operators will discern them? If so then we have the most complete account of PII in many-particle quantum mechanics: a pair of identical particles is indiscernible just in case they are mutually (anti)symmetrized.

References

Butterfield, J. (1993). 'Interpretation and identity in quantum theory'. *Studies in the History and Philosophy of Science*, **24**, 443–76.

French, S., and Redhead, M. (1988). 'Quantum physics and the identity of indiscernibles'. *British Journal for the Philosophy of Science*, **39**, 233–46.

Greenberg, O. W. (1991). 'Particles with small violations of Fermi or Bose statistics'. *Physical Review D*, **43**, 4111–20.

Greiner, W., and Muller, B. (1989). *Quantum Mechanics: Symmetries*. New York: Springer-Verlag.

Hartle, J. B., Stolt, R. H., and Taylor, J. R. (1970). 'Paraparticles of infinite order'. *Physical Review D*, **2**, 1759–60.

Redhead, M., and Teller, P. (1991). 'Particle labels and the theory of indistinguishable particles in quantum mechanics'. *British Journal for the Philosophy of Science*, **43**, 201–18.

14

Handedness, parity violation, and the reality of space

OLIVER POOLEY

1 Introduction

This paper is about asymmetry rather than symmetry. More specifically, it is concerned with the failure of *parity* to be a symmetry of elementary particle physics. The parity operation is spatial reflection through the origin: $(t, \vec{x}) \mapsto (t, -\vec{x})$. In the context of quantum field theory, it is closely connected to two other discrete transformations, namely *time reversal* (temporal reflection through the origin) and *charge conjugation* (the interchange of matter and antimatter). Although all three fail to be symmetries of the Standard Model, it is a theorem that their combination, CPT, is a symmetry of every local, Lorentz invariant quantum field theory.[1]

To understand parity violation (the failure of parity to be a symmetry) it is useful to consider another type of asymmetry: the sort of spatial asymmetry exemplified by human hands. Hands lack any plane of mirror symmetry. As a result they come in two varieties: left hands and right hands. Similarly we talk of left-handed and right-handed screws, left-handed and right-handed molecules, left-handed and right-handed coordinate systems, and so on.[2] Objects of opposite handedness that are otherwise qualitatively identical are 'mirror-images' of each other. Kant was the first philosopher to see something interesting in such objects. He called them *incongruent counterparts*. They clearly differ in some way. For example, a glove which might be a perfect fit for a right hand will not fit on its left-handed incongruent counterpart.

[1] See for example Streater and Wightman (1964). Currently there is no evidence that CPT fails to be a symmetry of our world. It should be noted, however, that some of the quantum field theory axioms used in proofs of the CPT theorem might in fact be false. If they are, CPT might not be a symmetry. For recent experimental and theoretical consideration of this possibility see Kostelecky (2002).

[2] Note that it is not required that the objects lack every sort of spatial symmetry. Screws, for example, can have (discrete) rotational symmetry if their threads are of the correct pitch.

Such objects are pertinent to parity violation because (in spaces of an odd number of dimensions) parity maps handed objects onto their incongruent counterparts.[3] If parity is a symmetry of a theory, it will always map physically possible states of affairs onto physically possible states of affairs. Hence if a particular handed object or process is physically possible, parity conservation implies that its incongruent counterpart will also be physically possible. Conversely, if parity fails to be a symmetry, then there will be at least *some* cases where it maps a physically possible state of affairs onto one that the law prohibits. If spatial translations and rotations *are* symmetries of a theory, then these will be cases where there is a type of handed object or process of one handedness that is physically possible and yet its incongruent counterpart is not. (For probabilistic theories, parity also fails to be a symmetry if *different* probabilities are assigned to a pair of counterpart yet incongruent processes.)

In the next section of this paper I review what philosophers have had to say about three questions: (i) What is the ground of the difference between incongruent counterparts? (ii) What is it to be a handed object? (iii) What is it to be of one particular handedness rather than the other? Kant first raised these questions in the context of a philosophical dispute, still very much alive today, that goes by the name of the substantivalist–relationalist debate (Kant, 1992a; German original, 1768).[4]

I will side with most – although admittedly not all – philosophers in defending an account of incongruent counterparts according to which they are *intrinsically identical*.[5] Moreover, I will defend a *relational* account of handedness according to which the difference between incongruent counterparts is grounded in their relations to each other and to other material objects. Kant thought that there were reasons to reject such an account. Initially he concluded that the difference between left and right hands did indeed come down to a difference in their relational attributes, but that these involved relations to 'universal space as a unity' (Kant, 1992a, p. 365). Not long after reaching this conclusion, he also rejected this substantivalist account

[3] Consider the Cartesian coordinates of a left hand relative to an arbitrary (left- or right-handed) set of axes. Now consider a passive parity inversion: a point that originally had coordinates (x, y, z) is assigned new coordinates $(-x, -y, -z)$. The coordinates of the left hand with respect to the new coordinate system are the coordinates of a possible *right hand* with respect to the original coordinate system.

[4] The debate concerns the ontological status of space. Its original protagonists were, on one side, Newton and Samuel Clarke and, on the other side, Leibniz. Substantivalists follow Newton in seeing space (or, in the context of relativistic physics, spacetime) as some kind of substance. It is as real, and as fundamental, as the material objects and events that exist in it. Relationalists follow Leibniz in denying that space is a fundamental entity. They do not deny that material objects are spatially extended. Nor do they deny that material objects stand in determinate distance relations from one another. But they hold that such facts are basic. They deny that space exists apart from such facts; that space has a reality of its own, independent of material objects and their spatial properties.

[5] Harper (1991) and Walker (1978) are two exceptions.

of handedness. Instead he now believed that the difference between incongruent counterparts was fundamentally incomprehensible: that it could only be grasped in perception, through a 'pure intuition,' and not by any 'characteristic marks intelligible to the mind through speech' (Kant, 1992b, p. 396; German original, 1770).

Consideration of Kant's arguments sets the stage for discussion of a more recent argument against relationalism due to John Earman. In *World Enough and Space-Time* (1989), he argues that the fact that our world violates parity and thus displays a *law-like* left–right asymmetry poses a serious challenge to relational accounts of handedness. This paper investigates whether viewing incongruent counterparts as intrinsically identical is compatible with parity-violating laws of nature, and also whether the fact that Nature distinguishes left and right has implications for the substantivalist–relationalist debate.

2 Incongruent counterparts

Imagine that you are given a model of a left hand and a perfect mirror-image (i.e. right-handed) duplicate of it. The distance between the tip of the thumb and the index finger will be the same for both hands. Similarly, the angles that the thumbs make to the planes of the palms will be identical in both cases. The two hands are perfectly identical in terms of the distances between their corresponding parts. Kant's way of making this point was to note that a complete description of one hand in terms of the positions of the parts relatively to one another will also be true of its mirror-image (Kant, 1992a, p. 370). For this reason he called them *counterparts*.

Yet despite this similarity, the two hands are nevertheless *incongruent*: they cannot be made to coincide – they cannot be superposed – by any rigid motion. Kant's own description of the incongruence runs: 'the limits of the one cannot also be the limits of the other' (1992a, p. 369). A little later he makes the same point by noting that the surface which encloses the physical space of one hand 'cannot serve as a boundary to limit the other, no matter how that surface be twisted and turned' (*ibid.*, p. 371). It is worth stressing that the relevant notion of possibility here is not that of physical possibility. It is physically impossible to superpose a left hand and its perfect *left-handed* duplicate if they are both solid material objects. Rather we must abstract from such physical limitations and consider whether it is *mathematically* possible for the distances between the two objects to be changed continuously in such a way that the two objects eventually coincide. By restricting ourselves to *rigid* motion, we are only considering changes of the total set of distances that preserve the internal distances between the parts of the two objects.

It is time to consider the question: *how is it possible that two objects which are counterparts can nevertheless be incongruent?* Answers will fall into one of two categories, within which there are further divisions. One's answer might involve the claim that left- and right-handed varieties of an object differ intrinsically, and then go on to exploit these differences in explaining their incongruence. Within this category one might view these intrinsic differences as primitive and unanalysable (Kant's later transcendental idealist position is perhaps of this type), or one might view the intrinsic differences as resulting, for example, from a difference in the way the parts of the objects are related to each other.[6] Alternatively one's answer might involve the claim that incongruent counterparts do not differ intrinsically in any way, and then seek to explain their incongruence in terms of something that is external to each object taken by itself. Kant's earlier substantivalist position is of this type: the difference is to be explained in terms of the objects' different relations to substantival space. But so too is the most economical, purely relational account according to which incongruence is explicable using only the resources of relative distances.

Kant's transcendental idealism aside, the view that left-handed and right-handed objects differ intrinsically has not been popular. There are at least three strong objections to such a view, one of a general nature and two that are specific. The general objection is that it is entirely unclear that we have any conceptual grasp of what such intrinsic differences involve.[7] In particular, no one has provided an illuminating account of how such intrinsic differences connect with incongruence, or with relevant practical abilities; for example, that I can without difficulty tell of a left hand that it is left-handed and distinguish it from a right hand.

The two specific objections are variations on a theme.[8] The first exploits the fact that two hands will never be incongruent if embedded in a non-orientable space.[9] So can hands in such spaces instantiate different intrinsic properties (primitive or otherwise), or not? If they can, one will be able to move a 'left-handed' object into the space occupied by a 'right-handed' one. Does a hand originally exemplifying the property of being 'left-handed' come to lose this basic property and acquire that of being 'right-handed' merely as a result of such a motion? Although they are said to differ in some primitive property, this property appears redundant, making no contact with any other spatial facts about the hands. But if hands in such spaces cannot instantiate the primitive properties of being left- or right-handed, then how is one meant to understand the dependence of whether or not an object can instantiate such a property on the type of space in which it is embedded?

6 Van Cleve calls this latter position 'internalism' (Van Cleve, 1991, p. 22). Clearly such relations must include more than just the relative distances and relations that are reducible to them.

7 This objection does not tell against Kant's later position, for he was at pains to deny that we have a conceptual understanding of the left–right distinction; it is supposedly only grasped in experience.

8 The following type of considerations are emphasized by Nerlich (1994, pp. 51–3), but see also Frederick (1991, p. 8) and Van Cleve (1991, pp. 22–3).

9 Or, to adopt relationalist language, if the spatial relations between them are such that the lowest-dimensional spaces in which they are embeddable are all non-orientable. Some believe that failure to provide an account of the orientability of space might show that relationalism is ultimately untenable. I disagree, for reasons elaborated on later in this section.

The other objection runs along the same lines, but this time exploits the fact that the incongruence of two hands in part depends on the dimensionality of the space in which they are embedded. One could move a left hand into the space occupied by a right hand if there was an extra spatial dimension through which it could be moved,[10] just as the letter 'F' can be brought into coincidence with its mirror-image on the page, ꟻ, if one is allowed to lift it off the page. Does whether or not hands have primitive left-handed or right-handed properties depend on the dimensionality of space?

These considerations give us more than enough reason to see whether incongruence can be explained without recourse to the postulation of intrinsic differences. I start by considering the purely relational account. Can we get by merely with relative distances?

The incongruence of left and right hands shows that they differ in some respect. It is surely reasonable to call this difference a purely *spatial* difference. That left and right hands match in terms of the distances between their parts shows that this difference is not grounded in *these* distances. However, the relationalist is not committed to the view that every spatial difference between two objects supervenes on a difference in the spatial arrangement of their parts. He can also appeal to the distance relations that hold between the two objects, and between them and other objects. Once this is acknowledged, there would seem to be no reason why the relationalist cannot view the incongruence of two counterparts as grounded in such external relations.

Now, nothing in my original definition of incongruence precluded relationalism.[11] In fact, one might even think that incongruence has been defined in purely relational terms: rigid motion is defined in terms of the constancy of the distances between the parts of an object, and coincidence (the occupancy of the same boundaries) is defined in terms of the distances between (the corresponding parts of) the two objects. Even when we seek the *ground* of such incongruence, it seems that the relationalist has no reason to be embarrassed. It is simply a mathematical fact, *and a comprehensible one*, that, when constrained to obey the algebraic relationships of Euclidean geometry, some numbers (the possible distances between two congruent counterparts) can be continuously altered so as to vanish while others (the distances between two incongruent counterparts) cannot.

The claim that the relationalist can not only accommodate but can *explain* incongruence is significant. As already noted, Kant came to the view that the incongruence of counterparts was, in a certain sense, fundamentally incomprehensible. He

[10] Although string theorists would have us believe there are in fact six or more such dimensions, they are 'too small' to permit the required motion.

[11] The definition deliberately does not follow others that can be found in the literature; see for example Earman (1989, chapter 7), Brighouse (1999), and Huggett (2000).

thought that it could be grasped only in experience. In his Inaugural Dissertation he writes (1992b, p. 396):

> between solid bodies which are perfectly similar and equal but incongruent, such as left and right hands (in so far as they are conceived only according to their extension), or spherical triangles from two opposite hemispheres, there is a difference, in virtue of which it is impossible that the limits of their extension should coincide – and that, in spite of the fact that, in respect of everything which may be expressed by means of characteristic marks intelligible to the mind through speech, they could be substituted for one another. It is, therefore, clear that in these cases the difference, namely the incongruity, can only be apprehended by a certain pure intuition.

But this is simply a *non sequitur*. Suppose that incongruent counterparts *are* intrinsically identical; that they do not differ *in themselves* in any way. So *a fortiori* we cannot, restricting ourselves to just the internal distances between the parts of two hands, understand or explain the hands' incongruence. Nevertheless we can both understand and explain their incongruence in terms of the different ways any two hands can be related to each other. It just does not follow from the fact that we cannot intellectually grasp an *intrinsic* difference between left and right hands that we can have no intellectual grasp of the basis of their incongruence, or that this incongruence is manifest only in experience.

Those whose intuitions lead them still to side with Kant at this point need to respond to the following challenge: suppose, for the sake of argument, that the relationalist is correct in asserting that all spatial facts are reducible to facts about relative distances between material objects. The relationalist will insist that any two coexisting hands stand in some quite determinate distance relations from one another. How can it be denied that the possibility of incongruent counterparts is already secured?[12] When things are put this way, surely the burden of proof is now on someone who wishes to assert that the incongruence or otherwise of the two hands is not determined *despite* the various facts about the distances between the them.

So far we have seen that relative distances alone are sufficient to ground the incongruence of two handed objects that are otherwise identical. However, there are two other questions about handedness that have exercised philosophers. The first is: in virtue of what is an object handed? The second is: what accounts for the particular handedness of a handed object – what makes it, say, a *left* hand?

My initial characterization of a handed object suggested that its key feature was that it lacked any plane of mirror symmetry. This is, again, a feature highlighted by Kant, who noted that a handed object cannot consist of 'two halves which are

[12] I have been tacitly assuming that the relative distances involved are those of an infinite N-dimensional Euclidean space. Things are obviously more complicated when one considers more general sets of relative distances. For some discussion, see Pooley (2002, chapter 6).

symmetrically arranged relatively to a single intersecting plane' (Kant, 1992a, p. 370). One might think that this is a characteristic that is reducible to facts concerning the relative distances between the parts of the object.[13] However, as Nerlich notes, more is needed. He defines an *enantiomorph* as follows (Nerlich, 1994, p. 51):

> each reflective mapping of [an enantiomorph] differs in its outcome from every rigid motion of it.

Otherwise the object is a *homomorph*. Nerlich's principal contention is that 'whether a hand . . . is enantiomorphic or homomorphic depends on the nature of the space it is in. In particular it depends on the dimensionality or the orientability, but in any case on some aspect of the overall connectedness or topology of the space' (*ibid.*, p. 53). Nerlich's claim, then, is that an object's being handed is not reducible to facts about the relative distances between material objects. It also depends on the dimensionality and orientability of the space in which the object is embedded.

Even if one were to agree with Nerlich about this, it is not clear that this observation can be used as an argument against a relational account of space, and for two reasons. The first is that we should ask why the relationalist about space is under any obligation to offer an equivalent, relationally pure definition of enantiomorphy. He believes that spatial facts are exhausted by the catalogue of relative distances between material points (and the fact that these must obey certain constraints). We have seen that this is enough to allow for the possibility of incongruent counterparts. If it turns out that substantivalism underwrites properties, such as enantiomorphy, which are not well defined by the lights of the relationalist's ontology, then so much the worse for enantiomorphy. Nothing in our experience of objects such as hands forces us to admit the existence of such additional properties, just as (so the relationalist would like to maintain) nothing in our theorizing about motion forces us to admit the reality of space.

Secondly, it has yet to be shown that the relationalist cannot provide a definition of enantiomorphy. Nerlich's observations might suggest that the relationalist needs to come up with a surrogate definition of orientability, and this is indeed the strategy that most have pursued (Brighouse, 1999; Huggett, 2000). Unfortunately for the relationalist, it has not been entirely successful. For example, although an object's being multiply related is a necessary condition of its being embeddable in a non-orientable space (for all non-orientable spaces are multiply connected), it is not a sufficient condition. Kant's own example of triangles on a sphere is precisely an example of multiply related yet enantiomorphic figures. Huggett suggests that what is needed is a 'general representation theorem of the form "space *S* is *orientable*

[13] Carol Brighouse, however, worries that talk of a plane of symmetry and of lines intersecting it at right-angles is not obviously relationally acceptable (Brighouse, 1999, pp. 56–8). The relationalist strategy that I outline below sidesteps Brighouse's worries.

iff relations of type ___ are instantiated" ' (Huggett, 2000, p. 225). No such theorem has been forthcoming. The relationalist also needs an account of the dimensionality of space.

I wish to propose that the relationalist has a way of sidestepping some of these difficulties. First, in order to be able to *exploit* (rather than explain) the fact[14] that the exact nature of an object's multiple relatedness (if it is multiply related) can determine whether or not the spaces in which it is embeddable are orientable or not, the relationalist does not need to have the type of representation theorem to which Huggett alludes. Second, as Huggett notes, the relationalist can talk freely of embedding the particular relative distances between the parts of some material object in a space, so long as the operation is understood to be a purely mathematical exercise (Huggett, 2000, p. 224).

So now suppose, additionally, that the relationalist has an account of the dimensionality of space.[15] He can then define the enantiomorphy of a material object by adopting Nerlich's definition, but now with respect to all abstract embedding spaces of the specified dimension. For example, if it is the case that according to an empirically adequate relational theory space has three dimensions, the relationalist can claim that an object is an enantiomorph iff, with respect to every possible abstract three-dimensional embedding space, each reflective mapping of the object differs in its outcome from every rigid motion of it. On this definition, planar objects count as homomorphs as do three-dimensional hands that are multiply related so as to be embeddable only within non-orientable three-dimensional spaces. Hands that are embeddable only within orientable three-dimensional spaces count as enantiomorphs, even though they are, of course, embeddable in spaces of higher dimensions.

I now return to the second question raised earlier: what accounts for the particular handedness of a handed object? To see that, once again, nothing more than relative distances is required, it will prove useful to consider the anti-relationalist argument of Kant's 1768 paper.

In this paper, Kant explicitly characterizes his aim as that of providing a 'clear proof that: *Absolute space, independently of the existence of all matter... has a reality of its own*' (1992a, p. 366). In other words he sets out to vindicate Newton's

[14] If it is a fact. This is something that the relationalist will want to prove. However, the mere possibility that it is a fact is enough, at this stage of the dialectic, to save the relationalist. The onus is now on the substantivalist to prove that the exact nature of an object's multiple relatedness does *not* fix the orientability of the lowest-dimensional embedding spaces. Thanks to Jeremy Butterfield and Carl Hoefer for saving me from overstating the relationalist's case.

[15] This might be fixed by the adoption of some specific relational dynamical theory. Relational theories typically simply assert, via the choice of some relative configuration space for example, that the relative distances between material objects are constrained to be embeddable in, say, a Euclidean space of no more than three dimensions; see, for example, Barbour and Bertotti (1982). Note that this is also likely to fix the orientability of space directly.

substantivalist conception of space over Leibniz's relationalist conception.[16] His argument does not challenge the claim that the relationalist can account for the *incongruence* of left and right hands. Rather it suggests that the difference between left and right goes beyond the relational facts so far cited.

After rehearsing the various definitional facts about incongruent counterparts and after noting that their incongruence cannot be grounded in a difference in how their parts are related, Kant makes the following claim (1992a, p. 371):

> ... imagine that the first created thing was a human hand. That human hand would have to be either a right hand or a left hand. The action of the creative cause in producing the one would have of necessity to be different from the action of the creative cause producing the counterpart.

Kant rightly notes that this is incompatible with relationalism:

> ... there is no difference in the relation of the parts of the hand to each other, and that is so whether it be a right hand or a left hand; it would therefore follow that the hand would be completely indeterminate in respect of such a property. In other words, the hand would fit equally well on either side of the human body; but that is impossible.

How should the relationalist respond to this particular challenge? He can simply deny Kant's initial premise, that every hand in an otherwise empty universe is necessarily either a right or a left hand. Certainly Kant is wrong to suppose that the lone hand's being of indeterminate handedness entails the absurdity that it can fit on *both* sides of a human body. For suppose that one is given a relational description of a hand and also a relational description of a handless human body that has various internal asymmetries involving the heart and other organs. One might then ask on which side of this body does the hand (properly) fit: the side on which the heart is, or on the other side?

The relationalist certainly should not answer 'both'. Rather he will deny that the question makes sense independently of a specification of the relative distances between the body and the hand. There are two incompatible ways in which a body satisfying the relational description and the lone hand could coexist in a single universe. According to one such way, the hand will fit on the side of the body that

[16] Some time prior to 1768, Kant is generally acknowledged to have held a Leibnizian, relational view of space. This, however, is a matter of some controversy amongst Kant scholars. Some argue that he is better seen as advocating some kind of compatibilism. Things are further complicated by the fact that, as already discussed, just two years after apparently arguing for a Newtonian view of space, Kant published his first 'critical' work, the Inaugural Dissertation of 1770, is which he rejects *both* substantivalism and relationalism, arguing instead that space is in some sense 'in us', a form of our intuition. The seeds of Kant's transcendental idealism about space are already discernible in the 1768 essay. However, the extent to which incongruent counterparts by themselves led Kant to transcendental idealism is again a matter of some controversy.

What should be stressed for the purposes of the present discussion is that it is evident that by 1768 Kant believed that his argument from incongruent counterparts provided a decisive reason to reject a purely relational account of handedness. Nowhere in his subsequent writings does Kant retreat from this claim.

the heart is on. According to the other way, the situation is reversed; the hand fits on the other side of the body to that on which the heart is. But either way, the hand will fit determinately on one, and only one, side of the body. And which side it fits is determined by the distances between the hand and the various parts of the body, i.e. by purely relational facts.

Although the relationalist's contention that the difference between left and right hands supervenes on the distances between them and between other material objects does not entail a patent absurdity, one might still wonder whether it is not in tension with our evident ability to recognize, for example, left hands as *left*. However, a little reflection suggests that the account the relationalist must give of our practical abilities and linguistic practices – of how we teach the meanings of 'left' and 'right' and of the fact that we are often prone to confuse left and right – is far more plausible than any account which postulates our recognizing an intrinsic difference between incongruent counterparts or recognizing that the hand bears some relation to (invisible) space. In fact, so far as I know, no one has attempted to give a genuine account of our abilities that postulates our recognizing either of these things.

The basic elements of a relational account have been outlined many times.[17] We have seen that, despite holding that left and right hands are intrinsically identical, the relationalist will also acknowledge that they fall into two equivalence classes defined, roughly speaking, by the relation of congruence. But it is then straightforward to understand how a practice of distinguishing members of these classes might involve all the hands of one class being given one 'name' ('left'), and all hands of the other being given another ('right'). Causal links between the speakers who are party to this practice, and between the speakers and actual hands, will ensure that the practice remains consistent. Together with an ability to recognize a hand as congruent or incongruent to hands with which one has previously been presented and has been told are left or right, these causal links are all that are required.

According to the relationalist account, therefore, the *only* facts about a left hand that make it left, is the fact that *we call it 'left'*, that it is congruent to every other hand that we in fact call 'left' and incongruent to every hand that we call 'right'.[18] Such an account of the meanings of 'left' and 'right' is, of course, very close to a causal theory of reference for proper names. And in certain respects the terms 'left' and 'right' are very much like names. What was it *about* Immanuel Kant, for example, that made it correct for his contemporaries to call him 'Immanuel Kant'?

[17] See, for example, Earman (1989, chapter 7, section 2), Gardner (1990, chapter 17), Huggett (1999, pp. 209–12), Hoefer (2000, section 3) and Saunders ('Incongruent counterparts: a Leibnizian approach', section 3.3, unpublished manuscript). These accounts, of course, differ from each other, and from my own, in minor ways.

[18] The question 'in virtue of what is a left hand left' is thus rather misleading. I should perhaps stress that my favoured relational account of handedness is not part of a general nominalism according to which the instances of any general term 'X' have in common only the fact that we call them all 'X'. It is only the left–right distinction in the 'which is which' sense, not their incongruence, that is purely nominal.

Nothing, other than the fact that he was actually known as 'Immanuel Kant', that there was a practice of calling him 'Immanuel Kant' and so on.

Can this really be all there is to the left–right distinction? I believe that it is. Such a point of view receives indirect support from what Jonathan Bennett calls the 'Kantian hypothesis' (Bennett, 1970).[19] This is the claim that chiral terms such as 'left' and 'right' cannot ultimately be explained without ostensively demonstrating, for example, a left hand. Various chiral terms can be explained in terms of each other. For example, one can define 'left' in terms of 'clockwise' and other, related, notions. But to break out of a rather *tight* circle, one must ultimately *show* what one means by 'clockwise' or by 'left.' Non-chiral words will never be enough.

This thesis can be put in the form of a predicament that Martin Gardner calls the *Ozma problem* (Gardner, 1990, chapter 18). Suppose that we are in radio contact with some extra-galactic civilization. Gardner's Ozma problem is (*ibid*., p. 167):

> Is there any way to communicate the meaning of 'left' by a language transmitted in the form of pulsating signals? By the terms of the problem we may say anything we please to our listeners, ask them to perform any experiment whatever, with one proviso: *There is to be no asymmetric object or structure that we and they can observe in common.*

If Bennett's 'Kantian hypothesis' is correct, we cannot manage the task without some asymmetric observable object in common with our alien friends. Appealing to the side of the body on which the heart is won't help, for example, because aliens' hearts, if they have hearts, might be on the right.[20] If the mechanism in virtue of which the terms 'left' and 'right' refer is indeed what I have suggested it is, the difficulty of explaining their meanings within the constraints of the Ozma problem are readily understandable.

Although Kant concludes in favour of substantivalism in his 1768 essay, he appears to do so very much by default. There is no explanation in his essay of *how* substantival space is able to ground that which relationalism supposedly cannot: the incongruence of counterparts. *If* substantivalism and relationalism represent two genuinely exhaustive alternatives, then an argument against one would be an argument for the other. But as I noted earlier, Kant quickly came to the view that they are not jointly exhaustive, and instead opted for the *tertium quid* of transcendental idealism.

However, Hoefer (2000) has recently pointed out that there is one way in which the postulation of substantival space *can* be used to secure Kant's intuition that a

[19] The reason for attributing this hypothesis to Kant is Kant's insistence, noted earlier, that the difference between left and right cannot be made intelligible through concepts.

[20] Actually, as will become apparent, we have known since the 1950s that we could exploit the fact that the laws of nature violate parity. One might worry that, since we are communicating through photons, we cannot rule out the possibility that our alien correspondents live in an *antimatter* galaxy. Were the laws CP invariant, their following our instructions and carrying out an experiment illustrating parity violation would lead them to conclude that 'left' meant right. Fortunately we can appeal to CP violating experiments to overcome this potential problem. Moreover CPT invariance is not a problem because our communicating at all presupposes that we agree about 'before' and 'after'!

hand in an otherwise empty universe is necessarily either a left or a right hand. One is to imagine that the universe contains a single hand and that the space in which the hand exists is the substantival space of our *actual* world. One would then appear to be able to appeal to facts of the following sort: in the imagined possible world, the lone hand is either determinately congruent to the hand-shaped region of space that is actually and currently occupied by my left hand, or it is determinately congruent to the hand-shaped region of space that is actually and currently occupied by my right hand. In the first case the lone hand is left-handed, in the second it is right-handed (cf. Hoefer, 2000, p. 241). I wish to make five observations about this substantivalist account of a lone hand's determinate handedness.

First, as Hoefer is keen to stress, such an account is only open to the substantivalist who believes that there are *primitive* facts about which points of space or spacetime in two different possible worlds count as the 'same' point. In terms that will be more familiar to philosophers, it is not enough that one be a substantivalist; to give such an account, one must also be a *haecceitist*. Since the issue was brought into focus by Earman and Norton's version of Einstein's 'hole argument', many philosophers have concluded that commitment to such primitive identities, and the corresponding haecceitistic differences between possible worlds, is not an obvious concomitant of a belief in the fundamental reality of space or spacetime (see, especially, Brighouse, 1994; Rynasiewicz, 1994; Hoefer, 1996).[21]

Second, the account only works for possible worlds the space(time)s of which have the same global topology as that of the actual world. This is because it is not clear what transworld identity relations, primitive or otherwise, could hold between two non-diffeomorphic spaces.

In fact, that the two spaces have the same global topology is not even sufficient. Let us assume substantivalism and primitive transworld identity for the sake of argument. There will nevertheless be spacetime points that are the location of some instantaneous stage of my left hand in this world but that form a perfect sphere in some other possible world. All that is required is that the region they constitute is topologically identical to my left hand. The handedness of the hands of this possible world will thus be undetermined for they will all be equally (in)congruent to the space actually occupied by my left hand.[22]

Third, although *these* complications do not tell against Hoefer's reconstruction being faithful to Kant's thinking – Kant and his contemporaries implicitly assumed that the spaces of all possible worlds were isometric to E^3 – the reconstruction certainly does not do justice to Kant's assertion that having a particular handedness

[21] I should also mention that not all philosophers agree. Belot and Earman, for example, argue against substantivalists who reject haecceitism, whom they brand 'sophisticated substantivalists' (see Belot and Earman, 2000 and 2001). For a response to their arguments, see Pooley (2002, chapter 8).

[22] The restriction to points underlying an *instantaneous* stage of *my* left hand is incidental. There are possible worlds in which the spacetime points forming the worldtubes of every actual hand form worldtubes of perfect spheres, or of objects whose handedness changes over time, etc.

is a matter of having the correct relation to space *as a unity*. According to Hoefer's Kant, it consists in having the right relation to *particular* regions of space, for example, the region which is actually the location of my left hand. This is something that Kant denies (Kant, 1992a, p. 365).[23]

Fourth, it is evident that the account is surprisingly close to the relationalist account of handedness just given. In particular, note how the account 'explains' what it is to be a left-handed hand-shaped region of space. This is held to be merely a matter of congruence to the actual material hands that we in fact, *actually*, call 'left'. If one believes in primitive identity, one can exploit the fact that a particular hand-shaped region of space exists in a large class of possible worlds to secure the handedness of material hands in all these world. One is effectively securing a *vicarious* congruence between material hands in two different possible worlds by way of particular hand-shaped regions of space that are supposed to exist in both. If this is all there is to the substantivalist's explanation of the handedness of the hands in one-hand worlds, then the relationalist's assertion that such hands do not have a determinate handedness starts to look decidedly less exceptionable.

The final, and related, observation is that the account surely does not connect with our epistemological situation. We certainly do not recognize hands as left in virtue of recognizing their congruence to particular regions of space. Such regions are invisible. This underlines the fact that the account is effectively a marriage of a relational account of handedness with substantivalism and haecceitism, so as to secure the determinate handedness of hands in other possible worlds.

3 The challenge of parity violation

The conclusion of the first part of this paper is that objects of opposite handedness that are otherwise identical, such as idealized left and right hands, do not differ intrinsically in any way and, furthermore, that their opposite handedness is a matter of their external spatial relations to each other (and to a language using community that has assigned quite arbitrary labels to the two incongruent classes of such objects). John Earman (1989, chapter 7) has suggested that, while the relational account outlined above may be able to deal with incongruent counterparts, the fact that a law of nature violates parity poses a more recalcitrant problem.

In fact, he sees parity violation as having implications for the substantivalist–relationalist debate in much the way Kant initially thought that incongruent counterparts had. The reason is that he believes the substantivalist *can* ground the left–right

[23] Hoefer (2000, p. 243) does point out that 'no particular points, lines, rays or regions [of space] are the ones that have to be mentioned'. However, this hardly makes it the case that being of a particular handedness is a matter of a hand's relation to space as a *unity*, rather than, say, to space as a *plurality*. Kant is explicit in his denial that handedness involves a hand's relation to places (and hence, presumably, to sets of these).

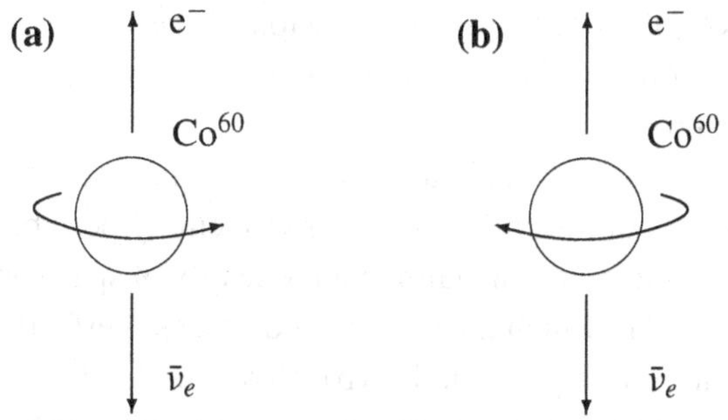

Figure 1. $Co^{60} \rightarrow Ni^{60} + e^- + \bar{\nu}_e$.

asymmetry exhibited in processes governed by parity-violating laws whereas the relationalist cannot. I shall shortly question this assumption.

Earman's example of a process that exemplifies such a law involves the decay of neutral hyperons that was experimentally investigated by Crawford *et al.* (1957) as a test for parity violation. An example that may be more familiar is the β-decay of radioactive cobalt atoms, the subject of the first experimental confirmation of parity violation (Wu *et al.*, 1957). In such a decay the electron and its antineutrino are preferentially emitted along the axis of nuclear spin. Given this, there are two, mirror-image possibilities, depicted in figure 1. In (a) the electron is emitted in the same direction as the spin of the cobalt nucleus, in (b) the electron is emitted in the opposite direction.

The weak interaction, which governs this decay process, fails to be symmetric under parity inversion. The decay (a) is, it turns out, much more probable than (b). In Wu's experiment, a sample of Co^{60} was cooled to near absolute zero and then subjected to a magnetic field to align the nuclear spins: many more electrons were detected emerging in the direction of nuclear spin than in the opposite direction.

In terms of this example, here is how Earman puts the challenge to the relational account of handedness defended above (1989, p. 148):

> The failure of mirror-image reflection to be a symmetry of laws of nature is an embarrassment for the relationist account sketched... for as it stands, that account does not have the analytical resources for expressing the law-like left–right asymmetry for the analogue of Kant's hand standing alone. If we may put some twentieth-century words into Kant's mouth, let it be imagined that the first created process is a [$Co^{60} \rightarrow Ni^{60} + e^- + \bar{\nu}_e$] decay. The absolutist has no problem in writing laws in which [(a)] is more probable than [(b)], but the relationist... certainly does, since for him [(a)] and [(b)] are supposed to be merely different modes of presentation of the same relational model. Evidently, to accommodate the new physics, relational models must be more variegated than initially thought.

Without doubt, Earman has put his finger on something. But one might wonder whether the full scale of the problem has been stated. Two things are worth saying immediately. First, given the conclusion of the first part of this paper, is it *obvious* that the absolutist (the substantivalist) has no problem 'in writing laws in which [a] is more probable than [b]'?

Modulo the qualifications made above, Hoefer's haecceitist substantivalist can secure the handedness of lone hands in otherwise empty possible worlds. In particular, Hoefer's substantivalist can claim that the reality of space grounds the genuine distinctness of a world in which the first created process perfectly resembles (a) and a world in which it perfectly resembles (b). However, he does so by claiming that processes of type (a) stand in different relations to *particular* bits of space to those in which processes of type (b) stand. It seems doubtful that such a substantivalist will want to write relations to particular bits of space into the laws. As Hoefer says (2000, p. 253): 'It seems wrong for a law of nature to contain reference to a particular, contingent physical object. But it seems (to me) at least as wrong for a law of nature to contain reference to particular bits of space . . . '.

Second, Earman's way of setting up the challenge, in terms of the 'first created process', suggests a relationalist response that echoes Herman Weyl's response to Kant's argument. Weyl wrote (1952, p. 21): 'Had God, rather than making first a left hand and then a right hand, started with a right hand and then formed another right hand, he would have changed the plan of the universe *not in the first but in the second act*, by bringing forth a hand which was equally rather than oppositely oriented to the first created specimen.' Similarly, perhaps the relationalist can maintain that whether the first created process is a typical decay governed by the weak interaction, or whether it is a possible but atypical decay, will depend on its incongruence or otherwise to the majority of subsequent similar decays. This, I think, is ultimately what the relationalist has to say. One aim of the rest of this paper is to highlight some of the costs involved.

The Weyl-style relationalist is obviously allowed the relationally acceptable distinction between a world where the first decay process is typical (i.e. congruent to the majority of subsequent decays) and one where it is atypical. He is also allowed the distinction between (parity-violating) worlds where the majority of decays are handed in the same way and (parity-symmetric) worlds in which decays of opposite handedness occur with equal frequency. The fundamental problem faced by any account of handedness according to which incongruent counterparts do not differ intrinsically is, in those worlds where the majority of decays are handed in the same way, how can this asymmetry be *explained*? If the decay modes (a) and (b) are intrinsically identical, what could *ground* their different likelihoods?

The challenge posed by parity violation is thus well put by Van Cleve, who anticipates Hoefer's unease with laws that make reference to particulars (1991, pp. 21–2):

God could no doubt *see to it* that certain kinds of particles always decay into configurations of the same handedness. But we need to be able to suppose that the result in question comes about through law rather than divine supervision. How can it be law that particles always... display decay modes of one orientation rather than another, if orientation is not intrinsic? If one particle has decayed in left-handed fashion, how does the next particle 'know' that it should do likewise? Its instruction cannot be to trace a pattern of a certain intrinsic description; it can only be to do what the first particle did.

The problem here is not 'action at a distance', though perhaps that will trouble some. It is rather that the required laws would make ineliminable reference to particular things, whereas it is generally supposed to be of the essence of laws that they state relations of kind to kind.

In a moment I shall suggest that in one respect Van Cleve is wrong; the problem *is* action at a distance and not ineliminable reference to particular things. But to see how reference to particular things – whether they be particle decays or regions of space – can be avoided, we need to review some of the details of the law that describes parity-violating processes.

4 A relational account of parity violation

That Nature treats left- and right-handed varieties of certain processes differently is puzzling. And yet we have an extremely well-confirmed physical theory describing how it does so. How does the mathematics of the theory work? Does it do so by revealing that left-handed and right-handed varieties of handed objects differ intrinsically after all?

The most fundamental description of parity-violating interactions so far formulated is that given by the Weinberg–Salam gauge field theory, part of the Standard Model.[24] The Weinberg–Salam theory treats elementary particles, such as electrons and quarks, as excitations of Dirac quantum fields interacting via gauge boson fields (the photon, the W^+ and W^-, and the Z). Parity is not a symmetry of the theory because it assigns different properties to the 'left' and 'right' chiral components of the same Dirac field; they couple to the interaction fields in different ways. In particular, 'right-handed' particle fields do not couple to the W bosons at all.

So what does it mean to call a component of a field 'left-handed'? The Dirac field can be thought of as the sum of two component fields, the left and right chiral components, which, in the zero mass limit, correspond to particles of definite and opposite *helicity*. The helicity of a particle is the projection of its spin in its direction of motion. Helicity eigenstates of a spin-$\frac{1}{2}$ particle involve the spin being either aligned or anti-aligned with the particle's direction of motion. These constitute two incongruent, 'handed' objects. By definition, *left-handed* massless particles

[24] I review some relevant details of the Weinberg–Salam theory in an appendix.

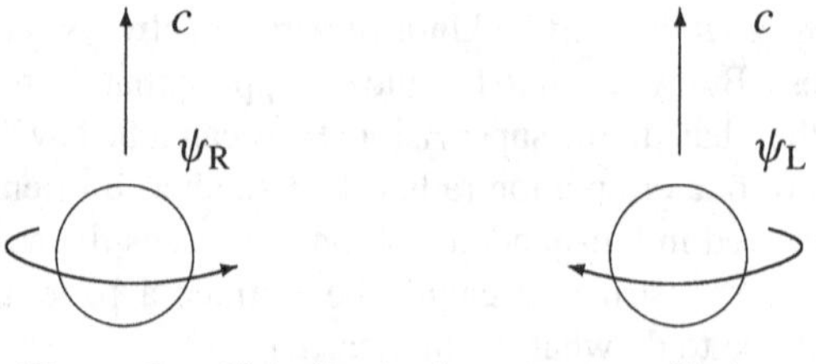

Figure 2. Chirality in the zero mass limit.

are particles of negative helicity (their spin is opposite to their direction of motion), while *right-handed* particles are particles of positive helicity (see figure 2).

Helicity and chirality are not quite the same thing, however. A helicity eigenstate of a massive particle will involve both left- and right-handed pieces. Moreover, while the chirality of a particle is Lorentz invariant, the helicity of a massive particle is not. For example, one can Lorentz boost by a large enough velocity in the direction of the particle's motion so as to reverse that direction of motion while leaving the direction of spin unchanged. This cannot be done, of course, if the particle is massless and thus travelling at the speed of light. The helicity of a *massless* particle *is* invariant under the (restricted) Lorentz group.

A spinning object defines an axis: that about which it is spinning. For a given axis there are then two possibilities involved: if one looks along the axis of spin from a given direction, the object will appear to be spinning either clockwise or anticlockwise. These two possibilities are represented by associating each with a direction: the spin vector points along the axis of spin away from the point of view from which the spinning appears clockwise. This is equivalent to the definition of angular momentum vector $\mathbf{l}$ as $l^i = \varepsilon^{ijk} x^j p^k$ where ε^{ijk} is completely antisymmetric in its three indices, $\varepsilon^{123} = +1$, and the components of $\mathbf{x}$ and $\mathbf{p}$ are given with respect to a conventional, *right-handed* set of Cartesian axes.

In the last analysis, in each case, the convention for associating a direction with a spinning object can be specified only via ostension. We can explain what we mean by left-handed and right-handed particles in terms of their relations to right-handed sets of axes, or in terms of their relations to typical clocks. But, if Bennett's 'Kantian hypothesis' is correct, the meanings of these terms cannot be conveyed without ostension.

Turning now to the case of negative- and positive-helicity particles, one sees from the conventionality of the definition of the direction of spin, and from the need for ostension in specifying this direction, that to explain the difference between negative and positive helicity one must also ultimately appeal to ostension. The parallel with left and right hands is obvious. An *intrinsic* description true of a negative-helicity particle will also be true of a positive-helicity particle. Yet we can understand why there are *two*, incongruent, types in purely relational terms (i.e. in terms that do not presuppose an intrinsic difference). For two spin-$\frac{1}{2}$ particles of definite helicity travelling in the same direction they can be spinning in the same sense, or the opposite sense.

It turns out that the mathematics of parity violation does not involve treating being left-handed and being right-handed as different substantive and intrinsic properties. As is explained in the appendix, left and right components of the fields are distinguished in terms of their differing congruence relations to the right-handed coordinate systems with respect to which the theory is standardly written. One then goes on simply to assert of these two components that they interact differently.

This difference in the way they interact is not further explained in terms of different intrinsic properties possessed by the particles. Although I note in the appendix that the left and right components of the field are assigned different values of 'weak hypercharge' and 'weak isospin', these are not properties that *explain* their particular couplings to the gauge fields. For two varieties of particle to have the particular values of weak hypercharge that they have, for example, *just is* for them to couple to the gauge field B_μ with the relative strengths that they do. In fact, it seems plausible that, quite generally, particle varieties are individuated only in terms of their particular place in a network of differently interacting particles.

Moreover, it cannot be the helicity of a particle alone that determines which type of interactions it can undergo. The reason is that the 'left-handed' Dirac field component $\psi_L(x)$ – and its adjoint, $\overline{\psi_L}(x)$ – are associated with left-handed particles and *right-handed* antiparticles. (In the massless limit these correspond to negative- and positive-helicity eigenstates respectively.) Therefore the theory not only violates parity P, it also violates *charge conjugation* C, the interchange of particles with their antiparticles. However, the fragment of the theory described in the appendix is invariant under the *combined* transformation CP. Just as only left-handed electrons and quarks couple to the Ws, only right-handed positrons and antiquarks couple to them.[25] So a description of the left-handed β-decay of a cobalt atom in intrinsic, relational terms will be equally true of a right-handed decay of an antimatter cobalt nucleus.

Does this fact, by itself, vindicate a relational account of handedness? Simon Saunders[26] has suggested that it does. His point is that the parity-violating law does not, after all, 'pick out' a particular handedness. Anything that is possible for a particle of one handedness is possible for particles of the opposite handedness, although it *may* be possible for the oppositely handed particles only if they are also antiparticles of the first. Just as the relationalist seeks to identify putatively distinct possible worlds containing nothing but single hands that supposedly differ solely

[25] The full Standard Model, involving all three generations of quarks, violates CP symmetry. However, as mentioned in the introduction, it is CPT-symmetric.

[26] See Saunders, 'Incongruent counterparts: a Leibnizian approach', unpublished manuscript, and this volume.

in the sense of their handedness, Saunders urges that we should treat models of a parity-violating but CP-symmetric theory that are related by a global CP-transformation as different representations of the *same* state of affairs. If we describe a world in which the first created process is a decay of a cobalt atom, there is no fact of the matter whether it was a decay of a matter nucleus emitting a right-handed antineutrino, or the decay of an antimatter nucleus emitting a left-handed neutrino. (Of course, if *per impossibile* the first created process of the actual world was a cobalt decay there would be a fact of the matter. But this would be a *relational* fact: did the decay involve the emission of a neutrino of the same handedness as particles that we in fact call 'right-handed' and a particle of the same charge as particles that we in fact call 'electrons'; or did it involve the emission of a neutrino of the same handedness as particles that we in fact call 'left-handed' and a particle of the same charge as particles that we in fact call 'positrons'?)

Unfortunately, CP (or even CPT) symmetry does not by itself save the relationalist. Nor, indeed, is it even a necessary component of the relationalist's account. To see this, consider the following toy models. The first involves a possible world whose fundamental objects are hand-shaped and come in two varieties: 'red' and 'green'. Red hands are never created, but they can 'decay' into green hands. Now let us suppose that only red 'left' hands can decay into green 'left' hands. Red 'right' hands never decay and no green 'right' hands exist at all. This decay law clearly violates parity.

We can, however, extend our example so that it involves a P-violating but 'CP'-symmetric law. In this second possible world, we now imagine that the red hands are 'charged' in that they attract or repel one another: similarly charged hands repel each other whereas oppositely charged hands attract each other. Now both left and right red hands can be 'negatively' and 'positively' charged. However, suppose that it is the case that only 'negatively' charged red left hands can decay into green (left) hands whereas 'positively' charged right hands can decay into green right hands. The law is now CP symmetric: if a particular hand can decay, then so can its incongruent, charge-reversed counterpart.

Is this CP-symmetric law really more susceptible of a relationalist interpretation than the first, P-violating law? The problem that the relationalist still faces is to explain why negatively charged red right hands cannot decay into similarly oriented green hands when negatively charged red left hands (i.e. hands that are identical apart from the sense of their handedness) can. Relationalism denies that there is some intrinsic difference between the two types of hand that can ground and explain their different interactions. The fact that *positively* charged red right hands can decay into similarly oriented green hands does nothing to ameliorate the problem. Similarly, the relationalist is at a loss to explain why only left-handed *electrons* couple to W bosons. How can it be that left-handed and right-handed electrons interact differently if the relationalist is correct in his deflationary account of what their being left- or right-handed consists in? The fact that right-handed *positrons* can couple to W bosons is of little comfort.

There is another reason for being wary of invoking CP (or CPT) symmetry to save a relational account of handedness: it fails to secure for the relationalist all that he desires. In his correspondence with Clarke, Leibniz, the arch-relationalist, insisted that two putatively distinct possible worlds differing solely over where the material universe was located in space were really just two ways of differently describing a single possibility. It is the homogeneity of space – that translations are a *symmetry* of Newtonian mechanics – that means that if one of these worlds is *physically* possible, then so is the other. Similarly, the relationalist will wish to see two models of a P-violating but CP-symmetric theory that are related by a global CP transformation as but two ways of representing a single state of affairs.

What, though, should they say about the model one obtains from one model from such a pair by performing a global parity transformation *without* also interchanging matter with antimatter? In the case of our second toy model, we obtain a world where it is negatively charged *right* hands and positively charged *left* hands that can decay. In the case of a model of the Standard Model, we describe a world where *right-handed electrons* and *left-handed positrons* couple to W bosons. The relationalist who sees the existence of a symmetry as a prerequisite of being able to identify the possibilities represented by models related by a non-trivial, global transformation must deny that these situations are equivalent to the original ones. He must even deny that they obey the same laws!

Something surely has gone wrong here. In what way does the parity-imaged model differ from the original? We call the quarks that couple to the Ws in the first model 'left-handed', and call the quarks which couple to the Ws in the second model 'right-handed', but what is the difference between them? In terms of the functional roles they play within the models, they are indistinguishable. If the original model was detailed enough to include human experimenters referring to left-handed quarks, then their counterparts in the second model will call the supposedly right-handed quarks 'left-handed' too. The relationalist intuition is surely that the left-handed quarks of the first model should be *identified* with the so-called right-handed quarks of the second.

Despite the fact that the worlds described by the two models display a law-like asymmetry between left and right, and despite the fact that the models are *nominally* the mirror-images of each other, they should be regarded as solutions of a *single theory*, and the two models should be judged to describe a single possibility.

So let us now return to Earman's claim that the relationalist lacks the 'analytical resources' to describe a law that can embrace both models. The relationalist must eschew terms such as 'left' or 'right', and he must not rely on a formulation that makes *implicit* use of such terms by relating the physics to, for example, right-handed coordinate systems. It seems, however, that the relationalist can indeed provide such a law.

Consider the first toy model again. The claim that all red hands which decay into green ones are *handed in the same way* embraces both the original possible world and its supposedly distinct parity-image. What is there to stop the relationalist claiming that this is a *law-like* statement and, moreover, that it is primitive: it cannot be further explained in terms of more fundamental laws? Turning to the second, CP-symmetric, toy model, we can state that red hands that *repel each other* and which can decay into green hands are handed in the same way; red hands that attract each other and which can decay into green hands are handed in the opposite way.

Saunders[27] offers a similar schematic law for β-decay:

PC charge-conjugate β-decay processes are oppositely oriented.

He stresses that this a PC-invariant statement of P violation. What is important for the present discussion is that it is *also* a P-invariant statement, even though it is an expression of P violation. The statement is true both of a model where 'left-handed' electrons couple to W bosons *and* of a model where 'right-handed' electrons do so. Of course, the law needs to be extended to include all varieties of fields – in particular, it must say something positive about the interactions which the electron-positron pair that cannot be produced in β-decays can undergo – but the outlines of how this is to be done are clear enough.

5 Orientation fields

In characterizing the relationalist's position in this way, I am very close to Carl Hoefer. He writes (2000, p. 253):

> The correct perspective, for either relationists or substantivalists, is this: P-violating laws mandate *that there shall be a certain, qualitative, spatial asymmetry* in events. They do not explain the asymmetry or how it arises . . . Bringing in enantiomorphic objects allows one to 'anchor' the asymmetry descriptively, but is in no way explanatory of the asymmetry, nor do such objects become 'referred to' in the laws by being so used.

In the case of the Standard Model, the enantiomorphic objects in question are actual left and right hands that are linked to the law, as we have seen, via the conventions that define what we mean by *right-handed* coordinate systems.

However, Hoefer also claims that parity-violating laws are 'purely phenomenological' and that this should, at least in part, ease any worries that we might have over the fact that the law-like asymmetry in phenomena is not ultimately explained.

[27] See Saunders, 'Incongruent counterparts: a Leibnizian approach,' unpublished manuscript.

This seems to me to be contestable. The Standard Model is not a 'purely' phenomenological law. There is a world of difference between the early descriptions of the weak interactions in terms of V–A currents, and the Standard Model together with the understanding that it provides of these truly phenomenological laws. The asymmetry in the phenomena *is* explained, albeit only in terms of a deeper asymmetry that is not. We postulate fundamentally handed particles (corresponding to massless spin-$\frac{1}{2}$ Dirac fields) and attribute different sets of interactions to oppositely handed fields. *Why* oppositely handed fields of the same particle type differ in this way is not explained. But there is no reason why the fact that they do should be seen as a phenomenological, rather than as a fundamental, fact.

Lack of a further explanation should not *per se* be seen as a problem for the relationalist. However, there is one feature of the relationalist's story that some will object to, and it is a feature that can lead back to substantivalism.

Recall that Van Cleve alleged that the problem faced by the relationalist was ineliminable reference to particulars rather than 'action at a distance'. We have seen that there is no ineliminable reference to particulars. However, the basic form of our proposed law – that all objects or processes of a certain relationally specifiable type are *handed in the same way* – is, in a certain specific sense, highly *non-local*. For the relationalist, being handed in the same way just is for two things to stand in certain spatiotemporal, and quite possibly space-like, relations.

It is time to consider the recent 'reconstruction' of Earman's argument offered by Nick Huggett (Huggett, 2000). One of Huggett's central claims is that rather specific geometrical structures are involved in a proper formulation of parity-violating laws. He illustrates this claim with his own toy model quantum theory, involving two particles in one dimension, coupled by the following asymmetric potential:

$$V(x_1, x_2) = \lambda(x_1 - x_2) + \mu(x_1 - x_2)^2. \quad (1)$$

In this theory the two directions in its one-dimensional space are not on a par. On measurement of their positions, the probability of finding particle 1 'to the left of' particle 2 (i.e. with a more negative position coordinate, relative to the coordinate expression of the potential given above) is greater than finding them the other way around.[28]

Huggett's claim is that the theory is not well defined until an 'arrow of space' has been given, enabling us to say whether $x_1 < x_2$ or $x_2 < x_1$ 'in absolute terms, not just relative to some arbitrary coordinates' (Huggett, 2000, p. 233). He links this

[28] Note that one unfortunate aspect of Huggett's example can be ignored. His theory violates both parity symmetry *and permutation symmetry*. It is not that one is more likely to find a particle of one *type* to the left of a particle of another type; rather one is more likely to find particle 1, *that particular particle*, to the left of particle 2. The fact remains that it is solely the theory's violation of parity that requires the introduction of an orientation field when expressed in a coordinate-free way. Similarly the coordinate-free expression of the Standard Model requires the introduction of an orientation field, even though this theory does not violate permutation symmetry.

claim to the fact that the coordinate-free expression of the potential will involve the explicit introduction of an *orientation field* (in the case of the one-dimensional theory, this field is simply a normalized 1-form). Similarly, if we were to express the equations of the Standard Model in a coordinate-free way, we would need to introduce an orientation field explicitly, and distinguish left- and right-handed components of fields by their relations to it, rather than to a standard coordinate system.

Huggett then puts the argument for substantivalism as follows, deliberately paraphrasing Earman's version of why the postulation of inertial structure to ground the distinction between absolute and relative motion licenses the move to substantivalism (Huggett, 2000, p. 236; cf. Earman, 1989, p. 125):

> ... the 'absolutist' asserts that 'the scientific treatment of motion ... requires some absolute quantities ... such as handedness. To make these quantities meaningful requires the use of an orientation, and this structure must be a property of or inhere in something distinct from bodies. *The only plausible candidate for the role of supporting the nonrelational structures is the spacetime manifold*.'

The parallel drawn here is suggestive and worth pursuing. Pure inertial motion can be thought of as manifesting rather noteworthy non-local correlations. Assuming the correctness of Newtonian mechanics, one can, from the relative motions of just three force-free bodies, construct a spatiotemporal coordinate system with respect to which all three of these bodies are moving uniformly and in straight lines.[29] This is already a highly non-trivial fact that one might feel calls for explanation, given that, as force-free bodies, the three bodies are supposed to be moving quite independently of each other. What is perhaps more striking is that every other force-free body is also moving inertially with respect to the coordinate system defined by the first three. The substantivalist offers a *local* explanation of these non-local correlations. According to him, the laws of motion constrain the motions of such bodies at each point of spacetime to be geodesics, as defined at each point by the affine connection.

Similarly, by reifying the orientation field, we can offer a local explanation of the non-local correlations between β-decays: that all neutrinos emitted in such decays are handed in the same way. According to Huggett's substantivalist, these correlations follow from the fact that the laws postulate that, at each point of spacetime, only quarks standing in one of the two possible relations to the orientation field *at that point* can couple to W bosons.

If the introduction of a real orientation field provides a genuine and local explanation of the congruence of all β-decays at no cost, we should surely admit such

[29] An elegant demonstration of this fact was given by Tait (1884). For a nice account of it, see Barbour (1999, chapter 6).

a field into our ontology. The question is, of course, whether the explanation is genuine, and whether there are costs. One might also wonder whether such a field supports spacetime substantivalism: does such a field obviously represent *space-time structure* (as does the metric field of relativistic theories[30]) rather than just another real, physical field in spacetime?

Hoefer has objected to the move on different grounds. He holds that it amounts to nothing more than writing reference to particular bits of space into the laws (Hoefer, 2000, p. 253):

> It seems wrong for a law of nature to contain reference to a particular, contingent physical object. But it seems (to me) at least as wrong for a law of nature to contain reference to particular bits of space... That this is what is going on may be masked by talk of 'absolute structures' or 'a preferred *n*-form defined at all points', or something of this nature. But such terminology, while not literally incorrect, really only disguises the dependence on primitive identity to make the distinctions between orientations for us.

The reader might be surprised at this assertion. Where, in the foregoing discussion of orientation fields, was reference made to particular bits of space? The reification of an orientation field does not entail a commitment to primitive identity. Care is needed, however, as is illustrated by the following quote from Huggett (2000, pp. 234–5), which follows his explicit introduction of an orientation field into his toy–model theory:

> At this point it is worth noting for clarity that there is also a conventional aspect to such handed theories. For suppose the arrow of space now runs in the opposite sense; if V_A remains the potential, then it will have the opposite handedness in space (compared to the original, or compared to some external bodies) and the system will behave differently. But if the potential also changes, $V_A \rightarrow -V_A$, then of course the dynamics will be as before. Thus, it does not make sense in this situation to ask in which direction the arrow of space runs, independently of a given Hamiltonian, and likewise it makes no sense to ask which sign of V_A is correct, independently of an arbitrary choice of arrow. Thus the two possible arrows and two possible Hamiltonians only allow two distinct theories not four. This point acknowledged, we can talk of *the* arrow and *the* Hamiltonian and bear in mind the freedom this actually leaves.

I agree with Huggett that there are certainly no more than two possible theories, not four. However, two readings of this passage are possible. And read in one way, precisely the wrong identifications are being advocated.

The orientation field is either supposed to be a real, physical field, or is supposed to represent some genuinely asymmetric structure of space or spacetime itself. If this is the case, then one *cannot* identify a theory that assigns a certain probability to the vector from particle 1 to particle 2 being aligned with the arrow defined by

[30] This claim is, perhaps, controversial. Here I am siding with, for example, Maudlin (1989, p. 318) and Hoefer (1998, pp. 459–60) against Rovelli (1997, pp. 193–4). For further discussion, see Pooley (2002, chapter 4).

the orientation field, with a theory that assigns precisely that probability to the case where the two vectors are in the *opposite* alignment. Similarly, a theory that asserts that all electrons which are 'congruent' to an orientation field couple to W bosons and those which are 'incongruent' do not, cannot be identified with a theory that predicts the same phenomena by asserting that all electrons which are *in*congruent to an orientation field couple to W bosons.

However, the anti-haecceitist substantivalist who follows Hoefer in denying primitive identity relations between the spacetime points of different possible worlds will not be able to distinguish worlds that involve the same relations between orientation field and the matter fields, but that differ solely in terms of the relations that all of these fields bear to particular points of spacetime. But this is exactly the distinction that Huggett might appear to be upholding when he talks, for example, of the 'arrow of space running in the opposite sense.' It seems that we are here being asked explicitly to imagine the orientation field bearing a different relationship to particular points of space.[31]

So we can introduce an orientation field to ground a local explanation of the non-local asymmetries that the relationalist must postulate as brute, law-like facts, in a way that does not involve an implicit commitment to haecceitism and primitive identities. Nevertheless, it does appear to involve an unavoidable commitment to the reality of differences that are unobservable in principle: the theory that has only electrons 'congruent' to such a field coupling to W bosons and the theory that has only electrons 'incongruent' to such a field coupling to W bosons must be regarded as distinct theories, even though they are observationally indistinguishable. This problem is not merely the result of viewing the orientation field as a physical field that is distinct from spacetime. If one were to insist that it is spacetime itself that has an intrinsically asymmetric structure, and that the orientation field is just a mathematical device to encode such structure, one still would not be able to collapse the four formally distinct theories down to one. As Huggett notes (on the intended reading of the above quote), one is free to choose either of the two possible orientation fields as encoding such structure. But this does not tell against there being a genuine metaphysical distinction between worlds where electrons bearing one type of relation to this structure interact in a certain way and worlds in which electrons bearing the opposite relation interact in an identical way. The only way to avoid postulating such a distinction is to adopt the relationalist's account of parity violation, together with its brute, law-like, non-localities.

[31] The alternative reading of Huggett's passage involves no such commitment to haecceitism. Instead one is merely noting that, *relative to a fixed coordinate system*, one can represent the asymmetric structure attributed to spacetime in two equally good ways. Although the coordinate system is kept fixed, one is contemplating a *passive* transformation that results in the orientation field of the mathematical model bearing a different relation to the physical orientation field (or to the asymmetric spacetime structure). Huggett has indicated that it was this freedom he was intending to highlight (private communication).

Acknowledgements

I am grateful to Harvey Brown, David Wallace, Mauricio Suarez, James Ladyman, Leah Henderson, Jeremy Butterfield, Carl Hoefer, Nick Huggett, Graham Nerlich, and Bill Child for comments on material for this paper and, particularly, to Simon Saunders for many discussions of handedness and parity violation. I would like to thank the editors for the invitation to contribute to this volume and for helpful comments on earlier drafts. Work for this paper was partly supported by the Arts and Humanities Research Board of the British Academy.

Appendix: The weak interaction

I should start by stressing that there are many reasons for thinking that the Standard Model is far from a final theory, not the least of which is the large number of unexplained parameters involved that must be fixed by hand for the theory to tally with experiment. There is also, as yet, no consensus on the correct way to extend the Standard Model so as to include neutrino mass. Indeed, this caveat is of relevance to our topic because some suggestions involve theories which are fundamentally parity symmetric. The standard electroweak theory results from parity symmetry being *spontaneously broken* in such theories.

I propose to set these issues to one side. Treating the Weinberg–Salam theory as a fundamental theory poses the most severe challenge for any account of handedness according to which handedness is an extrinsic, rather than an intrinsic, matter. So if such an account can be defended in this context, *a fortiori* it should be defensible in others.

Central to the Standard Model are the left-handed and right-handed components of the various Dirac fields.[32] Mathematically they are described in terms of the γ-matrices. In particular, for a Dirac field ψ, one has:

$$\psi_L = \frac{1}{2}(1 - \gamma^5)\psi, \quad \psi_R = \frac{1}{2}(1 + \gamma^5)\psi, \tag{2}$$

where $\gamma^5 = \gamma^0\gamma^1\gamma^2\gamma^3$. The γ-matrices might appear to be defined independently of our coordinate conventions: all that is required is that they obey the anti-commutation relations:

$$\{\gamma^\mu, \gamma^\nu\} = 2g^{\mu\nu}I, \tag{3}$$

and $g^{\mu\nu}$ is left–right symmetric. However, they are tied to the coordinate system through the Dirac equation:

$$(i\gamma^\mu\partial_\mu - m)\psi(x) = 0. \tag{4}$$

[32] In the following, I have drawn upon Peskin and Schroeder (1995) and unpublished lecture notes by I. T. Drummond and H. Osborn.

If we reverse the sense of our coordinate system, say by relabelling the x and y axes, we switch the roles of γ^1 and γ^2. Since, from equation (3), these anti-commute, $\gamma^5 \mapsto -\gamma^5$, *the mathematical description of the left-handed component of a fermionic field with respect to a right-handed set of axes is exactly the same description one gives to the right-handed component with respect to a left-handed set of axes*. The standard mathematical descriptions of the chiral components of a fermionic field thus clearly relate them, more or less explicitly, to the the handedness of the Lorentz chart with respect to which the physics is formulated. There is nothing in the standard mathematical description of a left-handed field that is not equally suited, given different conventions, to describing a right-handed field.[33]

Having introduced the left-handed and right-handed components of the Dirac field, we can see how parity violation is implemented in the Weinberg–Salam theory. The theory is based on a Lagrangian density involving the Dirac lepton fields, the gauge fields, and a scalar field, the Higgs field. The Lagrangian is invariant under certain local $SU(2) \times U(1)$ gauge transformations. The standard prescription followed in constructing a gauge field theory is to start with a Lagrangian that is invariant under some set of global gauge transformation and then create a locally gauge-invariant theory by replacing the derivative operators in the original Lagrangian with 'covariant derivatives' involving compensating gauge fields to ensure the required invariance. In this case, the original Lagrangian is the standard (parity-symmetric) Dirac Lagrangian for *massless* fields (for simplicity I consider only interactions involving electrons, positrons, and their associated neutrinos):

$$\mathcal{L}_{kin}(x) = \overline{\psi_e} i\gamma.\partial\psi_e + \overline{\psi_\nu} i\gamma.\partial\psi_\nu. \tag{5}$$

No mass terms have been included because to construct a parity-violating gauge theory, symmetries under which the left and right chiral components transform *differently* must be gauged. This prohibits mass terms, which mix left and right components and are thus not invariant under such transformations.

We now rewrite the Lagrangian as:

$$\mathcal{L}(x) = \bar{L}(x) i\gamma.\partial L(x) + \bar{R}(x) i\gamma.\partial R(x), \tag{6}$$

where L is a *doublet* involving the left-handed components of the neutrino and electron fields. Writing $\psi_\nu(x) = \nu(x)$ and $\psi_e(x) = e(x)$:

$$L(x) = \begin{pmatrix} \nu_L(x) \\ e_L(x) \end{pmatrix}. \tag{7}$$

R is a singlet involving the right-handed component of the electron field: $R(x) = e_R(x)$.[34]

[33] I am indebted to David Wallace for this way of seeing how the handedness of field components is defined in terms of the handedness of the coordinate chart.

[34] Although I have not included the right-handed component of the neutrino field, the fact that neutrinos have

$\mathcal{L}$ is invariant under the $SU(2)$ transformations:

$$L(x) \to e^{\frac{1}{2}i\alpha.\tau}L(x), \quad \bar{L}(x) \to e^{-\frac{1}{2}i\alpha.\tau}\bar{L}(x), \quad R(x) \to R(x), \tag{8}$$

where τ are the 2×2 Pauli matrices. It is also invariant under independent $U(1)$ phase transformations of L and R separately. In particular, it is invariant under the separate phase transformations $L \to e^{-i\frac{1}{2}\chi}L$ and $R \to e^{-i\chi}R$ which may be written $\psi \to e^{i\chi Y}\psi$ where $Y = -\frac{1}{2}$ for L and $Y = -1$ for R. These transformations are taken as the $U(1)$ transformations in an $SU(2) \times U(1)$ global symmetry group in order that the resulting gauge theory is invariant under local gauge transformations generated by the electric charge, and electromagnetism is recovered.

A local $SU(2) \times U(1)$ transformation is then written:

$$\psi \to e^{i\alpha(x).\mathbf{T}+i\chi(x)Y}\psi(x), \tag{9}$$

where $\mathbf{T} = \frac{1}{2}\tau$ acting on L and $\mathbf{T} = \mathbf{0}$ acting on R. A gauge-invariant Lagrangian is obtained by replacing the derivatives in equation (6) with the covariant derivatives

$$D_\mu = \partial_\mu - ig\mathbf{A}_\mu(x).\mathbf{T} - ig'B_\mu(x)Y\ . \tag{10}$$

The result is:

$$\mathcal{L} = \mathcal{L}_{kin} + g\bar{L}\gamma^\mu\frac{1}{2}\tau L.\mathbf{A}_\mu + g'\left(\frac{1}{2}\bar{L}\gamma^\mu L + \bar{R}\gamma^\mu R\right)B_\mu. \tag{11}$$

The consequences of gauging a symmetry group under which left and right field components transform differently are now manifest. Since R is a singlet under the chosen $SU(2)$ transformations, it does not couple to the three component fields of $\mathbf{A}_\mu$ at all. That the left and right fields are assigned different values of 'weak hypercharge' Y means that the strengths of their coupling to the $U(1)$ gauge field B_μ is different.

The quantum field theory derived from this Lagrangian, together with the appropriate Lagrangian for the free gauge fields, is not yet empirically adequate because the four gauge fields all correspond to *massless* gauge bosons. It was known experimentally that there is only one massless gauge boson involved in electromagnetic and weak interactions, the photon. The solution is to exploit the mechanism of 'spontaneous symmetry breaking'. A scalar field with a gauge-invariant potential term that has a minimum for non-zero values of the field is postulated. As a result of their coupling to this so-called Higgs field, all but one of the gauge fields acquire a mass and, with the right transformation properties under the gauge transformations assigned to the Higgs field, the massless field corresponds to the $U(1)$ gauge group of electromagnetism. The details need not concern us. All we need is that the

mass means that the Standard Model must be extended to include it. The simplest way of doing so would be to include the field as a separate singlet term. However, as mentioned above, there are many rival proposals which experiment has yet to decide between.

'physical' gauge fields corresponding to gauge bosons of definite mass are given by

$$W_\mu = \frac{1}{\sqrt{2}}(A_{1\mu} - iA_{2\mu}),$$
$$Z_\mu = \cos\theta_W A_{3\mu} - \sin\theta_W B_\mu,$$
$$A_\mu = \sin\theta_W A_{3\mu} - \cos\theta_W B_\mu, \tag{12}$$

where the Weinberg angle θ_W is defined by $\tan\theta_W = g'/g$. Making the identification $e = g\ \sin\theta_W$, equation (11) can be rewritten as:

$$\mathcal{L} = \mathcal{L}_{kin} + \frac{g}{2\sqrt{2}}(J^\mu W_\mu + J^{\mu\dagger} W_\mu^\dagger) + ej^\mu_{em} A_\mu + \frac{g}{2\cos\theta_W} J^\mu_n Z_\mu\,, \tag{13}$$

where the weak, electromagnetic and weak neutral currents are defined as follows:

$$\begin{aligned} J^\mu(x) &= \bar{L}(x)\gamma^\mu(\tau_1 + i\tau_2)L(x) = \bar{\nu}(x)\gamma^\mu(1-\gamma_5)e(x) \\ j^\mu_{em} &= \bar{L}\gamma^\mu\frac{1}{2}(\tau_3 - 1)L - \bar{R}\gamma^\mu R = -\bar{e}\gamma^\mu e \\ J^\mu_n &= \bar{L}\gamma^\mu(\cos^2\theta_W\tau_3 + \sin^2\theta_W 1)L - 2\sin^2\theta_W\theta_W\bar{R}\gamma^\mu R \\ &= \frac{1}{2}[\bar{\nu}\gamma^\mu(1-\gamma_5)\nu - \bar{e}\gamma^\mu(1-\gamma_5 - 4\sin^2\theta_W)e]. \end{aligned} \tag{14}$$

Here again we see that only the left-handed fields couple to the W bosons and that the coupling strengths of the left- and right-handed fields to the Z are different. Only the left- and right-handed couplings of the electron/positron field to the photon A are symmetric between left and right.

Let us now return to our original example of a parity-violating decay: the β-decay of cobalt atoms, $Co^{60} \to Ni^{60} + e^- + \bar{\nu}_e$. To model this in the electroweak theory outlined above, quarks need to be included. In addition to the doublet under local $SU(2)$ transformations, comprising the left-handed electron and electron-neutrino fields, there are weak isospin doublets containing the muon and muon-neutrino, the tau and tau-neutrino, and three left-handed quark doublets. The first of these comprises the up quark field $u_L(x)$ and a linear combination of the down and strange quark fields $d_{\theta L}(x) = [\cos\theta_C d_L(x) + \sin\theta_C s_L(x)]$. θ_C is the so-called Cabibbo angle and its significance is not important for our discussion. The β-decay in question involves the decay of a neutron within the cobalt nucleus into a proton with the emission of an electron and its antineutrino. This in turn is to be understood as the decay of a down quark bound within the neutron into an up quark ($n \sim udd \to p \sim uud$). The interaction term responsible for this decay is $J^{\mu\dagger} W^\dagger_\mu$ where $J^{\mu\dagger}$ contains the term $2\overline{e_L}\gamma^\alpha\nu_{eL} = \bar{e}\gamma^\alpha(1-\gamma_5)\nu_e$ as before and also $2\cos\theta_c\overline{d_L}\gamma^\alpha u_L = \cos\theta_c\bar{d}\gamma^\alpha(1-\gamma_5)u$. The down quark decays into an up quark emitting a W^- boson which decays into an electron and its antineutrino. But since only left-handed quarks can couple to the W boson, only left-handed quarks are involved in such decays and only left-handed electrons and positive helicity, right-handed antineutrinos are observed as a result.

References

Barbour, J. B. (1999). *The End of Time: The Next Revolution in Our Understanding of the Universe*. London: Weidenfeld & Nicholson.

Barbour, J. B., and Bertotti, B. (1982). 'Mach's Principle and the structure of dynamical theories'. *Proceedings of the Royal Society of London A*, **382**, 295–306.

Belot, G., and Earman, J. (2000). 'From metaphysics to physics'. In *From Physics to Philosophy*, ed. J. Butterfield and C. Pagonis, pp. 166–86. Cambridge: Cambridge University Press.

(2001). 'Pre-Socratic quantum gravity'. In *Physics Meets Philosophy at the Planck Scale*, ed. C. Callender and N. Huggett, pp. 213–55. Cambridge: Cambridge University Press.

Bennett, J. (1970). 'The difference between right and left'. *American Philosophical Quarterly*, **7**, 175–91. Reprinted in Van Cleve and Frederick (1991), pp. 97–130.

Brighouse, C. (1994). 'Spacetime and holes'. In *Proceedings of the 1994 Biennial Meeting of the Philosophy of Science Association*, Vol. 1, ed. D. Hull, M. Forbes, and R. Burian, pp. 117–25. East Lansing, MI: Philosophy of Science Association.

(1999). 'Incongruent counterparts and modal relationism'. *International Studies in Philosophy of Science*, **13**, 53–68.

Crawford, F. S., Cresti, M., Good, M. L., Gottstein, K., Lyman, E. M., Solmitz, F. T., Stevenson, M. L., and Ticko, H. K. (1957). 'Detection of parity nonconservation in Λ decay'. *Physical Review*, **108**, 1102–3.

Earman, J. (1989). *World Enough and Space-Time: Absolute Versus Relational Theories of Space and Time*. Cambridge, MA: MIT Press.

Frederick, R. E. (1991). 'Introduction to the argument of 1768'. In Van Cleve and Frederick (1991), pp. 1–14.

Gardner, M. (1990). *The New Ambidextrous Universe*, revised edn. New York: W. H. Freeman and Company.

Harper, W. (1991). 'Kant on incongruent counterparts'. In Van Cleve and Frederick (1991), pp. 263–313.

Hoefer, C. (1996). The metaphysics of space-time substantivalism. *Journal of Philosophy*, **93**, 5–27.

(1998). Absolute versus relational spacetime: for better or worse, the debate goes on. *British Journal for the Philosophy of Science*, **49**, 451–67.

(2000). Kant's hands and Earman's pions: chirality arguments for substantival space. *International Studies in the Philosophy of Science*, **14**, 237–56.

Huggett, N. (1999). *Space from Zeno to Einstein: Classic Readings with a Contemporary Commentary*. Cambridge, MA: MIT Press.

(2000). Reflections on parity non-conservation. *Philosophy of Science*, **67**, 219–41.

Kant, I. (1992a). ' Concerning the ultimate ground of the differentiation of directions in space'. In *The Cambridge Edition of the Works of Immanuel Kant: Theoretical Philosophy, 1755–1770*, trans. and ed. D. Walford and R. Meerbote, pp. 365–72. Cambridge: Cambridge University Press.

(1992b). 'On the form and principles of the sensible and the intelligible world'. In *The Cambridge Edition of the Works of Immanuel Kant: Theoretical Philosophy, 1755–1770*, trans. and ed. D. Walford and R. Meerbote, pp. 377–416. Cambridge: Cambridge University Press.

Kostelecky, A.V. (2002). *Proceedings of the second meeting on CPT and Lorentz Symmetry*, Bloomington, IN, 15–18 August 2001. River Edge, NJ: World Scientific.

Maudlin, T. (1989). 'The essence of space-time'. In *Proceedings of the 1988 biennial meeting of the Philosophy of Science Association*, Vol. 2, ed. A. Fine and J. Leplin, pp. 82–91. East Lansing, MI: Philosophy of Science Association.

Nerlich, G. (1994). *The Shape of Space*, 2nd edn. Cambridge: Cambridge University Press.

Peskin, M. E., and Schroeder, D. V. (1995). *An Introduction to Quantum Field Theory*. Cambridge, MA: Perseus Books.

Pooley, O. (2002). 'The reality of spacetime'. D.Phil. thesis, University of Oxford.

Rovelli, C. (1997). 'Halfway through the woods: contemporary research on space and time'. In *The Cosmos of Science*, ed. J. Earman and J. Norton, pp. 180–223. Pittsburgh, PA: University of Pittsburgh Press.

Rynasiewicz, R. A. (1994). 'The lessons of the hole argument'. *British Journal for the Philosophy of Science*, **45**, 407–36.

Streater, R. F., and Wightman, A. S. (1964). *PCT, Spin and Statistics, and All That*. Reading, MA: Benjamin/Cummings.

Tait, P. G. (1884). 'Note on reference frames'. *Procceedings of the Royal Society of Edinburgh*, session 1883–4, 743–5.

Van Cleve, J. (1991). 'Introduction to the arguments of 1770 and 1783'. In Van Cleve and Frederick (1991), pp. 15–26.

Van Cleve, J. and Frederick, R. E., eds. (1991). *The Philosophy of Right and Left*. Dordrecht: Kluwer.

Walker, R. C. S. (1978). *Kant*, pp. 44–51. London: Routledge and Kegan Paul. Reprinted as 'Incongruent counterparts' in Van Cleve and Frederick (1991), pp. 187–94.

Weyl, H. (1952). *Symmetry*. Princeton, NJ: Princeton University Press.

Wu, C.-S., Ambler, E., Hayward, R. W., Hoppes, D. D., and Hudson, R. P. (1957). 'Experimental test of parity conservation in beta decay'. *Physical Review*, **105**, 1413–15.

15

Mirror symmetry: what is it for a relational space to be orientable?

NICK HUGGETT

1 Introduction

I want to take issue with the definition of enantiomorphy that Pooley gives in his paper in this volume. His account goes something like this:

(a) Suppose that the relationist has an account of the dimensionality of space, according to which space is n-dimensional.
(b) The relations – especially the multiple relations – between the parts of a body determine whether it is geometrically embeddable in n-dimensional spaces that are either (only) orientable or (only) non-orientable.
(c) Then 'an object is an enantiomorph iff, with respect to every possible abstract [n]-dimensional embedding space, each reflective mapping of the object differs in its outcome from every rigid motion of it.'

This account depends on the truth of (b). Suppose that a body were embeddable in both orientable and non-orientable spaces of n dimensions. Then it might fail to be an enantiomorph, not because any of its possible reflections in physical space was identical to a rigid motion of the body, but because in some abstract space a reflection and a rigid motion of its image are identical. Pooley (in note 14) makes this point, but claims that the burden of proof falls on the opponent of his account to show that (b) is false. I wish first – in section 2 – to take up this challenge and, if not deliver such a proof, give convincing grounds for throwing the burden of demonstrating (b) back on Pooley's relationist. I do not, however, wish to follow Nerlich (1994, chapter 2) in arguing that the relationist project founders on these reefs; instead I want to offer a different proposal for how the relationist should understand orientability – and, as I shall explain, topology and geometry more generally.

Note. The topics of this paper are developed in far greater detail in Huggett, 'Geometry and topology for relationists' (in progress). That paper in turn owes a lot to discussions with Mihai Ganea. This material is based upon work supported by National Science Foundation Grant SES-0004375.

I asked whether (b) was true, but I might as well have asked for the details of the account of dimension presupposed by (a), since the account depends on that too – indeed, dimension and orientability are both topological properties, and what is really at stake is how such properties are understood by the relationist. Pooley suggests (in note 15) that a theory of mechanics might do the job, and indeed – at least in the example he gives – that it might also specify the geometry of space. If this means that the relationist simply takes facts about the topology and geometry of space to be brute, irreducible facts, among the facts about dynamical laws, then I think that it seriously undermines the interest of the relationist project – it seems to be to admit that there are more spatial facts than facts about relations. (It is in fact worth noting that to their detriment, discussions of relationism typically do simply assume that a full Riemannian geometry – typically Euclidean – is given.) If, however, Pooley is pointing out that a full theory of mechanics requires an account of the geometry of space (and spacetime) and that such a theory is determined by the relations of bodies alone, then I concur. However, in this case the substantivalist will no doubt demand to hear something more concrete about how such a programme is to work – in the next section we will consider some of the problems facing the programme, and then in section 3 I will propose a relationist account of space that resolves the difficulties. Interestingly, we can make considerable progress in this direction largely independently of considerations concerning dynamics.

2 The problems

Giving an account of orientability is the easiest of a series of problems for the relationist: the hardest is to give an account of the geometry of space, which I take to mean explaining how the relations between bodies determine a Riemannian geometry for the space; then easier, since a geometry has a topology but not vice versa, is the problem of the topology of a relational space; and then finally, because it is just one topological property, is the issue of (non-)orientability. I want to go through this series showing – or giving reasons to believe – that relations alone do not in general determine the features in question.

It is easy to produce examples to show that geometry doesn't supervene on relations: consider, for example, a system of bodies that is embeddable in a finite region of an n-dimensional Euclidean space. It is also embeddable in a space that is Euclidean everywhere except for some finite 'hole' in which a non-Euclidean region is smoothly joined, since the system of bodies can be embedded in such a way that it is entirely away from the hole. This example and its kind are embarrassing for the relationist – the two spaces could lead to physically different situations if, for example, bodies later head towards the hole.

The relationist has no better luck with the 'easier' problem of topology. Consider a sphere with a finite system of bodies scattered over it. If we cut an open hole in this space in such a way that it doesn't cut any geodesics joining two different (non-polar) bodies then we end up with an embedding of the same bodies and relations in a topologically distinct space: one topologically identical to the plane. (If we do cut a geodesic between two bodies then they are no longer multiply related, and so we don't have the same system of relations.)

You can probably see from these examples how the challege to orientability is going to go. Consider again a system of bodies embedded in a finite region of Euclidean space, then cut a hole in a region that they do not occupy and paste a non-orientable space into it – the result is a non-orientable space, with the same bodies with their Euclidean relations embedded in it. Now, it may seem that this very simple example shows that (b) is false: it is apparently a recipe for finding a system of bodies and their relations that are embeddable into orientable and non-orientable spaces. But not quite: it is, after all, possible that the operation of pasting a non-orientable region into the hole introduces a new geodesic connecting the bodies and so leads to a system in which the bodies are related in a different way – namely in that two of the bodies have different numbers of distance relations between them. All the same, the hole can be as small, and as flat, and as far away from the system of bodies – and of course still have an effect on them later in their history – as you like; saving (b) requires that however the non-orientable region is inserted, the result is a space with new geodesics between bodies. It seems to me unlikely that this is so, and I am absolutely certain that the onus is now on Pooley's relationist to prove that it is so.

Assuming that such a proof is not forthcoming, let's consider how the relationist might respond. He might, against my earlier advice, take it that there is some primitive fact, not reducible to the relations between bodies, about the geometry of physical space. And indeed there is a way in which such a position has been advocated – by the tradition of 'modal relationism', which traces back to Leibniz (e.g. Leibniz and Clarke, 1956, pp. 26 and 42). This position takes it that in addition to the occurrent facts about the relations between bodies, all the modal facts about what relations are possible in a world are also primitive. Then it is quite reasonable to suppose that the collection of all facts about whether systems of bodies and relations are possible or not is logically equivalent to the statement that a system is possible iff it is embeddable in some specified Riemannian geometry – the primitive geometry of relational space. As I already suggested, taking this step drastically weakens the relationist position (especially his empiricist credentials), but even so it is not enough to solve all his problems.

To see why, consider that for the substantivalist there are not just facts about geometry of space, but also concerning the location of matter in space (if the geometry is inhomogeneous). For instance, suppose that space had the (flat) geometry of $R^2 \times S^1$ – planar in two dimensions but rolled up in the third, with circumference, say, 10 m, except for a hole containing a non-Euclidean region (understood appropriately by the substantivalist and modal relationist). Suppose further that the only bodies are a spaceship and a probe that it can send out with a relative velocity of 10 m/s for 1 s after which it heads back to the ship at 10 m/s. Now consider the following scenario: at first the ship moves at a constant velocity such that if the probe is fired at 10 s intervals it always takes on non-Euclidean relations to the ship when it is 1 m away; later, after the ship accelerates by 1 m/s the same thing happens when the interval is only 5 s.[1] For the substantivalist the facts are that the ship is moving around the rolled up dimension first at 1 m/s and then at 2 m/s and is firing at just the right moments to shoot the probe into the inhomogeneity every time it goes past. According to him, whether or not a probe will hit the inhomogeneity is determined by (among other things) facts about the position of the ship in space when the probe is fired. Now consider things from the point of view of the modal relationist.

At any moment while the probe is on the ship, according to the relationist the spatial facts are exactly the same: that the ship and probe stand in such-and-such actual Euclidean relations, and that they could stand in any relations embeddable in the given space. But there are two (and more) very different relative motions that are embeddable in that space: those that actually take place, in which the probe passes through the non-Euclidean region, and those in which it is fired at the 'wrong' time and always misses – in the latter but not the former case the relations are always Euclidean. Hence the facts which the relationist allows at the moment a probe is fired do not suffice to determine what spatial facts will hold later. This situation is of course absurd: imagine the spaceship moving constantly and firing the probe 999 times once every 10 s, each time leading to non-Euclidean relations. After 10 s, at the start of the 1000th run the relational facts are just as they always are when the probe is on board, and so do not determine what will happen to the probe. Yet we would know by induction exactly what would result – non-Euclidean relations. But how could we possibly know something about the future if it were indeterminate? Clearly we can know something about the spatial facts beyond what relations are actual, and which are possible – something like where in the space of possible relations things are.

Modal relationists do indeed introduce such extra facts, though not necessarily for this reason. For instance, one of the best worked-out accounts was developed by Ken Manders (1982). His scheme has in its primitive vocabulary predicates not only for the relative positions of bodies but also to the effect that two bodies are

[1] I am assuming that Newtonian mechanics holds in the space, but similar examples could be cooked up for any kind of mechanics.

(or are not) in the 'same place', regardless of whether they are in the same relative positions or not. But if the modal relationist adopts this position then he is only a hair's breadth from substantivalism: for every point (up to isometries) that the substantivalist acknowledges, the relationist says that there is the possibility of a body being at a different place – effectively that places are 'permanent possibilities of location'. This is of course a position that one could adopt, but it seems to me to be rather far from the original intent of the relationist, and so I reject it.

3 The answer

So modal relationism is a dead end, even though it offered an account of geometry, topology and (non-)orientability. However, I want to propose what I believe to be a novel account of how these things should be understood by the relationist. This is a back-to-roots relationism: the only primitive spatial facts concern the actual relations between bodies, not their possible relations (though it respects the intuition that there are facts about what relations are and aren't possible, and gives an analysis of such facts). We know that the embeddability of actual relations won't pick out the properties of a space that we need, so let's make a fresh start on the problem.

Reconsider the example of a spaceship moving in $R^2 \times S^1$, regularly firing its probe so that the system takes on non-Euclidean relations every time. As we noted, after a number of repetitions the scientists on the ship are in a position to predict what will happen the next time they perform the experiment. Not only that, they could say what would happen if they were to perform the experiment, whether or not they actually do so. On what basis is this possible? On the basis of induction from past regularities in the way that relations evolved to possible or actual future events – namely that they will also fall under the regularities. Leaving the problem of induction to one side, what this observation indicates is that what is real about space for the relationist is not just the relations of bodies at any particular time, but *the way that relations evolve in regular ways over time*.

Thinking about things in this way connects the discussion to the topic of laws of nature, since they also concern the regular ways in which events occur. The analogy can be filled out in a number of ways, but here I just want to draw on an account of the nature of laws to help develop my relationism.

A fairly literal minded view of laws takes them to be statements of primitive natural necessities: for example, 'necessarily the gravitational force between any two bodies is proportional to their masses and inversely proportional to the square of their separation'. According to this view the fact that the actual force between two bodies is such-and-such holds because it must as a matter of necessity. A problem for this account though arises if – as is commonly supposed – there are incompatible sets of general statements compatible with any body of specific facts (as we saw many geometries compatible with a set of relations): if they are all true descriptions of the facts, how can one decide which is also *necessarily* true?

There is an account of laws that avoids this problem – and is appealing to empiricists (construed broadly), since it avoids any primitive necessities – namely the 'Mill–Ramsey–Lewis' (MRL) account.[2] According to this view, one does not take the laws as primitive, necessitating the regular ways in which events actually unfold, but instead takes the events and regularities as fundamental, determining the laws. Of course we've just said that if logical compatibility were the only constraint then the laws would be underdetermined by the events, so according to the MRL theory, a law of nature is, by definition, any member of the *strongest and simplest* set of true statements.[3] And that is all a law is, not a statement that is in addition 'necessarily true' in some primitive, independent sense – how could we hope to know that? As Earman (1984, p. 212) puts it: 'A world *W* is a world of non-modal facts' (and so our modal talk is unpacked as talk about which truths are simplest and strongest).

I believe we should take a similar approach to space. The only spatial reality of a world concerns the relations between bodies and the regular ways in which they evolve. Any talk about the geometry of space does not concern a substantial space, but instead is talk about those relations and regularities – and in particular the geometry of space is, by definition, the *simplest* one in which the entire history of relations is continuously embeddable.[4] That is, we adopt the MRL insight that law-talk is just the most convenient way to describe the regularities, and say that space-talk is just the most convenient way to talk about the regular ways in which relations evolve. According to this view the spatial part of a world is made up only of relations between bodies and the regular ways that those relations evolve, not of any further facts about geometry. Then when we make predictions or counterfactual statements about what will or could happen, all we are really talking about is the simplest geometry compatible with the history of relations – what we believe about what will or could happen is grounded in our inferences from the part of the history that we have seen to those that will be.

For obvious reasons I will call this the 'regularity account' of relational space, and we can give it some more bite by considering just what 'simplicity' means.

[2] See Earman (1984) for a very clear exposition and references.

[3] That the set is strongest means, roughly, that we value the most logically powerful statements describing the regular ways in which events unfold – thus we find that laws are typically universal generalizations. That the set is the simplest is a constraint in the other direction – the strongest set of true statements is the set that describes all actual events individually, but when it comes to laws we seek statements that summarize the facts as concisely as possible.

[4] Not strongest as well, since embeddability already ensures that all actual relations are captured.

I propose three criteria for simplicity in decreasing order of importance, though I suspect that the list is as yet incomplete. First of all, the simplest space is the one with the lowest possible number of dimensions in which the relational history is embeddable.[5] Next, of two spaces of the same dimension, the simpler is the one in which the geometry varies in a regular way – if a relational history is embeddable in a space in which the metric is a periodic function, and a space just like it except in some finite region, then the regular space is simpler (as with laws, we value regularities). Finally, if two spaces are otherwise equally simple, then the simpler is the one whose curvature varies in the smoothest way.

One might object that given any list of simplicity there will still be an underdetermination problem – many equally simple spaces into which a relational history is embeddable. But if so, then this is a problem for the substantivalist not the relationist, since he now takes space to have a definite but unknowable geometry. The relationist should respond to this situation by saying that there are geometric facts which are indeterminate. Specifically, the geometric facts are anything that is true of *all* the simplest geometries; so if, for instance, the simplest geometries disagree on the volume of some region, then the volume is in fact indeterminate. And don't think that this is fishy because the inhabitants of the world could measure the volume as accurately as they like – that is just to say that the relational history could have been different, in which case of course a different space could have been the simplest in which the history was embeddable.

And that is that: the regular ways in which relations evolve determine a Riemannian space with its geometry, topology and, crucially for the topic of mirror symmetry, (non-)orientability – conversely these properties are nothing more than succinct and powerful ways of expressing those regularities. This account also provides an acceptable resolution to the problem for the modal relationist, since it explicitly involves an embedding of the relational history into the space in question, and hence a fact about the embedding at any time – even though this too is just a fact about the history as a whole.

Except of course that is not quite that. The limitation of this account is that it has focused on merely embedding relations into space, when we know that in fact we also have to consider what geometric structures for space and time and spacetime are required by dynamics: An affine connection? Anisotropy? An orientation field? An arrow of time? Extra dimensions? I do not have room to address these complexities here, except to point out that the account I have given is amenable to development in such directions – the correct space is not only the simplest in which the relations

[5] Now, there might be other reasons to suppose that the number of dimensions is greater than embedding the relations requires – evidence for a Kaluza–Klein theory perhaps. True, but here we are trying to consider how relations, independently of dynamical considerations (as far as possible), determine the geometry of space. I will consider briefly how to integrate dynamical considerations below.

are embeddable, but also that required by dynamics, itself understood in relational terms (for instance as defended in Huggett, 1999). What we have seen here is one part of the way in which the geometry of relational space gets determined.

References

Earman, J. (1984). 'Laws of nature: the empiricist challenge'. In *D. M. Armstrong*, ed. R. J. Bogdan, pp. 191–223. Dordrecht: Reidel.

Huggett, N. (1999). 'Why manifold substantivalism is probably not a consequence of classical mechanics'. *International Studies in the Philosophy of Science*, **13**, 17–34.

Leibniz, G. W., and Clarke, S. (1956). *The Leibniz–Clarke correspondence*, ed. H. G. Alexander. Manchester: Manchester University Press.

Manders, K. L. (1982). 'On the space-time ontology of physical theories'. *Philosophy of Science*, **49**, 575–90.

Nerlich, G. (1994). *The Shape of Space*, 2nd edn. Cambridge: Cambridge University Press.

16

Physics and Leibniz's principles

SIMON SAUNDERS

Leibniz's principles made for an elegant and coherent philosophy. In part metaphysical, in part methodological, they addressed fundamental questions – in the treatment of symmetry, in the relationship of physics to mathematics, in logic – that are if anything even more pressing today than they were in Leibniz's time. As I shall read them, they also expressed a distinctive and uncompromising form of realism, a commitment to the adequacy of purely descriptive concepts. This doctrine has been called 'semantic universalism' by van Fraassen (1991), and the 'generalist picture' by O'Leary-Hawthorne and Cover (1996): it will become clearer in due course just what it entails.

The principles that I shall consider are the Principle of Sufficient Reason (PSR) and the Principle of the Identity of Indiscernibles (PII). In the first instance I shall take them both to be methodological principles. The former I shall read as requiring that the concepts of physics be entirely transparent. Analysis and explanation are to proceed without any limits. The perspective is impersonal: any epistemological limitation, to do with our human situation or perceptual apparatus, is to be viewed as a purely practical matter, reflecting no fundamental constraint. This puts in place a part of the generalist picture.

The PSR clearly promotes the use of mathematical concepts in physics. The PII, in contrast, depends on a sharp distinction between *purely* mathematical concepts, and physical ones. Leibniz too made use of this distinction (between 'real' and 'notional' or 'ideal' concepts), but in his hands the principle depended heavily, though often tacitly, on his metaphysical theory of substance (and, with qualifications, on his philosophical logic). He was led to a restrictive formulation of it in consequence (Leibniz, 1714, section 9):

> There are never in nature two beings which are perfectly alike and in which it would not be possible to find a difference that is internal or founded upon an intrinsic denomination.

In due course we shall come across two possible candidates for what Leibniz called 'intrinsic' denominations. But there have been major changes in logic since Leibniz's time, and a purely metaphysical theory of substance is unlikely to command much assent today: we shall find little reason to plump for either of them.

There is a widespread view that, apart from the trivialization of the PII whereupon identity is made out in terms of predicates that themselves involve identity, there are straightforward exceptions to every version of this principle. It is a mistaken view; as we shall see, there is a natural analysis of identity available for any formal language that is immune to the usual counter-examples; the principle is not, I hold, in any difficulties from this quarter. The problem, rather, concerns the *justification* for the PII – why embrace such a principle? What is wrong with identity taken as primitive?

In the most general context, I see nothing wrong with identity. But in physics – specifically identity as it figures in physical *theory* – there are special reasons to view it as derivative. I take it that we are concerned with physical objects in the logical sense, as objects of predication. I suggest it is through *talk* of objects, in the light of mathematical theories and experiments, that we achieve a clear interpretation of these theories and experiments in terms of physical objects – our understanding of what objects there are, I am suggesting, is clearest in our use of simple declarative sentences. And it is here that purely formal, logical considerations come into play; Quine and the logical empiricists had something important to say in this respect. But my suggestion is not that physical theories should be *reconstructed* in a formalized language (they should not be *rewritten*, as a construction in set theory). What I have in mind is *description*, as informed by theory, in predicative terms. (This puts in place the other part of the generalist picture.)

Taking this route, our first concern is with syntax. Here, I suggest, the non-logical symbols of the language – for simplicity I shall consider only finite, first-order languages – can be derived more or less directly from the physical theory. They are to be interpreted in terms of the real physical functions, properties, and relations. Our guide here, as for Leibniz, lies in the measurable quantities. Not so identity, as the relation that every physical object has to itself and to no other. It would be hard to imagine a quantity whose measurement could tell us about this directly. Nor is the identity relation itself under investigation in physical theorizing, unlike measurable properties and relations (in this sense it is not treated as a *physical* relation at all). From a formal point of view, the mathematics used in physics is far away from set theory, and still further from formal logic: the identity sign, as it figures in extant physical theories, signifies only the equality or identity of mathematical expressions, not of physical objects.

In summary, we may read off the predicates of an interpretation from the mathematics of a theory, and, because theories are born interpreted, we have a rough and ready idea of the objects that they are predicates of. But there is nothing systematic to learn from the formalism to sharpen this idea of object. It is plausible, in this situation, that we should look to a purely logical aid.

It is the fact that there is an essentially unique prescription for how to use the identity sign, available for any formal language whose predicates do not involve identity, that now is really telling: this is the chief selling-point for the PII as I shall understand it. Indeed, given a finite lexicon, this prescription even generates an explicit *definition* of identity.

I shall first sketch the details, and then show how the principle fares in the face of the usual scenarios offered as counter-examples to Leibniz's principles. The generalist picture is also supposed to be in trouble in these contexts; that too will need some defending. To proceed from that, we shall need the rudiments of Leibniz's theory of possible worlds; as we shall see, it can be taken over for our purposes with little change. The most important question, from that point on, is how we are to distinguish the real physical quantities of a theory from the purely mathematical ones. Leibniz too needed this distinction, but he fell back on a fairly crude form of verificationism. Measurable quantities will be our guide as they were for Leibniz, but they are only the starting point of our analysis: symmetries are the essential tool for going beyond them. At the end I will return to the PSR, and its relationship to the PII.

1 Identity

How, in the interpretation of a physical theory, is identity to be analysed in terms of other properties and relations, which do not themselves involve identity? There is a canonical answer to this question. Given the simplest case of a language with only finitely many predicates, for each $n \leq N$, let there be K_n n-ary predicate symbols $P_1^n, P_2^n, \ldots, P_{K_n}^n$. Now let '$s$' and '$t$' be terms (variables, names, or functions of such). The familiar axiom scheme for identity is:

$$s = s$$

$$s = t \rightarrow (Fs \rightarrow Ft) \tag{1}$$

where F is any predicate of the language – expressing, essentially, the substitutivity of identicals. As Gödel showed, a complete proof procedure for the predicate calculus without identity, supplemented by this scheme, yields a complete proof procedure for the predicate calculus *with* identity (Gödel, 1930, Theorem VII). It is therefore enough, from the point of view of completeness, to take the conjunction of the RHS of every instance of (1) as *sufficient* for identity. In the case of 1-place predicates, we obtain formulae of the form

$$P_i^1 s \leftrightarrow P_i^1 t \tag{2}$$

for $1 \leq i \leq K_1$. If it is right to read '1-place predicate' for 'intrinsic denomination', this would clearly do as a formalization of the Principle of Identity as Leibniz stated it (equation (2) is often called the *strong* version of his principle). In the case of 2-place predicates, generalizing on the free variable that remains in instances of (1) we obtain

$$\forall z_1((P_i^2 s z_1 \leftrightarrow P_i^2 t z_1) \wedge (P_i^2 z_1 s \leftrightarrow P_i^2 z_1 t)) \tag{3}$$

for $1 \leq i \leq K_2$. Likewise, for 3-place predicates:

$$\forall z_1 \forall z_2 (P_i^3 s z_1 z_2 \leftrightarrow P_i^3 t z_1 z_2) \text{ and permutations}$$

for $1 \leq i \leq K_3$. And so on, up to predicates in N variables. Call such formulae *identity conditions* for s and t. The conjunction of all these identity conditions is to serve as our definition of identity, where the question of which conditions in fact hold, for given terms, is to be settled by appeal to the physical theory from which the non-logical vocabulary is derived.

This principle is the *only* analysis of identity that is really workable from a modern logical point of view, embracing as it does every deductive consequence of the axioms of identity. It was first proposed as such by Hilbert and Bernays in the *Grundlagen der Mathematik* in 1934. It was subsequently defended by Quine in the above, definitional sense, for any first-order language with a finite lexicon (in *Set Theory and its Logic*, in *Word and Object*, and in *Philosophy of Logic*). Quine's interest in the principle was that it allowed him to extend his view of logical truth, as truth by virtue of grammatical form alone, to truths involving identity. As such he considered it an account of identity sufficient for mathematics as well.

There are plenty of reasons to be sceptical of Quine's programme, understood to have the generality that he intended for it, but there is no need to consider them in any detail; the proposal I am making is quite different. I do not suppose that there is anything wrong with identity, taken in an irreducible sense; whatever objects there are, we know what the identity relation is among them; given objects, identity can look after itself. Neither are we concerned here with the ground for logical truth. The proposal, rather, is that in a situation in which we *do not know* what physical objects there are, but only, in the first instance, predicates and terms, and connections between them, then we should tailor our ontology to fit; we should admit no more as entities than are required by the distinctions that can be made out by their means. The most common objection to Quine's proposal is that a language-relative notion of identity cannot possibly do – what we end up with is not *really* identity.[1] But in the present context that is to call into question either the correctness of the underlying physical theory (we may not have the right vocabulary or identity conditions), or the method of interpreting it in terms of objects (certain identities may be negated in an unanalysable sense). Certainly, theory or method may be wrong; both of course are defeasible; but just for that reason, neither can be rejected *a priori*.

There is certainly plenty of evidence in favour of the principle when it comes to ordinary physical objects.[2] Following Quine (1960, p. 230), call two objects *absolutely discernible* if there is a formula with one free variable true of one of them but not of the other. With the obvious extension of this terminology to sets of objects, it is clear that ordinary solid objects are all absolutely discernible: no two solid objects can occupy the same spatiotemporal position, and given an asymmetric distribution of such objects each will satisfy different spatiotemporal relations with every other (referred to by bound variables). Given any countable set of absolute discernibles, for each there will exist a finite formula in one free variable that applies to it uniquely; call it an *individuating predicate* for that object. It follows that every solid object in an asymmetric universe has an individuating predicate.

If the PII identified any two objects not absolutely discernible, we would have what is usually called the *weak* version of Leibniz's principle.[3] But this is only one category of discernibility, according to the PII. Call two objects *relatively discernible* if they are not absolutely discernible and there is a formula with two free variables that applies to them in only one order. For an example, consider the instants in time in an empty Newtonian spacetime; they are all relatively discernible (of any two, one will be earlier than the other, but not vice versa).

There is a third and final category. An identity condition may fail even when objects have exactly the same properties and exactly the same relations to all other objects *and* exactly the same relations to each other; equation (3) will be false if s and t satisfy only an *irreflexive* relation A (for then $\exists z_1 \sim (Asz_1 \longleftrightarrow Atz_1)$, namely when $z_1 = s$ or when $z_1 = t$). Call objects not absolutely or relatively discernible, that satisfy an irreflexive relation, *weakly discernible*;[4] if none of absolutely,

[1] See for example Wiggins (2001), chapter 6. Having considered Quine's original proposal, Wiggins goes on to consider a supervenience thesis of identity (*ibid.*, pp. 187–8) that is closer to what I am proposing, but at this point he puts the lessons that should have been earlier learned to one side (the counter-examples he adduces against the supervenience thesis are examples of relative or weak discernibles – see below).

[2] Quine has offered an extensive account of how concepts of objects are first acquired, according to which predications (and connections among predicates) will precede the full-blown notion of object involved in the use of the identity sign (Quine, 1974). Whether or not he is right on this, it can hardly be denied that the simplest ways in which we discriminate among ordinary objects is by qualitative, sensible differences.

[3] As a reading of Leibniz, this is to include relations involving bound variables as among the intrinsic denominations of a substance. (For a defence of the view that Leibniz was prepared to countenance relations not reducible to monadic predicates, see Ishiguro, 1972.)

[4] This category went unnoticed in Quine (1960), where the terms 'absolute' and 'relative' indiscernibles were introduced. Quine remarked on it later (Quine, 1976), but there he introduced a different terminology (in terms of 'discriminability'). It is true that weakly discernible objects are indiscernible under the strong or weak versions

relatively, or weakly discernible, *indiscernible*. Using these definitions the Hilbert–Bernays principle is precisely the Principle of the Identity of Indiscernibles. (This is what I shall mean by the PII from this point on.)

For an example of weakly discernible objects, consider Black's two iron spheres, one mile apart, in an otherwise empty space (Black, 1952; see this volume). The irreflexive relation A is '... one mile apart from ...'. It is *because* this relationship holds that we may say that there are two – that it is intuitively evident that there are two. The example was intended as a counter-example to Leibniz's original principle, in either the strong or weak form, and so it is; what went unnoticed is that it is the PII that sanctions the example, and shows us its logical form – and with which, of course, it is not in contradiction.

There are plenty of realistic examples of weak discernibles. Consider the spherically-symmetric singlet state of two indistinguishable fermions. Each has exactly the same mass, charge, and other intrinsic properties, and exactly the same reduced density matrix. Since the spatial part of the state has perfect spherical symmetry, each has exactly the same spatiotemporal properties and relations as well, both in themselves and with respect to everything else. But an irreflexive relation holds between them, so they cannot be identified (namely '... has opposite direction of each component of spin to ...'). Since symmetric, they are weakly, not relatively discernible. Indeed, indistinguishable fermions are *always* at least weakly discernible; an irreflexive relation exists between any pair of fermions, whatever their state.[5] There has been plenty of discussion of the bearing of Pauli's Exclusion Principle on principles of identity; the prevalent view is that none can be secured by it under the standard, minimial interpretation of quantum mechanics.[6] But the relations that I have made use of follow from the eigenvector–eigenvalue link, and are not in any doubt.

One might conclude that the PII is so weak that it can never be compromised, but that view too is mistaken. Indistinguishable elementary bosons may all exist in exactly the same state, and satisfy no irreflexive physical relation. It was argued by

of Leibniz's principle – the traditional ones in the philosophy literature – but then so, usually, are relative discernibles; I see no reason to follow Quine in this shift in terminology.

5 The most general antisymmetrized 2-particle state is of the form $\Psi = \frac{1}{\sqrt{2}}(\phi \otimes \psi - \psi \otimes \phi)$, where ϕ and ψ are orthogonal. Analogues of operators for components of spin can be defined as $S = P_\phi - P_\psi$, where P_ϕ, P_ψ are projections on the states ϕ, ψ. Each of the two particles in the state Ψ has opposite value of S, but no particle can have opposite value of S to itself. (For a general theory of the state in terms of systems of relations, see Mermin, 1998.)

6 As first argued by Margenau (1944), Leibniz's principle must be rejected as the reduced density matrix for each fermion in any antisymmetrized state is exactly the same. A more general argument was given by French and Redhead (1988) and by Butterfield (1993); the theorems proved there, like those proved by Huggett (this volume, Part II, chapter 13), apply only to either the strong or the weak version of Leibniz's principle, not to the PII. (For a criticism of such methods from a rather different perspective, see Massimi, 2001; for a commentary on interpretation-dependent treatments, see Castellani and Mittelstaedt, 2000.)

Cortes (1976) that photons are a counterexample to Leibniz's principle; free photons are certainly a counterexample, even to the PII. Does it follow that the principle should be abandoned? But the argument can be turned on its head. The stable constituents of ordinary matter are all fermions.[7] Apart from the Higgs particle – not so far observed – all elementary bosons are gauge quanta; they all mediate forces between fermions. The number of elementary bosons all in exactly the same state may better be thought of as the excitation number of a certain mode of a quantum field. It is the discrete measure of the strength of dynamical couplings, dependent on the mode, between the genuine physical objects of the theory, whether fermions or other modes of quantum fields.

Schrödinger argued very early on for such a view (Schrödinger, 1926). For a recent proposal of this sort, but with a somewhat different motivation, see Redhead and Teller (1992). Our conclusion, however, is that only boson numbers should be viewed as properties of things, not fermion numbers as well; the PII treats fermions quite differently. Given the contrast between the two, as gauge fields and sources respectively, it is a merit of the principle that it does.

2 The generalist picture

The immediate difficulty is not that the PII identifies indiscernibles that are not even weakly discernible; it is that it does not identify those that are. On familiar, Strawsonian lines, such highly symmetric situations call into question the adequacy of the generalist picture (of what he called 'descriptions-in-general-terms'), for given two weakly discernible objects, individual reference can be made to neither of them. Likewise in the case of relative discernibles. Yet confronted with two objects of this sort, there could be no obstacle to the use of indexicals; indexicals would do better here than any purely predicative description;[8] the generalist picture is therefore incomplete – and so the PSR is also in question.

There is an obvious flaw in this argument. The use of indexicals presupposes the existence of an observing agent, but introduce such an agent and the symmetry is broken. Each of two weakly discernible objects, once related to something as highly

[7] It is unclear to French and Rickles (this volume, note 23) 'why the metaphysics should follow the physics in this particular way, or at all'. Perhaps it need not. But if I am concerned with metaphysics at all, it is *descriptive* metaphysics, in Strawson's sense, as an aid to the interpretation of physics, and to that end I aim to preserve a good part of established practice. Ordinary objects had better turn out to be objects, on any account, and so they do on mine; it is as an extension from this that their stable constituents had better turn out to be objects as well. With the rest there is more latitude.

[8] It should be clear, here as elsewhere, that although the PII as stated permits the use of proper names (0-ary functions), it is contrary to the spirit of our programme to invoke them; certainly no general physical theory makes use of them explicitly. (They may do so tacitly; indexicals are obviously unavoidable in practice; the question, from the point of view of the PSR, is whether they are avoidable in principle.)

asymmetric as, say, a functioning human being, becomes absolutely discernible. And there is little point in envisaging a perfectly symmetric observer, so long as indexicals are tracking perception: attention to one rather than the other object, by whatever perceptual means, will surely break the symmetry. But for all that there is a difficulty, even concerning asymmetric observers; for it is easy to imagine a *large-scale* exact symmetry (for example, a spacetime containing a plane of exact mirror symmetry), where the observer too has a symmetric duplicate. It need not be the perception of an observer of *two* weakly discernible objects that creates the problem, it is enough, given that she has an exact duplicate elsewhere, that she sees only one of them.

The scenario is fanciful; one can deny that it is a genuine physical possibility. We are not concerned with defending the generalist picture or the PII in the face of any conceivable physics. For example, a physical theory that is explicitly a first-order formal theory (with physical objects as values of variables) is at least conceivable, however remote from the theories that we have; in such a case one might have as a law a sentence involving identity in which the identity sign cannot be treated as a defined term (using the PII) without contradiction. But these are challenges we do not have to meet.

Nevertheless, I think we should grasp this nettle.[9] We may grant a certain limitation to the generalist picture. But it is not that by the use of indexicals one can provide something *more* than what is there available, it is that one can provide something *less* – one provides less than a complete account of what there is. For suppose – the example is due to Adams (1979) – that each of Black's identical globes is inhabited, in such a way that the symmetry between them is preserved.[10] The inhabitant of each globe refers to his own uniquely. But *that* fact is perfectly well described in the generalist picture. Nothing is left out of it. That is the point of relative and weak discernibles in the generalist picture: one can only describe the part in terms of the whole.[11]

[9] Here I follow Pooley (2002, chapter 2). Yet another alternative is to embrace Hacking's strategy (Hacking, 1975), as recently endorsed by Belot (2001); I do not believe this strategy can be implemented with the generality Hacking claimed for it (see French, 1995), but it may be it can for the special case of large-scale spacetime symmetries, that gives rise to the present difficulty for the generalist picture.

[10] See Black (1952), extracts in this volume.

[11] Another challenge to the generalist picture arises from Kant's argument from incongruent counterparts. Pooley (this volume) may be read as insisting on a principled sense in which ostensive definition is required, in the context of a relational account of handedness. According to him, what objects we call 'left' can only, in the final resort, be shown. Against this – assuming spatial inversion is a symmetry (the relevance of this will become clear shortly) – I would maintain that a description of the universe which depicts a handed object as congruent to the hand on the side of the heart of a typical human body *describes* that object as left-handed (in other words, that the causal processes to which Pooley refers can themselves be described in the generalist picture).

Pooley rightly remarks that such descriptions do not solve the Ozma problem, but then this problem is not a difficulty for the generalist picture *per se* (concerning, as it does, the question of whether orientation can be *locally* defined).

3 Possible worlds

A similar challenge arises from a different quarter. What of possible worlds? Might there be possible worlds which are only weakly discernible?[12] In particular, might there be a possible world that is only weakly discernible from our own? We surely do refer uniquely to actual physical objects and to the actual world.

It should be evident that here the strategy just canvassed will hardly do. It is one thing if, in the generalist picture, in highly symmetric cases, we can only describe the entire world, but it is quite another if even in the physically realistic case we can do no better than describe a set of possible worlds – that we cannot describe the actual without describing all its possible simulacra as well.

In fact there is no such difficulty. We have been talking all along of *real* physical properties and relations. Objects can only be discernible, yet fail to be absolutely discernible, if they bear relations to each other – real, irreducible physical relations, relations that are not deductive consequences of their properties. If there are none such, the PII reduces to the identity of objects not absolutely discernible (and the latter in turn to the strong principle, equation (2)). What real physical relation can one possible world, a possible physical universe, bear to another? A world, in Leibniz's philosophy as in modern cosmology, is a system which is physically closed. For every real relation, from spatiotemporal and causal relations to quantum correlations, the relata always have to be included together to arrive at the closed physical system.[13]

There is of course a difficulty in saying just what are the real physical relations, as opposed to the purely nominal, mathematical ones – we shall come on to that in a moment. But we do not have to settle this question to justify the claim that possible worlds may bear no real relations to one another. I take this point to follow from the definition of physical closure: if we cannot make sense of the difference between real physical relations and mathematical ones, we will be equally hard put to say what physical closure really means. If the notion of 'real relation' is too vague then so is the notion of 'world'. And I suggest there is no good example in which the idea of closure under physical relationships is really in doubt. Even admitting exotica such as cosmic wormholes, or spacetimes with topological change, it is clear whether or not one is dealing with a closed physical system if only because one is considering a *single* solution to Einstein's field equations (or a single extension of

[12] The PII can hardly be applied to *all* possible worlds, since among them will be worlds governed by different physical laws; here I consider only possible worlds with the same physics as ours (and, naturally, the problem arises only for these).

[13] Teller's 'liberalized relationism', therefore, is not an option for us (Teller, 1991). Possible objects may, of course, be physically related, if they are described as such by a physical theory (as in the Everett interpretation of quantum mechanics): in that case they do not belong to distinct possible worlds, in Leibniz's sense (and they may well be only weakly discernible).

a solution).[14] Of course this is not to rule out comparisons of solutions to Einstein's equations. It clearly makes sense to talk of the mean matter density of one spacetime model in relation to another. But these relations are reducible to properties; they are deductive consequences of the properties of their relata.

Given that possible worlds bear no physical relations to one another, it follows from the PII that numerically distinct worlds will be absolutely (and in fact strongly) discernible. A world is surely an individual substance, in Leibniz's original sense of the term, even if nothing else is. But it follows from this that any object which is absolutely discernible from every other in one possible world, will be absolutely discernible from any other in any other possible world – for we have only to take the conjunct of its individuating predicate in the one possible world with the individuating predicate of the possible world to which it belongs; that will absolutely discern it from any possible object in any other world.[15]

When it comes to possible worlds, we not only obtain a form of the PII that is recognizably Leibniz's, we recover a part of his original motivation for it too. For the alternative to an analysis of identity in terms of predicates – taking identity as unanalysable – always amounted to a purely extensional account of it, in terms of whatever objects there are (as that relation that every object bears to itself and to no other). That is how Lewis, agnostic as to the nature of the full space of logical possibility, could declare himself agnostic on the principle: whether there exist indiscernible possible worlds depended, for him, on whatever possible worlds really exist. But for Leibniz, as for those of us who do not believe that possible worlds exist independent of us, that account will hardly do.

4 Leibniz Equivalence

I come back to the distinction between real and mathematical properties and relations in its more general setting. As we have just seen, transworld relations are either reducible or purely mathematical; our concern is with properties and relations internal to worlds. Here, as remarked earlier, our chief guide is experiment: properties and relations defined in terms of directly measurable quantities are certainly real. But to restrict ourselves to these would be a crude form of verificationism.

[14] If there is a difficulty it seems likely to lie in quantum cosmology or, classically, in regions interior to Cauchy horizons (in the neighbourhoods of singularities). The two topics are connected. In these cases we do not, properly speaking, have a serviceable theory at all. They offer no real threat to the view that a real physical relation is a part of reality, to be solved for along with all the other physically meaningful quantities, rather than a link to something beyond.

[15] It is true that the individuating predicate we end up with in this way will no longer be finite in length (for there the number of possible worlds is surely uncountable), but we are familiar with this from Leibniz's philosophy: there, in comparisons across worlds, the individual concept of a substance must be infinitely complex (this played an important role in his theory of contingency).

Symmetries are the key to moving beyond them. As I shall understand it, any exact symmetry of a system of equations is a transformation that leaves its *form* unchanged (under which the equations are *covariant*). It is their mathematical *form*, I take it, that has real physical meaning. Such symmetries therefore leave all the physically real quantities unchanged – among them the measurable ones. In the first instance we look for symmetries of these quantities; in the second to a theory or theory-formulation which respects these symmetries – which has a corresponding invariant structure or form. Under such a symmetry, those elements of the formalism that *are* transformed will have no direct physical significance, and their associated properties and relations that are likewise modified will not be physically real. But if there is no such theory formulation, that is an indication that there are *further* properties and relations, which are modified by these transformations, that are physically real – whether or not they are measurable. (This, as it were, *explains* why the transformations in question are not after all symmetries.) We then move to a smaller group of transformations, with respect to which the structure of the theory *is* preserved: the invariant quantities under these are the ones we count as real.

An example will illustrate the procedure. Relative distances between particles in Newton's theory of gravity (NTG) are surely measurable; they are invariant under translations, rotations, and boosts to reference frames with constant velocities. Since the equations of motion are form-invariant under these transformations – they are indeed symmetries – it follows that (absolute) positions and (absolute) velocities, as properties, are not physically real. What now of (absolute) linear accelerations? They are not directly measurable; but here the equations of NTG prove to be uncooperative. It turns out that they do not preserve their form under boosts to linearly accelerating frames. So the latter are not symmetries, and the quantities modified by these transformations – absolute accelerations, that are invariant under the symmetries of NTG – should also be counted as physically real.

This procedure, relying as it does on the details of a theory formulation, has its risks: even if the theory is substantially correct, a different formulation of it may come to light leading to a different conclusion. So it was with NTG: it turns out that there is a reformulation of the theory that does count boosts to accelerating frames as symmetries (where the accelerations are arbitrary functions of time, but are independent of position).[16] Absolute accelerations are no longer invariant, so they go the way of absolute velocities. Only relative accelerations and relative velocities, we learn from this theory, are invariants, along with relative distances,

[16] This was clear from Cartan's reformulation of NTG as a diffeomorphic-covariant system of equations, but nothing so elaborate is needed: the clue to it was already evident in the *Principia*. There Newton showed that his equations yielded the same results for the relative motions when referred to linearly accelerating frames (and needed to, to apply his principles to the Jupiter system); see Corollary VI, Book 1.

under the full symmetry group that this theory allows. Only they should be added to the list of real physical quantities.

But the story does not stop there. It turns out that there is an empirically adequate alternative to NTG according to which certain relative accelerations and relative velocities – those associated with non-zero total angular momentum – are necessarily zero. According to this theory, due to Barbour and Bertotti (1982), the affine structure of spacetime has a purely dynamical origin, deriving from a *geodesic* principle on the *relative* configuration space. It follows that masses and relative distances alone are fundamental; every other real physical quantity (including every time-dependent quantity and all temporal relations) can be defined in terms of these. The symmetry group of this theory is correspondingly greatly enlarged.

This example makes it clear that whilst we must start with measurable quantities, they may appear in a very different light at the end. Getting the right expression for measurable quantities in terms of theoretical ones has proved almost as difficult as moving beyond them. It is now a familiar story how this played out in the case of the diffeomorphism symmetries of the General Theory of Relativity (GTR). This symmetry group appeared to be a step too far, given the then standard interpretation of coordinates in a physical theory. It proved a considerable difficulty for Einstein to appreciate that observable quantities could in fact be coded into the theory in a diffeomorphic-invariant way.[17]

Physically real quantities are invariant under exact symmetries – this is the general lesson. It has been long in coming. Here there is a potential for confusion which it would be well to dispel. It is sometimes said that every symmetry transformation has a 'passive' and an 'active' interpretation. This is not the right distinction: what matters is the difference between transformations that are defined by their action on physical quantities that are not themselves modelled in the equations, as opposed to those that are, whose physical meaning, if any, has to be expressed by those equations themselves. Call them *extrinsically* as opposed to *intrinsically* defined transformations. In the former case, obviously, one is concerned only with a subsystem of the universe: the transformation in question will alter real physical relations between that subsystem and the rest (the change in the distance to the shore, when Galileo's ship is set smoothly in motion, is perfectly real). It is by means of these real physical quantities that one goes on to interpret the equations. Extrinsically defined symmetries, then, can usually be given an operational meaning. It is quite otherwise when the system of equations – *the very same* equations – is used to provide a model of the entire universe, or of a part of it (the Solar System, say, in NTG), *without* reference to the rest. The correspondence between active and

[17] Consensus on this now appears to have been reached: see Renn *et al.* (2000). (For further background see Norton, this volume.)

passive interpretations applies only to the former sort, to transformations that are extrinsically defined, not to intrinsic ones.[18]

Extrinsic transformations, that at least in some applications can be given a clear operational meaning, have also served as a guide to establishing the symmetries of a theory – the transformations under which its form should be preserved – and hence the theoretical, intrinsic symmetries too, but one misses some of the most important ones; one is not thereby led to local symmetry groups (symmetries which, viewed as Lie groups, are infinite dimensional). One is not even led to all the finite-dimensional ones (the symmetry of the Barbour–Bertotti theory under transformations to rotating frames of reference cannot be realized operationally, applying as it does only to the universe as a whole). What similarities there are disguise the fundamental difference: extrinsically defined symmetries, viewed from an active, operational point of view, transform real physical properties and relations, whereas symmetries that are intrinsically defined never do. Our concern is with the latter. Only equations and transformations that can be given an intrinsic physical meaning can be used to model the world as a whole.

The consequences of the PII for such transformations are then immediate. These are intrinsically defined symmetries; they therefore leave all the real physical quantities unchanged. The world thus arrived at does not differ, in respect of any real physical property or relation, from the world with which one begins. So they are numerically the same.

In the case of diffeomorphic spacetime models in GTR, this thesis has been called *Leibniz Equivalence*. But the thesis is quite general. It applies equally to any symmetry of a physical theory, when applied to the world as a whole, and to any transformation that can be only intrinsically defined. Some of these applications remain controversial. In the case of gauge theory, and specifically electromagnetism, the invariant quantities are the electromagnetic fields: the scalar and vector potentials A_μ, that are transformed by the gauge symmetry, do not directly correspond to any real physical magnitudes. Local (differential) relations among them are real – as exhibited by the gauge-invariant 2-form $\partial_\nu A_\mu - \partial_\mu A_\nu$ – but not the potentials themselves. This case is controversial because an explanation of certain effects – notably the Aharonov–Bohm effect – in terms of gauge-invariant quantities must be non-local, in contrast to an account of it in terms of the potentials. Another controversial case is the canonical approach to GTR (the constrained Hamiltonian

[18] The contrast has been put in a rather different way by Stachel (1993), who invokes the distinction between a theory interpreted in terms of a non-dynamical individuating field, and those interpreted in terms of a dynamical one. According to Stachel, GTR is unique in requiring the latter. It is true that there are special reasons why the symmetries of GTR must be intrinsically defined, but the option is important elsewhere as well; certainly the use of a dynamical individuating field was historically important to NTG (see Saunders, 2002; there I spoke of 'internally' defined symmetries, but since 'internal symmetry' is already in use and means something quite different, 'intrinsic' is better).

formalism). There the invariant quantities, the orbits of the group of transformations generated by the Hamiltonian constraint, are equivalence classes of 3-geometries. In fact it follows that no quantities preserved by the constraints can be functions of the time (there follows the 'frozen time formalism', as advocated by Rovelli, 1991).

Of the discrete symmetries, consider first permutation symmetry. This is as much a symmetry of classical statistical mechanics as of quantum theory, although in the former context it has received very little attention. By Leibniz Equivalence, permutations act as the identity – classical atoms are just as 'indistinguishable' as quantum ones, in the usual physicists' sense of the term.[19] In both cases, it is to this that the extensivity of the entropy function can be traced, as required of the classical thermodynamic entropy.[20] The failure of extensivity was a puzzle in the early days of classical statistical mechanics, when it was thought that permutations of particles should yield a physically distinct state of affairs, Gibbs' protests to the contrary notwithstanding. The puzzle was quickly overwhelmed by another, the discovery of quantum statistics. It was only much later that it was realized that permutation symmetry could be treated in exactly the same way in classical statistical mechanics as in quantum mechanics.[21] The point has yet to achieve broad acceptance by the physics community, let alone among philosophers, but it is a direct consequence of Leibniz Equivalence.[22] (An obvious question then arises: whence then the difference between classical and quantum statistics? But the answer here is clear enough, at least from the point of view of phase-space methods: all the differences can be traced to the use of a discrete measure on phase space rather than a continuous one.)

Finally, consider spatial inversion. Were this a symmetry, then applied to the world as a whole it would follow that only quantities and relations invariant under the transformation would be real: two spacetime models, the one the spatial inversion

19 I hestitate to call either classical particles or fermions 'non-individuals', in the light of French and Rickles' use of the term (for such particles satisfy at least one of their criteria for individuals, namely option (d), that a version of Leibniz's principle applies to them (French and Rickles, this volume, sections 2 and 5.2)).

20 In particular, to yield zero entropy of mixing for samples of the same gas. Against this, van Kampen (1984) has claimed that the extensivity of the classical entropy function is only a convention. Taking this line, Huggett (1999) has concluded that nothing physical hangs on Leibniz Equivalence in the case of the permutation group (that, as he put it, there is no physical basis to favour the abandonment of 'haecceistic' phase space, a view subsequently endorsed by Albert (2000, pp. 45–7)). It may be that the entropy of mixing for samples of the same gas is not directly measurable (although it may be viewed as the limiting case of entropies that are), but it hardly follows from that that the issue must be settled by *convention*.

21 For a history of this early controversy see Jammer (1966). For an account of the role of permutation symmetry in classical statistical mechanics see Hestines (1970).

22 French and Rickles (this volume, section 2) are clearly sympathetic to the view that Leibniz Equivalence, as applied to permutations, is incompatible with classical physics (equivalently, that classically one is committed to the use of haecceistic phase space); that Ehrenfest was right to criticize Planck's removal of the factor $N!$ as *ad hoc*. But I say that the use of the PII here, as applied to possible worlds, has an obvious pedigree in classical physics and classical metaphysics, and that there is nothing *ad hoc* about Leibniz Equivalence as I have derived it. Why insist that what was wanted was new physics, rather than a better interpretation of the old? (The same applies in the quantum case. The central question that they raise in section 5 – 'What is the ground of permutation invariance?' – is answered the same: it is a consequence of Leibniz's principles.)

of the other, would describe the same world, and the same handed objects within it. The hand considered in itself, in an otherwise empty space, would be neither left- nor right-handed.[23] This interpretation of global spatial inversion was first advocated by Weyl (1952). It was defended, with qualifications, by Earman (1989), and it has more recently been argued for by Hoefer (2000). Of course it has turned out that spatial inversion is *not* a symmetry of the Standard Model, but an argument to a similar effect remains: the combination of matter–antimatter inversion, space, and time (CPT symmetry) is demonstrably a symmetry of any relativistic quantum theory.[24] From Leibniz Equivalence, it follows that the world does not have one CPT orientation rather than the other. Its mirror image, on inverting matter and antimatter and the arrow of time, is one and the same.[25]

5 The Principle of Sufficient Reason

I have had more to say about the PII than the PSR; let me close with a remark on their relation. Despite the changes in the former principle, they still function in tandem.

Recall Clarke's criticism of the PSR (Leibniz and Clarke, 1956, pp. 20–21):

> Why this particular system of matter, should be created in one particular place, and that in another particular place; when, (all place being absolutely indifferent to all matter,) it would have been exactly the same thing vice versa, supposing the two systems (or the particles) of matter to be alike; there could be no other reason, but the mere will of God.

Leibniz did not respond to this directly; he surely agreed with Clarke that atomism is inconsistent with the PSR – but only given the further presupposition, common to them both, that *it must be possible to refer to a substance uniquely, independent of its relationships with other substances*. It would be odd to make such a claim today: it is a piece of metaphysics without any basis in modern logic; there is no reason to believe in it from the point of view of any physical theory; it is more contentious than any of the principles so far considered.

[23] Kant clearly saw this implication of Leibniz's principles, although initially he thought it confined to Leibniz's views on the nature of space (Pooley, this volume, section 2). Two years later, in the *Inaugural dissertation* of 1770, he had rejected the generalist picture as well, and with that much broader principles of Leibniz's philosophy.

[24] Pooley (this volume, section 4) denies that CPT symmetry is of any relevance to a relational account of handedness. But my claim is that it tells against absolutism: as I understand it, *only* transformations that are symmetries provide a sufficient condition for a quantity to be counted as unphysical. It is only insofar as absolute spatial orientation is changed by the CPT transformation, if a symmetry, that one can conclude it is not physically real. In the absence of such a symmetry I see no reason to suppose that absolute 3-dimensional spatial orientation is not a perfectly respectable real physical quantity in its own right, Pooley's arguments to the contrary notwithstanding.

[25] Of course there remain a number of open questions about the arrow of time, which cannot be addressed independent of the interpretation of quantum mechanics. (For example, CPT inversion can hardly remain a symmetry of state-reduction theories.)

Abjuring metaphysics of this order, there is no longer a conflict with the PSR. Leibniz Equivalence applies here as to any other symmetry – for given Clarke's assumption that the particles are exactly alike, particle permutations are surely symmetries. The permutation does not therefore lead to a possible world numerically distinct from the actual one. No decision as to which particle is to be placed in which position needs to be made.

A second example shows better how the novel features of the PII – the existence of relative and weak discernibles – works in tandem with the PSR. Consider the location of a material system in space. Were space a real entity, then, according to Leibniz, it would again follow that its parts must be individuated uniquely, without reference to anything else. But since they are qualitatively alike, there could be no reason to situate the material system in one place rather than another – a problem for the PSR. How does this case fare under the PII?

The points of space, independent of their relations to matter, unlike particles of matter, independent of their spatial relations, are in fact discernible. If, now, it were possible to refer to one point of space rather than another (without reference to matter), it would make sense to ask at which of the two points the material system is to be placed, leading to the same difficulty with the PSR. But in fact the points of space are only *weakly* discernible, so we cannot refer to any one point rather than another, and the difficulty does not arise. Evidently insofar as we can view the parts of a highly symmetric entity, such as a homogeneous space, as objects in their own right (as discernibles), without reference to anything else, it is essential – consistent with the PSR – that they *not* be absolutely discernible from one another.

It is also worth remarking that it is only in the generalist picture that there is no conflict with the PSR. For let us introduce into this space Adams' pair of identical globes, each with identical observers. The material system is to be placed adjacent to one globe rather than the other – which? In the generalist picture, there is no choice to be made, for neither can be referred to uniquely. From the standpoint of the two observers, one will see the system appear nearby, but not the other – again, this fact can be reported without any difficulty. But the one who sees it appear nearby will consider the event to be wholly arbitrary, and contrary to the PSR. Here, one might say, is *chance*, where in the generalist picture everything is deterministic.

6 Close

I have spoken throughout of the interpretation of theories in terms of objects, but it is object in the logical sense, the sense that Frege was concerned with. And, whilst I maintain that certain doubts have been laid to rest concerning it – for example, as expressed by Quine (1990, pp. 35–6), on whether the concept of object must crumble altogether in the face of particle indistinguishability in quantum mechanics – it is object in a very thin sense that is secured. In strongly interacting high-energy physics, it is doubtful that objects as individuated (using the PII) by the invariant properties and relations definable in quantum field theory will be quanta at all (although, in the light of asymptotic freedom in quantum chromodynamics, in the ultra-relativistic limit quarks as objects presumably remain). In classical field theories, obviously, one obtains little more than field values (or, given diffeomorphism symmetry, relations between field values). These are objects as events.

Coincidences of field values, and complexes of relations among them – this is a world understood in terms of structural descriptions, a world as graph, not a collection of things that evolve in time. It is a structuralist account, too, by virtue of its reliance on mathematical form, as the key to the distinction between the physical and the merely mathematical. But I see no reason, deriving from special relativity and quantum theory, to deny that the descriptions one ends up with, in the interpretation of physical theories, give the properties and relations among objects, in the logical sense of the term. The conservative notion of object that I have been concerned with, thin as it is, is not in jeopardy. It is not so thin as to be governed by an unanalysable notion of identity.

What is not so certain is that the current framework will be preserved in quantum gravity. In the quantum canonical approach, unlike in the classical theory, it is not so clear that quantities preserved by the constraints are sufficient to build up an account of change (not on any approach to the problem of measurement). And here there are avenues being explored – causal sets, for example – that, if successful, may well lead to a more direct account of identity than the logical one that I have given. They are speculative, and if contrary to the PII, they will have violated the methodology that I have been advocating; but the PII as I understand it is no *a priori* truth.

References

Adams, R. (1979). 'Primitive thisness and primitive identity'. *Journal of Philosophy*, **76**, 5–26.

Albert, D. (2000). *Time and Chance*. Cambridge, MA: Harvard University Press.

Barbour, J., and Bertotti, B. (1982). 'Mach's Principle and the structure of dynamical theories'. *Proceedings of the Royal Society (London)*, **382**, 295–306.

Belot, G. (2001). 'The Principle of Sufficient Reason'. *Journal of Philosophy*, **98**, 55–74.

Black, M. (1952). 'The identity of indiscernibles'. *Mind*, **61**, 153–64.

Butterfield, J. (1993). 'Interpretation and identity in quantum theory'. *Studies in History and Philosophy of Science*, **24**, 443–76.

Castellani, E., and Mittelstaedt, P. (2000). 'Leibniz's Principle, physics, and the language of physics'. *Foundations of Physics*, **30**, 1587–604.

Cortes, A. (1976). 'Leibniz's Principle of the Identity of the Indiscernibles: a false principle'. *Philosophy of Science*, **45**, 466–70.

Earman, J. (1989). *World Enough and Space-Time*. Cambridge, MA: MIT Press.
French, S. (1995). 'Hacking away at the identity of indiscernibles: possible worlds and Einstein's Principle of Equivalence'. *Journal of Philosophy*, **91**, 455–66.
French, S., and Redhead, M. (1988). 'Quantum physics and the identity of indiscernibles'. *British Journal for the Philosophy of Science*, **39**, 233–46.
Gödel, K. (1930). 'Die Vollständigkeit der Axiome des logischen Funktionenkalküls'. *Monaatshefte für Mathematik und Physik*, **37**, 349–60. Translated as 'The completeness of the axioms of the functional calculus of logic'. In *From Frege to Gödel* (1967), ed. J. van Heijenoort. Cambridge, MA: Harvard University Press.
Hacking, P. (1975). 'The identity of indiscernibles'. *Journal of Philosophy*, **72**, 249–56.
Hestines, D. (1970). 'Entropy and indistinguishability'. *American Journal of Physics*, **38**, 840–5.
Hilbert, D., and Bernays, P. (1934). *Grundlagen der Mathematik*, Vol. 1. Berlin: Springer-Verlag.
Hoefer, C. (2000). 'Kant's hands and Earman's pions: chirality arguments for the substantival space'. *International Studies in the Philosophy of Science*, **14**, 237–56.
Huggett, N. (1999). 'Atomic metaphysics'. *Journal of Philosophy*, **96**, 5–24.
Ishiguro, H. (1972). *Leibniz's Philosophy of Logic and Language*. Ithaca, NY: Cornell University Press.
Jammer, M. (1966). *The Conceptual Development of Quantum Mechanics*. New York: McGraw-Hill.
Leibniz, G. W. (1714). 'Monadology'. In *Monadology and other philosophical essays*, translated by P. Schrecker and A. Schrecker (1965). Indianapolis, IN: Bobb-Merrill.
Leibniz, G. W., and Clarke, S. (1956). *The Leibniz–Clarke Correspondence*, ed. H. G. Alexander. Manchester: Manchester University Press.
Margenau, H. (1944). 'The exclusion principle and its philosophical importance'. *Philosophy of Science*, **11**, 187–208.
Massimi, M. (2001). 'Exclusion principle and the identity of indiscernibles: a response to Margenau's argument'. *British Journal for the Philosophy of Science*, **52**, 303–30.
Mermin, D. (1998). 'What is quantum mechanics trying to tell us?' *American Journal of Physics*, **66**, 753–67.
O'Leary-Hawthorne, J., and Cover, J. A. (1996). 'Haecceitism and anti-haecceitism in Leibniz's philosophy'. *Noûs*, **30**, 1–30.
Pooley, O. (2002). 'The reality of spacetime'. D. Phil. thesis, University of Oxford.
Quine, W. V. (1960). *Word and Object*. Cambridge, MA: Harvard University Press.
(1974). *The Roots of Reference*. La Salle, IL: Open Court.
(1976). 'Grades of discriminability'. *Journal of Philosophy*, **73**. Reprinted in *Theories and Things* (1981). Cambridge, MA: Harvard University Press.
(1990). *Pursuit of Truth*. Cambridge, MA: Harvard University Press.
Redhead, M., and Teller, P. (1992). 'Particle labels and the theory of indistinguishable particles in quantum mechanics'. *British Journal for the Philosophy of Science*, **43**, 201–18.
Renn, J., Sauer, T., Janssen, M., Norton, J., and Stachel, J. (2000). *The Genesis of General Relativity: Sources and Interpretation*, Vol. 1: *General Relativity in the Making: Einstein's Zurich Notebook*. Dordrecht: Kluwer.
Rovelli, C. (1991). 'What is observable in classical and quantum gravity'. *Classical and Quantum Gravity*, **8**, 297–304.
Saunders, S. (2002). 'Indiscernibles, general covariance, and other symmetries: the case for non-eliminativist relationalism'. In *Revisiting the Foundations of Relativistic*

Physics: Festschrift in Honour of John Stachel, ed. A. Ashtekar, D. Howard, J. Renn, S. Sarkar, and A. Shimony. Dordrecht: Kluwer.

Schrödinger, E. (1926). 'Zur Einsteinshen Gastheorie'. *Physicalische Zeitschrift*, **27**, 95–101.

Stachel, J. (1993), 'The meaning of general covariance'. In *Philosophical Problems of the Internal and External Worlds: Essays Concerning the Philosophy of Adolf Grünbaum*, ed. A. Janis, N. Rescher, and G. Massey. Pittsburgh, PA: University of Pittsburgh Press.

Teller, P. (1991). 'Substance, relations, and arguments about the nature of space-time'. *Philosophical Review*, **100**, 363–97.

van Fraassen, B. (1991). *Quantum Mechanics*. Oxford: Clarendon Press.

van Kampen, N. (1984). 'The Gibbs paradox'. In *Theoretical Physics*, ed. W. Parry, pp. 305–12. Oxford: Pergamon Press.

Weyl, H. (1952). *Symmetry*, Princeton, NJ: Princeton University Press.

Wiggins, D. (2001). *Sameness and Substance Renewed.* Cambridge: Cambridge University Press.

Part III

Symmetry breaking

17

Classic texts: extracts from Curie and Weyl

On symmetry in physical phenomena, symmetry of an electric field and of a magnetic field

PIERRE CURIE

I. Introduction

I think that there is interest in introducing into the study of physical phenomena the symmetry arguments familiar to crystallographers.

For example, an isotropic body may have a rectilinear or rotational motion. If liquid, it may have turbulence. If solid, it may be compressed or twisted. It may be in an electric or magnetic field. It may carry an electric or thermal current. It may be traversed by unpolarized or linearly, circularly, elliptically, etc. polarized light. In each case a certain characteristic asymmetry is necessary at each point of the body. These asymmetries are even more complex if one assumes that several phenomena coexist in the same medium or if they take place in a crystallized medium, which already possesses, by its constitution, a certain asymmetry.

Physicists often utilize symmetry conditions, but generally neglect to define the symmetry in a phenomenon, for sufficiently often these symmetry conditions are simple and quite obvious *a priori.*

When teaching physics, it would be better to state these questions openly: In the study of electricity, for example, one should state almost immediately the characteristic symmetry of the electric field and of the magnetic field. One can then use these notions to simplify proofs.

From a general point of view the idea of symmetry can be linked to the concept of *dimension*: These two fundamental concepts are respectively characteristic of the *medium* in which a phenomenon occurs and the *quantity* which serves to evaluate its intensity.

Two media with the same asymmetry have a particular link between them, from which one can draw physical conclusions. A connection of the same kind exists between two quantities of the same dimension. Finally, when certain causes produce

Note. Extract from: 1982, *Symmetry in Physics: Selected Reprints*, ed. J. Rosen, trans. J. Rosen and P. Copié, Melville, NY: American Association of Physics Teachers, pp. 17–25.

certain effects, the symmetry elements of the causes must be found in the produced effects. Similarly, when a physical phenomenon is expressed as an equation, there is a causal relation between the quantities appearing in both terms and the two terms have the same dimension....

IV. Characteristic asymmetry of physical phenomena

Let us now consider any point of a medium in any physical state...

We state the following propositions.

The characteristic symmetry of a phenomenon is the maximal symmetry compatible with the existence of the phenomenon.

A phenomenon can exist in a medium which possesses its characteristic symmetry or that of one of the subgroups of its characteristic symmetry.

In other words, certain symmetry elements can coexist with certain phenomena, but they are not necessary. What is necessary is that some symmetry elements be missing. *Asymmetry is what creates a phenomenon.*

It is much more logical to call a plane of asymmetry any plane that is not a plane of symmetry, to call an axis of asymmetry any axis that is not an axis of symmetry, and so on, and in general to list the operations which are not recovery operations in the system. These are the operations indicating an asymmetry and therefore a possible property of the system. But in the groups we considered there are an infinite number of nonrecovery operations and in general a finite number of recovery operations: It is thus much simpler to list the latter operations.

One can also see that when several different phenomena are superimposed in the same system, their asymmetries add. The symmetry elements remaining in the system are only those that are common to all phenomena taken separately.

When certain causes produce certain effects, the symmetry elements of the causes must be found in their effects.

When certain effects show a certain asymmetry, this asymmetry must be found in the causes which gave rise to them.

In practice, the converses of these two propositions are not true, i.e., the effects can be more symmetric than their causes. Certain causes of asymmetry might have no effect on certain phenomena or at most an effect too weak to be discerned, which amounts to the same as no effect, for practical purposes....

VII. Conclusion

The characteristic symmetries of phenomena are of incontestable general interest. From the point of view of applications we see that the conclusions we can draw from symmetry arguments are of two kinds.

The first are firm but negative conclusions. They correspond to the incontestably true proposition: *there is no effect without causes.* Effects are phenomena which always require a certain asymmetry in order to arise. If this asymmetry does not exist, the phenomena are impossible. This often prevents us from searching for unrealizable phenomena.

Symmetry arguments also permit us to state a second kind of conclusions, which are of positive nature but do not offer the same certainty of their results as those of negative nature. They correspond to the proposition: *There is no cause without effects.* Effects are the phenomena which can arise in a medium possessing a certain asymmetry. One has here precise directions for the discovery of new phenomena, but the predictions are not as precise as those of thermodynamics. One has no idea of the order of magnitude of the predicted phenomena. One has only an imperfect idea of their precise nature. This last remark shows that one should beware of drawing an absolute conclusion from a negative experiment

Let us consider, for example, a tourmaline crystal which possesses a symmetry which is a subgroup of the electric field symmetry. We conclude that the crystal may be electrically polarized. We place the crystal in an electric field, its axis oriented 90° to the field; no polarization is observed in any way. One has no measurable torque acting on the crystal. One would be tempted to think that the crystal is not polarized, or that, if there is a polarization, it is less than what one can observe. Yet, there is a polarization and to make it appear it is necessary to modify the experiment, to heat the crystal uniformly, for example, which does not change its symmetry.

Symmetry

HERMANN WEYL

This is a special case of the following general principle: If conditions which uniquely determine their effect possess certain symmetries, then the effect will exhibit the same symmetry. Thus Archimedes concluded *a priori* that equal weights balance in scales of equal arms. Indeed the whole configuration is symmetric with respect to the midplane of the scales, and therefore it is impossible that one mounts while the other sinks. For the same reason we may be sure that in casting dice which are perfect cubes, each side has the same chance, $\frac{1}{6}$. Sometimes we are thus enabled

Note. Extract from: 1952, Princeton, NJ: Princeton University Press, pp. 125–6.

to make predictions *a priori* on account of symmetry for special cases, while the general case, as for instance the law of equilibrium for scales with arms of different lengths, can only be settled by experience or by physical principles ultimately based on experience. As far as I see, all *a priori* statements in physics have their origin in symmetry.

18

Cross fertilization in theoretical physics: the case of condensed matter and particle physics

GIOVANNI JONA-LASINIO

The following text is part of a talk entitled 'Cross Fertilization in Theoretical Physics: the Case of Condensed Matter and Particle Physics' given at the Young Researchers Symposium held on the occasion of the XIII International Congress on Mathematical Physics, London, 17–22 July 2000. One of the purposes of the symposium was the communication of personal experiences and points of view on the part of older researchers and this is reflected in the style of exposition.

The full text appears in *Highlights in Mathematical Physics*, edited by A. Fokas, J. Halliwell, T. Kibble, and B. Zegarlinski, and published in 2002 by the American Mathematical Society.

1 Breaking gauge and chiral invariance

Spontaneous breakdown of symmetry (SBS) is a concept that is applicable only to systems with infinitely many degrees of freedom. Although it pervaded the physics of condensed matter for a very long time – magnetism is a prominent example – its formalization and the recognition of its importance has been an achievement of the second half of the twentieth century. Strangely enough, the name was adopted only after its introduction in particle physics: it is due to Baker and Glashow (1962). I think that concepts acquire a proper name only when they attain their full maturity: this emphasizes the relevance of this case that took place over forty years ago. What is SBS? In condensed matter physics, SBS means that the lowest energy state of a system can have a lower symmetry than the forces acting among its constituents and on the system as a whole. As an example consider a long elastic bar on top of which we apply a compression force directed along its axis. Clearly there is rotational symmetry around the bar which is maintained as long as the force is not too strong: there is simply a shortening according to the Hooke's law. However when the force reaches a critical value the bar becomes inflected and we have an infinite number of equivalent lowest energy states which differ by a rotation.

Heisenberg (1960) was probably the first to consider SBS as a possibly relevant concept in particle physics but the idea was really appreciated only after Nambu and the present writer developed a specific model of relativistic field theory (see Nambu and Jona-Lasinio, 1961a; 1961b). To appreciate the innovative character of this concept in particle physics one should consider the strict dogmas which constituted the foundation of relativistic quantum field theory in the late fifties. One of the dogmas stated that the lowest energy state, the vacuum, should not possess observable physical properties and all the symmetries of the theory, implemented by unitary operators, should leave it invariant. While the first of these requirements may appear basic for the interpretation of the theory, the second one is far less obvious: in fact even if SBS is inobservable in the vacuum, its consequences may affect the structure of measurable quantities such as the particle spectrum.

Let me recall some steps in the transfer of SBS to particle physics.

The theory of superconductivity of Bardeen, Cooper, and Schrieffer which appeared in 1957 (see Bardeen *et al.*, 1957) provided the key paradigm for the introduction of SBS in relativistic quantum field theory. The spontaneously broken symmetry in this case is gauge invariance. Nambu (1960a) reformulated BCS theory in a field-theoretic language in which the analysis of gauge invariance became particularly transparent. He derived an expression for the matrix elements of the electromagnetic current between quasi-particle states of momentum $p + \frac{q}{2}$ and $p - \frac{q}{2}$ from which one could see that in the superconducting state, in order to keep gauge invariance and satisfy current conservation, a quasi-particle must be surrounded by an excitation field with energy spectrum proportional to q^2. The excitation field manifests itself through a pole in the current matrix elements. Similar results had been derived almost simultaneously (or even before) by other people including Anderson, Bogoliubov, and Rikayzen, but the approach in terms of propagators and vertex functions used by Nambu was such that a formal transfer to relativistic field theory was easy.

In the $V - A$ theory of weak interactions, if one makes an assumption of strict axial vector current conservation, one gets a structure for the matrix elements of the axial current between nucleon states, a quantity entering in the neutron β-decay, which exhibits a pole corresponding to a zero mass pseudo-scalar particle. If one identifies the nucleons with the quasi-particles and their mass m with the gap Δ and the pole with a forerunner of the pion, the similarity in structure between the electromagnetic current in the BCS theory of superconductivity and the axial vector current in the theory of β-decay becomes striking.

However, in β-decays the chiral current is only partially conserved. Nambu (1960b) made the assumption that due to the violation of chiral symmetry the pole at $q^2 = 0$ moves to $q^2 = -\mu^2$, the mass of the pion. From this he was able to derive easily the famous Goldberger–Treiman relation for the pion decay constant. In the same paper he mentioned a possible analogy with BCS theory.

When I arrived in Chicago in September 1959 as a research associate, Nambu was playing with these ideas and I became immediately involved.

We concentrated on two major problems. On the one hand it was necessary to check whether the analogy between gauge and chiral invariance leading to an unconventional derivation of the Goldberger–Treiman relation was not purely formal and could be given a physical content by deriving some other testable consequence. A second and more difficult question was the possibility of constructing a model of elementary particles where the mechanisms leading to the gap in superconductivity would find a counterpart in a relativistic quantum field theory.

It was experimentally known that the ratio between the axial vector and vector β-decay constants $R = g_A/g_V$ was slightly greater than 1 and ≈ 1.25. The following two hypotheses were then natural (see Nambu and Jona-Lasinio, 1961b):

(1) Under strict γ_5 invariance there is no renormalization of g_A.
(2) The violation of the invariance gives rise to the finite pion mass as well as to the ratio $R > 1$ so that there is some relation between these quantities.

Under these assumptions we calculated what we believed to be the main contribution to R and found a value close to the experimental one. More than the numerical value, the relevant result was that the renormalization effect due to a positive pion mass went in the right direction. This was quite encouraging as it showed that the picture made sense and we pursued the ambitious goal of constructing a model.

In a superconductor a basic fact is the attractive nature of electron–electron interaction, due to phonon exchange, near the Fermi surface. In a picture of the vacuum of a massless Dirac field as a sea of occupied negative energy states, an attractive force between particle and antiparticle should have the effect of producing a finite mass, the counterpart of the gap. At this point the choice of the model became important. In a relativistic theory interactions are usually associated with the exchange of bosons but due to the novelty of the approach a choice was far from obvious.

Heisenberg and his collaborators (see Dürr *et al.*, 1959) had developed a comprehensive theory of elementary particles based on a non-linear spinor interaction: the physical principle was that spin-$\frac{1}{2}$ fermions could provide the building blocks of all known elementary particles. Heisenberg was, however, very ambitious and wanted at the same time to solve in a consistent way the dynamical problem of a non-renormalizable theory. This made their approach very complicated and not transparent but it contained for the first time the idea of SBS in a field theoretic context. Nambu considered Heisenberg theory very formal but the 4-spinor interaction was attractive due to its simplicity and analogy with the many-body case. I had a more enthusiastic attitude. Shortly after my graduation in Rome, in the two years before going to Chicago, I had been exposed several times to the non-linear spinor theory – in a meeting in Venice where a very interesting discussion between Heisenberg and Pauli took place, and in Rome which Heisenberg visited just to explain his theory. Bruno Touschek in Rome was deeply interested in these ideas and I was impressed by comprehensive '*fundamental*' theories.

Following BCS, Nambu and I studied a self-consistent field approximation for a 4-spinor chiral invariant Lagrangian: from this an equation almost identical with the gap equation was obtained for the fermionic mass. The square of the coupling constant times the invariant cutoff played the same role of the energy density at the fermi surface. The introduction of the cutoff, a weakness from the standpoint of a deep dynamical treatment, turned out to be an important reason for the success of our model. In fact in this way the strict analogy with BCS made the physical mechanism leading to the SBS quite transparent and was understood by the elementary particle community. It was remarkable how, once a certain integral was assumed to be convergent, a rich mass spectrum followed. A most important conclusion was the existence of a zero mass pseudo-scalar particle, the first example of what later became known as the Goldstone theorem. It was also shown in our second paper (Nambu and Jona-Lasinio, 1961b) that a very small term in the interaction breaking explicitly chiral invariance induced a reasonable value for the pion mass.

Our work has been influential in the physics of the second half of the twentieth century. It provided the first example of a physically relevant relativistic model based on SBS and this concept became crucial in elementary particle physics. The model, suitably reinterpreted in terms of quarks, is still a main reference in the study of chiral symmetry, a subject not yet completely understood. The theory of Heisenberg and collaborators, in spite of the new ideas it contained, did not have an important follow up: this, I believe, was due in part to the obscurities connected with the attempt to solve too many problems at the same time. It was difficult to disentangle the different new concepts from the programme of reproducing the whole world of elementary particles, and analyse the merit and scope of each of them. The use of an indefinite metric in Hilbert space to master the dynamics added an exotic character which rendered the theory less accessible.

The self-consistent field approximation in our model was not very satisfactory: after my return to Italy and in a subsequent stay at CERN I started thinking whether one could give, within quantum field theory, a more satisfactory formulation of SBS. The outcome of this reflection is the subject of the next section.

2 The effective action

A characteristic feature of statistical mechanics, both classical and quantum, is the existence of variational principles determining the stable states of a system.

A quantum field and a many-body system both have infinitely many degrees of freedom: it was therefore natural to look for analogies at a very general level. Variational principles in quantum statistical mechanics had been introduced by Lee and Yang (1959; 1960) followed by Balian, Bloch, and De Dominicis (1961a; 1961b). The variables appearing in these principles were typically average occupation numbers and the corresponding stationary functionals required a rather complicated construction. On the other hand, Goldstone in his well-known paper (1961) had introduced a function of the homogeneous vacuum expectation value of the scalar field of his model, which was stationary at the physical values. Of this function, which later became known as the effective potential, he gave a perturbative construction. I was not completely satisfied by the perturbative approach and by the fact that the effective potential did not provide a complete description of the dynamics. Then a paper by De Dominicis and Martin (1964) appeared where they derived variational principles for the many-body problem in which the full time-dependent Green's functions were the natural variables. The key ingredient to the derivation of variational principles was the functional Legendre transform with respect to spacetime dependent potentials. At this point the analogy with quantum statistical mechanics was obvious to me and I introduced the effective action, a *c*-number action functional for quantum field theory whose arguments are the vacuum expectation values of the fields. The equilibrium states of the system are determined by its critical points (see Jona-Lasinio, 1964). This has become the standard approach to SBS in quantum field theory and can be found in textbooks. I learnt much later that the effective action had appeared for the first time in the mid-1930s in a paper by Heisenberg and Euler (1936) and had been considered also in unpublished notes by Schwinger.

From the standpoint of the dynamical evolution the effective action leads to equations for the expectation values of the fields. The effective action splits naturally into two parts: the classical and the quantum. The quantum part of the equations has a structure different from the usual Lagrange equations. In fact it is non-local in time and the whole history of the system is involved. Therefore an interpretation of the dynamical equations in terms of an initial value problem is not possible. This is a manifestation of the deep difference between classical and quantum dynamics. For a recent analysis of this aspect see Cametti *et al.* (2000).

References

Baker, M., and Glashow, S.L. (1962). 'Spontaneous breakdown of elementary particles symmetries'. *Physical Review*, **128**, 2462–71.

Balian, R., Bloch, C., and De Dominicis, C. (1961a). 'Formulation de la mechanique statistique en terme de nombres d'occupation (I)'. *Nuclear Physics*, **25**, 529–67.

(1961b). 'Formulation de la mechanique statistique en terme de nombres d'occupation (II)'. *Nuclear Physics*, **27**, 294–322.

Bardeen, J., Cooper, L. N., and Schrieffer, J. R. (1957). 'Microscopic theory of superconductivity'. *Physical Review*, **106**, 162–4.

Cametti, F., Jona-Lasinio, G., Presilla, C., and Toninelli, F. (2000). 'Comparison between quantum and classical dynamics in the effective action formalism'. In *New Directions in Quantum Chaos*, ed. G. Casati, I. Guarneri, and U. Smilansky, pp. 431–48. Amsterdam: IOS Press.

De Dominicis, C., and Martin, P. C. (1964). 'Entropy and renormalization in normal and superfluid systems I'. *Journal of Mathematical Physics*, **5**, 14–30.

Dürr, H. P., Heisenberg, W., Mitter, H., Schlieder, S., and Yamazaki, K. (1959). 'Zur Theorie der Elementarteilchen'. *Zeitschrift für Naturforschung A*, **14**, 441–85.

Goldstone, J. (1961). 'Field theories with "superconductor" solutions'. *Nuovo Cimento*, **19**, 154–64.

Heisenberg, W. (1960). 'Recent research on the nonlinear spinor theory of elementary particles'. In *Proceedings of the 1960 Annual International Conference on High Energy Physics in Rochester*, pp. 851–7.

Heisenberg, W., and Euler, H. (1936). 'Folgerungen aus der Diracschen Theorie des Positrons'. *Zeitschrift für Physik*, **98**, 714–32.

Jona-Lasinio, G. (1964). 'Relativistic field theories with symmetry breaking solutions'. *Nuovo Cimento*, **34**, 1790–4.

Lee, T. D., and Yang, C. N. (1959). 'Many-body problem in quantum statistical mechanics. I'. *Physical Review*, **113**, 1165–77.

(1960). 'Many-body problem in quantum statistical mechanics. IV'. *Physical Review*, **117**, 22–36.

Nambu, Y. (1960a). 'Quasi-particles and gauge invariance in the theory of superconductivity'. *Physical Review*, **117**, 648–63.

(1960b). 'Axial vector current conservation in weak interactions'. *Physical Review Letters*, **4**, 380–2.

Nambu, Y., and Jona-Lasinio, G. (1961a). 'Dynamical model of elementary particles based on an analogy with superconductivity, I'. *Physical Review*, **122**, 345–58.

(1961b). 'Dynamical model of elementary particles based on an analogy with superconductivity, II'. *Physical Review*, **124**, 246–54.

19

On the meaning of symmetry breaking

ELENA CASTELLANI

Symmetries can be attributed to physical states or to physical laws. The focus of this volume is on the symmetries of physical laws, that is the physical symmetries postulated by means of invariance principles. Accordingly, the focus of this review paper is on symmetry breaking in the case of physical laws. In this case, there are two different forms of symmetry breaking – 'explicit' and 'spontaneous' – the spontaneous symmetry-breaking case being the more interesting from a physical as well as a philosophical point of view.

After some general preliminary remarks on symmetry breaking, we start by examining how symmetry breaking was first considered in the literature, and then turn to the main subject of this paper, that is the physical and philosophical meaning of symmetry breaking of the laws of nature.

1 Preliminaries – I

A symmetry can be exact, approximate, or broken. Exact means unconditionally valid; approximate means valid under certain conditions; broken can mean different things, depending on the object considered and its context.

Our concern here is the breaking of physical symmetries. In physics, symmetry properties may be attributed to physical laws (equations) or to physical objects/phenomena (solutions). As the contributions to this volume clearly show, the two cases must be distinguished when considering the meaning and functions of physical symmetries and, accordingly, of their breaking.[1]

Another preliminary distinction is needed. The expression 'symmetry breaking' is in fact ambiguously used. On the one hand, it is taken to indicate the process by means of which the considered symmetry is broken, and is therefore usually

[1] Although there is of course a connection between the symmetries of physical laws and the symmetries of physical states, as will be seen in more detail later, especially when discussing the case of spontaneous symmetry breaking.

ascribed a 'dynamic' character in the literature (in contrast with the 'static' character attributed to a situation of symmetry). On the other hand, it is also taken to indicate the result of a symmetry-breaking process, that is, a broken symmetry situation (or simply the fact that a symmetry is not there). Failure to keep these two meanings distinct can lead to confusion.[2]

In all cases, the general questions raised by symmetry breaking are of the following three types: (i) why a symmetry breaks or is not there; (ii) how it breaks; (iii) what the effects or consequences of its breaking (or of its not being there) are.

Finally, let us emphasize once more the importance and usefulness of group theory. As is known, the symmetry of a 'something' (a figure, an equation, ...) is defined in terms of its invariance with respect to a specified transformation group, its symmetry group. Generally, the breaking of a certain symmetry does not imply that no symmetry is present, but rather that the situation where this symmetry is broken is characterized by a lower symmetry than the situation where this symmetry is not broken. In group-theoretic terms, this means that the initial symmetry group is broken to one of its subgroups. It is therefore possible to describe symmetry breaking in terms of relations between transformation groups, in particular between a group (the unbroken symmetry group) and its subgroup(s). As is clearly illustrated in the 1992 volume by I. Stewart and M. Golubitsky, starting from this point of view a general theory of symmetry breaking can be developed by tackling such questions as 'which subgroups can occur?', 'when does a given subgroup occur?'

2 Symmetry breaking and Curie's analysis

Symmetry breaking was first explicitly studied in physics with respect to physical objects and phenomena. This follows naturally from the developments of the theory of symmetry, at the origin of which are the visible symmetry properties of familiar spatial figures and everyday objects.[3] Moreover, examples of broken symmetries pervade the physical world, in particular the world of our everyday experience.

Any symmetry we can perceive (albeit in an approximate way) is indeed the result of a higher order symmetry being broken.[4] This can actually be said of any symmetry which is not the 'absolute' one (i.e. including all possible symmetry transformations). But we can say even more: in a situation characterized by an absolute symmetry, nothing definite could exist, since absolute symmetry means total lack of differentiation. For the presence of some structure, a lower symmetry

[2] Another source of confusion is due to unclear terminology in current literature, where usually no distinction is made between the result of a symmetry-breaking process ('broken symmetry'), the absence of one of the possible symmetries compatible with the situation considered ('non-symmetry', or 'dissymmetry' as it was termed in the nineteenth-century literature, notably by Louis Pasteur in his works on molecular dissymmetry), and the absence of all the possible symmetries compatible with the situation considered ('asymmetry').

[3] See the introduction to this volume, sections 1 and 2.

[4] The 'order' of a symmetry being the order of the symmetry group – that is, the number of independent symmetry operations.

than the absolute one is needed: in this sense, symmetry breaking is essential for the existence of structured 'things'. Crystals, the natural objects 'structured' *par excellence*, offer an exemplary illustration of this function of symmetry breaking, their many and striking symmetries being the result of the breaking of the symmetries of the initial medium from which they originated.[5] More precisely, the case of crystals illustrates how, given a physical medium as much undifferentiated as physical constraints allow, structured patterns arise when some of the possible symmetries are broken for some physical causes (how this breaking actually occurs is another question, as we will see in the next section). The resulting pattern is thus characterized by the presence of some symmetries and the absence of others; or, in other words, by specified symmetries and specified 'dissymmetries' (the term 'dissymmetry' meaning that some of the possible symmetries compatible with the physical constraints are not present).[6]

The symmetries (and dissymmetries) of crystals, besides being at the origin of the systematic study and classification of the possible types of symmetric patterns, also represent the starting point for the first explicit analysis of the role of symmetry breaking in physics. This analysis is due to Pierre Curie, who towards the end of the nineteenth century devoted a series of works – in particular his famous 1894 paper, extracts from which are reprinted in this volume – to examining the role of symmetry and symmetry breaking in physical phenomena, especially electric and magnetic.

As mentioned in the introduction to this volume, Curie's analysis was centred on the following question: which phenomena are allowed to occur in a given physical medium having specified symmetry properties? His studies of such properties as the pyro- and piezo-electricity of crystals had in fact persuaded him of the importance of the relationships between the symmetry of a physical medium (for example, a crystal) and the symmetry of the phenomenon occurring in it. By applying the methods and results of the theory of symmetry groups used in the crystallography of his times to the study of a number of physical phenomena, he arrived at definite conclusions known in the literature as 'Curie's principles'. Actually these principles amount to just one, Curie's conclusions clearly not being independent of one another. In this section we focus on the following group of propositions (called by some 'Curie's first principle'):

(a) 'The characteristic symmetry of a phenomenon is the maximum symmetry compatible with the existence of the phenomenon.'

[5] A hot gas of identical atoms has a very high symmetry, the equations describing it being invariant under all rigid motions as well as under all permutations of the atoms. As the gas cools down, the only mathematical solution maintaining these symmetries, which is the state where all the atoms are in the same place, is ruled out by the physics. In fact, the initial symmetry breaks down and the physical system takes up a stable state with less symmetry, that is the crystal lattice. On this point, and on the general role of symmetry breaking in the formation of Nature's patterns from the smallest scales to the largest, see in particular Stewart and Golubitsky (1992, chapter 3).

[6] See note 2.

(b) 'A phenomenon may exist in a medium having the same characteristic symmetry or the symmetry of a subgroup of its characteristic symmetry.'
(c) 'In other words, certain elements of symmetry can coexist with certain phenomena, but they are not necessary. What is necessary, is that certain elements of symmetry do not exist. Dissymmetry is what creates the phenomenon.'[7]

Curie makes very clear what he means with (a), (b), and (c) by discussing in his paper a number of concrete examples. These range from the determination of the characteristic symmetries of such phenomena as the Newtonian gravitation field, the electric field and the magnetic field (together with the symmetries of the media in which they originate), to the explanation – on the grounds of the symmetry groups implied and their relationships – of the possibility of such physical effects as the 'Wiedemann effect' and the 'Hall effect'.

In the case of the Wiedemann effect (the longitudinal magnetization of an iron cylindrical wire when a longitudinal electric field and an asymmetrical torque are applied), for example, the original symmetry group of the wire before the application of the electric field and the torque is the symmetry group of a cylinder. For the magnetization of the wire (the 'effect') to occur, some elements of this group must be absent, that is, the group must be broken to a lower order group (which is what happens under the combined action of the electric field and the torque).[8]

According to Curie, symmetry breaking thus has the following role: for the occurrence of a phenomenon in a medium, the original symmetry group of the medium must be lowered (broken, in today's terminology) to the symmetry group of the phenomenon (or to a subgroup of the phenomenon's symmetry group) by the action of some cause (the electric field and the torque in the above example). In this sense symmetry breaking is what 'creates the phenomenon'.

3 Preliminaries – II

A physical system is described in terms of its states, its observables, and its equation of motion (the states being solutions to this equation). All these components together form the physical theory of the system.[9]

The symmetries playing a central role in today's physics are the symmetries of physical theories. By this expression physicists mean the transformation groups leaving invariant (in form) the dynamical equations, the so-called physical laws (these invariance properties having then crucial implications for the physical states and observables).[10]

[7] Curie (1894), my translation; see also this volume.
[8] For details and comments on the discussion of Wiedemann effect by Curie, see for example Radicati (1987, pp. 199–200).
[9] The terminology used in this and the following section is, for brevity's sake, admittedly rough.
[10] See, for example, the introduction to this volume and Martin, this volume.

In the previous section we discussed symmetry breaking in the case of physical states. We now want to consider the meaning of symmetry breaking in the more general case of physical laws.

A preliminary question is the following: what is the relationship between the symmetry/asymmetry properties of states and the symmetry/asymmetry properties of laws? Since physical states are solutions to dynamical equations, it is natural to expect a relationship of some sort between the symmetry breaking of a state and the symmetry properties of the equation to which it is a solution. If we focus only on symmetry breaking of particular solutions (as in the previous section), nothing definite can be said about the symmetry properties of the equations. Depending on the case considered, asymmetric states can be solutions to symmetric as well as to asymmetric dynamical equations. In fact, the symmetry of a dynamical equation is not necessarily the symmetry of the individual solutions, but rather the symmetry of the whole set of solutions (in the sense that the symmetry of the dynamical equation transforms a given solution into another solution). Note that this explains, in particular, the role of asymmetric initial conditions in symmetry breaking of physical states (often emphasized in the literature): even if the considered equation is symmetric, an asymmetric solution at an initial time (i.e. asymmetric initial conditions) implies asymmetric solutions at later times. There remains the question of the origin of the asymmetric initial state (how do the laws give rise to asymmetric solutions?), which will be the subject of the next section.

4 Symmetry breaking of physical laws

Symmetries of physical laws can be broken in two ways: *explicitly* or *spontaneously*. We examine the two cases separately.

4.1 Explicit symmetry breaking

Explicit symmetry breaking indicates a situation where the dynamical equations are not manifestly invariant under the symmetry group considered. This means, in the Lagrangian (Hamiltonian) formulation, that the Lagrangian (Hamiltonian) of the system contains one or more terms explicitly breaking the symmetry. Such terms can have different origins, as the following cases illustrate.

(a) Symmetry-breaking terms may be introduced into the theory by hand on the basis of theoretical/experimental results. An illustration is offered by the case of the quantum field theory of the weak interactions, which is expressly constructed in a way that manifestly violates mirror symmetry or *parity* (see Pooley, this volume). The underlying result in this case is parity non-conservation in the case of the weak interaction, first predicted in the famous (Nobel-prize winning) 1956 paper by T. D. Lee and C. N. Yang (and then confirmed by others in the following year) considering the case of the decays K $\rightarrow 3\pi(\tau)$ and K $\rightarrow 2\pi(\theta)$ of the kaon K, known at the time as the '$\tau - \theta$ puzzle'.[11]

(b) Symmetry-breaking terms may appear in the theory because of quantum mechanical effects. One reason for the presence of such terms – known as *anomalies* – is the following: in passing from the classical to the quantum level, because of possible operator ordering ambiguities for composite quantities such as Noether charges and currents, it may be that the classical symmetry algebra (generated through the Poisson bracket structure) is no longer realized in terms of the commutation relations of the Noether charges.[12]

In practice it is not always possible to master the commutation rules of fields, and the alternative way is to check whether the Noether currents are conserved order by order in perturbation theory after an appropriate renormalization.[13] The use of a 'regulator' (or 'cut-off') required in the renormalization procedure to achieve actual calculations may itself be a source of anomalies. It may violate a symmetry of the theory, and traces of this symmetry breaking may remain even after the regulator is removed at the end of the calculations. Historically, the first example of an anomaly arising from renormalization is the so-called chiral anomaly, that is the anomaly violating the chiral symmetry of the strong interaction, 'discovered' in connection with the problem of understanding the observed decay rate of the neutral pion (see Weinberg, 1996, chapter 22).

(c) Finally, symmetry-breaking terms may appear because of non-renormalizable effects. As mentioned also in other contributions to this volume (see Martin, Part I, and Castellani, Part IV), physicists now have good reasons for viewing current renormalizable field theories as *effective field theories*, that is low-energy approximations to a deeper theory (each effective theory explicitly referring only to those particles that are of importance at the range of energies considered). The effects of non-renormalizable interactions (due to the heavy particles not included in the theory) are small and can therefore be ignored in the low-energy regime. It may then happen that the coarse-grained description thus obtained possesses more symmetries than the deeper theory. That is, the effective Lagrangian obeys symmetries that are not symmetries of the underlying theory. These 'accidental' symmetries, as Weinberg has called them, may then be violated by the non-renormalizable terms arising from higher mass scales and suppressed in the effective Lagrangian (see Weinberg, 1995, pp. 529–31).

[11] For a detailed history of parity violation, see in particular Telegdi (1987).

[12] See for example J. Govaerts, 'The quantum geometer's universe: particles, interactions and topology', 2002, hep-th/0207276, section 4.2.

[13] For details, see for example Itzykson and Zuber (1980, pp. 510–11).

4.2 Spontaneous symmetry breaking

Spontaneous symmetry breaking (SSB) indicates a situation where, given a symmetry of the equations of motion, solutions exist which are not invariant under the action of this symmetry *without the introduction of any term explicitly breaking the symmetry* (whence the attribute 'spontaneous').[14] How can this happen?

A situation of this type can be first illustrated by means of simple cases taken from classical physics. Consider, for example, the case of a linear vertical bar with a compression force applied on the top and directed along its axis, discussed also by Jona-Lasinio in this volume. The physical description is obviously invariant for all rotations around this axis. As long as the applied force is mild enough, the bar does not bend and the equilibrium configuration (the lowest energy configuration) is invariant under this symmetry. When the force reaches a critical value, the symmetric equilibrium configuration becomes unstable and an infinite number of equivalent lowest energy stable states appear, which are no longer rotationally symmetric but are related to each other by a rotation. The actual breaking of the symmetry may then easily occur by the effect of a (however small) external asymmetric cause, and the bar bends until it reaches one of the infinite possible stable asymmetric equilibrium configurations.[15]

In substance, what happens in the above kind of situation is the following: when some parameter reaches a critical value, the lowest energy solution respecting the symmetry of the theory ceases to be stable under small perturbations and new asymmetric (but stable) lowest energy solutions appear. The new lowest energy solutions are asymmetric but are all related through the action of the symmetry transformations. In other words, there is a degeneracy (infinite or finite depending on whether the symmetry is continuous or discrete) of distinct asymmetric solutions of identical (lowest) energy, the whole set of which maintains the symmetry of the theory.

SSB occurs in classical and quantum physics. In what follows, we focus on the meaning of SSB in quantum physics.

The first distinction to be drawn is between *finite* and *infinite* physical systems. In the case of finite systems, SSB actually does not occur: tunnelling takes place

[14] Historically, the name 'spontaneous breakdown of symmetry' appeared first in Baker and Glashow (1962), to indicate the fact that starting with a model possessing the higher symmetry $SU(3)$, 'without the introduction into the Lagrangian of any symmetry-breaking terms, solutions exist which have only the lower symmetries of isotopic spin and hypercharge'.

[15] Another example from classical physics which is often used in the literature to illustrate SSB is the case of a ball moving with no friction in a hoop constrained to rotate with a given angular velocity. The theory has a discrete inversion symmetry which is respected by the lowest energy solution when the velocity is below a specified critical value; when the velocity reaches the critical value, the symmetric equilibrium configuration becomes unstable and two lowest energy stable configurations appear, each one being asymmetric but related to the other by the action of the symmetry of the theory. This case, typically used as a simple model for the bifurcation theory of relative equilibria and its connection with dynamic stability theory in textbooks on classical mechanics, has been recently discussed in detail by C. Liu, 'The meaning of spontaneous symmetry breaking (I): From a simple classical model', 2002, PITT-PHIL-SCI00000563.

between the various degenerate states, and the true lowest energy state or 'ground state' turns out to be a unique linear superposition of the degenerate states. In fact, SSB is applicable only to infinite systems – many-body systems (such as ferromagnets, superfluids and superconductors) and fields – the alternative degenerate ground states being all orthogonal to each other in the infinite volume limit and therefore separated by a 'superselection rule'.[16]

Historically, the concept of SSB first emerged in condensed matter physics. The prototype case is the 1928 Heisenberg theory of the ferromagnet as an infinite array of spin-$\frac{1}{2}$ magnetic dipoles, with spin–spin interactions between nearest neighbours such that neighbouring dipoles tend to align. Although the theory (the spin–spin interaction as well as the rest of the Hamiltonian) is rotationally invariant, below the critical Curie temperature T_c the *actual* ground state of the ferromagnet has the spins all aligned in some particular direction (i.e. a magnetization pointing in that direction), thus not respecting the rotational symmetry. What happens is that below T_c there exists an infinitely degenerate set of ground states, in each of which the spins are all aligned in a given direction. A complete set of quantum states can be built upon each ground state. We thus have many different 'possible worlds' (sets of solutions to the same equations), each one built on one of the possible orthogonal (in the infinite volume limit) ground states.[17]

To use a famous image by Sidney Coleman, a little man living inside one of these possible asymmetric worlds would have a hard time detecting the rotational symmetry of the laws of nature (all his experiments being under the effect of the background magnetic field). The symmetry is still there – the Hamiltonian being rotationally invariant – but 'hidden' to the little man. Besides, there would be no way for the little man to detect directly that the ground state of his 'world' is part of an infinitely degenerate multiplet. To go from one ground state of the infinite ferromagnet to another would require changing the directions of an infinite number of dipoles, an impossible task for the *finite* little man.[18] As said, in the infinite volume limit all ground states are separated by a superselection rule.

The same picture can be generalized to quantum field theory (QFT), the ground state becoming the *vacuum state*, and the role of the little man being played by ourselves. This means that there may exist symmetries of the laws of nature which are not manifest to us because the physical world in which we live is built on a vacuum state which is not invariant under them. In other words, the physical world of our experience can appear to us very asymmetric, but this does not necessarily mean that this asymmetry belongs to the fundamental laws of nature. SSB offers a key for understanding (and utilizing) this physical possiblity. As stated by Baker

[16] For details see for example Weinberg (1996, pp. 164–5).
[17] For this image see Aitchison (1982, pp. 75–7).
[18] See Coleman (1975, pp. 141–2).

and Glashow (1962, p. 2463), 'it is thus made plausible that the intricacies of the physical world are not reflected in an equally intricate fundamental theory'.

How the concept of SSB was in fact transferred from condensed matter physics to QFT in the early 1960s, thanks especially to works by Y. Nambu and G. Jona-Lasinio, is described in the other contributions to Part III of this volume (see Jona-Lasinio, Earman, and Morrison). In particular, Jona-Lasinio's paper offers a first-hand account of how the idea of SSB was introduced and formalized in particle physics on the grounds of an analogy with the breaking of (electromagnetic) gauge symmetry in the 1957 theory of superconductivity by J. Bardeen, L. N. Cooper, and J. R. Schrieffer (the so-called BCS theory).

The application of SSB to particle physics in the 1960s and successive years led to profound physical consequences and played a fundamental role in the edification of the current Standard Model of elementary particles. We just mention some of the main ideas here, referring to the other chapters of Part III of this volume (and the references therein) for further details.

In short, as it turned out, a number of field theories may display spontaneous breaking of internal symmetries. In the case of a *discrete symmetry*, this does not lead to especially interesting consequences. In the case of a *continuous symmetry*, what results from SSB can be summarized in the following points.

- *Goldstone theorem.* In the case of a *global* continuous symmetry, massless bosons (known as 'Goldstone bosons') appear with the spontaneous breakdown of the symmetry according to a theorem first stated by J. Goldstone in 1960. The presence of these massless bosons, first seen as a serious problem since no particles of the sort had been observed in the context considered, was in fact the basis for the solution – by means of the so-called Higgs mechanism (see the next point) – of another similar problem: that is, the fact that the 1954 Yang–Mills theory of non-Abelian gauge fields predicted unobservable massless particles, the gauge bosons (massless because of the gauge invariance).
- *Higgs mechanism.* According to a 'mechanism' established in a general way in 1964 independently by (i) P. Higgs, (ii) R. Brout and F. Englert, and (iii) G. S. Guralnik, C. R. Hagen, and T. W. B. Kibble, in the case that the internal symmetry is promoted to a *local* one, i.e. to a *gauge symmetry*, the Goldstone bosons 'disappear' and the gauge bosons acquire a mass; or, in other words, the Goldstone bosons are 'eaten up' to give mass to the (otherwise massless) gauge bosons, and this happens without (explicitly) breaking the gauge invariance of the theory.
- *Renormalizability.* The above mechanism for the mass generation for the gauge fields is also what ensures the renormalizability of theories involving massive gauge fields (such as the Glashow–Weinberg–Salam electroweak theory

developed in the second half of the 1960s), as first generally demonstrated by M. Veltman and G. 't Hooft in the early 1970s.[19]

Let us end this brief review of SSB by mentioning *dynamical symmetry breaking* (DSB). In such theories as the Glashow–Weinberg–Salam unified model of electroweak interactions, the SSB responsible for the masses of the gauge vector bosons is due to the symmetry-violating vacuum expectation values of scalar fields (the so-called 'Higgs fields') introduced *ad hoc* in the theory. For different reasons – first of all, the *ad hoc* character of these scalar fields, for which there is no experimental evidence (no 'Higgs particle' has been observed up to now) – increasing attention has been drawn to the possibility that the scalar fields could be phenomenological rather than fundamental, that is bound states resulting from a specified dynamical mechanism. SSB realized in this way has been called 'DSB'.

Note that SSB was in fact first introduced in the DSB form. In the BCS theory of superconductivity, as well as in the 1961 theory of broken chiral symmetry by Nambu and Jona-Lasinio, SSB is realized dynamically through a fermion condensate. In the BCS theory, for example, the gauge invariance of electromagnetism is spontaneously broken by pairs of electrons that condense – forming a bound state – in the ground state of a metal. Although DSB has not (so far) proved successful as an alternative route to the problem raised by the Higgs fields in the Standard Model, it has been applied with success to specific cases: for example, besides the already mentioned case of the BCS theory, the current quantum field theory of the strong interaction (quantum chromodynamics), in the approximation that quark masses are very small, possesses chiral symmetries that are spontaneously broken by a condensation of quark–antiquark pairs.[20]

5 Symmetry breaking and philosophical questions

Symmetry breaking raises a number of philosophical issues. Some of them relate only to the breaking of specific types of symmetries, such as the issue of the significance of parity violation for the problem of the nature of space (see Pooley and Huggett, chapters 14 and 15 of this volume). Others, for example the connection between symmetry breaking and observability, are particular aspects of the general issue concerning the status and significance of physical symmetries, discussed in Part IV of this volume (see also the introduction, section 3). Finally, there are issues raised by the role and nature of symmetry breaking in general; these are the subject of this concluding section.

[19] Their Lagrangians in fact involve only massless gauge fields coupled minimally to conserved currents, the mass of the gauge vector bosons and the non-conservation of the currents (whence the apparent non-renormalizability) being the result of SSB. It is a general result that the divergence structure of a renormalizable quantum field theory is not affected by SSB.

[20] In a similar way to the spontaneous breaking of the chiral symmetry of the strong interactions hypothesized in 1961 by Nambu and Jona-Lasinio, although of course not in terms of quarks and antiquarks.

From what we have seen so far, the main points about the general role of symmetry breaking in physics can be summarized as follows:

(i) Symmetry breaking is essential for the existence of something structured (for the presence of some structure, a lower symmetry than the absolute one is needed). In Curie's words, the absence of certain elements of symmetry or dissymmetry is what creates the phenomenon.

(ii) Asymmetric states can be solutions to symmetric as well as asymmetric equations (the symmetry of a dynamical equation not being necessarily a symmetry of the individual solutions but rather the symmetry of the whole set of solutions). In the case of symmetric equations, asymmetric initial conditions explain the presence of asymmetric solutions, but there remains the question of the origin of the asymmetric initial state.

(iii) Spontaneous symmetry breaking allows symmetric theories to describe asymmetric reality. The asymmetries of the physical world of our experience do not necessarily belong to the fundamental laws of nature: there may exist symmetries of the laws of nature which are not manifest to us because the physical world in which we live is built on a vacuum state which is not invariant under them. In short, SSB provides a way of understanding the complexity of nature without renouncing fundamental symmetries.

In the light of these points, the philosophical discussion has focused on methodological and epistemological aspects of spontaneous symmetry breaking, and in particular the following questions:

(A) If, according to point (i), the absence or breaking of some symmetry is not only common but even necessary for the existence of the phenomena, why search for a way (provided by the mechanism of SSB) of avoiding attributing these asymmetries to the fundamental laws?[21] The question is actually twofold.
  - Why should we prefer symmetric to asymmetric fundamental laws? We refer to the other contributions to this volume for this point, the answer being provided by the many relevant functions and consequences of symmetry principles in contemporary physics.
  - Why should we assume that an asymmetry needs an explanation? The underlying idea is that we can start with symmetries, but not asymmetries.[22] In other words, it is assumed that an observed asymmetry requires a cause, which can be an explicit breaking of the symmetry of the laws, asymmetric initial conditions, or SSB. Note that this idea is very similar to the one expressed by Curie in his famous 1894 paper. In section 2 we mentioned one

[21] On this sort of question see, for example, Earman (this volume, Part III) and Kosso (2000).

[22] Of course, there remains the fundamental issue of the origin of the symmetries, which is the subject matter of other papers in the volume (in particular, Parts I and IV).

of the forms in which Curie expressed his conclusions about the relationships between the symmetries of a physical medium and the symmetries of a phenomenon occurring in it. Another equivalent formulation but in terms of 'causes' and 'effects' – usually known as 'Curie's principle' *tout court* in the literature – is the one stating that the symmetries of the causes must be found in the effects; or, equivalently, the asymmetries of the effects must be found in the causes (see Curie, 1894, this volume). This formulation, when extended to include the case of SSB, is well appropriated to express the methodological principle considered here: an asymmetry of the phenomena ('the effects') must come from the breaking – 'explicit' or 'spontaneous' – of the symmetry of the fundamental laws ('the causes'). What the real nature of this principle is remains an open issue, at the centre of a developing debate (see the introduction to this volume, section 2).

For the sake of completeness, let us mention an aspect of the debate on the validity of Curie's principle that is in apparent contrast to the 'extended' way in which it has been presented above. The point is sometimes made in the literature that SSB appears to violate Curie's principle because of the absence of any asymmetric cause. Thus, when SSB occurs we have a situation where an asymmetry of the effect does not derive from an asymmetry of the cause; or, equivalently, a situation where the symmetry of the cause (the law) is not to be found in the effect (the phenomenon). That this is not the case is easily shown. It is true that SSB indicates a situation where solutions exist that are not invariant under the symmetry of the law (dynamical equation) without any explicit breaking of this symmetry. But the symmetry of the 'cause' is not lost, it is conserved in the ensemble of the solutions (the whole 'effect') (see section 4, and point (ii) in this section).[23]

(B) What is the epistemological *status* of a theory based on SSB? This general issue is currently addressed by tackling the following cluster of questions.

- What is the empirical stand of the 'hidden' symmetries of the fundamental laws (given that, in the SSB case, what is directly observed – the physical situation, the phenomenon, . . . – is asymmetric)? How can we know them, if they are hidden? The question is part of the general issue of the empirical evidence for the symmetries of physical laws, here addressed in the case where there is no direct correspondence between the symmetry of the laws and the symmetry of the phenomena (see for example Morrison, this volume, and Kosso, 2000).
- In the absence of direct empirical evidence, the above question then becomes whether and how far the predictive and explanatory power of theories based on SSB provides good reasons for believing in the existence of the hidden symmetries.

[23] Stewart and Golubitsky (1992), for example, speak of an 'extended Curie's principle' to indicate this situation.

This is in part a specific issue:

(a) In each of the particular cases in which SSB is applied, is the existence of the hidden symmetries of the laws the best explanation for the behaviour of the physical world? How successful is the specific SSB case considered in producing observable consequences?

And in part a general issue:

(b) Can the predictive and explanatory power of these theories be a basis for scientific realism with respect to the assumed hidden symmetries, and, more generally, with respect to all the (directly) non-observable features introduced for the SSB mechanism to function? In this respect, for example, Morrison (this volume) discusses two particular issues arising in the specific but paradigmatic case of the Glashow–Weinberg–Salam theory of electroweak interactions: that is, the question of the epistemological *status* of the Higgs particles, and the issue of the nature of the physical assumptions about the quantum field theory vacuum, which play a crucial role in the history of SSB.

Conclusions

Concluding this final section, it is worth underlining that the philosophical reflection on symmetry breaking (in particular in its 'spontaneous' form) is very recent in the literature. The discussion of the methodological and epistemological issues mentioned above, while important from both a physical and a philosophical point of view, has just started. Stimulating this discussion by offering an overview of its basic elements and possible developments is the aim of this review paper and the other contributions (Jona-Lasinio, Earman, and Morrison) to Part III of this volume.

Acknowledgements

Many thanks to Katherine Brading, Roberto Casalbuoni, Leonardo Castellani, and Giovanni Jona-Lasinio for helpful comments and suggestions.

References

Aitchison, I. J. R. (1982). *An Informal Introduction to Gauge Field Theories*. Cambridge: Cambridge University Press.

Baker, M., and Glashow, S. L. (1962). 'Spontaneous breakdown of elementary particles symmetries'. *Physical Review*, **128**, 2462–71.

Coleman, S. (1975). 'Secret symmetry: an introduction to spontaneous symmetry breakdown and gauge fields'. In *Laws of Hadronic Matter*, ed. A. Zichichi, pp. 138–215. New York: Academic Press.
Curie, P. (1894). 'Sur la symétrie dans les phénomènes physiques. Symétrie d'un champ électrique et d'un champ magnétique'. *Journal de Physique*, 3rd series, **3**, 393–417.
Itzykson, C., and Zuber, J.-B. (1980). *Quantum Field Theory*. New York: McGraw-Hill.
Kosso, P. (2000). 'The epistemology of spontaneously broken symmetries'. *Synthese*, **122**, 359–76.
Lee, T. D., and Yang, C. N. (1956). 'Question of parity conservation in weak interactions'. *Physical Review*, **104**, 254–8.
Radicati, L. A. (1987). 'Remarks on the early developments of the notion of symmetry breaking'. In *Symmetries in Physics (1600–1980)*, ed. M. G. Doncel, A. Hermann, L. Michel, and A. Pais, pp. 195–206. Barcelona: Servei de Publicacions.
Stewart, I., and Golubitsky, M. (1992). *Fearful Symmetry. Is God a Geometer?* Oxford: Blackwell.
Telegdi, V. L. (1987). Parity violation. In *Symmetries in Physics (1600–1980)*, ed. M. G. Doncel, A. Hermann, L. Michel, and A. Pais, pp. 433–49. Barcelona: Servei de Publicacions.
Weinberg, S. (1995). *The Quantum Theory of Fields. I*. New York: Cambridge University Press.
(1996). *The Quantum Theory of Fields. II*. New York: Cambridge University Press.

20

Rough guide to spontaneous symmetry breaking

JOHN EARMAN

Understanding spontaneous symmetry breaking is essential to understanding characteristic phenomena of solid-state physics, condensed matter physics, elementary particle physics, and cosmology. I will forego the details of physical applications in order to concentrate on the implications of spontaneous symmetry breaking for general issues concerning laws and symmetries and for the foundations of quantum field theory (QFT).[1]

1 Laws and symmetries

The most intriguing aspects of spontaneous symmetry breaking derive from features peculiar to QFT. But before turning to this theory, I want to rehearse some themes that are familiar from that innocent era that existed not only prior to the advent of QFT but also before the *verdammt* quantum complicated our world.

Theme 1. Consider equations of motion that are derived from an action principle that demands that the allowed motions extremize the action $\mathfrak{A} = \int_\Omega L(\mathbf{x}, \mathbf{u}, \mathbf{u}^{(n)})\, d^p\mathbf{x}$ (here $\mathbf{x} = (x^1, \ldots, x^p)$ stands for the independent variables, $\mathbf{u} = (u^1, \ldots, u^r)$ are the dependent variables, and the $\mathbf{u}^{(n)}$ are derivatives of the dependent variables up to some finite order n with respect to the x^i); the resulting equations of motion take the form of Euler–Lagrange equations. This is a mild restriction in the sense that it covers almost all of the live candidates for the role of fundamental equations of motion in modern physics. A Lie group $\mathcal{G}$ of transformations $\mathcal{G} \ni g : (\mathbf{x}, \mathbf{u}) \to (\mathbf{x}', \mathbf{u}')$ is called a *variational symmetry group* if the infinitesimal generators of $\mathcal{G}$ leave L form invariant. Such symmetries are necessarily symmetries of the Euler–Lagrange equations in that they carry solutions to solutions.[2]

[1] For a more detailed guide, see J. Earman, 'Spontaneous symmetry breaking for philosophers', unpublished manuscript. Good reviews of the physics of spontaneous symmetry breaking abound in the physics literature; I can especially recommend Aitchison (1982, chapters 5 and 6), Coleman (1985, chapter 5), and Guralnik *et al.* (1968).

[2] The converse is not true.

Furthermore, if $\mathcal{G}$ is an s-parameter ($s < \infty$) Lie group, then Noether's first theorem tells us that there are s conserved currents.[3] In the cases to be considered below, the transformations $\mathcal{G}$ are such as to produce a unit Jacobian for the transformation of the independent variables and, thus, produce strict numerical invariance of the action. Such symmetries are referred to somewhat imprecisely in the physics literature as 'symmetries of the Lagrangian'. I will follow this usage here.

Theme 2. Symmetries of the laws of motion need not be – and typically are not – exhibited in particular solutions or in particular states belonging to solutions.

Combining these two themes yields the moral that a symmetry of the Lagrangian may be 'broken' in that it may not be exhibited in the solution corresponding to the actual world. For example, the tables, chairs, and other objects in my neighbourhood do not exhibit rotational symmetry even though the Lagrangian that governs the motion of the particles that compose these objects is (presumably) rotationally invariant. Put in this light, symmetry breaking seems so commonplace and so innocuous that, at first blush, it is hard to see how it could hold any interest for philosophers of physics. However, cases of symmetry breaking do serve to underscore the point – emphatically emphasized by Eugene Wigner (1967) – that an appreciation of the role of symmetry principles in physics presupposes a distinction between laws and initial/boundary conditions since the symmetries of concern to the physicist are symmetries of the former that typically fail to be symmetries of the latter. Taken at face value, this point appears to be flatly inconsistent with the so-called 'no-laws' view of science, at least insofar as this view denigrates or downplays the concept of laws of physics. It remains to be seen whether it is incompatible with the more moderate no-laws view which does not denigrate the concept of laws of physics – in the sense of general principles fashioned by us in our attempts to understand the world – but denies that these principles capture the laws of nature after which philosophers have lusted – in the sense of objective facts about nomic necessities.[4] A second methodological issue arises from the first. If asymmetry is such a pervasive feature of the phenomena, why should physicists believe so strongly in the symmetry of the laws that govern these phenomena?[5] And if their reasons are good reasons, do they provide a basis for scientific realism with respect to the unobservables in terms of which the laws governing the phenomena are formulated?

I leave it to the philosophers to deal with these methodological issues, in order to concentrate on the subclass of cases that fall under the heading of spontaneous symmetry breaking.

[3] For details see Brading and Brown, this volume.
[4] Two rather different no-laws views are to be found in van Fraassen (1989) and Giere (1999).
[5] On this matter see Kosso (2000).

2 Spontaneous symmetry breaking in classical physics and QFT

What distinguishes cases of spontaneously broken symmetries from generic cases of broken symmetries? Weinberg (1996, p. 163) casts the net widely by counting as a case of spontaneous symmetry breaking any one where the ground state (lowest energy state) is degenerate and where the symmetry of the Lagrangian is not a symmetry of the ground states but carries one asymmetric ground state to another. A more common usage would restrict the label 'spontaneous symmetry breaking' to cases where, as the value of some parameter surpasses a critical value, the ground state becomes degenerate and, without any apparent asymmetrical cause, the system enters one of the asymmetrical degenerate ground states. Examples of this kind can be given in elementary classical mechanics – a rather nice one was constructed by Greenberger (1978).[6] On closer inspection, however, it turns out that the breaking of the symmetry is not really spontaneous and that the appearance of a spontaneous breaking arises from the combination of dynamical instability plus some asymmetrical factor which may not be macroscopically detectable.

I will not pause to enter the debate about how exactly the subclass of spontaneously broken symmetries is to be delimited from the general class of broken symmetries. For my main concern is with spontaneous symmetry breaking in QFT, and in this context the situation is quite different from that in classical mechanics or ordinary non-relativistic quantum mechanics (QM). For starters, as long as the system of interest is closed, there is no temporal evolution involved in spontaneous symmetry breaking in QFT since *every* physically relevant state of the system is asymmetric with respect to the symmetry of the Lagrangian. A better term than 'spontaneously broken symmetry' might be 'ubiquitously broken symmetry'. Furthermore, although some of the words describing spontaneous symmetry breaking in classical mechanics and ordinary QM carry over to QFT, their meaning is quite different. For instance, the role of the 'ground state' is played in QFT by the vacuum state in a Fock space, and in cases of spontaneous symmetry breaking the vacuum state becomes 'degenerate'. However, the relevant sense of degeneracy in QFT is wholly different from that in classical mechanics or ordinary QM. Indeed, the situation in QFT is so unfamiliar it can generate a sense of the cognitive dissonance. Here are a few of the puzzles that someone encountering the topic for the first time may experience:

- In cases of spontaneous symmetry breaking in QFT a symmetry of the Lagrangian is spontaneously broken by failing to be a symmetry of the vacuum state of the system. But if the vacuum state is a state of nothingness, how can such a state fail to share the symmetries of the Lagrangian?

[6] See also C. Liu, 'Spontaneous symmetry breaking (I): Its meaning from a simple classical model,' preprint.

- Since the said symmetry of the Lagrangian is not a symmetry of the vacuum state of the system, it carries this state to another state which, it can turn out, is also a vacuum state of the system. So in cases of spontaneous symmetry breaking in QFT, the vacuum state is degenerate. But how can this be, since in relativistic QFT the vacuum state is supposed to be the unique state picked out by Poincaré invariance and positivity of energy?
- Part of the answer to the second puzzle is that the said symmetry of the Lagrangian is not unitarily implementable, i.e. its action is not faithfully represented by a unitary operator on Hilbert space. But how can this be, since Wigner's theorem has taught us that a symmetry in QM is represented by a unitary transformation (or, as in the case of time reversal, an anti-unitary transformation)?
- The remainder of the answer to the second puzzle involves the fact that the 'degenerate' vacuum states belong to unitarily inequivalent representations of the algebra of observables. But what does this mean physically and why does it happen in QFT and not in ordinary QM?

To help resolve these puzzles I turn to the algebraic formulation of QFT.

3 Using the algebraic formulation of QFT to explain spontaneous symmetry breaking

The algebraic formulation of QFT has the virtue of making transparent the abstract mathematical structure of cases of spontaneous symmetry breaking in QFT. It points one to definitions, both natural and precise, of the relevant senses of symmetry and symmetry breaking, and it allows the formulation and proof of quite general statements about the conditions under which a symmetry is or is not unitarily implementable. Anyone who masters this formalism will not be subject to the puzzles enumerated above.

In the algebraic approach the basic object is an abstract algebra $\mathcal{A}$, usually taken to be a C^*-algebra with unit element.[7] An algebraic state is a normed positive linear functional $\omega : \mathcal{A} \rightarrow \mathbb{C}$, and the expectation value of an element $A \in \mathcal{A}$ is then $\omega(A)$. The familiar Hilbert space formalism is recovered in a representation of $\mathcal{A}$ in the form of a structure preserving map $\pi : \mathcal{A} \rightarrow \mathcal{B}(\mathcal{H})$ into the algebra of bounded operators $\mathcal{B}(\mathcal{H})$ on a separable Hilbert space $\mathcal{H}$, which might or might not be a Fock space. A fundamental result of Gelfand, Nimark, and Segal (GNS) guarantees that any state ω determines a cyclic representation $(\pi_\omega, \mathcal{H}_\omega)$ that is unique up to unitary equivalence.

In the algebraic setting a symmetry is given by an automorphism θ of the algebra $\mathcal{A}$.[8] An automorphism θ of $\mathcal{A}$ is said to be *inner* just in case there is a unitary $U \in \mathcal{A}$

[7] See appendix. A detailed presentation of the mathematics of algebraic QFT is to be found in Bratteli and Robinson (1987; 1996).

[8] Everything said here is easily generalized to groups of automorphisms.

such that $\theta(A) = UAU^{-1}$ for every $A \in \mathcal{A}$. Many automorphisms of C^* -algebras are not inner. And worse, they may even fail to be inner with respect to a state ω in that they fail to have the property that θ behaves like an inner automorphism 'under the expectation value', i.e. there is no unitary $U \in \mathcal{A}$ such that $\omega(\theta(A)) = \omega(UAU)$ for every $A \in \mathcal{A}$. Glimm and Kadison (1960) proved that θ is inner with respect to a pure state ω iff θ is unitarily implementable with respect to ω in that in the GNS representation $(\pi_\omega, \mathcal{H}_\omega)$ determined by ω, there is a unitary operator $\hat{U}$ on $\mathcal{H}_\omega$ such that $\pi_\omega(\theta(A)) = \hat{U}\pi_\omega(A)\hat{U}^{-1}$ for every $A \in \mathcal{A}$. Typical cases of spontaneous symmetry breaking in QFT turn out to be cases where there is a symmetry of the Lagrangian that induces an automorphism θ of the algebra $\mathcal{A}$ of field operators, but θ is not unitarily implementable with respect to any physically relevant state ω (see below).

Note that an automorphism θ of the algebra $\mathcal{A}$ can be thought of as acting on states: for any state ω, θ produces a new state $\widehat{\theta\omega}$ where $\widehat{\theta\omega}(A) := \omega(\theta(A))$ for every $A \in \mathcal{A}$.[9] If ω is θ-invariant, i.e. $\widehat{\theta\omega} = \omega$, then trivially θ is unitarily implementable with respect to ω. Since in spontaneous symmetry breaking we are concerned with cases where a symmetry θ is not unitarily implementable, $\widehat{\theta\omega} \neq \omega$. If ω is a vacuum state and θ commutes with the symmetry (e.g. Poincaré transformations), the invariance under which picks out a vacuum state, then $\widehat{\theta\omega}$ will be a different vacuum state. And, moreover, these different vacuum states belong to unitarily inequivalent representations of $\mathcal{A}$, for the Glimm and Kadison result can be extended to show that θ is unitarily implementable with respect to a pure state ω iff $\widehat{\theta\omega}$ and ω determine unitarily equivalent GNS representations of $\mathcal{A}$, in that there is an isomorphism $\hat{V} : \mathcal{H}_\omega \to \mathcal{H}_{\widehat{\theta\omega}}$ such that $\pi_{\widehat{\theta\omega}}(A) = \hat{V}\pi_\omega(A)\hat{V}^{-1}$ for every $A \in \mathcal{A}$ (see Arageorgis *et al.*, 2002).

A concrete toy example of spontaneous symmetry breaking in QFT starts from the Lagrangian density $L = \frac{1}{2}\partial_\mu\varphi\partial^\mu\varphi$ for a real-valued scalar field φ. L is invariant under the transformations of the field $\varphi \to \varphi' = \varphi + \chi$, where χ is an arbitrary real number. When the field is quantized this symmetry is not unitarily implementable. The treatment of this example from the perspective of standard QFT is to be found in Aitchison (1982, chapter 5), and the algebraic analysis is given in Streater (1965). In this particular example, as in many examples of spontaneous symmetry breaking, there is a nice connection with Noether's first theorem. In the present case, Noether's first theorem tells us that since the variational symmetries form a one-dimensional Lie group, there is a conserved current j^μ. The spatial volume integral of the time component j^0 is the Noether charge Q. If, upon quantization of the field, there were a well-defined self-adjoint charge operator $\hat{Q}$ corresponding to Q, it would be the generator of a 1-parameter family of unitary transformations that

[9] If the automorphism represents time evolution, the difference between the two points of view amounts to the difference between the Heisenberg and Schrödinger pictures.

implement the transformation of the quantum field $\hat{\varphi} \to \hat{\varphi}' = \hat{\varphi} + \chi$. But a reductio argument shows that there is no such $\hat{Q}$ if the vacuum state is (as normally assumed) translationally invariant and if $\hat{Q}$ commutes with translations (see Aitchison, 1982, pp. 71–2, and Fabri and Picasso, 1966).

To translate this no-go result back into the algebraic setting, call an algebraic state ω on a C^*-algebra $\mathcal{A}$ a *Fock state* if its GNS representation is unitarily equivalent to a Fock representation. Then what the reductio proof for the toy model shows is that if it is demanded that algebraic states be Fock states with translationally invariant vacuums, there is no state ω with respect to which the automorphism θ of $\mathcal{A}_L$ induced by the symmetry of L is unitarily implementable and, *a fortiori*, there is no state ω which is θ-invariant. Thus, if the physically relevant states are Fock states with translationally invariant vacuum states, it is precluded that the symmetry is broken by evolution of any sort from a symmetric state to an asymmetric state since there are no symmetric states in the relevant sense.[10] Of course, this no-go result applies only to closed systems. It remains a possibility that in open systems there is some interesting sense of temporal symmetry breaking. But if that open system is a subsystem of a larger system that itself is a closed system, then that sense is inoperative for the larger system.

The features of spontaneous symmetry breaking that have been emphasized above cannot arise in the ordinary quantum mechanics of a system with a finite number of degrees of freedom since the Stone–von Neumann theorem says, roughly, that in such cases the representation of the canonical commutation relations is unique up to unitary equivalence. With $U_m(s) := \exp(iq_m s)$ and $V_n(t) := \exp(ip_n t)$, the Weyl form of the canonical commutation relations $p_n q_m - q_m p_n = -i\delta_{nm}$, etc., is given by $V_n(t)U_m(s) = U_m(s)V_n(t)\exp(ist\delta_{nm})$ and $U_m(s)U_n(t) - U_n(t)U_m(s) = 0 = V_m(s)V_n(t) - V_n(t)V_n(s)$. When the ranges of m and n, are finite, the Stone–von Neumann theorem says that the only irreducible representation of these relations by continuous unitary groups on Hilbert space is unitarily equivalent to the Schrödinger representation. It is the breakdown of the Stone–von Neumann theorem when the number of degrees of freedom is infinite – in particular, for the case of fields with an uncountable number of degrees of freedom – that opens up the possibility of spontaneous symmetry breaking. In algebraic terms, the Weyl relations generate a special C^*-algebra called (naturally) a Weyl algebra,[11] and this algebra admits innumerably many unitarily inequivalent representations. The existence of such representations was long known. The discovery of spontaneous symmetry breaking can be seen as the discovery of a particular physical mechanism

[10] It is sometimes said that non-Fock representations are needed to describe interacting quantum fields. This claim is disputable. And in any case the use of non-Fock representations renders inapplicable the above style of argument for showing that a symmetry of the Lagrangian is not unitarily implementable. Thus, if such representations are to host spontaneous symmetry breaking, a much different style of argument is needed.

[11] For an explicit construction of the Weyl algebra for the scalar Klein–Gordon field, see Wald (1994).

for generating such representations. Characterized in this way spontaneous symmetry breaking seems like pretty small potatoes. However, this impression is belied by the fact that an apparent problem with spontaneous symmetry breaking led to the Higgs mechanism, which is an essential ingredient in the Standard Model of elementary particles (see section 5, below).

4 Resolving the puzzles

We are now in a position to resolve the puzzles raised in section 2. The first puzzle has a quick *modus tollens* 'resolution': spontaneous symmetry breaking underscores the point that the vacuum in QFT is not a formless nothingness but rather a state with an intricate structure. A full resolution obviously requires an extended discussion of the structure of the vacuum, but it cannot be provided here.

The second puzzle is quickly dissolved. The degeneracy of the vacuum that occurs in spontaneous symmetry breaking does not contradict the usual statement about the uniqueness of the vacuum: the latter refers to the existence of a unique state vector satisfying various conditions within a given Fock space representation of the algebra of field operators, whereas the former refers to the existence of multiple unitarily inequivalent representations, each with its own unique vacuum state.

The resolution of the third puzzle requires a bit more commentary. Call a map $\mathcal{S}:\Phi \to \Phi'$ of the rays of a separable Hilbert space $\mathcal{H}$ a *Wigner* or *unbroken symmetry* just in case it preserves probabilities in the sense that for all rays Φ and Ψ, $|\langle\Phi|\Psi\rangle| = |\langle\Phi'|\Psi'\rangle|$ where $|\Phi\rangle$ and $|\Psi\rangle$ are vectors belonging to Φ and Ψ respectively. Wigner (1959) proved that $\mathcal{S}$ is unbroken iff there is a unitary or antiunitary operator $\hat{U}$ such that $|\Phi'\rangle = \hat{U}|\Phi\rangle$. An unbroken $\mathcal{S}$ induces a transformation of operators $\mathcal{T}:\hat{A} \to \hat{A}' = \hat{U}\hat{A}\hat{U}^{-1}$. Obviously, if $\hat{U}$ is unitary then $\mathcal{T}$ is an automorphism of $\mathcal{B}(\mathcal{H})$. Conversely, any automorphism of $\mathcal{B}(\mathcal{H})$ takes this form for a unitary $\hat{U}$ – in the terminology of section 3, $\mathcal{T}$ is an inner automorphism of the C^*-algebra $\mathcal{B}(\mathcal{H})$. How then can a symmetry in the guise of an automorphism of a C^*-algebra $\mathcal{A}$ fail to be an unbroken or Wigner symmetry? The answer is that two different senses of 'broken symmetry' are in play. For a broken symmetry in the sense of spontaneous symmetry breaking, the C^*-algebra $\mathcal{A}$ is not isomorphic to $\mathcal{B}(\mathcal{H})$; indeed, a representation π of $\mathcal{A}$ is into rather than onto $\mathcal{B}(\mathcal{H})$, and there is no continuous extension of $\pi(\mathcal{A})$ to all of $\mathcal{B}(\mathcal{H})$. An automorphism θ of $\mathcal{A}$ is broken in the sense of spontaneous symmetry breaking not because it is broken in the Wigner sense in that it fails to preserve probabilities but because it is not an automorphism of $\mathcal{B}(\mathcal{H})$. At the level of Hilbert space representations, the action of the broken θ is best construed as moving between unitarily inequivalent representations.

The fourth puzzle has been resolved in the sense that the existence of non-unitarily equivalent representations has been explained in terms of the algebraic apparatus.

It remains to give more of a feeling for what these representations mean in terms of physics. If one tries to get a grasp on the meaning of unitarily inequivalent representations by insisting on thinking of the vacuum states from different representations as all belonging to one big Hilbert space, then the different vacuum states must lie in different superselection subspaces in that a genuine superposition of states from two different such subspaces is not physically meaningful.[12] This follows from the fact that for pure states ω_1 and ω_2 on a C^*-algebra $\mathcal{A}$, unitary inequivalence is the same as *disjointness*, the latter of which means intuitively that any vector from the GNS representation of ω_1 is 'orthogonal' to any vector in the GNS representation of ω_2 and vice versa.[13] In ordinary QM or QFT one can make a symmetric state by superposing asymmetric states, and 'measurement collapse' of a superposition onto one of its 'branches' can produce an asymmetric state from a symmetric one. But neither of these things can happen if an 'asymmetric state' is understood to mean not just that the state is not invariant under the relevant symmetry but also that the symmetry is not unitarily implementable with respect to the state.

5 Issues of interpretation

The above rough guide to spontaneous symmetry breaking provides an entry point into the subject without revealing the subleties and unresolved issues that a more thorough guide would unveil. Here I will simply mention three important issues that the more assiduous explorers will encounter, leaving it to them to investigate the details and to provide their own resolutions.

In section 3 I indicated that spontaneous symmetry breaking arises in cases of continuous symmetries falling under Noether's first theorem: if the symmetry were unitarily implementable, it would be generated by the global charge operator $\hat{Q}$, which is obtained by integrating over all space the time component $\hat{j}^0$ of the current operator corresponding to the conserved Noether current; but, given plausible assumptions such as the translation invariance of the vacuum, provably there is no such self-adjoint operator. Actually, the charge operator $\hat{Q}_V$ corresponding to a finite volume V of space is well defined. It is only the infinite volume limit $\hat{Q} = \lim_{V\to\infty} \hat{Q}_V$ that is ill defined. Since actual physical systems exhibiting spontaneous symmetry breaking – e.g. ferromagnets and superconductors – occupy only a finite volume, it needs to be asked whether or not the distinctive features of spontaneous symmetry breaking discussed above are merely artifacts of the infinite volume idealization.

12 And in the case of the above toy example where the cardinality of the degenerate vacuum states is that of the continuum, the Hilbert space won't be separable.

13 For a precise definition of disjointness, see Bratteli and Robinson (1987, pp. 370–1). As emphasized by Castellani (this volume, Part III), the existence of a superselection rule for ground states holds not only for QFT but more generally for quantum systems with an infinite number of degrees of freedom.

The second issue arises from Goldstone's theorem (see Goldstone *et al.*, 1962) which says, roughly, that in cases of spontaneous symmetry breaking arising from a finite parameter Lie group symmetry, the imposition of standard quantum field theoretic assumptions, such as Poincaré invariance and locality, implies the existence of massless scalar bosons. Since all of the experimental evidence indicates that such particles do not exist, the upshot is a dilemma: either spontaneous symmetry breaking does not occur or else the standard quantum field theoretic assumptions must go.

The response that most particle theorists have settled on is known as the *Higgs mechanism*. It passes through the horns of the dilemma by changing the problem. Additional fields are added in such a fashion that the new Lagrangian is invariant under an infinite-dimensional Lie group, where the group parameters are arbitrary functions of the independent variables. The situation now falls under Noether's second theorem, and the symmetries at issue are gauge symmetries. It is found that the gauge can be set in such a way that the Goldstone bosons are absent, which shows that in the changed environment these particles can be dismissed as gauge epiphenomena.

The Higgs mechanism has been incorporated into the Standard Model of particle physics not only because it suppresses Goldstone bosons but also because it provides a way to give the particles their masses. But because the new theory is a gauge theory, the upshot for the nature of spontaneous symmetry breaking remains opaque until the veil of gauge is removed and the theory is reduced to its gauge-independent content.[14] Suppose that this content is described by local fields which, when quantized, obey the standard assumptions of QFT, such as Poincaré invariance and locality – otherwise the implementation of the Higgs mechanism would necessitate a radical revision of QFT. On pain of resurrecting Goldstone bosons, the reduced theory cannot admit a finite parameter Lie group as a symmetry group. Goldstone's theorem does not apply if the reduced theory admits a discrete symmetry group, but neither does Noether's first theorem nor the argument sketched in the preceding section for the spontaneous breakdown of symmetry. There are heuristic arguments for the spontaneous breakdown of discrete symmetries in QFT, but as far as I am aware there are no demonstrations of even modest rigour for such a breakdown. In sum, the fate of spontaneous symmetry breaking in the Higgs model is unsettled.

[14] As discussed elsewhere in this volume, there is in principle a method for accomplishing this goal. If the variational symmetries consist of transformations that depend on arbitrary functions of the independent variables, then the Hamiltonian formulation contains constraints. The subclass of first-class constraints generates the gauge transformations on the Hamiltonian phase space. If various technical obstructions do not intervene, quotienting out the gauge orbits yields the reduced phase space whose canonical coordinates are gauge-invariant quantities.

The third issue also revolves around the gauge concept. In the case of the Higgs mechanism the concept of gauge is invoked in order to overcome an apparent failure of determinism that results when the action admits a variational symmetry group of transformations that depend on arbitrary functions of the independent variables. But apart from considerations of determinism, there can be other grounds for seeing gauge freedom at work. For example, there might be reasons for thinking that a complex-valued scalar field φ is not observable, whereas combinations such as $\varphi^*\varphi$ are and, consequently, that the transformations $\varphi \to \varphi' = \exp(i\alpha)$, $\partial_\mu \alpha = 0$, are gauge transformations in the sense that they connect different descriptions of the same physical state. While such a line is defensible, it leads to peculiar consequences in cases that would normally be described as cases of spontaneous symmetry breaking. In the first place, assuming that $\varphi \to \varphi' = \exp(i\alpha)$ is a symmetry of the Lagrangian, the stance under discussion implies that the symmetry isn't broken in the relevant sense: the world doesn't break the symmetry by choosing one from among an infinity of physically distinct states; it is rather that we make a conventional choice from among different gauges describing the same physical situation. In the second place, since different values of α label unitarily inequivalent representations, it follows from the stance under discussion that these different representations must be treated as physically equivalent. This is not an unknown position. For example, some of the pioneers of algebraic QFT promoted the concept of *weak equivalence* (see appendix) of representations as the explication of physical equivalence, and they proved that all faithful representations of a C^*-algebra are weakly equivalent. However, there are reasons to doubt that weak equivalence is sufficient for physical equivalence (see Arageorgis *et al.*, 2002) and, thus, reasons to doubt the current line on gauge.

6 Conclusion

In addition to its implications for the topic of laws and symmetries, spontaneous symmetry breaking is a marvellous vehicle for probing some of the deepest and most important problems in the foundations of QFT. But by the same token, since these problems are notoriously difficult and contentious, they collectively provide a barrier to a full understanding of the basis and ramifications of spontaneous symmetry breaking. This is, perhaps, a large part of the reason that there is little of substance in the philosophical literature about spontaneous symmetry breaking in QFT. If this rough guide to spontaneous symmetry breaking serves to encourage philosophers of science to explore the topic, it will have fulfilled part of its purpose. The other part will have been fulfilled if it serves to indicate to physicists how the algebraic formulation of QFT, though useless for calculations, helps to clarify foundational issues.

Appendix

A *C^*-algebra* $\mathcal{A}$ is an algebra, over the field $\mathbb{C}$ of complex numbers, with an involution $*$ satisfying: $(A^*)^* = A$, $(A + B)^* = A^* + B^*$, $(\lambda A)^* = \bar{\lambda} A^*$ and $(AB)^* = B^* A^*$ for all $A, B \in \mathcal{A}$ and all complex λ (where the overbar denotes the complex conjugate). In addition, a C^*-algebra is equipped with a norm, satisfying $\|A^* A\| = \|A\|^2$ and $\|AB\| \leq \|A\| \, \|B\|$ for all $A, B \in \mathcal{A}$, and is complete in the topology induced by that norm. It is assumed here that $\mathcal{A}$ contains a unit 1 such that $1A = A1 = A$ for all $A \in \mathcal{A}$. Observables are identified with self-adjoint elements of $\mathcal{A}$, i.e. elements A such that $A^* = A$. A *state* on $\mathcal{A}$ is a linear functional ω that is normed ($\omega(1) = 1$)) and positive ($\omega(A^* A) \geq 0$ for all $A \in \mathcal{A}$). ω is a *pure state* iff it cannot be written as $\lambda_1 \omega_1 + \lambda_2 \omega_2$ where $\omega_1 \neq \omega_2$ and $0 < \lambda_1 < 1$ and $0 < \lambda_2 < 1$.

A *representation* of a C^*-algebra $\mathcal{A}$ is a mapping $\pi : \mathcal{A} \to \mathcal{B}(\mathcal{H})$ from the abstract algebra into the concrete algebra $\mathcal{B}(\mathcal{H})$ of bounded linear operators on a Hilbert space $\mathcal{H}$ such that $\pi(\lambda A + \mu B) = \lambda \pi(A) + \mu \pi(B)$, $\pi(AB) = \pi(A)\pi(B)$, and $\pi(A^*) = \pi(A)^\dagger$ for all $A, B \in \mathcal{A}$ and all $\lambda, \mu \in \mathbb{C}$. A representation is *faithful* if $\pi(A) = 0$ implies $A = 0$. A fundamental theorem due to Gelfand, Naimark, and Segal (GNS) guarantees that for any state ω on $\mathcal{A}$ there is a representation $(\pi_\omega, \mathcal{H}_\omega)$ of $\mathcal{A}$ and a cyclic vector $|\Psi_\omega\rangle \in \mathcal{H}_\omega$ (i.e. $\pi_\omega(\mathcal{A})|\Psi_\omega\rangle$ is dense in $\mathcal{H}_\omega$) such that $\omega(A) = \langle \Psi_\omega | \pi_\omega(A) | \Psi_\omega \rangle$ for all $A \in \mathcal{A}$; moreover, this representation is the unique, up to unitary equivalence, cyclic representation. Two representations $(\pi_1, \mathcal{H}_1)$ and $(\pi_2, \mathcal{H}_2)$ of a C^*-algebra $\mathcal{A}$ are said to be *unitarily equivalent* just in case there is an isomorphism $\hat{U} : \mathcal{H}_1 \to \mathcal{H}_2$ such that $\hat{U} \pi_1(A) \hat{U}^{-1} = \pi_2(A)$ for all $A \in \mathcal{A}$.

The *kernel* $Ker(\pi)$ of a representation $(\mathcal{H}, \pi)$ of $\mathcal{A}$ is defined as $\{A \in \mathcal{A} : \pi(A) = 0\}$. Two representations are said to be *weakly equivalent* iff their kernels are the same. Unitary equivalence of representations implies weak equivalence, but not conversely. Obviously all faithful representations of a given C^*-algebra are weakly equivalent. It turns out that two representations $(\mathcal{H}_1, \pi_1)$ and $(\mathcal{H}_2, \pi_2)$ of $\mathcal{A}$ are weakly equivalent iff for any algebraic state ω_1 corresponding to a density matrix on $\mathcal{H}_1$ and any $A_1, \ldots A_n \in \mathcal{A}$ and any $\varepsilon_1, \ldots, \varepsilon_n > 0$, there exists a state ω_2 corresponding to a density matrix on $\mathcal{H}_2$ such that for all $i = 1, \ldots, n$, $|\omega_1(A_i) - \omega_2(A_i)| < \varepsilon_i$. This finite operational equivalence is arguably not sufficient for full physical equivalence.

References

Aitchison, I. J. R. (1982). *An Informal Introduction to Gauge Field Theories*. Cambridge: Cambridge University Press.

Arageorgis, A., Earman, J., and Ruetsche, L. (2002). 'Weyling the time away: the non-unitary implementability of quantum field dynamics on curved spacetime'. *Studies in the History and Philosophy of Modern Physics*, **33**, 151–84.

Bratteli, O. and Robinson, D. W. (1987). *Operator Algebras and Quantum Statistical Mechanics*, Vol. 1, 2nd edn. New York: Springer-Verlag.
(1996). *Operator Algebras and Quantum Statistical Mechanics*, Vol. 2. New York: Springer-Verlag.
Coleman, S. (1985). *Aspects of Symmetry*. Cambridge: Cambridge University Press.
Fabri, E., and Picasso, L. E. (1966). 'Quantum field theory and approximate symmetries'. *Physical Review Letters*, **16**, 409–10.
Giere, R. (1999). *Science Without Laws*. Chicago, IL: University of Chicago Press.
Glimm, J. M., and Kadison, R. V. (1960). 'Unitary operators in C^*-algebras,' *Pacific Journal of Mathematics*, **10**, 547–56.
Goldstone, J., Salam, A., and Weinberg, S. (1962). 'Broken symmetries'. *Physical Review*, **127**, 965–70.
Greenberger, D. M. (1978). 'Exotic elementary particle phenomena in undergraduate physics – spontaneous symmetry breaking and scale invariance'. *American Journal of Physics*, **46**, 394–8.
Guralnik, G. S., Hagen, C. R., and Kibble, T. W. B. (1968). 'Broken symmetries and the Goldstone theorem'. In *Advances in Particle Physics*, ed. R. L. Cool and R. E. Marshak, Vol. 2, pp. 567–708. New York: Interscience.
Kosso, P. (2000). 'The epistemology of spontaneously broken symmetry'. *Synthese*, **122**, 359–76.
Streater, R. F. (1965). 'Spontaneous breakdown of symmetry in axiomatic theory'. *Proceedings of the Royal Society of London, A*, **287**, 510–18.
van Fraassen, B. C. (1989). *Laws and Symmetry*. Oxford: Oxford University Press.
Wald, R. M. (1994). *Quantum Field Theory on Curved Spacetime and Black Hole Thermodynamics*. Chicago, IL: University of Chicago Press.
Weinberg, S. (1996). *Quantum Theory of Fields*, Vol. 2. Cambridge: Cambridge University Press.
Wigner, E. P. (1959). *Group Theory and its Application to Quantum Mechanics*. New York: Academic Press.
(1967). *Symmetries and Reflections*. Bloomington, IN: Indiana University Press.

21

Spontaneous symmetry breaking: theoretical arguments and philosophical problems

MARGARET MORRISON

1 Introduction

Arguments regarding the ontological status of symmetries typically involve questions such as the following: how does the mathematics of symmetry relate to the matter of the physical world and do we have good reasons for thinking that the symmetries inherent in the mathematical structure of our theories have a counterpart in the physical world? In cases where there seems to be a corresponding relation between the symmetries present in the physical system (e.g. rotational and translational symmetries) and the symmetries in the equations that govern this system, one might think the relation is relatively straightforward and that the former is simply an empirical manifestation of the latter.[1] But our questions are complicated by the fact that spontaneous symmetry breaking (SSB) is also a crucial feature of modern physics. In cases such as these the physical system displays none of the symmetry present in the equations that govern it. This symmetry is sometimes referred to as a hidden symmetry so the question, then, becomes one of determining whether the symmetry of the equation should be interpreted in a realistic way given that it seems to have no empirical manifestation.

But perhaps this notion of a 'hidden' symmetry should not raise philosophical worries, especially given that SSB lies at the foundation of some of the most successful theories in physics – superconductivity and quantum field theory (QFT) to name just two. However, unless we are strict hypothetico-deductivists or proponents of 'inference to the best explanation', there may be good reasons to think that the predictive power of our theories does not automatically license a realistic construal of their entire theoretical structure. The specific example I want to consider is the $SU(2) \times U(1)$ electroweak theory initially formulated by Glashow, Weinberg, and Salam (hereafter GWS) which unifies electromagnetism and the weak force via

[1] A good deal has been written recently about the empirical status of symmetries so I won't rehearse these arguments here; see for example Brading and Brown (in press), Budden (1997), and Kosso (2000).

SSB and the Higgs mechanism. Much of the argument surrounding the introduction of SSB into QFT was based on an analogy with superconductivity. But, like many analogical arguments, it isn't always clear that theoretical relations holding in one context are easily transferable to another. There are also issues specific to GWS that raise philosophical questions regarding the epistemological and ontological status of SSB. Those issues raise concern at two levels. The first involves particular difficulties associated with the Higgs vacuum while the second is the more general worry that the phenomenon of SSB rests on a foundation that is ultimately non-empirical, resulting primarily from assumptions about the vacuum that are not capable of being empirically confirmed.[2] Consequently, if the theoretical picture underlying SSB is itself called into question there will, as a result, be corresponding reasons for questioning any realistic interpretation of the hidden symmetry that is allegedly broken (in this case the $SU(2) \times U(1)$ symmetry) and the kinds of claims we can make about its status as a part of the physical world.[3]

It is perhaps important to point out here that one can give two types of arguments for a realistic interpretation of symmetries. The first deals with claims about correspondences between the mathematics of our theories and the world. Here we rely on evidence that the world in fact displays the mathematical relations embedded in our theories. This is achieved in the way I mentioned above, via some type of direct or indirect comparisons (like those between spacetime symmetries and objects that possess these symmetries), or by relying on the predictive power of the symmetries to produce observable consequences and then inferring the existence of the symmetries themselves. I will briefly mention these kinds of arguments in the following section. However, my goal in the paper is to provide a different kind of analysis of the problem. I want to focus on a second type of argument, one that raises questions about the overall theoretical picture of how the symmetries function and their character within the context of a *specific* theory. These kinds of issues I take to be problematic since they speak not simply to philosophical disagreements but also to fundamental theoretical issues that bear directly on the legitimacy of specific hypotheses relating to the symmetries.

2 Mathematics and the physical world

What exactly is involved in the claim that symmetries are part of the physical structure of the world? There are a few possibilities that come to mind. We can have empirical evidence that they exist by observing them or observing that a certain transformation has taken place; or symmetries can be a source of explanatory or

[2] I use the term 'Higgs vacuum' in the following way: we say that the electromagnetic vacuum is the ground state of the quantized electromagnetic field. In the electroweak case the Higgs field is postulated as existing *throughout* the physical vacuum, which is the analogue of the many-body ground state.

[3] Brown and Cao (1991) provide a nice account of the rediscovery and integration of SSB in quantum field theory.

predictive power. In other words, given certain effects or the existence of particular phenomena the only explanation of their occurrence or existence is the presence of certain kinds of symmetries. In other words, the existence of symmetries is presupposed in order to explain why the world behaves as it does. Typically we say a symmetry exists whenever the laws of physics are unaffected by a change in the reference frame. There are many different kinds of symmetries, all of which are amply discussed in various papers in this volume. As a result I will focus only on internal local symmetries, specifically the $SU(2) \times U(1)$ symmetry associated with SSB in GWS.

Local symmetries involve transformations that are variable and can differ at different points in space and time. A specific instance is the rotation of field components where the proton is identified as the up state of isospin-½ at one point in space and the neutron as the down state, with no guarantee that the up state will be the same at some other location. The unique aspect of all local symmetries is that their variability essentially destroys the symmetry of the Lagrangian, which is then recoverable by introducing a connection responsible for transmitting the effects of the local change from one part of the system to another. This connection has the effect of restoring invariance and in that sense it functions as a local variable that can specify the direction of the isospin.

What the famous Yang–Mills $SU(2)$ gauge theory (1954) succeeded in showing was that the isotopic spin connection and hence the potential acts like the $SU(2)$ symmetry group. For instance, if we want to relate 'up' states of a particle at different locations we can ask how much the up state at x needs to be rotated so that it is oriented in the same direction as the up state at y. So, if the particle is moved through the potential field from x to y, its isotopic spin direction will be rotated by the field. Since the components of isospin can be transformed into one another by the elements of the $SU(2)$ group the connection must be capable of performing the same isospin rotations as the group.[4] From the point of view of physical dynamics the interesting issue is how gauge invariance and the group-theoretic structure

[4] What exactly is the relation between the $SU(2)$ group structure and the isospin connection? The components of isospin are transformed by the elements of the group, consequently the connection (potential) is capable of performing the same isospin transformations as the transformations of the $SU(2)$ group. In order to see how the potential generates a rotation in the isospin space we need to first consider how the potential relates to a rotation. A three-dimensional rotation of a wavefunction is $R(\theta)\psi = e^{-i\theta L}\psi$ where θ is the angle of rotation and L is the angular momentum operator. It is important to note that the potential is not a rotation operator and instead is considered to be a 'generator' of a rotation. Consequently the potential must include three charge components corresponding to three angular momentum operators L_+, L_-, L_3. $A_\mu = \sum A^i_\mu(x)\, L_i$ is a linear combination of the angular momentum operators where the coefficients A^i_μ depend on spacetime position. So, the potential acts like a field in spacetime and an operator in the abstract isospin space. Each of the potential components behaves like an operator, as in the case of the raising operator L_+ which transforms a down state into an up state. This formal operation corresponds to the case of a neutron absorbing a unit of isospin from the gauge field and becoming a proton. The $SU(2)$ group has three generators so three gauge fields are necessary. So, in addition to the positive and negatively charged fields a third neutral component W_3 must be added to correspond to the isospin operator τ_3. We can see then how gauge invariance can dictate the form of the potential – the gauge field must, unlike the electromagnetic potential, carry electric charge. In addition, the requirement of local invariance dictates that the mass of the potential field be zero.

imposes constraints or bears on the field/particle interactions. In order to answer that question we need to know the status of the connection. In electrodynamics, where our interest is invariance under local phase transformations, the connection is associated with the electromagnetic potential A_μ which relates different phases at different points. Since A_μ is also identified with a gauge field from which the electromagnetic field can be derived it is commonly said that gauge theory, or more specifically the requirement of gauge invariance, generates a dynamics. But the sense in which a dynamics is 'generated' is rather intricate and not altogether straightforward.[5]

Very briefly, quantum-mechanical equations of motion (e.g. the Schrödinger equation) typically involve derivatives of the wavefunction $\psi(x)$ (as do many observables). These transform under local phase rotations in a way that involves a local gradient of the phase term ∂_μ which spoils local phase invariance. However, the equations of motion (and the observables) involving derivatives can be modified in a way that restores this invariance by replacing ∂_μ with the gauge covariant derivative $D_\mu \equiv \partial_\mu + ieA_\mu$ where e is the charge in natural units of the particle described by $\psi(x)$. The additional gauge field $A_\mu(x)$, which is also introduced here, transforms in a way that is precisely the form of a gauge transformation in electrodynamics, i.e.

$$A_\mu(x) \to A'_\mu(x) \equiv A_\mu(x) - (1/e)\partial_\mu\alpha(x). \tag{1}$$

Essentially gauge invariance is restored by combining the conventional derivative with a new vector field A_μ. But it is important to point out the sense in which the introduction of this field is necessary and not simply an add-on. Local phase invariance is not possible for a free particle waveequation, so in order to locally change the phase of a charged particle's wave function we need to introduce a field in which the particle moves. Or, in slightly less physical terms, local phase invariance is possible only if we have an *interacting* theory. However, we still don't have a physical dynamics for the field A_μ, but what the gauge covariant derivative does is prescribe the form of the coupling between the field and the particle. To get dynamical equations for A_μ we need to construct the appropriate Lagrangian density which requires the addition of a kinetic term involving the tensor form of the electric and magnetic fields such that $L_{em} = -1/4\, f_{\mu\nu} f^{\mu\nu}$. However, with just the Lagrangian density for ψ we get, via the gauge covariant derivative, an interaction term that shows how the charged particle will behave in the presence of the gauge field. The point here is to show that a *complete* dynamical theory is not achieved simply from the requirement of gauge invariance, but what we do get is the presence of an interaction field (term). Once the appropriate Lagrangian

[5] For additional discussion of these points see Martin, this volume.

density is constructed we then get a proper dynamical theory. As a concluding remark it is worth mentioning that in constructing the Lagrangian we need to make use of additional assumptions about what is and is not physically reasonable with respect to the field equations, but this in itself need not be problematic. If we distinguish what one 'gets' simply from gauge invariance and what the broadened notion of a gauge theory involves, we can view the additional assumptions as part of the overall conditions that go into constructing theories that are gauge invariant.

The connection between the group theoretical properties and the 'physical' aspects of the gauge theory thus turns out to be rather intricate. Although gauge invariance determines the form of the particle/field interactions, i.e. the form of the forces on the charged particle, the group-theoretic structure defines, as it were, the type of gauge theory one has. This in turn can dictate the number of fields that are required and whether they can combine with each other. In that sense the two are intimately connected, leaving us with what is seemingly a direct link between the physical dynamics of the theory and its mathematical structure; a situation that should supposedly provide us with strong evidence for the physical status of the symmetries. So, it isn't simply the case that the gauge invariance 'generates' a physical *dynamics*; rather, what gauge invariance does is demand the presence of an interaction field, on the basis of which we can go on to construct a gauge theory which may or may not be empirically adequate in its description of the phenomena. Its power lies in the fact that it provides, in some cases, a theoretical framework that could not have been constructed from the phenomenology of the physics alone.[6]

At this point it seems as though the evidence for a physical interpretation of symmetries might rest on whether the gauge theory is successful in describing/predicting the dynamics of the physical systems we are interested in. If so, one could then reason hypothetico-deductively and claim that we are indeed justified in interpreting the relevant symmetry as physically real, giving us reason to think that there exists an 'external' counterpart to the symmetry operations that take place in 'internal' space. However, there is much more to the story than this. Not only does one need to determine whether the dynamical predictions of a specific gauge theory are borne out, but in the case of the electroweak $SU(2) \times U(1)$ theory spontaneous symmetry breaking (SSB) complicates the matter by introducing a mechanism (Higgs) that breaks the symmetry at the level of the physical states while maintaining a symmetric Lagrangian. This seems to complicate the situation by severing any

6 Although this was not the case with electromagnetism (gauge theory added nothing to the theoretical structure that wasn't already in place) it was true of the electroweak theory. Because the bosons are massive particles while the photon is massless it is difficult to see how any theoretical unification would have been forthcoming on the basis of phenomenological considerations alone.

direct predictive or explanatory links between the symmetrical equations and any empirical manifestation of the symmetry. Instead, the Higgs mechanism is introduced as a separate and distinct cause of the symmetry breaking. Now the question of whether we should interpret this hidden symmetry in a realistic way is bound up with the legitimacy of the SSB story. Some criteria for evaluating its success are: Does the theory yield successful predictions? Are its assumptions coherent and independently testable/verifiable? While the former is partially satisfied I want to claim that the latter is clearly not. In order to argue that point I now turn to a discussion of SSB in GWS.

3 Spontaneous symmetry breaking: how it all began

The idea of a spontaneously broken symmetry needs to be distinguished from the more straightforward notion of a broken symmetry. In the latter case the symmetry is approximately true of the laws of nature or true for some interaction laws but not others. For example, isospin symmetry holds for weak interactions but is violated by electromagnetic forces. Here the term broken means that there are terms in the Hamiltonian that violate symmetry; neither the Hamiltonian nor the physical world exhibits symmetry. In the case of hidden or SSB the solutions to the equations of motion (the physical states) have less symmetry than the equations themselves. We say that the symmetry is spontaneously broken because once the theory develops degenerate vacua (non-unique vacuum states) the symmetry can be broken without external perturbations. Yet, there is also a sense in which the symmetry here is not really broken because the equations of motion remain symmetrical. We can see then why some might think that the straightforward idea of a broken symmetry is not inherently problematic from a philosophical point of view. We simply say that in some cases the symmetry holds and in some cases not. The spontaneously broken symmetry, however, presents us with a different concern. The empirical evidence points to an asymmetrical physical state while at the same time we want to affirm the existence of a hidden symmetry based on the form of the equations of motion. How, then, can we justify the ontological and epistemological status of hidden symmetry as a fundamental feature of the physical world?

Attempts to understand weak interactions in terms of a gauge theory (Schwinger, 1957; Glashow, 1961) were based on the idea that leptons carried a 'weak' isospin analogous to the strongly interacting particles discussed by Yang and Mills (1954). On this assumption weak interactions could be considered invariant under rotations in the isospin space and the corresponding Lagrangian would be invariant under the group of isospin rotations characterized by the $SU(2)$ group. By analogy with electromagnetism where the force is mediated by photon exchange, weak decay was supposedly mediated by the exchange of a spin-1 quantum called the W boson.

This had the added advantage of making it appear that weak and electromagnetic processes could be described using the same kind of theory.[7] The most serious problem with the W particle was that its mass prevented the theory from being renormalizable since renormalizability requires, minimally, that we have a gauge-symmetric Lagrangian. But gauge invariance allows only massless particles, so the W would effectively break the symmetry. Hence one obtained an asymmetric solution (massive W bosons) from a symmetric theory (gauge theory). In order for the theory to work it had to be possible for the gauge particles (W) to acquire a mass in a way that would preserve gauge invariance. In other words, the boson masses could not be added by hand because that would break the symmetry of the Lagrangian – there had to be some systematic way of introducing mass terms that didn't destroy the gauge invariance of the Lagrangian. Moreover, in order for electromagnetism and the weak force to be describable using the same theory, some account of the coupling of massive and massless particles was crucial. The answer to these problems was provided by a mechanism known as spontaneous symmetry breaking.

Hopes of achieving a true synthesis of weak and electromagnetic interactions came in 1967 with a paper on leptons by Steven Weinberg. Weinberg focused on leptons because they interact only with photons and the bosons that mediate the weak force. One of the most important assumptions of the gauge theory of weak interactions is that the lepton pairs can be considered weak isospin doublets like the neutron and proton. Weinberg extended the $SU(2)$ weak isospin symmetry group to form the larger $SU(2) \times U(1)$ group, an idea originally proposed by Glashow. This larger group structure allowed for the unification of weak and electromagnetic interactions and the mixing of the fields but the remaining problem was one of generating the appropriate masses for the $W^{\pm}$ and the Z^0 particles/fields. Weinberg's idea was that one could understand the mass problem and the coupling differences by supposing that the symmetries relating the two interactions were exact symmetries of the Lagrangian that were somehow broken by the vacuum. That is to say, the situation supposedly described by the symmetric Lagrangian does not hold of the physical system. Since the core of my argument centres on the difficulties associated with the vacuum in GWS I will discuss these issues at length below. In the meantime, however, let me mention briefly the main historical facts crucial to the story.[8]

Ideas about SSB were introduced into QFT from work in solid-state physics on superconductivity. Landau introduced the idea of a non-vanishing macroscopic

[7] There were, however, significant differences between the photon and the W boson. The boson had to exist in at least two charged states, W^+ and W^-, since all weak interactions involved the exchange of electric charge between the currents, and the boson had to be extremely massive in order to explain the short range of the weak force.

[8] See Jona-Lasinio, this volume. My discussion mentions only a few aspects of that history, enough to give the reader a sense of the analogy between superconductivity and QFT, and the plausibility of introducing fields with non-zero average values.

symmetry-breaking parameter in his work on phase transitions in the 1930s and again later in the phenomenological theory of superconductivity developed with Ginzburg (see Ginzburg and Landau, 1965). This idea of a parameter that measures the extent of symmetry breaking is also found in Heisenberg's work on the infinite ferromagnet. In the ferromagnetic case the basic interactions between the constituents of the system and the Hamiltonian describing the spin–spin interaction are rotationally invariant while below the ferromagnetic transition temperature the ground state is not; instead a particular direction in space is singled out as the preferred one due to the alignment of the spins. This is possible because the symmetry of the Hamiltonian does not require that the ground state be also symmetrical, unless it is unique or non-degenerate. The situation where the ground state configuration does not display the symmetry of the Hamiltonian is referred to as a symmetry that is spontaneously broken. Here again the symmetry is simply 'hidden' since there are no non-symmetric terms added to the Hamiltonian. In a case where, for example, an external magnetic field term is added, resulting in the selection of one specific direction, rotational invariance is explicitly broken. But in the case where any direction is equally good the symmetry still exists – it is merely hidden by the association of one of the possible directions with the ground state of the ferromagnet. The important point is that in the ferromagnetic case there exists an infinitely degenerate set of ground or vacuum states, each of which can support a complete set of quantum states that describe the same physics. The magnetization is considered the parameter that measures the extent of the symmetry breaking.

In his work on phase transitions Landau showed that spontaneous symmetry breaking occurs whenever different phases have different symmetry. He generalized the idea of a symmetry-breaking parameter by introducing the notion of a characteristic order parameter that vanishes in the symmetrical phase and is non-zero in the asymmetrical phase. In superconductivity this parameter is replaced by an effective wavefunction ψ as the characteristic function for the superconducting electrons. Like the order parameter, if ψ does not vanish SSB has occurred and the superconducting state characterized by ψ is asymmetrical. The microscopic Bardeen–Cooper–Schrieffer (BCS) theory interpreted ψ as an energy gap that exists just above the lowest energy value in the spectrum (see Bardeen *et al.*, 1957). This gap was thought to arise as the result of a phenomenon called Cooper pairing which involved the pairing of electrons with opposite momenta and spins. The gap was the finite amount of energy required to break the pairing.

These ideas were taken by Nambu who together with Jona-Lasinio extended them into the domain of particle physics, using, once again, Heisenberg's model of SSB developed to account for the breaking of isospin symmetry in ferromagnetism (see Heisenberg, 1956; Nambu and Jona-Lasinio, 1961). Although that model had originally been unsuccessful, Nambu and Jona-Lasinio used it as the basis for an analogy between the physical vacuum of a QFT and the ground state of an interacting many-body system – both of which are states of minimum energy.[9] The specific details of the analogy are not important for our purposes here; however, what *is* important are the physical assumptions made about the vacuum, assumptions that were crucial for the SSB idea to be successfully translated into QFT. The fundamental hypothesis that allowed Nambu to utilize Heisenberg's notion of a degenerate vacuum was that the vacuum was a plenum packed with many virtual degrees of freedom. The plenum idea was based on an analogy between the Fermi sea of electrons in a superconducting metal and the sea of electrons in the Dirac vacuum. In order to determine the legitimacy of these ideas and their role in QFT we need to say more about this notion of a plenum and the role it plays with respect to the Higgs mechanism.

4 At sea in the vacuum

The idea that the fermion vacuum was a 'filled sea' emerged out of an attempt to give a physical interpretation to Dirac's relavitistic wave equation. Essentially the difficulty was that certain solutions to the equation involved negative energies, which meant that matter would be essentially unstable. As a way of preventing the transition from positive to negative energy states, Dirac suggested that all the negative states were occupied thereby preventing, due to the Exclusion Principle, no further transitions. Despite the many problems with this picture it turned out to be a very useful heuristic (see Aitchison, 1985). For our purposes, however, the important point is the extension of the idea of a filled sea to the physical picture of the vacuum as plenum. Although the plenum concept can explain, among other things, electron–positron pair creation, there is no direct sense in which this picture can be said to have experimental support. Phenomena such as the energy fluctuations that accompany the Lamb shift and the Casimir effect are taken as evidence that the vacuum is not empty. Yet there is a significant difference between affirming the reality of these so-called vacuum effects and the notion of a plenum, which allows for the possibility of a *manifold* of vacuum states required for SSB. The question that emerges as significant is whether one should think of the Dirac vacuum as a mathematical device or a physical reality. Nambu certainly understood it in the latter way and indeed one of his main concerns was that the 'arguments for the existence of a multiplicity of vacua be made as convincing as possible' (see Nambu, 1997).

[9] Because it appeared impossible to construct a fully symmetrical vacuum state, Heisenberg suggested that it shouldn't really be considered a vacuum but rather a 'world state' which forms the substrate for the existence of elementary particles.

Although Nambu's idea was implemented with some success it had one major drawback. Goldstone, a high energy physicist, showed that cases of spontaneous symmetry breaking involve the production of a massless, spinless particle, that came to be known as the Goldstone boson (see Goldstone, 1961). For each degree of freedom in which the symmetry is spontaneously broken, a massless scalar field always appears. Since no such particles were thought capable of existing it was difficult to see how the idea of spontaneous symmetry breaking could be applied to field theory. But, once again, a clue was forthcoming from the field of superconductivity. In that context symmetry breaking does not give rise to Goldstone excitations, a result that is directly traceable to the presence of the long-range Coulomb interaction between electrons. The first indication of a similar effect in relativistic theories was provided by Anderson (1963), who showed that the introduction of a long-range field such as the electromagnetic field might serve to eliminate massless particles from the theory. Basically, what happens is that in cases of local symmetries that are spontaneously broken the zero-mass Goldstone bosons can combine with the massless gauge bosons to form massive vector bosons, thereby removing the undesirable feature of massless particles. This result was fully developed for relativistic field theories by Higgs (1964a; 1964b; 1966) and others who showed that Goldstone bosons could be eliminated by a gauge transformation so that they no longer appeared as physical particles. Instead they emerge as the longitudinal components of the vector bosons.

The Higgs model is based on the idea that even the vacuum state can fail to exhibit the full symmetry of the laws of physics. If the vacuum state is unique, meaning that there is only one state of lowest energy, then it must exhibit the full symmetry of the laws. However, the vacuum may be a degenerate (non-unique) state such that for each unsymmetrical vacuum state there are others of the same minimal energy that are related to the first by various symmetry transformations that preserve the invariance of physical laws. The phenomena observed within the framework of this unsymmetrical vacuum state will exhibit the broken symmetry even in the way that the physical laws appear to operate. The analogue of the vacuum state for the universe is the ground state of the ferromagnet. Although there is no evidence that the vacuum state for the electroweak theory is degenerate it can be made so by the introduction of the Higgs mechanism, which involves the artificial insertion of an additional field. This field has a definite but arbitrary orientation in the isospin vector space (thereby breaking the symmetry of the vacuum) as well as properties that would make the vacuum state degenerate. And, not only is the symmetry-breaking Higgs mechanism able to rid the theory of the Goldstone boson but in doing so it solves the mass problem for vector bosons while leaving the theory gauge invariant.

The Higgs field (or its associated particle the Higgs boson) is really a complex $SU(2)$ doublet consisting of four fields that are required to transform the massless gauge fields into massive ones. So, in addition to the four scalar quanta we also have four vector gauge quanta (three from $SU(2)$ and one from $U(1)$). A massless gauge boson like the photon has two orthogonal spin components transverse to the direction of motion while massive gauge bosons have three, including a longitudinal component in the direction of motion. Three of the four Higgs fields are absorbed by the W^+, W^-, and Z^0, and appear as the longitudinal polarization in each of the three quanta. Since longitudinal polarization in a vector field is equivalent to mass we obtain three massive gauge quanta with charges $+$, $-$, and 0. The fourth vector quantum remains massless and is identified as the photon. Of the four original scalar bosons, only the Higgs boson remains. The four gauge fields and the Higgs field undergo unique, non-linear interactions with each other and because the Higgs field is not affected by the vector bosons it should be observable as a particle in its own right.

5 What's wrong with this picture?

As a theoretical story the GWS model presents a coherent and predictively successful account of the unification of weak and electromagnetic interactions. SSB via the Higgs mechanism functions as the cornerstone of the theory, yet a proper assessment of its success requires that we look beyond the theory's predictions to the assumptions, particularly the vacuum assumptions, necessary for SSB to play this rather significant role. At this point one might object that although these vacuum assumptions were important *historically* for the transition of SSB from superconductivity to QFT, the approach now is much different. What is important is the choice of the Lagrangian, which then allows us to derive everything we are interested in, including ground state properties. So, in that sense hypotheses about the nature of the vacuum state are largely irrelevant. Although clearly correct at one level, the objection runs the risk of collapsing the distinction between the theoretical background and the methodology one uses in 'doing' the physics. In other words, it ignores the fact that in order for the SSB story as presented in GWS to work, one needs the underlying vacuum assumptions regarding the plenum and degeneracy as part of the 'physical' picture. In other words, these assumptions are still present although not part of the explicit methodology of the theory; and, insofar as that is the case, their status is subject to scrutiny.

We know the vacuum isn't really a 'thing' in the classical sense of the word but rather a state of minimum energy referred to as a ground state. That is, the vacuum state is one where the average value of the quantum fields is zero, which means that the vacuum contains no particles or excited states. Having said that, though, we must immediately make some qualifications. Although the *average value* of quantum fields in the vacuum is zero there are cases where the *mean square values* of the fields are non-zero, resulting in fluctuations that exert a measurable effect. Such is the case with the Casimir effect, the Lamb shift, and several other phenomena. The fluctuations are thought to be the result of the zero-point energy of the field. In classical mechanics the state of lowest energy would have zero energy but in quantum mechanics the uncertainty principle tells us that if a particle is confined to a particular region of linear dimension a then its momentum will have values of the order $p_0\hbar/a$ which means that its energy will be of the order $E_0 = p_0^2/2m \;\; \hbar^2/2ma^2$. In other words, quantization prevents us from removing all the energy from the system – even in the ground state there is some motion, i.e. zero-point vibrations or oscillations which result from the localization caused by the oscillator potential.[10] In that sense then, quantum mechanics tells us that we cannot have a vacuum that is empty of all mass and energy because that would entail that we had information about the motion and energy of the system at a given time – a violation of the uncertainty principle.

In the case of the Casimir effect we have a free (non-interacting) quantum field which is typically represented as a sum of an infinite number of modes. The zero-point energy of the field will depend on the mode frequencies, which in turn depend on the boundary conditions imposed on the field. Although the Casimir effect allegedly proves the existence of a base level of electromagnetic oscillation (or zero-point energy) in the vacuum, the physical explanation is not without its problems.[11] The primary difficulty is the presence of macroscopic objects whose function is to modify the boundary conditions. As far as I know no satisfactory understanding of the microscopic mechanism responsible for these boundary conditions has been forthcoming (see Milonni, 1994; Mostepanenko and Trunov, 1997). A quantum-mechanical account of the interaction between the radiation field and the atoms in the conducting surfaces is needed in order to fully understand the physical basis of zero-point energy. So, to that extent, the existence of the effect doesn't tell us much about the structure of the vacuum, except that there is evidence to suggest that it is non-empty.

But the presence of zero-point energy in the vacuum is not really the issue here since the kinds of assumptions required for the Higgs mechanism and SSB extend

[10] It is probably worth noting that Aitchison (1985) points out that the equality between the uncertainty argument and the exact value for the ground state energy of the oscillator is a fiddle; π and factors of two, etc., have been ignored.

[11] Milonni (1994, p. 250), for example, points out that in the case of the Lamb shift the interpretation of the Casimir force in terms of the vacuum field is largely a matter of taste: underlying that interpretation is a particular and arbitrary choice of ordering of field operators. Different orderings reveal that the vacuum-field picture is only one of many ways to describe the effect. An alternative intepretation in terms of source fields emerges from a different but normal ordering.

beyond fields whose *mean square value* is non-zero. Instead we are dealing with fields whose *average value* is non-zero, where the vacuum is said to have non-zero expectation value. This kind of field is called an order parameter and is characteristic of some specially ordered ground state, as in the case of the ferromagnet where the average value of the magnetization is non-zero. It is these latter kinds of fields that are required for the SSB/Higgs mechanism story.

The Higgs mechanism requires that we extend the Nambu analogy between the particle physics vacuum and superconductivity one step further to include the existence of a matter field as an analogue to the Cooper pairs. What this means, essentially, is that a superconductor functions as the model for the weak force vacuum. While this may at first seem a reasonable assumption given the initial success of the analogy, a closer look reveals the extent to which this hypothesis embodies a somewhat dramatic shift in our physical picture of the vacuum. In a superconductor currents are set up in the surface of the material producing a **B** field, which below critical temperature cancels the field in the body of the superconductor. These currents are produced by the charged current-carrying matter in the superconductor. The analogy therefore requires the existence of a weak current-carrying matter field, which must be present in the vacuum. It is important here to emphasize that this is not a fluctuation phenomenon of the sort observed in the Casimir effect; there the assumption was that the average value of all fields in the vacuum was zero. In the case of these matter fields, however, we have a non-zero vacuum value, that is, the vacuum state is one in which some field quantity has a non-zero average value (which we referred to above as an order parameter). Although this phenomenon is present in condensed matter physics in the form of the Cooper pair wavefunction and in the ferromagnet below its transition temperature (where spontaneous magnetization occurs), there is no evidence to suggest that the weak vacuum has these same properties. Unlike the zero-point energy phenomena, a hypothesis that can be derived directly from quantum mechanics, the existence of non-zero average value fields in QFT has no dynamical basis – the story is purely phenomenological.[12]

There are other difficulties both physical and philosophical that plague the Higgs hypothesis and its connection to the vacuum. Although the two kinds of concerns are not unrelated I want to focus specifically on the philosophical reasons why one might approach this theoretical picture with some scepticism, reasons that are not, strictly speaking, grounded in the fact that the Higgs particle has not been found. But first let me briefly mention the 'physical' issues.

[12] One possible way around the problem is dynamical symmetry breaking. However, the problem with these models is that there is, as far as I know, no way to construct a *phenomenologically* acceptable model that gives mass to the quarks and leptons. Castellani (this volume, Part III) provides a discussion of dynamical symmetry breaking.

Two theoretical problems associated with the Higgs field are the hierarchy problem and the cosmological constant problem.[13] The former relates to the different values of the boson masses introduced in the electroweak unification and those introduced in the unification of the weak, strong and electromagnetic forces in quantum chromodynamics (QCD). Without going into specifics, the different energy scales in the two types of interactions will induce mass shifts in particles causing all masses including the W and Z to be increased to that of the grand unified bosons. As yet there is no generally accepted satisfactory way for this hierarchy in scales to be maintained. Some separation of the electroweak and grand unified Higgs fields seems to be required in order to address the problem, but in order for the theory to be renormalizable there needs to be fundamental couplings of the two types of fields. One solution is to regard the Higgs field as a bound state of some hitherto unknown matter fields. This would mean it had a characteristic size, thereby providing a physical cut-off to the scale of corrections to its mass. Although this might seem a promising strategy, no acceptable model has been forthcoming.[14] The other problem has to do with the vacuum energy density of the Higgs field. Because of the non-zero values of some fields in the vacuum a term is added to its potential energy. This results in an energy density that would be constant throughout all of space and one that would have significant cosmological implications, especially with respect to inflation. However, at all events the fact remains that the vacuum value of V (the potential energy) must be nearly zero now.

Underlying these technical and theoretical difficulties is a fundamental philosophical problem: all of the physics we have described so far ultimately rests on crucial assumptions about the nature of the vacuum, yet these assumptions are, in a very significant sense, not subject to direct empirical confirmation. Consequently, their status becomes philosophically questionable at best. In the case of SSB we have a series of assumptions each of which builds on the previous one, culminating in a theoretical picture of what the vacuum must be like in order for the SSB story to work as GWS says it does. Beginning with Nambu's plenum hypothesis we move to the idea of a degenerate vacuum, neither of which enjoys any independent evidential support. The degeneracy hypothesis is then put on a slightly stronger foundation by introducing the Higgs mechanism, which acts to break the symmetry of the vacuum by having a definite but arbitrary orientation in space. Although the Higgs particle has not been found, its discovery would allegedly go some way towards lending credibility to the vacuum hypotheses and hence the entire SSB edifice. But that would not signal the end of the problem.

13 Most texts on QFT discuss the hierarchy problem; see for example Kaku (1993). See also Aitchison (1985; 1991). The cosmological constant problem is also widely discussed in texts on particle physics and cosmology. See for example Collins *et al.* (1989) and, more recently, Allday (2002).

14 Here again dynamical symmetry breaking affords a possible, though as yet ultimately unsatisfactory, explanation.

If we return for a moment to the superconductivity analogy we see that in that context there is a microscopic theory (BCS), which allows us to derive the phenomenological London account from which we get the notion of an order parameter.[15] The analogue of this, the Higgs field, differs from the Cooper pair wavefunction in that it is assumed to exist *throughout* the vacuum as a kind of many-body ground state. However, because neither the BCS theory nor the Higgs vacuum for QCD are based on perturbation theory it becomes extremely difficult to envision a BCS-type account for the weak vacuum that would justify the Higgs phenomenology.[16] In the BCS case we have a binding mechanism responsible for the Cooper pairing, but without a microscopic theory of the vacuum we have no way of knowing whether the Higgs field is a new elementary field or a composite, i.e. a bound system of more elementary fields. Simply finding the Higgs particle needn't provide direct answers to these kinds of questions. Indeed, even the most well-documented vacuum phenomena – zero-point energy and fluctuations – have been measured experimentally only in the electromagnetic case with no such experimental effects detected for the weak vacuum. The structure and nature of the latter remain fundamentally unknown and perhaps even unknowable.

What we have seen then is that the notion of spontaneous symmetry breaking in GSW, far from being simply an ontological question about the physical status of the $SU(2) \times U(1)$ symmetry, is one that is heavily embedded in a theoretical story about the nature of the vacuum; a story that, I want to claim, lacks the kind of independent evidence capable of supporting a realistic interpretation of SSB. My intention is not, however, to claim that this theoretical story is somehow illegitimate from the point of view of physical *methodology* – in many cases the process of theory construction relies on hypotheses that are not, themselves, well grounded. Instead my goal was to call attention to some of the hypotheses involved in the SSB story and cast doubt on a *realistic* interpretation of their claims. What I have tried to show is that in this particular context debates about the status of spontaneously broken (hidden) symmetries as compared to the more straightforward spacetime symmetries must extend beyond the kinds of direct arguments frequently offered in support of the latter and beyond the hypothetico-deductive arguments typically offered in support of theoretical entities/structure. Instead, a careful examination of the *theoretical framework* supporting the symmetry claims is required. And, as we have seen, in this particular case the various vacuum hypotheses which provide the necessary theoretical foundation are essentially problematic, for both physical and philosophical reasons. In view of this it would be folly to accept a robust physical interpretation of the SSB story and the accompanying $SU(2) \times U(1)$ symmetry on

[15] Leaving aside the problem that BCS is inadequate in accounting for high-temperature superconductivity.

[16] Aitchison (1985) points out the potential difficulty in making this connection between the Higgs phenomenology and a BCS account of the weak vacuum.

the predictive success of the GWS theory; even Maxwell himself was sceptical of the ether hypothesis from which he derived his field equations.

Acknowledgements

Support of research by the Social Sciences and Humanities Research Council of Canada is gratefully acknowledged. I would also like to thank the editors for their valuable suggestions and active participation, which is not to suggest that they agree with what I say.

References

Aitchison, I. J. R. (1985). 'Nothing's plenty'. *Contemporary Physics*, **26**, 333–91.

(1991). 'The vacuum and unification'. In *The Philosophy of the Vacuum*, ed. S. Saunders. Oxford: Oxford University Press.

Allday, J. (2002). *Quarks, Leptons and the Big Bang*. London: Institute of Physics Publishing.

Anderson, P. W. (1963). 'Plasmons, gauge invariance and mass'. *Physical Review*, **130**, 439–42.

Bardeen, J., Cooper, L. N., and Schrieffer, J. R. (1957). 'Theory of superconductivity'. *Physical Review*, **108**, 1175–204.

Brading, K., and Brown, H. R. (in press). 'Are gauge symmetry transformations observable?' *British Journal for the Philosophy of Science*.

Brown, L., and Cao, T. (1991). 'Spontaneous breakdown of symmetry: its rediscovery and integration into quantum field theory'. *Historical Studies in the Physical Sciences*, **21**, 211–35.

Budden, T. (1997). 'Galileo's ship and spacetime symmetries'. *British Journal for the Philosophy of Science*, **48**, 483–516.

Collins, P. D., Martin, A. D., and Squires, E. J. (1989). *Particle Physics and Cosmology*. New York: Wiley Interscience.

Ginzburg, V. L., and Landau, L. (1965). 'On the theory of superconductivity'. In *Collected Papers*, ed. L. Landau, pp. 546–8. New York: D. ter Haar.

Glashow, S. L. (1961). 'Partial symmetries of weak interactions'. *Nuclear Physics*, **22**, 579–88.

Goldstone, J. (1961). 'Field theories with superconductor solutions'. *Nuovo Cimento*, **19**, 154–64.

Goldstone, J., Salam, A., and Weinberg, S. (1962). 'Broken symmetries'. *Physical Review*, **127**, 965–70.

Heisenberg, W. (1956). 'Research on the non-linear spinor theory with indefinite metric in Hilbert space'. *Proceedings of the International Conference on High Energy Physics*, CERN, Geneva, pp. 119–22.

Higgs, P. W. (1964a). 'Broken symmetries, massless particles and gauge fields'. *Physics Letters*, **12**, 132–3.

(1964b). 'Broken symmetries and masses of gauge bosons'. *Physical Review Letters*, **13,** 508–9.

(1966). 'Spontaneous symmetry breaking without massless bosons'. *Physical Review*, **145**, 1156–63.

Kaku, M. (1993). *Quantum Field Theory*. Oxford: Oxford University Press.

Kosso, P. (2000). 'The empirical status of symmetries in physics'. *British Journal for the Philosophy of Science*, **51**, 81–98.
Milonni, P. W. (1994). *The Quantum Vacuum*. New York: Academic Press.
Mostepanenko, V. M., and Trunov, N. N. (1997). *The Casimir Effect and its Applications*. Oxford: Oxford University Press.
Nambu,Y. (1966). 'A 'superconductor' model of elementary particles and its consequences'. In *Proceedings of the Midwest Conference on Theoretical Physics*. Purdue University, West Lafayette, IN.
(1997). Panel session: 'Spontaneous symmetry breaking'. In *The Rise of the Standard Model*, ed. L. Hoddeson, L. Brown, M. Riordan, and M. Dresden. Cambridge: Cambridge University Press.
Nambu, Y., and Jona-Lasinio, G. (1961). 'Dynamical model of elementary particles based on an analogy with superconductivity'. *Physical Review*, **122**, 345–58.
Schwinger, J., (1957). 'A theory of fundamental interactions'. *Annals of Physics*, **2**, 407–34.
Weinberg, S. (1967). 'A model of leptons'. *Physical Review Letters*, **19**, 1264–6.
Yang, C. N., and Mills, R. (1954). 'Conservation of isotopic spin and isotopic gauge invariance'. *Physical Review*, **96**, 191–5.

Part IV

General interpretative issues

22

Classic texts: extracts from Wigner

Invariance in physical theory

EUGENE P. WIGNER

Initial conditions, laws of nature, invariance

The world is very complicated and it is clearly impossible for the human mind to understand it completely. Man has therefore devised an artifice which permits the complicated nature of the world to be blamed on something which is called accidental and thus permits him to abstract a domain in which simple laws can be found. The complications are called initial conditions; the domain of regularities, laws of nature. Unnatural as such a division of the world's structure may appear from a very detached point of view, and probable though it is that the possibility of such a division has its own limits,[1] the underlying abstraction is probably one of the most fruitful ones the human mind has made. It has made the natural sciences possible.

The possibility of abstracting laws of motion from the chaotic set of events that surround us is based on two circumstances. First, in many cases a set of initial conditions can be isolated which is not too large a set and, in spite of this, contains all the relevant conditions for the events on which one focuses one's attention. . . .

However, the possibility of isolating the relevant initial conditions would not in itself make possible the discovery of laws of nature. It is, rather, also essential that, given the same essential initial conditions, the result will be the same no matter where and when we realize these. This principle can be formulated, in the language of initial conditions, as the statement that the absolute position and the absolute time

Note. Extract from: 1967, *Symmetries and Reflections*, Bloomington, IN: Indiana University Press, pp. 3–13.

[1] The artificial nature of the division of information into 'initial conditions' and 'laws of nature' is perhaps most evident in the realm of cosmology. Equations of motion which purport to be able to predict the future of a universe from an arbitrary present state clearly cannot have an empirical basis. It is, in fact, impossible to adduce reasons against the assumption that the laws of nature would be different even in small domains if the universe had a radically different structure. One cannot help agreeing to a certain degree with E. A. Milne, who reminds us (*Kinematic Relativity*, Oxford: Oxford University Press, 1948, p. 4) that, according to Mach, the laws of nature are a consequence of the contents of the universe. The remarkable fact is that this point of view could be so successfully disregarded and that the distinction between initial conditions and laws of nature has proved so fruitful.

are never essential initial conditions. The statement that absolute time and position are never essential initial conditions is the first and perhaps the most important theorem of invariance in physics. If it were not for it, it might have been impossible for us to discover laws of nature.

The above invariance is called in modern mathematical parlance invariance with respect to displacement in time and space. Again, it may be well to remember that this invariance may have limitations. If the universe should turn out to be grossly inhomogeneous, the laws of nature on the fringes of the universe may be quite different from those which we are studying; and it is not impossible that an experimenter inside a closed room is in principle able to ascertain whether he is in the midst, or near the fringes, of the universe, whether he lives in an early epoch of the expansion of the universe, or at an advanced stage of this process. The postulate of the invariance with respect to displacement in space and time disregards this possibility, and its application on the cosmological scale virtually presupposes a homogeneous and stationary universe. Present evidence clearly points to the approximate nature of the latter assumption.

Invariance

What are the other laws of invariance? One can distinguish between two types of laws of invariance: the older ones which found their perfect, and perhaps final, formulation in the special theory of relativity, and the new one, yet incompletely understood, which the general theory of relativity brought us.

The older theories of invariance postulate, in addition to the irrelevance of the absolute position and time of an event, the irrelevance of its orientation and finally, the irrelevance of its state of motion, as long as this remains uniform, free of rotation, and on a straight line. The former theorems are geometrical in nature and appear to be so self-evident that they were not formulated clearly and directly until about the turn of the last century. The last one, the irrelevance of the state of motion, is far from self-evident, as all of us know who have tried to explain it to a layman. There would be no such principle of invariance if Newton's second law of motion read 'All bodies persist in their state of rest unless acted upon by an external force'; on the contrary, the scope of this invariance could be extended considerably if the bodies maintained their state of acceleration rather than their velocity in the absence of an external force. It is fitting that this principle was first enunciated, in full clarity, by Newton in his *Principia*.

The fact that the older principles of invariance are the products of experience rather than *a priori* truths can also be illustrated by our gradual abandonment of a very plausible principle, the principle of similitude. This principle, formulated perhaps most clearly by Fourier, demands that physical experiments can be scaled;

that the absolute magnitude of objects be irrelevant from the point of view of their behavior on the proper scale. The existence of atoms, of an elementary charge, and of a limiting velocity spelled the doom of this principle.

The formulae describing what I am calling the older principles of invariance were first given completely by Poincaré, who derived them from the equations of electrodynamics. He also recognized the group property of the older principles of invariance and named the underlying group after Lorentz. The significance and general validity of these principles were recognized, however, only by Einstein. His papers on special relativity also mark the reversal of a trend: until then, the principles of invariance were derived from the laws of motion. Einstein's work established the older principles of invariance so firmly that we have to be reminded that they are based only on experience. It is now natural for us to try to derive the laws of nature and to test their validity by means of the laws of invariance, rather than to derive the laws of invariance from what we believe to be the laws of nature.

The general theory of relativity is the next milestone in the history of invariance. The fact that it is the first attempt to derive a law of nature by selecting the simplest invariant equation would in itself justify the epithet.

The role of invariance principles in natural philosophy

EUGENE P. WIGNER

What is the role and proper place of invariance principles in the framework of the physical sciences?

. . . There is a strange hierarchy in our knowledge of the world around us. Every moment brings surprises and unforseeable events – truly the future is uncertain. There is, nevertheless, a structure in the events around us, that is, correlations between the events of which we take cognizance. It is this structure, these correlations, which science wishes to discover, or at least the precise and sharply defined correlations. . . .

If we look a little deeper into this situation we realize that we would not live in the same sense we do if the events around us had no structure. Even if our bodily functions remained unaltered, our consciousness could hardly differ from that of plants if we were unable to influence the events, and if these had no structure or if we were not familiar with some of this structure, we could not influence them. There

Note. Extract from: 1967, *Symmetries and Reflections*, Bloomington, IN: Indiana University Press, pp. 28–37.

would be no way our volition could manifest itself and there would be no such thing as that which we call life. This does not mean, of course, that life depends on the existence of the unbelievable precision and accuracy of the correlations between events which our laws of nature express, and indeed the precision of these laws has all the elements of a miracle that one can think of.

We know many laws of nature and we hope and expect to discover more. Nobody can foresee the next such law that will be discovered. Nevertheless, there is a structure in the laws of nature which we call the laws of invariance. This structure is so far-reaching in some cases that laws of nature were guessed on the basis of the postulate that they fit into the invariance structure.

It is not necessary to look deeper into the situation to realize that laws of nature could not exist without principles of invariance. This is explained in many texts of elementary physics even though only few of the readers of these texts can be expected to have the maturity necessary to appreciate these explanations. If the correlations between events changed from day to day, and would be different for different points of space, it would be impossible to discover them. Thus the invariances of the laws of nature with respect to displacements in space and time are almost necessary prerequisites that it be possible to discover, or even catalogue, the correlations between events which are the laws of nature. This does not mean, of course, that either the precision or the scope of the principles of invariance which we accept at present are necessary prerequisites for the existence of laws of nature. Both are very surprising indeed, even if not quite as amazing as the precision of some laws of nature which I am always tempted to add to Kant's starred sky above us, and the categorical imperative within us.

This then, the progression from events to laws of nature, and from laws of nature to symmetry or invariance principles, is what I meant by the hierarchy of our knowledge of the world around us. . . .

In fact, if the universal law of nature should be discovered, invariance principles would become merely mathematical transformations which leave that law invariant. They would remain, perhaps, useful tools for deriving consequences of the universal law, much as they were used to derive the qualitative rules of spectroscopy from the laws of quantum mechanics. However, if the universal law of nature should be discovered, the principles of invariance would lose their place in the hierarchy described before.

23

Symmetry as a guide to superfluous theoretical structure

JENANN ISMAEL AND BAS C. VAN FRAASSEN

Symmetries can be a potent guide for identifying superfluous theoretical structure. This topic provides a revealing illustration of the power of formal methods for illuminating the *contents* of our theories, and bears potentially on some very old philosophical problems. The philosophical and scientific literature contains a good many discussions of individual cases, but the treatment is rarely general and tends to be technically involved in a way that may bury the basic physical insight as well as making it inaccessible to philosophers. We wish to identify the sorts of symmetry that signal the presence of excess structure, and do so in a completely general way, applicable to all theories and all genres of theory.

1 What is superfluous structure?

For any entity whether concrete or abstract we distinguish its elements and its structure; the latter is specified by listing relations between the elements (equivalently, features of sets or sequences of elements). Whether or not some of its structure is superfluous is clearly an interest-relative question. A sowing machine has superfluous structure if some features of or relations between its elements are dispensable for sowing, although these may be quite relevant to it from an aesthetic or antique collectors' point of view. Each of two features may be dispensable for the given purpose, but they may not be both dispensable at once, namely if the machine has multiple features which can play each other's roles. Obviously then any machine at all – classified in terms of intended function and design – has superfluous structure. In the case of an abstract entity or intellectual product, classified in the same way, it may not be absurd to think of discarding all superfluous structure.

A physical theory provides us with descriptions and models that can be used to represent physical situations. We say that a theory has superfluous structure if it provides multiple representations for the same physical situation. Unfortunately, the way we describe and identify or distinguish physical situations tends itself to be quite theory-laden. So it is not usually possible to have also at hand an authoritative, adequate, theory-independent account of nature against which we can test for superfluous structure. What we can do is inspect the theoretical representations and look for internal evidence to suggest that several of them are so alike that to distinguish between them makes a distinction without a difference.

'So alike': thereby hangs a tale. If they are alike to that extent, they must differ only in ways that amount to physically superfluous structure. Logically, of course, there could be physical differences that correspond to no measurable or observable difference. Logic alone will not decide what is physically superfluous in a model. Metaphysical views, physical intuition, and empiricist preferences for empirical content over metaphysical distinctions will all play a role in what is identified as really superfluous. But symmetry is a guide to this identification in all cases, and we wish to display this in a sufficiently general framework for the study of theory structure.

Outside the theoretician's study our clues to physically significant differences are the observable or measurable differences. But these provide no easy guide, and as we will see, the distinction between 'observable' and 'measurable' itself turns out to matter. A theory is usually around long before we know whether it contains unmeasurable quantities, and which quantities those might be. The process of identifying unmeasurable quantities, i.e. of smoking them out of their hiding places in the theoretical apparatus, is long, hard, and highly non-trivial. It is a discovery, then, and not a happy one, that a theory contains unmeasurable quantities.

Why is there something amiss with a theory that contains unmeasurable quantities? It is not because we have any *a priori* guarantee that there are no such quantities, i.e. that our senses see right through to the bottom of things, but because they are not a proper part of the subject matter of theorizing. To isolate an unmeasurable quantity mathematically is to demonstrate that the theory to which it belongs has idle parts, wheels which turn without turning anything. But it requires a clear view of theories as well as of theory–phenomena relations to clarify this.

2 What is a theory?

A theory has two main ingredients: a *theoretical ontology* which specifies its initial (*metaphysical*) *possibility space*, and a set of *laws* which selects therefrom the *physical possibilities*. There is for any historically developed theory a great deal of leeway in how it is conceived and presented. In the case of classical mechanics, for example, we need only recall the names of Newton himself, Lagrange, Hamilton, Kirchhoff, Mach, Birkhoff, and Mackey to see the diversity possible in foundational reconstruction. Our own form of presentation is meant to help bring out, as perspicuously as possible, the roles of symmetry in conceptions of physical theory structure.

2.1 Theoretical ontology

The initial framework of a theory – its (*theoretical*) *ontology* – serves to delineate in very broad terms what will count as its initial (*metaphysical*) *possibility space.* Following common usage we will refer to the points in this space as (*possible*) *worlds.*

We will think of the ontology as specified by means of a *catalogue*, listing classes of entities, quantities, and relations which together determine the theory's parameters of representation. These serve as the 'supervenience base' for the description of nature: every possibility conceivable in this theoretical context can be conceived of entirely in terms of parameters that are either among or derivative from items in the catalogue. The items in the catalogue must be 'typed' (in the way that sets in type theory are, for example). Thus if A is one of the entities and Q a quantity, the specification that $Q(A) = r$ will make sense only if Q, A, and r are of the right type. All complete specifications of this sort identify points in the initial possibility space: the (*metaphysically*) *possible worlds.* This space is, in effect, the set of worlds obtained from the catalogue by means of any arrangement whatsoever of the theory's basic building blocks.[1]

As a first example think of the revival of ancient atomism in the sixteenth and seventeenth centuries. The world is conceived of as made up of atoms, whose number is the first basic quantity; each atom is characterized by means of the fixed list of primary qualities. Those primary qualities are actually quantities; their quantification was a crucial step for the new sciences. Initially at least, the 'mechanical philosophy' of the seventeenth century saw no need for more; any more clearly qualitative aspects of the world were merely derivative. What David Lewis and others have called a Principle of Recombination is clearly held in this context.[2] For if we take any part of the class of atoms of one world, combine it with some of those of another world, keeping in each case their primary qualities, then the result is a third (metaphysically) possible world.

If a theory is more holistic in its world picture, it is not equally straightforward to see it in this way. One way would be to think of the supervenience base as containing more complex quantities with the simpler ones as derivative. Another way, perhaps equally 'formal', would be to think of holistic properties of complex systems as

[1] In what follows, when we say 'possibility', we will always mean *metaphysical* possibility. Physical possibilities will always be explicitly identified as such.

[2] There are questions, here, about the metaphysical possibility of worlds that contain quantities or entities of kinds that are nowhere instantiated at our world (see Lewis, 1983), but these play no role in physical contexts.

external relations (i.e. relations that don't supervene on the intrinsic properties of their relata) between their parts.

Consider, for example, an *N*-body system in elementary quantum mechanics. The state of the whole is not determined by the states of its parts. In fact, as Schrödinger already pointed out, if the whole system is in a pure state, states ascribed to its parts can in general not be pure; and different pure states of the whole system are compatible with the same (mixed) states for its parts. Thus, one way to think of this is to take the state of the whole as basic and the rest as derivative. We can think here for example of taking all the states in the Schrödinger picture of *N*-body systems, for each *N*, as basic. With time then as an independent parameter, represented by the real number continuum, each possible world is a trajectory in one of the relevant state spaces. But it is also possible, in a formal sense, to think of the state of the whole as encoding the states of the parts plus certain non-supervening, non-spatial relations between its constituents that are not captured in their individual states.[3]

Even so, before any laws are introduced, and depending a bit on how much or how little we specify in the catalogue, many of these (metaphysically) possible worlds will be very strange and nothing like what we can think of as a quantum mechanically (physically) possible world. But what we will have at this point is a basic framework for which the laws can be formulated.

2.2 Laws and physical possibilities

Thus we can represent a theory as a structured set of possibilities, and this turns out to be a nice way of thinking of theories.[4] These structured sets of possibilities (heretofore, 'possibility spaces') relate in a straightforward way to spaces that physicists are accustomed to dealing with (phase spaces and configuration spaces, for instance, can be identified with subspaces of them), and they are familiar sorts of objects to philosophers.

We can think of the laws of theory as given in many different ways, but they have one simple role. They select a subset of points of the metaphysical possibility space: that selection is (or represents) the set of *physical possibilities*; we speak of these also as the *physically possible worlds*. If the laws are specified separately as conditions expressed by propositions or equations, for example, then the selected

[3] The case is complicated by the fact that the relations in question are relations between probabilities, i.e. correlations. See Mermin (1998, appendix A).

[4] Features that play an important role in some conceptions of theories (e.g. the language in which they are formulated or the mathematical form of their equations) are treated as incidental on this approach, relegated to the subsidiary role of picking out the worlds that are the real locus of interest. We don't set much store on debates about the 'right' way to represent theories; any way that contains enough information to reconstruct ontology and laws is adequate. It also doesn't much matter how we mark the physical possibilities, but it's convenient to keep the markings separate from the intrinsic structure of the space, so that structures intrinsic to the space represent only internal relations between possibilities.

worlds are precisely those points in the metaphysical possibility space that satisfy those conditions.

3 Empirical content

We must now introduce the ingredient that is external to theory structure, i.e. to the ontology and laws, namely the theory–phenomena relation. This enters in two contexts: when we are trying to decide what kind of epistemic attitude to take towards a theory, and when we are trying to interpret it.

3.1 Qualitative structure

To begin, we introduce the interpretative predicate 'qualitative'. A world has qualitative features as well as structure, but this is a distinction we impose from without, by relating its features to us, the epistemic community. Thus the distinction is not to be found inside the theory or its models; it pertains to the theory's (and its models') use. Qualitative aspects of a physical situation correspond in our terminology to parameters which characterize that situation, and are directly accessible to us through perception. When we specify implications of a theory that pertain to qualitative aspects of a situation we are linking the theory to the content of possible perceptions, hence specifying empirical content.

We realize that the equation of qualitative with perceptible features is contentious, since something more than a choice of terms may be at issue. So we will expand on this by relating these terms to each other and to further terms typically used in this context.

Traditionally the terms 'quality' and 'qualitative' did indeed have a different meaning. Thus Kant's Table of Categories lists *quality*, *quantity*, and *relation*, and distinguishes them in that traditional way. But for us today quality cannot contrast with quantity. Take any quality Q which an entity A may or may not have; we can think of Q as a quantity (a function that takes numbers or similar mathematical objects as values) such that $Q(A)$ equals 1 if A has Q and equals zero otherwise. Nor need quantities be real-number valued: a partially ordered set of determinations under a determinant can be represented by a function that maps entities into a partially ordered mathematical structure for its values. The distinction between metric and non-metric aspects of even a geometric structure is closer to the traditional quantity/quality distinction, but equally soft.[5]

[5] Coordinates can be introduced in synthetic projective geometry by construction, and a metric can be defined from these. This point was central to Russell's (1897) diatribe against those who in his view betrayed the real theory of space.

3.2 Qualitative vs. measurable

But the distinction between directly accessible and non-accessible quantities, with the former's link to observation and experience, is crucial to any evaluation of how well or poorly a theory manages to represent nature. This distinction often appears disguised in terms such as 'physical intuition', 'physically meaningful', 'physically significant', and the like; indeed, its name is legion. Thus in a recent paper by Belot (2001), which we shall discuss further in a later section, it is easy to find telling examples. In his discussion of Hertz's programme, for example, he writes (p. 72):

> in fact, it is all too easy to implement it technically. The problem is rather that known implementations have little discernible *physical* interest. The kinetic energy in question bears no straightforward relation to any physical notion.

It would be hard to find any reading for 'physical' in this context that does not imply at least 'measurable' or even 'observable' .

Like so many in the philosophical lexicon, 'observable' is an accordion term all too easily squeezed or stretched out. One use applies only to *entities* – objects, events, processes which may or may not be small enough, large enough, massive enough, etc., to be observable by us.[6] A second use, which we will adopt here, applies to quantities that can characterize a situation, distinguishable by even a gross discrimination of colour, texture, smell, and so on. These alone we shall here count as *qualitative*. They are accordingly to be distinguished from *measurable* quantities. The latter include only quantities whose values make some discernible impact on gross discrimination of colour, texture, smell, and so on, but it doesn't matter how attenuated the connection is, how esoteric the impact, or how special the conditions under which it can be discerned. It will do no harm to count qualitative quantities as measurable, but many measurable quantities will be non-qualitative. We observe therefore a tripartite distinction between qualitative, measurable but non-qualitative, and unmeasurable quantities.

Which quantities are qualitative we take to be part of our resources for fixing the interpretation of a theory. But then we may take that either to be a fact of nature, or to be contextual, with the qualitative/non-qualitative line drawn differently in different theoretical contexts. While we have views on this, the choice does not seem to affect our main discussion. The measurable/unmeasurable distinction, on the other hand, is certainly not such a brute given line, but must be determined using those resources, for each theory to be interpreted. This point relates directly to our main topic: superfluous structure will align with the presence of unmeasurable quantities in the theory's world picture.

[6] This is the preferred exclusive use of one of us (BvF), but is relaxed in the present context.

3.3 Measurable but non-qualitative features

The history of this subject was ably told and given a remarkable continuation in Glymour's *Theory and Evidence* (1980). Imagine a discussion between Newtonians and their rivals the Cartesians around 1700. All will consider the quantities of extension as directly given to observation, for observable objects. But the measurement of some of these quantities, measurable by means of rulers and clocks, already presupposes certain substantive assumptions – in the case of planetary motions, for example, the immediate observations do not suffice without some assumption about how light travels. So far, however, with respect to the kinematic quantities, the two parties share the minimal needed presuppositions. It is different for the Newtonians' new quantities of mass and force. Certainly they offer operational procedures for determining these, but the measurement of the masses and forces in a system is a calculation on the assumption that this system is a system of Newton's mechanics.

This was the main point of Poincaré's discussion of quantities in classical mechanics in his *Science and Hypothesis* (1952), but it followed on the nineteenth century's attempts to define the dynamical quantities in kinematic terms. These attempts cannot be successful (given our current precise notion of definition) but it is possible to state theoretical assumptions presupposed in the measurement of those dynamical quantities using only kinematic terms. The best known example is still Mach's 'definition' of mass in terms of mutually induced accelerations under special circumstances. Since the special circumstances usually do not obtain, the challenge is to see whether the kinematic behaviour of the system implies the values of the masses and forces. The trivial case of a body never accelerated refutes this as a general claim, even if we amend 'implies' to 'implies relative to Newtonian mechanics'. Mach's procedure is inadequate for less trivial cases also.[7] But in the next century Pendse proved that if the complete kinematic data are given for the elements of a point particle system at a sufficient number of times then the forces and masses will be uniquely determined for a large class of cases. The main point is of course that this large class of cases is certainly not exhaustive – not only the general validity of the theory (such as constancy of mass) but also dynamical isolation (with conservation of linear and angular momentum) are assumed in the measurement calculations.

Subsequently Sneed (1971) gave a very precise form to Poincaré's conclusion; Glymour, while severely critical of Sneed, actually shows how this point emerges also in such other writers as Hans Reichenbach, and relocates it in his concepts of testing and evidence. A theory is tested by means of measurements of quantities on the assumption that the theory is satisfied; for certain quantities nothing more basic is possible. These are quantities which are measurable and definitely not qualitative,

[7] For a clear discussion of this and also the following, see Jammer (1961, chapter 8).

in our terms. From an epistemological point of view, therefore, measurement is in effect a procedure conditioned on ambient theories to generate new observable phenomena set in a theoretical context. Hence our emphasis here on observation rather than measurement for the theory–phenomena relation.

To sum up then: we are going to connect superfluous structure with the presence of unmeasurable quantities. However, what is measurable/unmeasurable cannot be read off directly from a theory. We need to make use of what is observable in order to make this distinction, but what is observable/unobservable does not align with what is measurable/unmeasurable. How the two distinctions are related to each other is spelled out to some extent in the above capsule history, but there is clearly more work to be done on this topic.

4 Symmetries: kinds of structure-preservation

We will adopt here the following precise terminology. A *transformation* is a one-to-one mapping of its domain onto its range. In the present context the domain and range are the initial possibility space onto itself. A transformation may be definable by means of instructions for adding or removing objects from a world, changing their properties, or shifting them around in space and time, but in general it may not even be computable or definable in any sense.

Intuitively, *symmetries* are transformations that preserve structure, and are therefore to be classified in terms of the (kind of) structure that they preserve. The symmetries *of a world* have to be distinguished from the symmetries *of a theory*. The symmetries of a world are transformations that map the world to itself. Thus if M is a possible world and h is a transformation pertaining to it, h is a symmetry of M if and only if $hM = M$. Symmetries of a theory, by contrast, are transformations of the initial possibility space that preserve satisfaction of its laws, and also preserve non-satisfaction. (These definitions are to be kept clearly in mind when we discuss various symmetry groups that can be associated with a theory.) In the terminology adopted above, the symmetries of a theory take physically possible worlds into physically possible ones and physically impossible worlds into physically impossible ones. Hence we can equally well call them the symmetries of its laws. (Indeed, one way to specify the laws is simply to specify that symmetry group – not to be confused with narrower symmetry groups that may be under discussion in a specific context!) They need not preserve any particular characteristic of these worlds.

It would in general make no sense to suggest that only features preserved by (invariant under) the symmetries of a given theory are real. For to identify all possible worlds related by such symmetries (as representations of the same physical possibility) would be to claim that there is only one physically possible world. Note that *any* mapping of worlds to worlds that leads from 'legal' to 'illegal' ones are included; hence each physically possible world is the image of any other such world under some such mapping. In the case of Newtonian particle mechanics, for example, these mappings include ones that relate one system to another one with the same number of particles (but different trajectories), but also mappings that relate systems with different cardinalities.

We must look therefore to other symmetries, transformations that preserve not only satisfaction and non-satisfaction of the laws but other significant features as well. For example, in the construction of a theory one could begin with transformations that preserve some spatiotemporal structure, and perhaps some other quantities, and then ask under which among these the laws are preserved. Weyl (1952, pp. 26–7) made this general point, which can be illustrated with the example of Galilean transformations in classical mechanics. They are intimately related to the laws. Yet we cannot say that a feature invariant under all Galilean transformations is always a matter of law. For number – the number of particles in the system, say – is invariant, but the laws of classical mechanics do not impose any constraint on number. The laws constrain *relations between* quantities, but not, typically, the particular values they take. The nomologically contingent, but nonetheless real, features of a situation are precisely those that distinguish worlds related by a symmetry of the laws.

So in a diagnosis of what is real and what is superfluous structure the symmetries to focus on may belong to some intermediate class(es). At least one springs to mind: the class of *symmetries of the theory that also preserve all qualitative features of every model*. This is a class that can be defined in the same way for all theories, and therefore has the same generality as the notions of symmetry of a world and symmetry of a theory.[8]

5 The signs of superfluous structure

Consider a symmetry of theory T which maps each physically possible world to one that is qualitatively indistinguishable.[9] It will always be possible, without affecting

[8] The Galilean transformations do not preserve position or velocity, yet these are directly measurable quantities. That is an objection at first blush, but perhaps not on second thoughts. Position and velocity in a given frame of reference are not preserved, while distances and relative velocities are. But it is precisely the latter that are measured; the former result from a measurement plus some convention or paper-and-pencil operation. Recall from above that, as we draw the lines, the qualitative features are those directly observationally accessible to the observer, while measurement in general involves the deliverances of observation plus some theoretically conditioned calculation. Note that it may also be necessary to add an indexical element into this conception. A navigator on a ship may, for example, record the presence of an iceberg ten miles to the north-east. It is clear that such a report is to be understood as carrying the indexical 'from here' Such a report can be made in ignorance of the navigator's location on the globe, and its indexical content is independent of that. But we leave this aspect of observation to other occasions.

[9] This is essential; the transformations in question have to preserve the qualitative structure of all physical possibilities. See below for a further apparent disagreement with Belot.

the empirical content of theory, to interpret such a transformation as a symmetry of each of the worlds it relates. It will always be possible, that is to say, to interpret it as acting as the identity on the worlds in question, notwithstanding that it may not act as the identity on our *representations* of those worlds. If h is a qualitative-structure-preserving symmetry of the laws, and M_T and M_T^* are theoretical representations of worlds (descriptions or mathematical world-representing structures), such that $M_T^* = hM_T$ and $M_T^* \neq M_T$, it will always be possible to identify the worlds represented by M_T and M_T^* respectively, and to think of the structures that distinguish their representations as superfluous.

5.1 Trivial and non-trivial transformations

To interpret a transformation as *trivial* is to interpret it as permuting (at most) physically insignificant features of theoretical structure. As physically significant we count precisely (i) those features that are (jointly or individually) constrained by the laws, plus (ii) the qualitative features, regardless of whether they are constrained by the laws. *Hence we submit that it is precisely the qualitative-structure-preserving symmetries of the laws that are indicative of the presence of superfluous theoretical structure and should always be interpreted as trivial.*

We are distinguishing here three classes of transformations, with corresponding interpretational claims:

(a) Transformations that preserve qualitative structure but are not symmetries of the laws. These will turn at least some physically possible worlds into physically impossible ones, though there is no observable difference.
(b) Symmetries of the laws that are not qualitative structure preserving. These produce distinct physically possible worlds.
(c) Qualitative-structure-preserving symmetries. We submit that it is precisely these which suggest the presence of superfluous structure.

The symmetries in class (b) include non-geometric transformations that permute the values of dynamically relevant unobservable quantities (i.e. transformations that induce unobservable changes with potential but non-actual observable effects), and geometric transformations that are not symmetries of the laws (i.e. transformations that can change the state of motion of a system in a way relevant to their dynamical behaviour).[10]

5.2 A sophistical argument

Here is an argument that all theories will necessarily have superfluous structure by the above criteria. Consider even the very simple theory that says:

[10] We will discuss an example below, in connection with a claim made in Belot (2001).

(i) There are exactly two objects.
(ii) There is a red object and a black object.

Take a world that satisfies these laws; it has two members in its domain. Call the red one **a** and the black one **b**. There is another world, produced by the permutation of the two (a transformation that does not preserve colour, but certainly legality) in which **a** is the black one and **b** the red one. These two worlds are qualitatively indiscernible. But they are clearly distinct, for in the first one the property $(\lambda x)(x = \mathbf{a}\ \&\ \mathrm{Red}(x))$ is instantiated and in the other world it is not instantiated. (Notice that the Principle of the Identity of Indiscernibles does not get any purchase here: in each world each individual has a uniquely defining description.)

Various gambits are familiar from the philosophical literature to try and block this conclusion. Could we insist that non-trivial transformations must preserve the individuals' essential properties, and that the *identity properties* (such as $(\lambda x)(x = \mathbf{a})$) are essential? Bad move, for then only the identity transformation $f(x) = x$ will count as non-trivial. Could we suggest that '=' be dropped from our admissible language for constructing theories? Sounds like a bad move too, if we construe (i) in the usual way as 'There are x, y such that not $[x = y]$ but for all z, $z = x$ or $z = y$'. We should be able to express counting. However, it is possible to go higher order and have primitive number properties: 'in these worlds Red and Black are singly instantiated' and 'the property of being Red or not Red is doubly instantiated'. Then '=' can perhaps be dropped.

We see this as a sophistical problem, artificially generated. There are two obvious choices for philosophical bookkeeping. We can say that the argument succeeds in displaying superfluous structure for every theory, but that this is *trivial superfluous structure*, defining that notion precisely as 'superfluous structure that every theory has'. Or alternatively we can say that the displayed worlds are identical, that we only have two descriptions differing in an arbitrary labelling that has no physical significance. Our own preference is for the latter, but we do not see it as a substantive issue.[11]

In objection it may be suggested that each individual has its own haecceity, which we thereby ignore, and that this makes the issue substantive. But introducing haecceity, or any other hidden individuating factor, will just push the sophistical argument one step further. For if all non-trivial transformations must preserve haecceity, then only the identity transformation is non-trivial. If on the other hand haecceity need not be preserved, then there will be trivial but distinct variants on worlds once more produced by permutation of the individuals. This dialectic is then merely a useless replay of the same argument. In what follows we will sometimes just say 'ignoring the individuals' identity' to remind ourselves of this bit of

[11] Cf. van Fraassen (1991, chapter 12, section 4.2) and Huggett (1999).

philosophical book-keeping. But let us add a few more remarks to locate this issue properly.

There are two ways one can think about transformations. One can describe transformations in an interpreted metalanguage, with as much structure as one likes. In that case, for the purposes of describing their action on the class of structures under consideration, they should be grouped into equivalence classes insofar as they are indistinguishable by their effects on structures in the relevant class. Alternatively, one can describe the transformations in the same language in which the structures are described. In that case they are individuated more coarsely, so that there are no distinct transformations that are indistinguishable by their effects on the structures in question. This has the effect of tying the individuation of transformations to the structures in the class in such a way that when one identifies superfluous structure in the models, it is identified simultaneously in the class of applicable transformations. The second alternative is more natural, because it ascends a level, applying the same reasoning to the individuation of transformations as to the individuation of physical possibilities. Transformations that act in the same way on all structures in the class are identical, notwithstanding notational variation in their presentation. In application to the example, it has the consequence that talk of permuting colours over individual elements while leaving global colour distribution the same is nonsense.

6 Symmetries of the laws

The claim we have now introduced we have not so far supported, but by examining other accounts of the matter and examples that illustrate the differences we mean to make a plausible case. Our claim is in at least apparent disagreement with Belot, who writes (2001, p. 55):

> if there were a class of possible worlds whose shared geometry and laws were invariant under some set of symmetries, then . . . there would exist distinct worlds sharing all of their qualitative and relational properties.[12]

Belot's article, framed as a discussion of Leibniz's Principle of Sufficient Reason (PSR), is devoted to an exploration of the possibility of a trivial interpretation of certain symmetries of classical mechanics. Instead of focusing on spaces of possible *worlds*, Belot speaks in terms of spaces of possible *states*, and he focuses on spaces whose geometric structure determines the physically possible trajectories through it, effectively writing a theory's dynamical laws into the space. So what we call worlds are the trajectories in his discussion. That is just a different way of speaking, but there are reasons (good reasons, we think, that discourage otherwise seductive

[12] We readily admit that we are using this excellent and insightful article as a stalking horse, and that the apparent disagreements may not go very deep; they will still clearly bring out the contents of our claims.

confusions) to prefer talking, as we have, in terms of spaces of worlds with no more intrinsic structure than is given by the internal relations among their elements, and we will continue to do so.

Returning then to the quoted thesis, we note that Belot in effect omits the restriction to symmetries that preserve qualitative structure, allowing for simple counterexamples. Consider, for instance, the very simple theory that says that there are two types of thing in the world, triangles and squares, and that there are two laws:

(i) There are only three objects in the world.
(ii) Either everything is a black triangle, or everything is an orange square.

Ignoring the identity of the individuals (see above!) there are here only four physically possible worlds, distinguished by having 0, 1, 2, or 3 black triangles. Consider now the transformation **h*** that changes triangles into squares and squares into triangles, and changes black things into orange ones and orange into black. This transformation **h*** is a symmetry of (i) and (ii), and maps the set of physically possible worlds onto itself. But these worlds are all distinct in structure.[13] Physically realistic examples are easy to find. Whenever you've got a theory with only qualitatively distinguishable models, any automorphism of the set of physical possibilities will constitute an example.

For a contrasting example in which qualities are preserved and in which there are distinct, indiscernible worlds, take any theory, add to it a hidden, unmeasurable quantity, Q, and let **j** be the transformation that leaves everything else intact but permutes the value of Q. Since Q is unmeasurable, it is causally isolated from the values of all other quantities, and so permuting Q-values at a world is not going to affect the values of observable or measurable quantities. As long as we don't add new laws explicitly mentioning Q and relating them to observable or measurable quantities, **j** will be a symmetry of the theory, and among the theory's models will be worlds that share a geometry, and all qualitative and relational properties.

A more interesting example is provided by Glymour's example of the revised Newtonian theory that replaces Newtonian force with the complex quantity morce + gorce (1980, pp. 356–62). Let **k** be the transformation that adds to morce what it takes from gorce, leaving force unaffected (suppose morce and gorce can take negative values). The revised theory will be invariant under **k**, and world **w** will be (except in degenerate examples) distinct but indistinguishable from world **kw**.

Why would Belot have ignored such examples? Perhaps if we focus on geometric symmetries the distinction between qualitative-structure-preserving and non-qualitative-structure-preserving symmetries of the laws will not spring to mind. What distinguishes the geometric symmetries as a class is that they preserve

[13] To fill out the example we could add that (i) and (ii) are invariant under Galilean transformations, and that **h*** leaves the metrical structure of a world intact.

qualitative structure; but it is not in general true that symmetries of a theory preserve the qualitative structure of its models. It is only by contrast with geometric transformations that are *not* symmetries and symmetries that do *not* preserve qualitative structure, that what is special about geometric symmetries emerges. By considering other symmetries that figure prominently in the literature (e.g. gauge symmetries and permutation symmetry) we can get a handle on the general physical significance of symmetries. Transformations that preserve qualitative structure are not all symmetries of the laws, and transformations that are symmetries of the laws do not all preserve qualitative structure. Only those that have *both* features suggest the presence of superfluous theoretical structure. Only those that have *both* features permit a trivial interpretation.[14]

6.1 Qualitative indiscernibility with dynamic differences

Belot sometimes speaks as though Leibniz's Principle of Sufficient Reason (PSR) enjoins the identification of *any* pair of qualitatively indistinguishable models of a theory. He introduces the article, for instance, with the claim (2001, p. 55) that:

> a description of a set of possible worlds which includes pairs of worlds with identical qualitative structures can [and, according to PSR, *should*] always be taken to correspond to the sparser set of possibilities which arises when qualitatively identical worlds are identified.

But he can't mean that.[15] Consider the transformation that maps an empty rotating bucket in a Newtonian world onto a qualitatively indistinguishable bucket at rest, and maps every other world onto itself. Both are models of Newton's theory, and if neither bucket is filled, they are qualitatively indistinguishable. There are undoubtedly differences between the two according to Newton's theory, which would become manifest if measuring instruments (or water) were introduced into the buckets. But by hypothesis there are only the buckets, and they are therefore the same in all their qualitative features. PSR, however, ought not to be understood as instructing us to identify these worlds. It ought to be understood, rather, as counselling aversion to recognition of dynamical distinctions that have no *potential* qualitative effects. By 'potential qualitative effects' we mean here qualitative effects that show up in some *physically possible* circumstance, according to the theory (i.e. in some of the theory's physically possible worlds).[16]

[14] This provides some insight also into the conditions under which we recognize intrinsic geometric structure, given that geometric transformations all preserve qualitative structure; viz., when they have potential qualitative effects.

[15] And some of the other things that he says suggest that he doesn't, though there are still others that make it clear that he is not *simply* misspeaking.

[16] Belot discloses (in personal correspondence) that he means 'qualitative' in a different sense from ours: 'I use "qualitative" in the metaphysicians' sense [to refer simply to the intrinsic properties of an object]; it is, I think,

The difference is far from trivial. We get the space of metaphysical possibilities, in physical contexts, by unconstrained recombination from the entities, quantities, and relations that we take to be the building blocks of the actual world. If we took the state space of Newtonian mechanics, as suggested by Belot, and simply identified any pair of qualitatively identical worlds, we would give up combinatorial structure (the space would contain, for instance, worlds in which there are spinning water-filled buckets, but no worlds in which there are empty spinning buckets), and that isn't something we can surrender just because it happens to suit our theoretical purposes. One of the reasons why we *care* about the space of possibilities, one of the things that makes it an object of physical interest, is that it is related, in the way expressed by the Principle of Recombination, to the structure of every world *in* it. We care about (metaphysical) possibilities, in physical contexts, at least in part, because they relate in a principled way to the structure of actuality, and we can't abandon the relating principle without relinquishing (this aspect of) their significance.[17]

The physical intuition to which Leibniz's principle answers is that if we are recognizing a set of possible worlds with a great deal of qualitative redundancy, we may be recognizing more structure in the actual world than there is good reason to suppose, and if we can find a way of trimming away some of the fat without cutting into the meat (i.e. if we can find a way of doing without some of the non-qualitative structure without losing any qualitative distinctions), we should do so. But this business of identifying qualitatively indistinguishable possibilities won't tell us anything about the structure of the actual world if we give up Recombination. We don't understand what the actual world is *like* according to a theory that removes some of the qualitative redundancy unless we can place it against the background of a combinatorially structured space. We don't really understand what a theory says the building blocks of the actual world are unless we know how to take them apart and put them back together; that is, unless we know what the theory says the real degrees of metaphysical freedom are.

much weaker than your usage; and it is not directly founded on considerations involving perception . . . '. We have not emphasized the difference because it does not affect the counterexamples. We can simply stipulate in the orange square/black triangle example that the only real, intrinsic properties of the worlds described are the colour and shape, and there are no more any real, intrinsic properties to distinguish the rotating from the stationary empty Newtonian bucket worlds than there are qualitative ones. The weakened sense of 'qualitative' does make Belot's discussion less ambitious than one might have hoped; since which properties are qualitative in the weakened sense is something that is not determined independently of a theory's interpretation, Belot is not recommending a general, theory-neutral criterion for identifying superfluous structure. His discussion applies only after interpretive decisions have been made. See further, below.

[17] There may be a new space, with a combinatorial structure, containing all and only the worlds in the modified Newtonian one, but the worlds in that space would have a non-Newtonian structure. The claim is that we wouldn't understand the physical situations depicted by the modified Newtonian worlds – we wouldn't understand what their constituents were, what kinds of entities, quantities, and relations they were really made of – until we saw them in the context of the new space, until we saw how to take them apart and put them back together.

To review, then: symmetries of a theory, T, are transformations that map the set of physically possible worlds onto itself. Transformations that *aren't* symmetries of T, by contrast, sometimes take you from a world that is physically possible by T's lights, to one that is not, and can be understood as changing the world in some dynamically relevant way. Some of these dynamically relevant changes will be visible, but some will have visible effects only under certain conditions (e.g. the difference between empty Newtonian buckets in rotating and non-rotating universes). The difference between such worlds is qualitatively potential because if we filled the buckets with water then, *ceteris paribus* and keeping the laws fixed, qualitative differences would emerge. The PSR should be understood as a prohibition on invisible, dynamically irrelevant differences. It should be thought of as a ban on the recognition of invisible, intrinsic differences that do not have visible manifestations, under any physically possible conditions, actual or counterfactual.

Geometric transformations are special in that they preserve qualitative structure, and so recognizing differences between worlds related by geometric symmetries is always a violation of the PSR. But symmetries do not in general preserve the qualitative structure of their models, neither in our sense of 'qualitative', nor in Belot's weaker sense. If they did, and if we identified all physically possible worlds related by symmetries, no theory would have more than a single model. Belot has suggested, in personal correspondence, a restricted notion of symmetry:

> I count as symmetries of a theory only those permutations of its space of possible worlds which preserve the structure defining the dynamics of the theory (thus my symmetries are diffeomorphisms preserving, say, the Hamiltonian and the symplectic structure or Hilbert space structure, in typical physics cases).

There is nothing illegitimate about building restrictions into your definitions. But there is a cost to this. If you *define* the symmetries of a theory as those that preserve the structure defining the dynamics, you cannot then use considerations involving symmetry to see how much dynamical structure is needed to reproduce the empirical content of a theory, i.e. to reveal dynamical structure that is not really doing any empirical work. Symmetry is a mathematical notion, and we think it best to keep our definition of the symmetries of a theory uncontaminated by physics. We also consider it important that any notion of symmetry used in a special context should derive from general notions defined for any theory in the same way. Combine a purely mathematical notion of symmetry with a theory-neutral distinction (as indispensable precursor to interpretation) between qualitative and non-qualitative structure in the models, and you have a good guide to identifying superfluous structures, one whose epistemic motivation is plain, and whose application doesn't wait on the very interpretive decisions we want to use it to make.

6.2 Symmetries of a world, and identity of indiscernibles

The symmetries of a given world, as distinct from the symmetries of a theory, are just those transformations that map it into itself (its *automorphisms*). That is the sense in which a 360° rotation is a symmetry of any letter, while a 180° rotation is a symmetry of the letter O but not of the letter P. For these symmetries it would certainly make no sense to suggest that they detect superfluous structure in a particular world. Since they map the world into itself, they cannot be used to support any claim to the effect that this possible world is really the same as (represents the same physical situation as) some other world.

However, if a transformation is a symmetry of some worlds but not others, we can *raise the question* whether it does not perhaps preserve *all* significant structure, and thus relates only worlds that represent the same way a real world could be. If physical situations S and S' are represented as mirror-images of each other, for example, are there really two distinct physical possibilities being represented, or only one? The question is: does the mathematical operation correspond to a physical operation? In applying this transformation are you really reorganizing a world, or mapping it onto a duplicate, or just permuting insignificant bits of the representation?

There should be a strong suspicion of superfluous structure if two distinct worlds are related by a transformation that has some world as a fixed point. As an intuitive example, familiar from much literature, imagine that worlds w_1, w_2, and w_3 have in them respectively only a left hand, a right hand, and two hands (one right and one left, which are each other's mirror-image reflected through a central plane). Reflection will turn w_1 into w_2 and vice versa but turns w_3 into itself. This is the clear danger sign that makes us think that w_1 and w_2 do not represent two really distinct possible physical configurations but only one.[18]

This historical example has much about it that is questionable and has made it the topic of a large and diverse literature. Consider a more abstract example, the four-element group known as Klein's *Viertelgruppe*.[19] This is a commutative group with four elements e, a, b, c, in which e is the identity element ($ex = x$ for all x), each element is its own inverse, and if x, y, z are distinct elements other than e then $xy = z$:

	e	a	b	c
e	e	a	b	c
a	a	e	c	b
b	b	c	e	a
c	c	b	a	e

[18] Cf. Pooley, this volume.
[19] This example is used to advantage in Rynasciewicz (2001).

Here the element e is uniquely definable. But we cannot construct a uniquely identifying description for any of the other elements; they are structurally and qualitatively indiscernible. The automorphisms of this structure are precisely the permutations of its set of non-identity elements, and only what is invariant under the automorphisms can be definable. It would however make no sense to suggest that those elements should therefore be identified. The result would be a two-element group, and we would have the strange consequence that the *Viertelgruppe* does not exist (though there are objects that have this form, plus additional structure).

There are two points to be made here. The first is that the group is invariant under permutations of the three non-identity elements. Replace a by b and conversely, and the table is just

	e	b	a	c
e	e	b	a	c
b	b	e	c	a
a	a	c	e	b
c	c	a	b	e

and that is quite obviously the same table as before, written in slightly different order. This should raise the suspicion that if two worlds are related by such a permutation, they do not represent two distinct possibilities. But secondly, this permutation invariance is no basis for suggesting that this world, the *Viertelgruppe*, has been redundantly depicted in the above table.[20]

6.3 A new sophistical argument

We emphatically used the word 'suggest' above: each case has to be examined separately. Thus we distinguish this topic (symmetries of worlds) very emphatically from that of qualitative-structure-preserving symmetries of the laws and their clear implication of superfluous structure.

But there have certainly been suggestions, typically connected with Leibniz's Principle of the Identity of Indiscernibles (PII), to draw an exceedingly general moral. Consider world w_3 above consisting of two hands which are each other's mirror-image. That world is left unaffected by reflection through its central plane – conclude then that actually it contains only a single hand! Black's world consisting of two identical spheres provides the simplest (if not the most illuminating) example. Why not conclude that this world actually contains only a single sphere, and is here redundantly described, as in our own familiar duo of the Evening Star and the Morning Star?[21]

[20] Cf. French and Rickles, this volume.

[21] One of us did at best (on the most charitable reading) come exceedingly close to being taken in by this line of argument; see van Fraassen (1985, pp. 63–5).

These arguments are sophistical, trading on an untenable version of PII. There are undoubtedly cases in which a theory has distinct models that are related by a qualitative-structure-preserving symmetry of its laws and cannot represent distinct possibilities. That this is indeed so in a specific case may be suggested by the fact that this symmetry has some worlds as fixed points. But here, in this sophism, the inference goes in the opposite direction: to the conclusion that it is those fixed points which have superfluous structure in themselves, and hence are superfluous items in the theory's physical possibility space!

That it is a sophism the very examples of spheres and hands should already illustrate.[22] If they do not, it is because in such fanciful cases the cost of denying obvious possibilities may not seem so high. But first of all, the other examples we gave should make the cost quite clear, and secondly, we can put the matter quite abstractly.[23]

Suppose that for a certain two-place relation R the following is true in a given world:

There are objects x and y such that Rxy and not Rxx.

In that case the world contains at least two objects. This follows regardless of anything else that may be true in this world, and therefore regardless of whether the objects in question are differentiated in any describable way.

This simple point defeats many a naive version of PII. There are more sophisticated versions that do not fall so easily, but we will not hold them sacrosanct if they are not tautologous.[24]

The main point is also well illustrated by an example Belot cites in his discussion of whether indiscernibles in a theory's models can be identified. The logical way to do that would be to reduce each world *modulo* the indiscernibility relation (2001, p. 60):

> The upshot: whenever we have a structure that admits non-trivial symmetries, we can factor these out, constructing a quotient structure . . .

The example that halts this suggestion in its tracks (suggested to Belot by Kit Fine) is simple enough (*ibid.*):

> Consider two structures for a given countable set of objects; in one structure that set of objects is given an ordering isomorphic to the integers, in the other an order isomorphic to the rationals. The quotient of each structure is just: a single object, related to itself.

Belot notes simply that it is 'necessary to examine the relation between a structure and its quotient on a case-by-case basis' (*ibid.*). Rather disappointing, if one was

[22] See the discussion in van Fraassen (1991, pp. 454–6 and 459–65).
[23] *Ibid.* (p. 456, last paragraph of section 3.2).
[24] For a thorough discussion of the issue, with carefully nuanced distinctions, see Saunders, this volume.

hoping for a general method for wholesale elimination of putative superfluous structure![25] We can see now, however, what crucial distinction tends to be ignored in this area. Symmetries *of worlds* – of single structures meant to represent nature – can at best offer a suggestion of, or clue to, possible superfluous structure in the theory. They can certainly not imply that – the world could after all be symmetric in any way it likes! So a theory must be allowed to have models that have any conceivable kind of symmetry. It is not the symmetry of any given world, but the qualitative-structure-preserving symmetries *of laws*, that can definitively reveal superfluous structure.

7 The wider theoretical context

Formalisms with little superfluous structure are nice, of course, because they reflect cleanly the structure of what they represent; they have fewer extra mathematical hooks on which to hang the mental structures that we project onto the phenomena. But we want to conclude this general discussion of *theory structure* with a reflection on the *structure of theorizing*.

Methods for removing excess structure are much more than mopping up procedures. They are not something merely to be done *after* our representations have been crafted, like portraitists erasing stray pencil marks, or sculptors removing extra clay. Methods for removing excess structure are the very heart of theorizing; we figure out what the world is *like* by seeing what kinds of representations it supports. In theorizing one starts, that is to say, with the *representations*, and works one's way towards ideas about the intrinsic character of their common object by a kind of triangulation.

There are two stages in theory construction. The first is to generate a set of models rich enough to embed the phenomena, the second is to attempt to simplify those models by exposing and eliminating excess structure. Continuing in this way the structure of the models is pared down, being careful not to jeopardize their capacity to embed the phenomena. The whole class is thrown over only if a new significantly simpler set of models is found.[26] These inside-out procedures for identifying superfluous structure are indispensable, and the identification of qualitative-structure-preserving symmetries of the laws is paramount among them.

25 Something Belot, in correspondence, disavows: 'You might wonder, after all these qualifications, what the project of my paper really is? A modest one: to point out that there will always be available in philosophy of physics a trick which allows you to pass from a formulation of a theory that admits symmetries to a related one which does not; and to make a very modest start on assessing the interpretative merits and demerits of making this move in some classical cases.' Even this modest project, however, cannot proceed without a distinction between symmetries that are, and symmetries that are *not*, candidates for reinterpretation. Transformations that do not preserve qualitative structure, transformations that map observationally distinguishable worlds onto one another, even if they are symmetries of the laws, are not candidates for reinterpretation.

26 Or one whose models are demonstrably simpler than our good faith estimation of the potential for simplifying the ones we have.

If we interpret such transformations as trivial, we drain the structures that distinguish the representations they relate of significance, giving us simpler models at no empirical cost.[27]

This two-stage conception of theorizing has many historical illustrations. One familiar example is provided by the development of quantum mechanics.[28] At first one must be struck by the differences between Schrödinger's wave mechanics and Heisenberg's matrix mechanics. A first simplification came with von Neumann's insight into the shared Hilbert space structure, a second with Weyl's display of the still more basic group-theoretic structure behind the algebras of observables.[29]

A guide for identifying superfluous structure, however, is not a recipe for formulating a nice theory that does without it, that is to say, a local intrinsic description of the world that has all of the properties we like theories to have. There is no better illustration of this than the problems, amply chronicled in this volume, associated with the interpretation of gauge symmetry. Focusing on Yang–Mills theories, we have local symmetries of the generalized phases associated with the wave functions of the matter fields that show all the formal signs of revealing superfluous structure, but we can't simply excise the problematic structures without rendering the theory non-local. It seems that we need *something* in the region of space occupied by the gauge potentials to explain effects like the Aharanov–Bohm effect in a local manner. Redhead and Nounou (both this volume) explore the options.

8 Conclusion

We have been exploring the question of how symmetries can function, in the context of physical theory, as guides to the presence of superfluous structure. The philosophical lesson that can be taken away from the discussion is an insight into what has emerged as the most characteristic feature of modern physics. The ontologies of our most fundamental theories are not guided by physical intuition; they are not shaped by philosophical prejudices, but led, at their best, by the ideal of a kind of formal simplicity. The history of modern physics has been (to adapt a phrase from a recent book by Barbour)[30] 'a long, sustained effort to shed redundant concepts', and symmetries of the right sort, symmetries of the sort that we have been talking about, can act as beacons of redundancy.

27 The flip-side is, of course, that the larger the set of geometric symmetries of a theory's laws, the less dynamically significant spatiotemporal structure it recognizes.

28 We can be brief here; see for example Otavio Bueno's discussion of this development in the context of the heuristic value of symmetry oriented theorizing, in his 'Weyl and von Neumann: symmetry, group theory, and quantum mechanics', PITT-PHIL-SCI00000409.

29 See Bub (1981).

30 Barbour (1999). Barbour applies the phrase not to physics, but to the book itself.

References

Barbour, J. (1999). *The End of Time: The Next Revolution in Our Understanding of the Universe.* London: Weidenfeld & Nicholson.

Belot, G. (2001). 'The Principle of Sufficient Reason'. *Journal of Philosophy*, **98**, 55–74.

Bub, J. (1981). 'Hidden variables and quantum logic – a skeptical review'. *Erkenntnis*, **16**, 275–93.

Glymour, C. (1980). *Theory and Evidence*. Princeton: Princeton University Press.

Huggett, N. (1999). 'Atomic metaphysics'. *Journal of Philosophy*, **96**, 5–24.

Jammer, M. (1961). *Concepts of Mass in Classical and Modern Physics*. Cambridge, MA: Harvard University Press.

Lewis, D. K. (1983). 'New work for a theory of universals'. *Australasian Journal of Philosophy*, **61**, 343–77.

Mermin, D. (1998). 'What is quantum mechanics trying to tell us?' *American Journal of Physics*, **66**, 753–67.

Poincaré, H. (1952). *Science and Hypothesis*. New York: Dover.

Rynasciewicz, R. (2001). 'Definition, convention, and simultaneity: Malament's result and its alleged refutation by Sarkar and Stachel'. *Philosophy of Science*, **68**, S345–57.

Russell, B. (1897). *An Essay on the Foundations of Geometry*. Cambridge: Cambridge University Press.

Sneed, J. (1971). *The Logical Structure of Mathematical Physics*. Boston: Reidel.

van Fraassen, B. C. (1985). *An Introduction to the Philosophy of Time and Space.* 2nd edn. New York: Columbia University Press.

(1991). *Quantum Mechanics: An Empiricist View*. Oxford: Oxford University Press.

Weyl, H. (1952). *Symmetry.* Princeton, NJ: Princeton University Press.

24

Notes on symmetries

GORDON BELOT

These notes discuss some aspects of the sort of symmetry considerations that arise in philosophy of physics. They describe and provide illustrations of: (i) one common sort of symmetry argument; and (ii) a construction that allows one to eliminate symmetries from a given structure.[1] I hope that they suggest a unifying perspective.

1 Symmetries

It is helpful to begin with an abstract characterization of symmetries.

A *structure* consists of: a set, D, of *objects* together with a set, $\mathcal{R} = \{R_i\}_{i \in I}$, of relations defined upon D (no restrictions are placed on the cardinality of D or on that of the index set I). If $(D, \{R_i\}_{i \in I})$ and $(D', \{R'_i\}_{i \in I})$ are structures, then we say that a map $\phi : D \to D'$ *fixes* the n-ary relation R_i if: $R_i(x_1, \ldots, x_n)$ iff $R'_i(\phi(x_1), \ldots, \phi(x_n))$ for every n-tuple of objects in D. The *automorphisms* of the structure $(D, \{R_i\}_{i \in I})$ are bijections $\phi : D \to D$ that fix each $R_i \in \mathcal{R}$. The set of automorphisms forms a group under composition of functions.

There are two approaches to talking about symmetries.[2] Under the first, one identifies the symmetries of a structure with its automorphisms – we will call automorphisms symmetries *in the first sense*. Under the second, symmetries are taken to be permutations of names of objects rather than of the objects themselves. This works as follows. Suppose that we have a first-order language with a predicate symbol, $\mathfrak{R}_i$, for each relation, R_i, of our structure, and enough constants, $a, \ldots, c$, to serve as names for each object in D. A *nomenclature* is a bijection from the set of objects to the set of names. If we fix a nomenclature, then we can consider

[1] Some of this material is developed more fully in Belot (2001; and 'Dust, time, and symmetry', unpublished manuscript).

[2] The distinction below is related to that between active and passive symmetries, and coincides with it in some contexts.

the set of atomic sentences true of our structure under that nomenclature – that is, the set of formulae of the form $\mathfrak{R}_i(a, \ldots, c)$ true of our structure under our convention of associating names with objects. Call this set the *complete description* of our structure relative to the given nomenclature. Such a complete description determines the associated structure up to isomorphism. If we now compose the given nomenclature with a permutation of the set of names, we generate a new nomenclature; the permutation is a symmetry *in the second sense* when the complete description relative to the induced nomenclature is identical to that generated by the original nomenclature. The symmetries in the second sense relative to a given nomenclature again form a group under composition, and this group is isomorphic to the group of symmetries in the first sense. Because the two senses are so closely related, it is seldom necessary to decide which sense is in play in a given discussion.

Objects related by a symmetry occupy identical roles in the pattern of relations described by their structure – think of the identity of role of points in Euclidean geometry, or of congruent sides of an isosceles triangle. Below we will be interested in structures whose objects correspond to possibilia – typically, possible objects or worlds. We assume that only appropriately qualitative relations are represented in our structures – so objects related by symmetries will be qualitatively indistinguishable.

2 Symmetry arguments

Symmetry arguments have played a central role in natural philosophical debates, especially those concerning the nature of space, time, and motion. Many of them fall under the following argument form (illustrative examples appear in section 3).

- The point of departure is a given structure, held to provide a representation of the features under investigation that is taken to be (more or less) adequate for the purposes at hand. Very often this structure will be either a representation of (aspects of) a certain spatiotemporal world (in which case it will encode, for example, geometrical facts about the spatiotemporal relations between the parts of the given world) or a space of physical possibilities (which will also typically carry a geometrical structure – though, of course, a non-spatiotemporal one).
- In order to solve some outstanding problem, it is proposed to *extend* the given structure by supplementing its class of relations, yielding an enriched representation of the subject matter.
- We ask whether every symmetry of the original structure is a symmetry of the new structure.
- If not, then we can find a symmetry $\Phi : D \to D$ of the original structure, a new relation R of the extension, and objects $x_1, \ldots, x_n \in D$ such that $R(x_1, \ldots, x_n)$ but not $R(\Phi(x_1), \ldots, \Phi(x_n))$. In this case, the new relations make distinctions between objects which are qualitatively indistinguishable in the original structure.

To the extent that we are confident that the symmetries of the original structure are the 'correct' symmetries at the level at which we are working, the failure of the new relations to respect the symmetries of the original structure provides a reason to reject the proposed new structure – and the problem solution that it serves. The content and force of judgements of correctness will vary from case to case.

- If, on the other hand, one can show that the proposed extension is *invariant*, in the sense that it respects the symmetries of the original structure, then the proposed solution has met a minimum standard. If the extension can be shown to be the unique invariant extension of the sort under consideration, then one has reason to accept the extended structure as a (more or less) adequate representation of the features under investigation – to the extent that one is confident that the type of extension under consideration is indeed the best way to approach the problem at hand.

3 Examples

This section contains five examples of the argument form discussed above, ranging from the ancient to the modern. The first three provide examples in which the structure under investigation is a representation of a single world. In the last two examples the structures are more abstract: in the final example, the structure is the space of worlds possible relative to classical mechanics; in the penultimate example, the structure can be thought of as a coarse-graining of the space of two-particle collision worlds. This selection gives an indication of the range of application of the argument form within natural philosophy.

3.1 Platonic cosmology

The *Timaeus* includes a nice instance of our argument form.

> It is entirely wrong to suppose that there are by nature two opposite regions dividing the universe between them, one 'below,' toward which all things sink that have bodily bulk, the other 'above,' toward which everything is reluctant to rise. For since the whole heaven is spherical in shape, all points which are extreme in virtue of being equally distant from the center must be extremities in just the same manner; while the center, being distant by the same measure from all extremes, must be regarded as the point 'opposite' to them all. ... When a thing is uniform in every direction, what pair of contrary terms can be applied to it and in what sense could they be properly used? If we further suppose that there is a solid body poised at the center of it all, this body will not move toward any of the points on the extremity, because in every direction they are all alike ...[3]

Here our initial structure is a highly idealized representation of the Platonic cosmos: a sphere whose distinguished central point represents the Earth.[4] This

[3] 62*c*–63*a*. Translation from Cornford (1997, pp. 262–3).

[4] More precisely: our structure is a spherical subset of Euclidean space, with the usual betweenness and congruence relations defined on the set of points.

structure is invariant under any reflection through a plane through the central point. It follows that it is invariant under rotations (which arise as products of reflections), and hence that any two points on the surface of the sphere are related by a symmetry – and thus count as being qualitatively identical in this representation of the cosmos. Plato supposes that any definition of *down* would have to involve the choice of a distinguished inward-pointing normal to the cosmic sphere. But no such definition – amounting to the supplementation of the original structure by a property possessed only by the single distinguished point – would respect the original symmetries.

Plato moves immediately from this point about definition to a claim about dynamics. In both cases, the claim that the original representation is perspicuous is of course crucial – if we are allowed to bring into play asymmetries in the cosmos or in the central body then the problems Plato considers admit of easy solutions (*up* points towards Polaris; a central shoe moves *toe-wards*).

Here Plato expects his readers to grant as a matter of course that in the sort of investigation he is engaged in, it is only the largest-scale features of the world that should be taken into account. Perhaps to deny this is to mistake the sort of understanding he claims to offer.

3.2 Simultaneity in special relativity

One might run a similar argument to explain to a beginning student Minkowski's claim that with special relativity 'space by itself, and time by itself, are doomed to fade away into mere shadows, and only a kind of union of the two will preserve an independent reality' (1952, p. 75).

In Newtonian spacetime, one has binary relations 'at the same time as' and 'at the same place as' defined on the set of spacetime points. Suppose that we wanted to introduce such structures in Minkowski spacetime.[5] We should, at the very least, require any candidate for 'at the same time as' to be an equivalence relation with three-dimensional, connected, space-like equivalence classes, and require any candidate for 'at the same place as' to be an equivalence relation with one-dimensional, connected, time-like equivalence classes. But there are no such equivalence relations definable on Minkowski spacetime that are also invariant under the symmetries of the spacetime.[6] In this sense, the shift from Newtonian spacetime to Minkowski spacetime deprives us of fully autonomous notions of time and space.

Now, while every student ought to be exposed to this observation, it does little to settle the question of the nature of time in special relativity – for both

[5] Here and below the objects of differential geometry are taken to be structures in which the points of the space are the objects, and the topological and differential structures of the space, along with any tensors defined on it, are taken to be encoded in some appropriate fashion in the relations of the structure.

[6] The only invariant equivalence relations are the trivial ones: the one in which every point is related only to itself, and the one in which every point is related to every other. See Giulini (2001, Theorem 4).

reactionary metaphysicians and heretical philosophers of quantum mechanics deny that Minkowski spacetime provides a complete representation of spatiotemporal reality in the first place. This is quite typical: symmetry arguments are of little polemical value in situations where fundamental questions are at stake, since those are the cases in which there will be little agreement as to whether a given structure provides an acceptable point of departure for such an argument.

But note: in this setting *no one* is interested in using contingent features of the matter distribution to introduce asymmetries which make it (all too) easy to generate invariant simultaneity relations. The question is about the structure of spacetime, and in the context of special relativity it would be cheating to take matter into account.

3.3 Time in dust cosmology

The situation is quite different in general relativistic dust cosmology. There one conceives of spacetime as filled everywhere by dust motes (representing galaxies), which interact with one another only gravitationally – this provides a tractable (and relatively honest) idealization of the large-scale dynamics of the universe for all but the earliest (and, possibly, latest) times (at which other interactions must be taken into account). A solution to the equations consists of a spacetime geometry together with a congruence of time-like geodesics (the worldlines of the dust) and a scalar function (the matter density). The stress-energy of the dust serves as the source term in the Einstein equations.

Now, note that if one augments the structure of Minkowski spacetime by the choice of a privileged congruence of inertial observers at rest relative to one another – in effect imposing a relation 'at the same place as' – then there is a *unique* candidate for 'at the same time as' invariant under the symmetries of the augmented structure.[7] The associated equivalence classes are just the hypersurfaces everywhere orthogonal to the privileged congruence – and coincide with the surfaces of Einstein simultaneity associated with the privileged inertial observers.

Let us now consider a solution of dust cosmology. The dust congruence gives us an analogue of the congruence of freely falling observers in the augmented version of Minkowski spacetime. So it is natural to wonder whether we can define a decent relation 'at the same time as' in our dust cosmology. We look for invariant equivalence relations with connected, three-dimensional equivalence classes each of which intersects each dust worldline exactly once.[8]

[7] Giulini (2001, Theorem 5). See Malament (1977) and Stein (1991) for related results departing from the causal structure of Minkowski spacetime.

[8] The dust congruence and matter density of a dust solution are definable from the metric alone. So in dust cosmology, there is no difference between studying the symmetries of the structure 'spacetime geometry' and studying the symmetries of the structure 'spacetime geometry + material contents'. This equivalence fails for some general relativistic systems, including the Einstein–Maxwell field; see Kramer *et al.* (1980, p. 114).

Such equivalence relations exist in physically realistic dust cosmologies. Indeed, in any dust cosmology with a trivial symmetry group, any partition by space-like hypersurfaces will do – so there will be far too many candidates. But in highly symmetric models, like the Friedmann–Robertson–Walker solutions and the Einstein static universe, there is a *unique* relation of the desired type – the one whose equivalence classes are the hypersurfaces everywhere orthogonal to the dust congruence.

In the early decades of relativistic cosmology, this was taken to suggest that attention to astronomy restored what the local physics of special relativity had dissolved – an objective separation of spacetime into space and time.[9] Gödel was able, however, to construct a dust solution in which there is no invariant equivalence relation possessing the desired features.[10]

On its surface, this shows only that the Minkowski-style argument carries over to Gödel's solutions – a result of some interest, given that a dust congruence is a natural generalization of a structure sufficient to generate an invariant temporal slicing when added to Minkowski spacetime. Gödel himself thought the result had much greater significance.[11]

3.4 Huygens on collision

Here is a bloodless paraphrase of the opening moves in Huygens's analysis of collision (1977, pp. 574–8):

- We consider two bodies A and B. We assume that these bodies are perfectly elastic, of equal mass, and moving along the same line. We assign this line a sense – we call one end 'left' and the other end 'right'. The space of ordered pairs of real numbers is our *space of states*: the ordered pair (a, b) corresponds to a state where A has velocity a along the line while B has velocity b, employing the convention that positive velocities correspond to motion towards the right. This is an impoverished notion of state: we ignore the positions of the bodies on the line.
- We are interested in the dynamics of collision. We seek to define an irreflexive binary relation, $\rightarrow$, on the space of states, where $(a, b) \rightarrow (a', b')$ iff a system initially in state (a, b) eventually evolves into a distinct (a', b').
- We introduce three hypotheses.
  - **Hypothesis I.** The system remains in the initial state unless a collision occurs. So states of the form (a, b) with $a \leq b$ are dead-ends – they do not arrow anything.
  - **Hypothesis II.** For $a > 0$, $(a, -a) \rightarrow (-a, a)$.

[9] See Eddington (1920, p. 163) and Jeans (1936, p. 21ff.).

[10] Gödel (1949a, p. 447); (1949b, p. 560). It is possible to construct hypersurfaces of orthogonality only when the solution is non-rotating. But rotation alone does not suffice to rule out a relation of the desired sort – in Gödel's later, somewhat more realistic, expanding rotating solutions the surfaces of constant matter density have the properties required above of instants; see Gödel (1952).

[11] See Belot, 'Dust, time, and symmetry', unpublished manuscript.

- **Hypothesis III**. For any real number x, if $(a, b) \to (a', b')$ then $(a + x, b + x) \to (a' + x, b' + x)$.
- These hypotheses suffice to determine $\to$. In light of Hypothesis I, we need only investigate states (a, b) with $a > b$.
  - **Proposition I.** For $a > 0$, $(a, 0) \to (0, a)$, while for $b < 0$, $(0, b) \to (b, 0)$.
    Proof. Consider the first case. Let $x = -a/2$. Then $(a + x, 0 + x) = (a/2, -a/2)$. By Hypothesis II, $(a/2, -a/2) \to (-a/2, a/2)$. So by Hypothesis III, $(a/2 - x, -a/2 - x) \to (-a/2 - x, a/2 - x)$. That is, $(a, 0) \to (0, a)$. □
  - **Proposition II.** For $a > b$, $(a, b) \to (b, a)$.
    Proof. As above, choosing this time $x = -\frac{1}{2}(a + b)$. □

This analysis can be recast in our canonical form. Take as the set of objects our space of states and equip it with a binary relation, $\sim$, such that $(a, b) \sim (c, d)$ iff there is a real number x such that $(c, d) = (a + x, b + x)$.[12] Take as our problem the construction of a dynamics, to be encoded in an arrow relation satisfying Hypotheses I and II. Huygens shows that there is unique such relation that respects the symmetries of the original structure.

Hypothesis III, the principle of relativity of inertial motion, is the lynchpin of the analysis – and the element most likely to be challenged by Huygens' Cartesian contemporaries.[13] The success of this principle in solving the problem of collision, and others, ought to be the chief reason given for its acceptance.

In Huygens' proofs, the colliding objects are passed at the moment of collision from the hands of a sailor on a boat gliding down a river to the hands of a confederate at rest on the bank; the situation is arranged so that the solution of the problem for one party follows from Hypothesis II; the other party is then able to solve the problem by appeal to Hypothesis III. This suggests that Huygens viewed each dynamical state as representing the velocities of the colliding bodies relative to some observer, and viewed states related by boosts as descriptions of the *same* system from the point of view of different observers. From this perspective, the principle of relativity says that there is a single set of rules for predicting the outcome of collisions given the initial states, and that observers obtain the correct result by applying this rule to their own description of the initial state.

But there is a second way of interpreting this little theory – one more appealing, probably, to modern eyes than it would have been to Huygens.[14] We can interpret the

[12] This enforces Hypothesis III by ensuring that the symmetries of the original structure are the maps of the form $(a, b) \mapsto (a + x, b + x)$.

[13] Descartes employs the same basic framework in his laws of motion and collision (1991, Part II, sections 36–53) – but Huygens' Propositions I and II directly contradict Descartes' third and sixth rules of impact, while Huygens' Hypothesis I follows from Descartes' first and second laws of motion, and Huygens' Hypothesis II appears as Descartes' first rule of impact. So a Cartesian interested in upholding Descartes' analysis of impact against that of Huygens must locate the error of the latter in the principle of relativity (a principle that plays no role in Cartesian physics).

[14] Of course, some contemporary commentators prefer the original approach; see for example Brown and Sypel (1995). See also note 19, below.

states of the theory as encoding instantaneous absolute velocities of the particles existing in two-particle worlds. States related by boosts now represent distinct possibilities rather than distinct descriptions; and the principle of relativity tells us about the relation between the dynamics of collisions at sets of worlds related by boosts. Note that states represent sets of worlds rather than individual worlds, since they encode no information about the location of the particles – there will be many worlds corresponding to a state of the form (a, a).

3.5 Symmetries in classical mechanics

Suppose that we have n particles of equal mass moving in Euclidean space, subject to (inter-particle and/or external) forces that depend on the location of the particles in space. Then we can cast our dynamics in Hamiltonian form, writing a state of the system as (q, p) where the $3n$-vector q encodes the position coordinates of the particles while the $3n$-vector p encodes their momentum coordinates. We equip the *phase space* $T^*Q := \{(q, p)\}$ with a tensor, the *symplectic form*, $\omega := \sum dq^i \wedge dp_i$, and a scalar function, the *Hamiltonian*, $H := \frac{1}{2}|p|^2 + V(q)$.[15] The first term in the Hamiltonian is the kinetic energy, the second is the potential energy; the components of the forces on the particles are given by $-\frac{\partial V}{\partial q^i}$. There is a unique vector field, X_H, on T^*Q such that $\omega(X_H, \cdot) = dH$.[16] The flow generated by this vector field gives the dynamics of the theory. We write $(q, p) \rightarrow_t (q', p')$ when the state (q, p) evolves into the state (q', p') after t units of time.

We can study the symmetries of the structure 'phase space + symplectic form + Hamiltonian'. Because we employ smooth objects, a symmetry will be a diffeomorphism from the phase space to itself that preserves the symplectic form and the Hamiltonian.[17] Because a symmetry preserves the structures that determine the dynamics, the dynamics is also invariant under symmetries – that is, if $\phi : T^*Q \rightarrow T^*Q$ is a symmetry and $(q, p) \rightarrow_t (q', p')$, then $\phi(q, p) \rightarrow_t \phi(q', p')$.

If we restrict attention to forces, such as gravity, that depend only on the inter-particle distances (and not on the location of particles in absolute space) then transformations that correspond to shifting the system in Euclidean space,

[15] The symplectic form is a closed non-degenerate 2-form. The variety of dynamics defined below can be constructed whenever one has a manifold equipped with such a form.

[16] Note that the symplectic form on T^*Q induces the *Poisson bracket* – a Lie bracket satisfying Leibniz's rule – on the space of smooth function on T^*Q via the rule: $\{f, g\} := \omega(X_f, X_g)$. We can write the dynamics in terms of this bracket as $\dot{f} = \{f, H\}$. The advantage of this form is that it applies whenever we have a *Poisson manifold* – a manifold whose space of functions is equipped with a Lie bracket satisfying Leibniz's rule. The notion of a Poisson manifold is more general than that of a symplectic manifold – indeed, every Poisson manifold can be decomposed as a disjoint union of symplectic manifolds.

[17] The group of diffeomorphisms which preserve the symplectic form is immense – it is infinite-dimensional. But the symmetry group of the full structure will be much smaller – at most, of dimension $3n$ (in the case of an integrable system).

or reorienting it by a rotation, will be symmetries.[18] The fact that the operation of shifting the entire system in Euclidean space is a symmetry shows that there is no dynamically preferred origin in the Euclidean space the system inhabits – for the dynamical behaviour of a system whose centre of mass was at the hypothetical origin would be indistinguishable from that of an otherwise similar system that had been shifted (i.e. there is no invariant way to privilege a set of points in the phase space as representing the system as being located at the spatial origin).

So far we have been thinking of the points in our structure as representing possible complete dynamical instantaneous states of a system of particles. But given the determinism (modulo certain technicalities that we can ignore here) of classical mechanics, we might just as well think of them as representing complete possible physical histories of the particles – the specification of a state at a given time is enough to determine the entire history of the system. Thus we can take our phase space to be a space of physically possible worlds, carrying a geometrical structure and scalar which determine the dynamics – where now if $(q, p) \rightarrow_t (q', p')$ then the worlds (q', p') and (q, p) have the same sequence of instantaneous states, with these states occurring t units of time later in one world than in the other.[19]

4 Symmetries of solutions and symmetries of laws

These examples are pretty typical of those one comes across in philosophy of physics – most of the structures that arise in the course of symmetry arguments are either representations (generally, highly idealized ones) of a given spatiotemporal world, or spaces of such representations carrying a structure that encodes the dynamics of a physical theory.

Let's put it this way: suppose that we are interested in philosophical aspects of some physical theory; then we will spend some time contemplating the physics of individual solutions of the equations of the theory, and some time contemplating the content of the laws of the theory by studying (what physicists and mathematicians have to say about) the space of solutions to the equations.

I will make a few remarks about these two occupations and the relation between them.

[18] Galileian boosts are not symmetries in the present sense – they preserve the dynamical trajectories, but not the Hamiltonian.

[19] The statement in the text is tendentious – it is a matter of controversy whether two worlds can differ in this way. My view is that if one denies that the application of time translation (or any other symmetry) generates distinct physical possibilities, then one ought to prefer to the standard formulations of classical mechanics those in which the offending symmetry has been factored out. See section 5 below and Belot (2001).

4.1 Structuring the space of solutions

In the previous section we saw a couple of examples where we were *handed* structures whose objects represented (complete or partial) characterizations of possible worlds, and which came equipped with enough structure to single out a physically interesting group of symmetries. The case of classical particle mechanics, section 3.5, is entirely typical: the symmetries of classical physical theories are studied by studying the symmetries of joint structure composed of the space of states/solutions of the theory and the dynamics-determining structures defined on that space.[20]

Where do these latter structures come from?

This isn't a frivolous question – for if we just look at the differential equations of a theory or at the corresponding space of solutions, we will not have enough structure to pick out the physically interesting symmetries. If we take the latter route, and study the set of solutions as an unstructured set, then arbitrary permutations of the set will count as symmetries. If we base our analysis instead on the differential equations of the theory, we will get a bit further: the space of solutions will plausibly be equipped with a topological and a differential structure, arising out of the use of continuous variables; so the symmetries will at least be diffeomorphisms on the space of states. In either case, it will be possible to relate any pair of solutions by a symmetry – a solution of Newton's gravitational equations in which the planets are all falling into the Sun will count as 'equivalent' to one in which they are in stable orbits. This is a disaster.

Furthermore, the interesting connection between symmetries and conservation laws will be absent under such approaches. This connection can be established when the differential equations of the theory arise as the equations of motion for a Lagrangian or Hamiltonian formulation. In this case, the space of solutions is equipped with a natural geometric structure, the symmetries of the joint structure 'space of states + geometric structure + Lagrangian or Hamiltonian' has as its symmetries the intuitively correct symmetries of the theory, and each (continuous) symmetry of this structure is associated with a conservation law for the original equation(s) of the theory.[21]

[20] The spaces of states of quantum theories also come equipped with familiar structures defining the corresponding dynamics. For reasons of convenience I discuss only the classical case here. But it is interesting (and perhaps important) to note that there is a sense in which quantum theories are special cases of classical theories, as dynamical structures. (i) The space of rays of a Hilbert space carries a symplectic structure; the Schrödinger dynamics are given by solving for the vector field associated with the function on this symplectic manifold given by the expectation value of the Hamiltonian operator; see Landsman (1998, section I.2.5) or Ashtekar and Schilling (1999). (ii) More generally, the space of states of a C^*-algebra carries a dynamically relevant Poisson structure; see Landsman (1998, Proposition I.2.6.8 and Theorem I.3.8.1).

[21] If we set out from a Lagrangian formalism, then there is a natural way to equip the space of solutions of the associated Euler–Lagrange equations with a closed 2-form; this form will be non-degenerate (and hence symplectic) if the Euler–Lagrange equations have a well-posed initial value problem. See Deligne and Freed (1999, section 2).

That the setting of the theories of classical physics within the unifying framework of geometrical mechanics has proved so immensely fruitful suggests that we ought to take as basic, not the equations of motion of classical physics, but rather the Lagrangian or Hamiltonian formulations that give rise to them.[22] This is the implicit or explicit practice of most mathematicians and physicists. But what exactly this 'taking as basic' commits us to is a philosophical question that remains largely unexplored.[23] At first blush it is tempting to think that we are being told to take the set of worlds physically possible relative to some theory as carrying a structure determined by the physics itself, and that this ought to have some consequences for philosophical debates about, for example, the nature of laws, or for the nature of physical possibility more generally. But making out these claims is not a straightforward matter.

4.2 Symmetries of equations vs. symmetries of solutions

- In the case of Newtonian gravitating point particles we find, of course, that the symmetries of the laws – translation in time and the Euclidean symmetries – are not symmetries of typical solutions: only zero-particle solutions are invariant under translations in Euclidean space, and only static solutions are invariant under time translation. This is, in fact, a general feature of differential equations: generic solutions have less symmetry than do the equations that determine them. See Olver (1993, chapter 3).
- Some questions, like that of the existence of a preferred parity, admit of two construals – one focusing on properties of solutions, the other on properties of the laws. Thus we can ask whether there are types of objects or properties such that nature prefers one parity over the other (a question we can reconstrue as whether an imbalance between parities appears in solutions which are good representations of our world); or we can ask whether the laws are invariant under reversal of parity. And, of course, the latter question can be given an affirmative answer even if there are systematic preferences for one parity over the other in nature – although preferences in certain fundamental cases will suggest asymmetries in the laws.

If we set out from a strictly Hamiltonian formalism, then we begin with a space of states/space of initial data equipped with a symplectic form, which we can as usual pull back to the space of solutions by the isomorphism between the spaces induced by fixing a time at which the initial data are posed. More generally, we might work with a space of states equipped with a presymplectic form (a closed but possibly degenerate 2-form whose foliation by null manifolds has a well-behaved leaf-space) – in which case the usual construction equips the space of solutions with a similar form.

22 The structures on the space of solutions discussed in the previous paragraph need not be the end of the story – for example, often we will want to view this space as a cotangent bundle over a configuration space, and view the kinetic term in the Lagrangian or Hamiltonian as arising from a Riemannian metric on the configuration space.

23 See, however, J. Butterfield, 'Solving all problems, postulating all states: some philosophical morals of analytic mechanics', unpublished manuscript.

Now, in the case of parity, the laws version of the question is taken to be much deeper than the particular-solution version of the question. Why should that be? These days, physicists are acutely aware that their present theories are, strictly speaking, false – and much of the most creative and influential work in physics is directed towards creating new theories. The result is that the question 'What do present theories tell us about the world?' becomes 'What hints do present theories contain about future physics?' And while the physics of particular solutions undoubtedly looms large in answers to the first question, it is swamped by considerations relating to structural features of the laws in approaches to the second question. (For further discussion of parity, see Pooley, this volume.)

- Curie's principle is often taken to forbid the evolution of a system from a symmetric state into an asymmetric state. (For the original formulation, see Curie, this volume.)

The principle is true for a large class of theories.[24] Suppose that we are given a space, X, of dynamical states, along with a deterministic dynamics – i.e. for each $x \in X$ and $t \in \mathbb{R}$, there is a unique x' that x evolves into after t units of time; we write $x \rightarrow_t x'$. Suppose, further, that if a state counts as symmetric, it is in virtue of being left invariant by a non-trivial, physically relevant transformation $\Phi : X \rightarrow X$. Finally, suppose that every Φ arising in this way is a symmetry of the dynamics, in the sense that $x \rightarrow_t x'$ implies $\Phi(x) \rightarrow_t \Phi(x')$. Then if x is a symmetric state, so is each x' such that $x \rightarrow_t x'$: by the symmetry of the dynamics, $\Phi(x) \rightarrow_t \Phi(x')$; and by the symmetry of x, $\Phi(x) = x$; so $x \rightarrow_t x'$ and $x \rightarrow_t \Phi(x')$; so by the determinism of the dynamics, $x' = \Phi(x')$.

So in the classical realm, Curie's principle holds so long as the symmetry operations performable on states are also symmetries of the dynamics. This appears to hold true for realistic systems. But it is easy enough to violate for artificial examples.

Consider a Hamiltonian theory of three point particles. Let us say that a state is *equilateral* if it represents the particles as forming an equilateral triangle, with momenta of equal magnitude directed in the same sense along the angle bisectors. Let us say that a state is *scalene* if it represents the particles as forming a scalene triangle. Let (q, p) be some equilateral state; and let Φ be a transformation on the phase space corresponding to the action of reflection in Euclidean space through one of the angle bisectors. Φ permutes the states of our theory, but leaves (q, p) invariant – and it is in virtue of being invariant under such Φ that an equilateral state counts as symmetric. If gravity is the only force acting on our particles, then each such Φ, being a Euclidean symmetry, is a symmetry of our dynamics – and hence leaves (q', p') invariant if $(q, p) \rightarrow_t (q', p')$, so that Curie's principle is satisfied. But we can also consider deviant theories,

[24] See Ismael (1997); J. Earman, 'Spontaneous symmetry breaking for philosopers', unpublished manuscript, and the references therein.

constructed by messily altering the expression for the gravitational potential energy outside of some open neighbourhood of (q, p), so that Φ is no longer a symmetry of the dynamics. In such a theory we have $(q, p) \rightarrow_t (q'', p'')$ with, in general, $(q'', p'') \neq (q', p')$ and (q'', p'') a scalene state. So we have a violation of Curie's principle – the initial state is symmetric with respect to the configuration variables, momenta, and forces, but it evolves into an asymmetric configuration.

Enthusiasts of the principle may be tempted to dismiss such counter-examples. After all, our equilateral state evolves into a scalene state – so one of the three originally congruent sides of the triangle ends up being the longest. In the simplest cases, this will be because the potential of the new theory encodes forces which differentiate between directions in Euclidean space, or treat, say, the first particle differently from the second and the third. It is tempting, perhaps, to think that more complex cases are just variants on these two options. And if this is so, then the example under consideration ought not to be viewed as a counter-example to Curie's principle, for the initial equilateral state is not genuinely symmetric – reflection or interchange of particles is not a *real* symmetry of the initial state, given the sort of information that the potential function takes into account.

Now there is something funny about this objection, since by construction the potential is identical to the Newtonian gravitational potential on an open neighbourhood of the initial equilateral state. So, in effect, we are told the initial state is not symmetric because . . . it later evolves to an asymmetric state.

In any case, the presupposition of the objection – that asymmetries in our final state must be grounded in the potential's caring about directions in space or the identity of the particles – is mistaken. We could have first eliminated Euclidean and permutation symmetries from the Newtonian theory (see section 5.2 below) before perturbing the potential. The result would have been a theory in which the particle states are characterized by relative distances and velocities (so the particles live in a relational space, without absolute directions), and in which the state of the system is characterized by a set of three-particle states rather than by an ordered triple of particle states (so that there is no longer a question of which particle is which). Nonetheless, it is possible to rig the potential so that an initial state in which the three relative distances are equal and the three relative velocities are equal evolves into a state in which the particles form a scalene triangle. Only great stubbornness could lead someone to insist that such an initial state should not count as symmetric.

5 Quotienting out symmetries

For every structure $\mathcal{S} = (D, \{R_i\}_{i \in I})$, we can define the associated *quotient structure* $\bar{\mathcal{S}}$, which arises by factoring out the symmetries of $\mathcal{S}$. We define an equivalence relation, $\sim$, on D by declaring $x \sim y$ whenever there exists a symmetry of $\mathcal{S}$,

$\phi : D \rightarrow D$ with $y = \phi(x)$. The equivalence class of x under this relation is denoted $[x] := \{y \in D : x \sim y\}$. $\bar{\mathcal{S}}$ has as its set of objects the set $[D] := \{[x] : x \in D\}$, of equivalence classes of $\sim$. For each n-ary relation R_i of $\mathcal{S}$, $\bar{\mathcal{S}}$ has an n-ary relation $[R_i] := \{([x_1], \ldots, [x_n]) : (x_1, \ldots, x_n) \in R_i\}$.

- If $\mathcal{S}$ admits no non-trivial symmetries, then $\bar{\mathcal{S}}$ and $\mathcal{S}$ are isomorphic. If $\mathcal{S}$ is homogeneous – if every pair of its objects are related by a symmetry, as in the case of the order structure of the integers or the rationals – then the quotient has a single object, related to itself by the counterparts of all the non-vacuous relations of $\mathcal{S}$. The interesting cases lie in the intermediate region.
- We can use the same language to describe both structures (using the same predicate symbol, '$\mathcal{R}_i$', as a name for both R_i and $[R_i]$). Choosing names for our new objects generates a complete description of the quotient structure.

  The complete descriptions of the two structures will be closely related. A complete description of the original structure can be transformed into a complete description of the quotient structure by taking names of objects related by symmetries to name identical objects.

  We can also consider the relations between theories describing the two structures. For instance, any constant-free sentence that employs only one-place predicates or that is free of negation symbols will be true of the quotient if it is true of the original structure.[25] But sentences combining multi-place predicates and negation symbols need not be true: let $\mathcal{S}$ be the countable structure whose sole relation, R, gives it the order structure of the integers; then $\bar{\mathcal{S}}$ has a single object, $[x]$, with $[R]([x], [x])$. So the sentence $\forall x \sim \mathcal{R}(x, x)$ is true in $\mathcal{S}$ but false in $\bar{\mathcal{S}}$.
- More generally, we can quotient out by the action of a subgroup of the full symmetry group – declare two objects to be equivalent if related by an element of the chosen subgroup, go on to take equivalence classes, etc.

5.1 *Quotienting solutions*

If one has a description of a possible world – in the form of a solution of a differential equation, for instance – one can ask whether it admits any symmetries. If it does, then one can consider the distinct description that arises as the quotient of the original.

Advocates of the Principle of the Identity of Indiscernibles will want to deny that any description admitting symmetries corresponds directly to a possible world, while granting that the related quotient (*ceteris paribus*) does so.[26] Indeed, they can go on to insist that a description admitting symmetries is merely a misdescription

[25] Indeed, something stronger is true: if a constant-free sentence is in negation normal form (so that any negation symbols apply to atomic formulae) and each of its negation symbols applies to a one-place predicate, then its truth in the original structure implies its truth in the quotient structure. This follows from Hodges (1997), Theorem 8.3.3(a), since the map $x \mapsto [x]$ is a surjective homomorphism that fixes one-place relations.

[26] For this attitude, see Hacking (1975) and L. Smolin, 'The present moment in quantum cosmology: challenges to the arguments for the elimination of time', PITT-PHIL-SCI 00000153.

of the corresponding quotient, under the strange convention according to which some objects are given multiple names.

To most, this will appear unmotivated at best.[27] Like Black (this volume), many suppose that there could be a Euclidean world, otherwise empty except for two identically constituted iron spheres. What motivation is there for saying that the only possibility in the neighbourhood is the uglier world, containing a single such sphere, related geometrically to itself in its strange non-Euclidean space? Even in cases where there is no gain in awkwardness in passing from the symmetric description to its quotient – as in passing from the covering spacetime of a non-simply connected spacetime to the non-simply connected spacetime itself – many will still feel that there are two genuine possibilities in the neighbourhood.

5.2 Quotienting the space of solutions

If we take the space of solutions of a classical physical theory to be equipped with the sort of rich structure discussed in section 4.1 above, then taking the quotient by the action of a group of symmetries often leads to an interesting result – sometimes with interpretative implications. There exists a large mathematical literature on this technique.[28] I think that the techniques and results of this literature promise to offer a unifying perspective on a number of classic problems in philosophy of physics (the relation between the nature of space and the nature of motion in Newtonian physics, identical particles, the nature and significance of gauge freedom and general covariance).

Let us return to our example of n gravitating point particles.

- Example: Euclidean symmetries. Consider the theory of n gravitating Newtonian point particles. The symmetries of Euclidean space – translation, rotations, reflections, and their products – are symmetries of this theory (action on initial data by one of these symmetries transforms the dynamical trajectory by the action of same symmetry). Call this group $E(3)$. Translations and rotations are generators of the continuous symmetries of the theory, and the corresponding conserved quantities are the total linear and (centre of mass) angular momentum.

  Let δ be the set of points which represent the particles as forming symmetric configurations in Euclidean space. Let Δ be the set of collision points of the phase space, representing states in which two or more particles occupy the same point of Euclidean space. These are sets of measure zero which we excise from the phase space.[29] Call the resulting phase space M. We are also interested in

[27] Hacking's motivation, I take it, stems from his views on the nature of logic; see Hacking (1978; 1979).

[28] For introductions to the mathematics, see Marsden (1992), Marsden and Ratiu (1994), and Singer (2001).

[29] The excision of δ is a convenience that allows us to sidestep complications in the construction of the quotient theory; see Belot (2003, section 10) for discussion and references. The excision of Δ is more essential, since certain types of collision singularity are intractable.

a second space, M_0, the subspace of M corresponding to states in which the system has vanishing linear and angular momentum. Since M_0 is a dynamically closed subspace (being defined by the vanishing of conserved quantities), we can consider the dynamical theory defined upon it by structure inherited from M. We are interested in the quotient structures, $\bar{M}$ and $\bar{M}_0$, that arise when we take the quotient of these spaces by the action of $E(3)$. Each of these encodes a mathematically well-behaved physical theory.[30]

$\bar{M}_0$ gives a theory closely related to the Barbour–Bertotti form of relational dynamics (Barbour and Bertotti, 1982). The points of the quotient space are parameterized by the relative distances and relative velocities of the particles. The dynamics is such that specifying an initial point determines a dynamical trajectory – which gives the same evolution of the relative distances and relative velocities as one gets if one chooses an initial point in the full Newtonian theory with the same pattern of relative distances and relative velocities for the particles and with vanishing angular momentum, then reads off the subsequent values of the relative distances and relative velocities from the Newtonian evolution. Thus $\bar{M}_0$ is an attractive relational theory of motion: it is, mathematically, of the same form as the Newtonian theory (one does not, for instance, have to specify higher derivatives in order to get a well-posed initial value problem); and the relative distances and relative velocities between the particles form a dynamically closed set, whose evolution is deterministic; furthermore the predictions for the empirically accessible variables match those of the Newtonian theory on the cosmologically relevant non-rotating sector. It is also a theory which is naturally set in the space of a classical relationalist about Euclidean space: the theory recognizes no difference between the state of a system located *here*, and one differing only by being shifted over *there*.

The story with $\bar{M}$ is a bit more complicated. We started in M with $6n$ variables (corresponding to the components of the position and momentum of each particle), then we eliminated six of these by identifying points related by the six-dimensional group of Euclidean symmetries. $3n - 6$ of the remaining variables fix the relative distances of the particles and $3n - 6$ of them fix the relative velocities between the particles. Three further variables correspond to the velocity of the centre of mass of the system – and can simply be dropped from the theory, since they are dynamically inert. Informally speaking, the final three variables encode information about the angular momentum of the system – these variables stand in the way of the most straightforward sort of relativist/relationalist interpretation.[31]

30 Each inherits a Hamiltonian from M, and carries a geometric structure adequate to determine a dynamics – $\bar{M}$ is a Poisson manifold while $\bar{M}_0$ is a symplectic manifold.

31 These variables correspond to the components of the angular momentum in a frame rotating with the system (rather than in the usual spatially fixed frame). The magnitude of the corresponding vector is preserved, but its direction evolves in time.

- Example: permutation symmetries. Let's alter the notation slightly: now we write the states as $((\vec{q}_1, \vec{p}_1), \ldots, (\vec{q}_n, \vec{p}_n))$ where $\vec{q}_i$ is the position vector of the ith particle and $\vec{p}_i$ is the corresponding momentum. Now consider the transformation $\pi_{(12)} : ((\vec{q}_1, \vec{p}_1), \ldots, (\vec{q}_n, \vec{p}_n)) \mapsto ((\vec{q}_2, \vec{p}_2), (\vec{q}_1, \vec{p}_1), (\vec{q}_3, \vec{p}_3), \ldots, (\vec{q}_n, \vec{p}_n))$, and the corresponding transformations $\{\pi_{(ij)}\}$ for $1 \leq i < j \leq n$. Each of these is a symmetry of our theory, so the group that they generate, S_n, is a symmetry group of the theory.[32] This group permutes the identity of the particles (see French and Rickles, this volume, for further discussion). Let us denote by $\hat{M}$ the quotient of M by the action of S_n. This is again a well-behaved physical theory.[33] We can think of its states as consisting of sets $\{(\vec{q}_1, \vec{p}_1), \ldots, (\vec{q}_n, \vec{p}_n)\}$ of particle states rather than ordered tuples of particle states – the theory keeps track of how many particles are in each state, rather than which particle is in which state. It is an interesting fact that M and $\hat{M}$ underwrite the same statistical theories (see ter Haar, 1995, section 5.9; Huggett, 1999). The reason is as follows. In classical statistical mechanics, one is interested in the ratio of phase space volumes. Now the volume measures in play in M and its subspaces are S_n-invariant, deriving ultimately from the S_n-invariant symplectic form. And the physically interesting subsets of M are also S_n-invariant. The volume of the image of such a subset in $\hat{M}$ will just be $\frac{1}{n!}$ of its volume in M ($n!$ counts the number of elements in S_n). So we will get the same answer whether we measure the ratio of two such regions in M, or measure the ratio of their images in $\hat{M}$.

  Now, the difference between Maxwell–Boltzmann statistics and Bose–Einstein statistics is often informally cashed out in terms of ways of counting possible outcomes of double coin-flips: under the first set of statistics, there are two ways to get {H,T}, under the second, only one. Does the argument above show that classical particles obey Bose–Einstein statistics after all? *No* (see Huggett, 1999, pp. 16–17). A better way of describing the difference between the two sorts of statistics is to say that under one {H,T} is twice as likely as either {H,H} or {T,T}, while under the other these three alternatives are equiprobable. But recall that in constructing M we eliminated all points in which there were particle collisions, including all states in which distinct particles share the same position and momentum – so we cannot make the comparison of the likelihood of classes of states fixed by permutations with classes of states not fixed by permutations in M, or the corresponding comparisons in $\bar{M}$. [34]

[32] Note that while it was a mere notational convenience to take the particles to be of equal mass in previous examples, it is here a necessity.

[33] It is a symplectic manifold, equipped with the projected Hamiltonian.

[34] What happens if we restore to our phase space the subset of Δ consisting of the physically bizarre states fixed by elements of S_n, in which two or more particles share the same position and momentum? On the one hand, not much: this set and its image under the projection to $\hat{M}$ are of measure zero in their respective spaces – so we don't get any difference between the relative weight assigned to these two sets in their respective spaces. On the other hand, there is the following suggestive fact: for a generic (permutation symmetry-free) state there

Whenever we compare two spaces of possible worlds, one the quotient of the other, we are contrasting two ways of counting possibilities. In the two examples discussed above, and in other examples that arise naturally in philosophy of physics, we are in essence faced with the difference between a relatively haecceitistic means of counting possibilities and a relatively anti-haecceitistic means of counting possibilities.[35] In M, but not in $\bar{M}$, if it is possible for the particles to be *thus and so* and to be *here*, then it is also possible for them to be *thus and so* and to be *there* – with the two states qualitatively identical, and differing only as to which spacetime points are occupied. In M, but not in $\hat{M}$, if it is possible for the particles to be *thus and so* with *this* particle playing a certain role, then it is also possible for them to be *thus and so* with *that* particle playing the given role – with the two states qualitatively identical, and differing only in their distribution of roles to particles.

Note that while in the case of the Euclidean symmetries, it was natural to think of (some versions of) the quotienting procedure as leading to the elimination of spacetime points, there is no such straightforward ontological purge in the identical-particles case. For even after the quotient has been taken, the space of states is rich enough to associate with each dynamically possible history a set of continuous spacetime trajectories, labelled by features such as mass, etc. – so there is little impetus to say that anything has been eliminated from the ontology of the worlds described by the theory. Why the difference? Well, the existence of spacetime points is closely tied up with questions of counting of possibilities – so they are vulnerable to elimination in the transition from a haecceitistic means of counting to an anti-haecceitistic one. But in the case of particles we have much more to hang on to.

There is also a crucial technical distinction between the two cases. Taking the quotient of a phase space by the action a discrete group of symmetries yields a phase space that is almost everywhere locally equivalent to the original. This is not so if one works with a continuous group: the presence of continuous symmetries indicates the inclusion of dynamically irrelevant variables in the space of states; the elimination of such variables results in an interestingly distinct theory (the new phase space will be of smaller dimension than the original, and (unless one restricts attention to a subspace) will be a Poisson space even when the original phase space is symplectic).[36]

are $n!$ states upstairs for every state downstairs; but if we look at states in which exactly two particles share the same position and momentum, this factor goes down to $\frac{n!}{2}$.

35 Hence while there may perhaps be more temptation to apply the quotienting procedure across the board in the case of spaces of solutions than in the case of individual solutions, familiar haecceitistic modal intuitions – see Adams (1979) and Lewis (1986, section 4.4) – provide a countervailing force.

36 Gauge theories provide an especially vivid example. One starts with a theory in many ways analogous to that of M_0 above, then quotients out the action of the infinite dimensional group of gauge transformations. For details, see Belot (2003), and Earman (this volume, Part I), and Redhead (this volume).

Acknowledgements

I would like to thank the editors of this volume for their patience, encouragement, and helpful advice.

References

Adams, R. (1979). 'Primitive thisness and primitive identity'. *Journal of Philosophy*, **76**, 5–26.

Ashtekar, A., and Schilling, T. (1999). 'Geometrical formulations of quantum mechanics'. In *On Einstein's Path*, ed. A. Harvey. Berlin: Springer-Verlag.

Barbour, J., and Bertotti, B. (1982). 'Mach's principle and the structure of dynamical theories'. *Proceedings of the Royal Society of London A*, **382**, 295–306.

Belot, G. (2001). 'The principle of sufficient reason'. *Journal of Philosophy*, **98**, 55–74.

(2003). 'Symmetry and gauge freedom', *Studies in History and Philosophy of Modern Physics*, **34**, 189–225.

Brown, H., and Sypel, R. (1995). 'On the meaning of the relativity principle and other symmetries'. *International Studies in the Philosophy of Science*, **9**, 235–53.

Cornford, F. (1997). *Plato's Cosmology*. Indianapolis, IN: Hackett.

Deligne, P., and Freed, D. (1999). 'Classical field theory'. In *Quantum Fields and Strings: a Course for Mathematicians*, Vol. 1, ed. P. Deligne *et al.* Providence, RI: American Mathematical Society.

Descartes, R. (1991). *Principles of Philosophy*. Dordrecht: Kluwer.

Eddington, A. (1920). *Space, Time, and Gravitation*. Cambridge: Cambridge University Press.

Giulini, D. (2001). 'Uniqueness of simultaneity'. *British Journal for the Philosophy of Science*, **52**, 651–70.

Gödel, K. (1949a). 'An example of a new type of cosmological solution of Einstein's field equations of gravitation'. *Review of Modern Physics*, **21**, 447–50.

(1949b). 'A remark about the relationship between relativity theory and idealistic philosophy'. In *Albert Einstein: Philosopher–Scientist*, ed. P. Schilpp. New York: Harper and Row.

(1952). 'Rotating universes in general relativity theory'. *Proceedings of the International Congress of Mathematicians, Cambridge, MA, 1950*. Providence, RI: American Mathematical Society.

Hacking, I. (1975). 'The identity of indiscernibles'. *Journal of Philosophy*, **72**, 249–56.

(1978). 'On the reality of existence and identity'. *Canadian Journal of Philosophy*, **8**, 613–32.

(1979). 'What is logic?' *Journal of Philosophy*, **76**, 285–319.

Hodges, W. (1997). *A Shorter Model Theory*. Cambridge: Cambridge University Press.

Huggett, N. (1999). 'Atomic metaphysics'. *Journal of Philosophy*, **96**, 5–24.

Huygens, C. (1977). 'The motion of colliding bodies', translated by R. J. Blackwell. *Isis*, **68**, 574–97.

Ismael, J. (1997). 'Curie's principle', *Synthese*, **110**, 167–90.

Jeans, J. (1936). 'Man and the universe'. In *Scientific Progress*, ed. E. Appleton *et al.* New York: MacMillan.

Kramer, D., Stephani, H., Herlt, E., and MacCallum, M. (1980). *Exact Solutions of Einstein's Field Equations*. Cambridge: Cambridge University Press.

Landsman, N. (1998). *Mathematical Topics Between Classical and Quantum Mechanics*. Berlin: Springer-Verlag.
Lewis, D. (1986). *On the Plurality of Worlds*. Oxford: Blackwell.
Malament, D. (1977). 'Causal theories of time and the conventionality of simultaneity'. *Noûs*, **11**, 293–300.
Marsden, J. (1992). *Lectures on Mechanics*. Cambridge: Cambridge University Press.
Marsden, J., and Ratiu, T. (1994). *Introduction to Mechanics and Symmetry*. Berlin: Springer-Verlag.
Minkowski, H. (1952). 'Space and time'. In *The Principle of Relativity*, ed. A. Einstein *et al.* New York: Dover.
Olver, P. (1993). *Applications of Lie Groups to Differential Equations*. Berlin: Springer-Verlag.
Singer, S. (2001). *Symmetry and Mechanics*. Basel: Birkhäuser.
Stein, H. (1991). 'On relativity and openness of the future'. *Philosophy of Science*, **58**, 147–67.
ter Haar, D. (1995). *Elements of Statistical Mechanics*, 3rd edn. Oxford: Butterworth-Heinemann.

25

Symmetry, objectivity, and design

PETER KOSSO

1 Introduction

Symmetry has an undeniable heuristic value in physics, as is demonstrated throughout this volume. To see if the value is more than instrumental, that is, to see if the symmetries are somehow in nature itself, we should ask a transcendental question: what must the world be like such that symmetry would be so effective in understanding it?

I will describe two apparent conditions of symmetry: objectivity and design. It will turn out that only one of these, objectivity, can be securely linked to symmetry. But symmetry does not serve as evidence for design in nature. In fact, the very aspects of symmetry that link it to objectivity suggest that it is not the result of design. The two apparent implications of symmetry are incompatible, and there is clear reason to retain the notion of objectivity and give up design.

2 Symmetry and objectivity

The accomplishment of knowledge involves keeping track of relations between the permanent and the ephemeral. Sensations keep changing while the relevant categories to describe them stay the same, and we have empirical knowledge of the world when we can accurately associate the fleeting sensations with their more permanent concepts. Coordinate systems remain fixed as an object's position changes, and the science of kinematics is useful insofar as it can describe the variable positions in terms of the stable reference frame. In general, knowledge is intimately involved in the interplay between what changes and what doesn't.

Étienne Klein and Marc Lachièze-Rey (1999, p. 11) make this point specifically about physics: 'The very goal of physics is to understand what changes in terms of what is permanent'. But perhaps they have the order wrong, targeting what changes as the goal and using what stays the same as the means. In fact, the importance of symmetry in physics suggests that the roles are reversed. Symmetries put the focus on invariance under transformation,[1] that is, understanding what is permanent in terms of what changes. This may be just a matter of emphasis, in the sense that the goal should really be to understand both the variable and the constant, each in terms of the other. But using symmetry as the analytic tool unavoidably highlights what does not change.

Emphasis on invariance was made early on by Hermann Weyl. Weyl saw the importance of symmetry as rooted in its ability to separate objective facts from the conventions of description. He associated invariance with objectivity, arguing that 'objectivity means invariance with respect to the group of automorphisms' (1952, p. 132). By noting what is invariant in our changing perspectives on the world, Weyl suggests that we can discover what is really in nature and separate it from the artifacts of description.

Not surprisingly, this effectiveness of symmetry has spread beyond the discipline of physics. Robert Nozick applied symmetry as a general epistemic tool in his presidential address to the American Philosophical Association, declaring that 'An objective fact is invariant under various transformations' (1998, p. 21). This idea is expanded as the central premise in Nozick's book, *Invariances: The Structure of the Objective World* (2001).

There has been a lot of informative philosophical analysis of the concept of objectivity.[2] I do not offer any precise characterization other than that the objective components of knowledge are those that reveal information from nature itself, in contrast to the components of information added by us. Nature's contribution to our description, that is, nature's constraint on our theorizing, will be shown as those aspects of the description over which we have no control. These objective aspects of knowledge will be those that stay the same through all our various, subjective ways of describing the world. This constraint, independent of us, reveals something about nature rather than about us. Thus objectivity has epistemic value as a link to reality, and is expressed as permanence under change of perspective.

By separating the ephemeral effects of perspective from the enduring reality, symmetries are a key epistemic tool, facilitating the step from appearance to reality. Assuming that what is real is perspective-independent, and broadening the concept of perspective to include all kinds of descriptive systems, symmetries amount to the discovery of reality in the appearance by understanding what is permanent in terms of what changes.

[1] I start by using the common description of symmetry as invariance under transformation, but it will soon be clear that in associating symmetry with objectivity, covariance, rather than invariance, is the more appropriate term for what remains the same.

[2] A sampling of recent work includes Newell (1986), Daston and Galison (1992), Galison (1999), and Nozick (2001).

The role of symmetries in physics is thus relevant to the question of the epistemic limits of physics as characterized by the disagreement between Bohr and Einstein. Regarding the potential of physics and its realistic limitations, Bohr cautioned that 'It is wrong to think that the task of physics is to find out how nature is. Physics concerns what we can say about nature' (quoted in Petersen, 1963, p. 12). The implication is that we are stuck with appearance, with no objective access to the way nature really is. This is famously countered by Einstein's realism (Einstein, 1949, p. 81): 'Physics is an attempt conceptually to grasp reality as it is thought independently of its being observed. In this sense one speaks of physical reality.' Symmetry, with its link to objectivity, can be important evidence in this debate.

The question of realism that divides Bohr and Einstein is an issue of epistemology. It is about our ability to know about nature, not directly about the way things are in nature. And this is the appropriate context in which to consider the objectivity associated with symmetries in physics. Discovering that there is invariance under transformation is demonstrating that there are ways in which we can change our perspective on the world without effecting a change in all aspects of the appearance. There are constraints, natural constraints on the evidence. In this way, our description of the natural world is in some ways subjective, given some freedom to choose a perspective, but in other ways objective, given the independence of some aspects. Thus we can know some aspects of nature that are independent of our perspective. Symmetries play a role in responding to Hans Reichenbach's epistemological challenge (1958, p. 37): 'The only path to objective knowledge leads through conscious awareness of the role that subjectivity plays in our method of research.'

The subjective role in choosing a perspective and a reference frame is similar in some respects to the subjective role in employing a language to describe experience. A subjectively chosen language can deliver objective information about the world. Individually arbitrary terms for things come together in naturally constrained relations. This aspect of language helps clarify an epistemic value of symmetry and the link between symmetry and objectivity.

In our day-to-day interactions with the world and with each other, we are free to choose a language for description and communication. Not only is there some choice in which language to use, English, Greek, or whatever, but each of these languages has been formed by convention. The words, their link to objects in the world, and the rules of grammar are all subjective. They are influenced by imagination, creativity, and other forms of human self-indulgence. And they are changeable by human will. Most informative for our purposes is the ability to change the description of nature from one language to another. The analogy is to a change from one reference frame to another. The transformation is not haphazard or capricious, since once the particular frame is chosen the values of properties are fixed, just as the choice of a language fixes the descriptive terms and their reference.

The important feature of the linguistic analogy is this. Once a particular language system is chosen, nature itself constrains the possible combinations of words. We are free to choose English (or any other language) as our descriptive tool. We are free, that is, to choose 'Sun' to refer to the bright, hot disc in the daytime sky, and choose 'Moon' and 'Earth' as we do. But once this choice is made we have no control over the proper relations among these particular words. We have to say, for example, that the Moon is closer to the Earth than is the Sun. The linguistic conventionality and the subjective influence ends at the level of making propositions. There is a kind of propositional objectivity.

To push this analogy in the direction of symmetry and physics, the ideas of propositional objectivity and word-choice subjectivity lead to a kind of linguistic covariance. In different languages, features of the world are described by different words. That is, in the transformation from one language to the next, from English to German, say, individual linguistic values change. Under transformation from English to German, the word 'Moon' is not invariant; it changes to 'Mond'. And there is no reason to expect invariance of translation for individual words, since the choice of a sound and sequence of letters to represent a particular thing in the world is highly subjective. Adapting Weyl's assessment of simultaneity as having no 'objective significance' (1952, p. 130), we can say that the particular word used to refer to the waxing and waning white disc in the sky has no objective significance either.

But combinations of words do have objective significance, and this is demonstrated by the linguistic covariance. While transformations from one language to another change values of terms, these values must covary in a way that leaves the propositional content, the meaningful relations among terms, unchanged. This is the translational covariance. It is not that grammatical structure is preserved; rather it is the propositional content that stays the same. In English we describe the local astronomical situation by saying that the Moon is closer to the Earth than is the Sun. Under transformation to German, 'Moon' goes to 'Mond', 'Earth' goes to 'Erde', and 'Sun' goes to 'Sonne'. These I had to look up, because of the subjectivity inherent in languages. There is no algorithm for translation, and in this way the analogy between language and reference frame is flawed. There are algorithmic transformations between different reference frames. But even in the looser linguistic case, there is no lingering subjectivity in the proper combination of terms to describe the astronomical arrangement. *Der Mond ist näher als die Sonne zur Erde.* This propositional covariance results from the astronomical facts of the matter. Nature constrains the relation among terms, and this covariance reveals what is objectively significant in the description.[3]

[3] Note that the covariance would be trivially easy and no indication of the state of nature if we allow meanings of terms to vary without restriction. Any proposition can be made true if we allow *ad hoc* re-interpretation of terms. The covariance is only meaningful when interpretation of individual terms remains fixed. This restriction is

There are other noteworthy differences between the use of language and the role of symmetry in physics. Most important is that transformations of reference frames are constrained in a way that languages are not. A foundational accomplishment of the Special Theory of Relativity, for example, is the discovery that nature allows Lorentz-transformed coordinate systems but not Galilean-transformed systems. As John Norton (this volume) points out, such restrictions on allowable coordinate systems contribute to the 'physical content' of the theory. There are no analogous restrictions in natural languages.

Even with these limitations, the linguistic analogy highlights relevant aspects of the role of symmetry in physics. Language is necessary for describing nature, and language is inescapably subjective, but the subjective language can nonetheless be an objective tool and can provide objective access to nature. Furthermore, the objectivity is revealed by restrictions on the way that terms can be put together. These are not our restrictions; they are nature's. In this sense the covariance, the constancy in the semantic relations among individual terms from one description to another, indicates what is physical fact rather than descriptive artifact. The link is between covariance and objectivity, and it applies to relations among descriptive terms. It is not directly relevant to terms taken individually.

To apply this analogy to the role of symmetry in physics, generalize the idea of a reference frame to be any descriptive system. The most natural idea of a reference frame involves spacetime coordinates, and it is commonplace to consider the symmetries of Lorentz frames and transformations of rotation or translation. But choosing a quantum-mechanical observable is also specifying a reference frame. We can use a position representation or a momentum representation, and it is possible to transform the description from one to the other just as we transform from one spacetime reference frame to another. Choosing a gauge representation is another kind of reference frame. And so on. Once these representational choices are made, the natural system is projected onto the frame. Different reference systems render different property values, but the relations among the values are the same. The laws are covariant, and this is expressed as a symmetry. The laws of nature are preserved under transformations from one reference frame to another. And this is why covariance, rather than invariance of individual properties, is the primary link between symmetry and objectivity.[4]

It is worth repeating that in physics, unlike in translations between natural languages, only certain transformations between reference systems are allowable.

relevant to the Kretschmann objection to Einstein's accomplishment of a generally covariant theory of relativity. (See Norton, this volume.)

4 There is an important question of the epistemic or ontological import of invariance. Max Born (1953) argues persuasively that relative properties are no less real than invariant properties. But it is not uncommon to see reality reserved for the invariant (see, for example, Lange 2001, p. 225). These discussions are about ontology. Clarification is also needed regarding the epistemic role of invariance and its link to objectivity.

Discovering which transformations are allowed gives empirical content to particular symmetries. In physics, it is not symmetry in general that is important, but specific symmetries such as Lorentz symmetry. And from an empirically founded global symmetry such as Lorentz symmetry, there is the prescribed technique of localizing the transformation to find the associated dynamical symmetry. This is the so-called 'gauge argument'. (See Norton, Earman, and Martin, this volume, Part I.) The important point is that the particular symmetries in physics introduce explicit restrictions on allowable transformations. We are not entirely free to choose a reference system in the way we are free to choose a language.

Some choice does remain to us though, and this gives us the opportunity to note what always remains the same against this changing background. This is the link between the covariance of laws and objectivity. A principle of covariance can be restated to say that the laws of nature must be objective. They must be the same in all allowable representational systems, showing their independence from the vagaries of human perception and description. This is the link between symmetry and objectivity.

Every symmetry principle is thus a principle of objectivity. But it is important to note that this link has only been applied to relations among properties, not to properties themselves or to objects themselves. The argument here is not that values of properties are invariant and hence objective; rather, it is relations among the values that are covariant and hence objective. It is about covariance, not invariance. Eugene Wigner points this out (1967, p. 19): 'It is good to emphasize at this point the fact that the laws of nature, that is, the correlations between events, are the entities to which the symmetry laws apply, not the events themselves.'

This emphasis on the covariance of laws over the invariance of individuals complements the conclusion reached by French and Rickles in their analysis of permutation symmetry in this volume. Though my perspective is epistemological while theirs is ontological, both arguments provide, as they put it, 'a natural home for the structural understanding' of symmetries in physics. On this understanding, physics is primarily about relations in nature, and the success of a theory is in terms of its accurately describing nature's underlying structure. It is not about identifying the individual constituents.

Henri Poincaré put the same emphasis on accuracy in describing the basic relations in nature rather than getting the individuals right.[5] Describing the individual components of Fresnel's and Maxwell's theories of electromagnetism, Poincaré (1905, p. 161) explains that:

[5] I acknowledge the wonderful paper by John Worrall (1989) on structural realism for pointing out Poincaré's introduction of the idea.

these are merely names of the images we substituted for the real objects which Nature will hide for ever from our eyes. The true relations between these real objects are the only reality we can attain, and the sole condition is that the same relation shall exist between the images we are forced to put in their place.

In other words, we can label the pieces as we choose. That part is subjective and there is no reason to think it is descriptive of nature. We cannot access the piecemeal level of reality. But we can know about the underlying structure, and the evidence is in the persistent relations among things. It is the covariant relations that reveal nature's constraint on human theorizing.

The point is this. The link between symmetry and objectivity is primarily about laws, not about individual events, objects, or properties. Covariance of laws shows our lack of control over the situation. We can choose a reference system and, given the physical restrictions specified by the particular symmetry, invoke a transformation from one system to another. Within the allowable group of transformations, we can design whatever reference frame suits us, whether it is the spacetime frame of a laboratory or a passing train. To this extent, the description of nature is our choice and under our control. But we cannot control the resulting organization of the terms, the covariant laws. This is nature's constraint. In this way the symmetry, in particular the covariant results, show an immunity from our influence and our intentions.

3 Symmetry and design

Symmetries in nature have been cited as evidence of design, and there is much intuitive appeal to this. The argument often has an aesthetic aspect. Symmetry is a sign of beauty and precision, and the regularity shows precise planning. Symmetry is cited as evidence of the intentions and creative influence of a mindful designer. Anthony Zee makes this connection throughout his book *Fearful Symmetry* (1986), starting in chapter 1, 'Symmetry and design', and continuing with the search for the 'Ultimate Designer' (pp. 3 and 281). Zee begins with the assumption that 'Nature, at the fundamental level, is beautifully designed' (p. 3), and the 'beauty means symmetry' (p. 13). Thus, 'we do not doubt that symmetry will light our way in our quest to know His mind' (p. 283).

The intuition behind this link between symmetry and design is most likely rooted in the geometric symmetries of physical objects. An amorphous lump of stuff can happen by accident, but a symmetric shape like a snowflake or a paper doll only happens on purpose. Give a monkey paper and scissors and he'll make a mess. Give me the same tools and materials and I can return a bilaterally symmetric paper doll. The symmetry won't have come about spontaneously; I did it intentionally. I folded the paper and cut out half a doll. The symmetry is there by design, and so symmetry seems to be evidence of design.

To cite symmetry in nature as evidence of design would be, naturally enough, a kind of argument from design. As such, it employs an analogy between the products of human design and the natural world. The world is like a great machine, like a watch, and so the world must have been brought into being by a great watchmaker. Symmetry only shows up in our products when we put it in on purpose, by design, and so the best explanation for symmetries discovered in nature is that they too were put in on purpose, by design.

The link between symmetry and design goes through the analogy to human intentions and human influence over their creations. But in the previously described link between symmetry and objectivity we saw that symmetries reveal aspects of nature that are *not* influenced by human intentions. Symmetry indicates objectivity because it shows a lack of control and an independence from any purposeful, intentional choices that would be associated with design. Creative influence is blocked by the symmetry in the laws of nature, and that is exactly how symmetry is valuable for showing objectivity. This suggests a tension between objectivity and design. Objectivity is about the lack of intentional control; design is about having intentional control.

This raises the question of whether we can have both design and objectivity as implications of symmetry. And there are more fundamental reasons besides this apparent incompatibility with objectivity to argue that symmetry does not indicate design. Symmetry, we will see, does not serve as evidence in an argument for design, and the reason for this is found directly in the definition of symmetry as invariance under transformation.

It is often said that a perfect symmetry is unobservable, since flawless and complete invariance makes no physical impact on the world. John Emmerson (1972, p. 164) points out that 'any useful symmetry must be broken in order to give it physical content'. This has epistemic ramifications, as Heinz Pagels (1986, p. 202) explains, specific to gauge symmetries, 'if the Yang–Mills symmetry is exact, then the symmetry remains completely hidden'. A system that experiences no changes whatsoever under any and all transformations has no discernible structure. It is the antithesis of a well-ordered system. In fact, it is the least ordered systems in nature that are the most symmetric, since a truly random arrangement transformed in any way is still a random arrangement. It is well known that by increasing the order in a system, one actually decreases the symmetry. As Ian Aitchison and Anthony Hey (1982, p. 205) put it, '*ordering* here entails *loss* of symmetry!' So if design is linked to putting order in nature, design must result in decreased symmetry.

Here is a homespun example to bring our intuitions up to speed on severing the connection between symmetries and design. Imagine, as a child, separately visiting your two sets of grandparents. The two households are markedly different. The first house is tidy and orderly, with everything in its designated place. There is nothing lying around, no stray magazines or empty glasses. This house is carefully neatened by a meticulous and watchful housekeeper. The second house is clean but disorganized. Furniture and knick-knacks are haphazard in their placement. Newspapers lie about. Old magazines lean in disorderly stacks. No one seems to care about the appearance or organization here.

As a child, it is probably more fun to visit the second house. It is certainly less stressful, since you can move things around the living-room, even take things out or add to the clutter without anyone noticing. You can get away with anything here, or at least with more unruly behaviour than you could in the first house. In the language of symmetry, you can perform more transformations that leave the appearance of the room invariant, more in the messy house than in the tidy one. The messy house has more symmetry. In the first house, the neat one, the careful planning, the strict design, has left fewer options for undetectable rearrangements. There is less rearrangement invariance. In this sense, the greater design has reduced the symmetry. And conversely, the greater symmetry of the second house is a result of indifference. Symmetry is not evidence of close attention to detail and conscious design.

The first household wasn't always so neat and asymmetric. When these grandparents first moved in things were scattered and stacked randomly by the moving-men. These guys are great for putting symmetry into your possessions. Only a natural disaster such as an earthquake or a tornado can do it better. Your grandparents then actively broke the symmetry. From the chaos of moving in, they designed and implemented the careful structure of their living-room. Your other grandparents in the second household also did some symmetry breaking by consciously and purposefully arranging their things, but they did notably less of this than their more attentive counterparts.

The point in these examples is that it is the symmetry breaking, not the symmetry, that is the product of the design. It is symmetry breaking, not symmetry, that warrants an explanation. Weyl (1952, pp. 25–6) makes a similar point in discussing the spherical Earth: 'The feature that needs explanation is, therefore, not the rotational symmetry of its shape but the deviations from this symmetry as exhibited by the irregular distribution of land and water and by the minute crinkles of mountains on its surface.' Symmetries, being too easily the result of equilibrium (or moving-men), are the default structure from which the notable features of nature are formed.

Steven Weinberg (1996, p. 164) has a similar example, closer to the concerns of physics. Building a chair, he points out, is an act of symmetry breaking. Each atom in the chair is rotationally symmetric, but the chair itself is not. The design and construction of the chair didn't add to the symmetry of the system; it decreased it. Design broke the symmetry.

It would be premature at this point to conclude that broken symmetries in nature are evidence of design. The argument and the examples are intended only to discredit the use of symmetry as evidence of design. We see that symmetry is correlated to a lack of purpose and intentional control. This is the case with Weinberg's chair, where the influence of design broke the symmetry in the raw materials, the unstructured heap of molecules. Adding the creative, purposeful information to nature reduces the symmetry. Similarly, getting information from nature involves suppressing the creative and intentional input. The objective knowledge, the covariant structures revealed by symmetries, show where we have no control. Symmetry, in both cases, is about indifference.

This argument is relevant to the debate between Leibniz and Newton regarding the nature of space. The space of Newtonian physics has several important symmetries such as rotational and translational symmetries. The laws of motion are the same here as they are there, and they are the same in this direction as in that. They are invariant, that is, under transformation of position or orientation. But the absolute space of Newton's philosophical description of nature breaks these symmetries. Absolute space distinguishes location in a way that the results of mechanical experiments do not. There is a difference between here and there in an absolute space. Thus, the formation and placement of the universe where it is, rather than a metre to the left, broke the symmetry by selecting one position, similar to the way that magnetizing a bar magnet selects one particular orientation out of all the possible orientations. But why was *this* particular position the one? That is, why was the symmetry broken in this particular way? Leibniz cannot see any sufficient reason for one choice over another. Assuming, as Leibniz does, that there is a designer, there must be a reason for the designer to break the symmetry of space to make it an absolute space. Without that reason, there is no absolute space. In other words, without the intervention of the designer (there being no sufficient reason to intervene) the symmetry remains unbroken and space is not absolute.

But notice that it is symmetry *breaking* that would show the influence of design. The unbroken symmetry of space is evidence of non-intervention by a designer. And so it is generally. Symmetry is the result of indifference to detail and of a lack of creative control. It cannot be evidence of design.

4 Conclusion

There are a lot of reasons given for the effectiveness and importance of symmetry in physics. Two of them have been discussed here: symmetry indicates objectivity and symmetry indicates design. But these two implications are at odds and we are forced to give up one of them. We must abandon the idea that symmetry is evidence of design. A symmetry, after all, reveals an invariance, a non-happening. It shows the aspects of nature that are not influenced by conscious decisions. This is what makes a symmetry a valuable sign of objectivity. It is also what severs the link between symmetry and design. What is uninfluenced by conscious decision cannot be the product of design.

Recall the debate between Bohr and Einstein, the debate about realism. In particular, recall Bohr's advice that 'Physics concerns what we can say about nature.' There is actually a hint of realism here, in the suggestion that there are some things we *can* say about nature and some things we *cannot* say. What would constrain the description in this way other than physical reality itself?

We get to the point about realism and objectivity more quickly by noting what we *cannot* say about nature. But symmetry reveals that once the choices are made as to how to describe nature, once the reference frame is chosen by us, there are relations among the properties that are fixed by nature. The laws are covariant, and we cannot say otherwise. It is in noting these constraints, noting what we cannot say about nature, that we find the physical reality.

So, what must the world be like such that symmetry would be so effective in understanding it? It does not have to have been brought about on purpose or by design. On the contrary, it must be, in important ways, independent of purpose and design. And it must be, in some ways, independent of our subjective ways of describing it.

Acknowledgements

I am grateful to Katherine Brading, Elena Castellani, and Mauricio Suarez for helpful comments and encouragement.

References

Aitchison, I., and Hey, A. (1982). *Gauge Theories in Particle Physics*. Bristol: Adam Hilger.

Born, M. (1953). 'Physical reality'. *Philosophical Quarterly*, **3**, 139–49.

Daston, L., and Galison, P. (1992). 'The image of objectivity'. *Representations*, **40**, 81–128.

Einstein, A. (1949). 'Autobiographical notes'. In *Albert Einstein: Philosopher–Scientist*, ed. P. Schlipp. La Salle, IL: Open Court.

Emmerson, J. (1972). *Symmetry Principles in Particle Physics*. Oxford: Clarendon Press.

Galison, P. (1999). 'Objectivity is romantic'. *American Council of Learned Societies*, Occasional paper, no. 47.

Klein, É., and Lachièze-Rey, M. (1999). *The Quest for Unity*. Oxford: Oxford University Press.

Lange, M. (2001). 'The most famous equation'. *Journal of Philosophy*, **98**, 219–38.

Newell, R. (1986). *Objectivity, Empiricism and Truth*. London: Routledge & Kegan Paul.

Nozick, R. (1998). Presidential address, *Proceedings and addresses of the American Philosophical Assocation*, **72**, 21–48.
(2001). *Invariances: The Structure of the Objective World.* Cambridge, MA: Harvard University Press.
Pagels, H. (1986). *Perfect Symmetry: The Search for the Beginning of Time.* New York: Simon & Schuster.
Petersen, A. (1963). 'The philosophy of Niels Bohr'. *Bulletin of the Atomic Scientist*, **19**, 8–14.
Poincaré, H. (1905). *Science and Hypothesis.* New York: Dover.
Reichenbach, H. (1958). *The Philosophy of Space and Time.* New York: Dover.
Weinberg, S. (1996). *The Quantum Theory of Fields*, Vol. 2. Cambridge: Cambridge University Press.
Weyl, H. (1952). *Symmetry.* Princeton, NJ: Princeton University Press.
Wigner, E. (1967). *Symmetries and Reflections.* Bloomington, IN: Indiana University Press.
Worrall, J. (1989). 'Structural realism: the best of both worlds'. *Dialectica*, **43**, 99–124.
Zee, A. (1986). *Fearful Symmetry.* New York: MacMillan.

26

Symmetry and equivalence

ELENA CASTELLANI

As other contributions to this volume also testify, the notions of symmetry and equivalence are closely connected. This paper is devoted to exploring this connection and its relevance to the symmetry issue, starting from its historical roots. In fact, it emerges as an essential and constant feature in the evolution of the modern notion of symmetry: at the beginning, as a specific relation between symmetry and equality; in the end, as a general link between the notions of symmetry, equivalence class, and transformation group.

1 Symmetry and equality

Weyl's 1952 classic text on symmetry starts with the following distinction between two common notions of symmetry:

> If I am not mistaken the word *symmetry* is used in our everyday language in two meanings. In the one sense symmetric means something like well-proportioned, well-balanced, and symmetry denotes that sort of concordance of several parts by which they integrate into a whole. . . . The image of the balance provides a natural link to the second sense in which the word symmetry is used in modern times: *bilateral symmetry*, the symmetry of left and right . . .

Bilateral symmetry is in fact a particular case of the scientific notion of symmetry, the symmetry being defined as invariance with respect to a transformation group (in the case of bilateral symmetry, the group of spatial reflections). This is the symmetry we are interested in here: the notion which, thanks to its group-theoretic characterization, has proved so successful in twentieth-century physics. But how did it develop? We know that the symmetry of the Greeks and Romans, current until the end of the Renaissance, was the one grounded on proportion relations. But we also know, thanks to a 1673 text by the French architect and physician

Claude Perrault (the brother of the more famous Charles), that the second notion of symmetry (not yet in its group-theoretic formulation, of course) became commonly used among his contemporaries. In fact, in presenting his French translation of Vitruvius' *De Architectura*, Perrault introduced a very similar distinction to the one stated above, i.e. the distinction between an *ancient notion* of symmetry, based on proportions (and used for example by Vitruvius), and a *modern notion*, defined as a relation of equality between parts that are opposed – the most familar case being the relation of equality between the right and left parts of a figure (Perrault, 1673, p. 10).

We thus know that symmetry, in its initial modern sense, was closely related to spatial equality. What exactly is the nature of this relationship? In particular, how does it change in the conceptual development of symmetry from its initial meaning (a relation of equality between opposed parts) to its group-theoretic definition? This section and the next are devoted to exploring this issue.

The equality of the parts is indeed the distinguishing feature of the modern notion in its original version: it marks the difference with respect to the ancient notion (the previous role of symmetry being that of 'according' parts of different size and form by means of proportions). At the same time, the equality of the parts is what made possible the mathematical development of the symmetry notion. Parts that are equal can be exchanged or substituted: it thus became possible to introduce mathematical operations (such as translations, rotations, and reflections) for describing how, in fact, the parts could be exchanged.

The equality of the parts is, of course, not enough. Symmetry is something more than simple repetition. It is characterized by the sort of regularity inherent in the arrangement of equal parts. In its modern sense, symmetry is always a property of the whole: in the case of a spatial arrangement of equal parts, a property of the whole arrangement. The relationship between symmetry and equality is therefore to be understood with respect to the whole: the parts of a symmetric arrangement are each equal to one another and with respect to the whole; that is, the entire configuration does not change when the parts are exchanged. The symmetric character of an arrangement of equal parts has thus to do with the fact that the whole figure does not change when the parts are exchanged by means of some operations. These operations are then identified as the 'symmetry operations' of the figure, and the type of the symmetry (translational, rotational, etc.) is given by the type of symmetry operations considered (translations, rotations, etc.).

Symmetry could thus be characterized in terms of the invariance of the figure studied under specified operations. The next step, in its conceptual history, was of a purely mathematical nature: the nineteenth-century introduction of the algebraic concept of a *group* and the subsequent development of the theory of transformation groups. The symmetry operations ('symmetry transformations') of a figure were found to satisfy the conditions for constituting a transformation group.[1] This led to the final stage in the evolution of the modern notion of symmetry: its general mathematical definition as 'invariance under a group of transformations'.

2 Symmetry, equivalence, and group

What about the initial equality of parts? We started with the familiar case of a spatial arrangement of equal parts, and arrived at the abstract group-theoretic notion of symmetry applied in modern science. But the substance has not changed: the 'equality of parts' is still there, but in the form of an *equivalence relation*, i.e. the equivalence relation between the elements that are connected to each other by the symmetry transformations of the 'whole'.

To see better what this means, let us stay on familiar ground by taking again the case of a symmetric arrangement of parts that are related to each other by an equality relation. For a relation of this nature we expect that it is

(i) *reflexive* – each part is equal to itself;
(ii) *symmetric* – if part A is equal to part B, then B is equal to A;
(iii) *transitive* – if part A is equal to part B and part B is equal to part C, then A is equal to C.

Equality is in fact the most obvious example of an *equivalence relation* (i.e. a relation that is reflexive, symmetric, and transitive). Now, this has the following interesting implications for the symmetry operations of the whole figure (the operations exchanging the equal parts in such a way that the whole remains invariant):

(i) the existence of the identity operation, corresponding to the reflexive property of the relation between the parts;
(ii) the existence of the inverse operation, corresponding to the symmetric property of the relation between the parts;
(iii) the closure of the product between any two operations, corresponding to the transitive property of the relation between the parts.

In substance, the fact that the parts are related by means of an equivalence relation (which is at the same time *the* equivalence relation between the parts, ensuring their interchangeability in the considered context) corresponds to the fact that the family of operations transforming the parts into each other while leaving the whole invariant satisfies the conditions for constituting a group (i.e. the existence of the identity and inverse operations, and the closure of the product).[2] In other words,

[1] A group is defined to be a set G with a product $\cdot$ which assigns to each pair (g_1, g_2) of elements in G one element $g_1 \cdot g_2$ in G, and is associative, has a neutral element, and for which each element has an inverse.
[2] For a more detailed discussion see for example Yaglom (1988, pp. 114–6).

the two main features for capturing the essence of a symmetry – the invariance of the whole and the equivalence of the parts (at the same time between each other and with respect to the whole) – are mathematically mirrored in the group properties. It is worth underlining that not any equivalence relation will do, but only *the* equivalence relation which is such as to ensure the invariance of the whole when the equivalent parts are exchanged. This is the difference between a symmetry and a bare repetition. There are conditions regulating the way in which the parts are equivalent: namely, the conditions dictated by the invariance of the whole under the transformations relating the equivalent elements, mathematically translated into the group conditions for the symmetry transformations.

The link thus obtained between the group nature of the symmetry operations and the fact that they exchange elements that are equivalent is in fact valid in general, independently of whether we are considering the symmetry properties of familiar spatial figures or of abstract mathematical relations. What the equivalent elements are, how they are related and in which way the symmetry of the whole is obtained of course depend on the specific situation considered. But the connection that can be established between the notions of symmetry, equivalence, and group is a general feature.

More can be said, in fact. We know that, given an equivalence relation on a set A, the set of all the elements equivalent to a given element x of A forms the *equivalence class* of x. A lemma says that if two elements are equivalent, they have the same equivalence class. Then we have a theorem stating that, if we have an equivalence relation on A, the family of all the equivalence classes is a *partition* of A.[3] Proceeding the other way round, we may begin with a partition and end up with an equivalence relation. It is thus legitimate to conclude that the two concepts of *equivalence relation* and *partition* are 'twin aspects of the same structure on sets' (Pinter, 1982, p. 120). Now, as we have seen, the elements exchanged with one another by the symmetry transformations of a figure (or whatever the 'whole' considered is) are connected by an equivalence relation. They thus form an equivalence class. More generally, we can then say that the presence of a symmetry group in a given context is related to a partition into equivalence classes.[4] The symmetry transformations, by permuting the elements in the same equivalence class, leave this structure invariant. At the same time, they leave those features invariant that are common to all the elements constituting the same equivalence class. Note that these features are exactly the ones necessary to characterize the class as such within the considered context. Two fundamental aspects of the notion of symmetry clearly emerge here: (i) its generality – what remain invariant under symmetry transformations are properties

[3] For details see for example Pinter (1982, chapter 12).

[4] For a slightly different way of presenting the connection between the concepts of equivalence relation, partition, and group transformations, see van Fraassen (1989, chapter 10, section 3), who arrives at the conclusion that these three concepts 'amount really to the same concept' (p. 246).

of classes, not of individual elements;[5] and (ii) its 'structuring' role – symmetry at the same time defines and preserves structures.[6]

3 Equivalence and irrelevance

A symmetry, we have seen, corresponds to a situation of equivalence with respect to a given context: elements that are connected with one another by symmetry transformations form an equivalence class. How does this apply to physics? A physical symmetry corresponds to the equivalence of a number of elements with respect to the physical theory considered. The contributions to this volume offer a good variety of examples: permutation symmetry and the equivalence of so-called identical particles (French and Rickles, Huggett (chapter 13), Saunders), the equivalence of spacetime points connected with relativity principles (Norton), and the equivalence of phase space points lying in the same gauge orbit (Redhead, Earman (chapter 8)), to mention some. Note that the role of the 'whole' is now played by the physical theory; more precisely, by its dynamical equations. The theory gives the same description – that is, the fundamental dynamical equations do not change – when the equivalent elements are exchanged with one another by the transformations of the symmetry group. This is the received view in the physics literature.

Different questions naturally arise at this point. This paper focuses on the following one: how is the equivalence of the elements to be understood? Generally speaking, it is quite common to think that if some elements are equivalent from the viewpoint we are considering, we do not need to take all of them into account. One of the equivalent elements can be representative of the others. In which way representative and with respect to what needs to be clarified, of course. But the moral is general: in some way, equivalence seems to go with *irrelevance* or *redundancy*, the presence of equivalent elements suggesting the presence of irrelevant or redundant features in the context considered.

A concrete example of this moral is the standard attitude among physicists towards the symmetries postulated by special relativity, that is the global symmetries of the spacetime continuum. Take, for example, the case of (spatial) translations and the corresponding symmetry (homogeneity of space): positions are equivalent to one another from the viewpoint of the physical description, the dynamical equations not varying in form under translations. Now, this is usually taken to indicate the irrelevance of an absolute position to the physical description: all space points are equivalent as far as the physics (i.e. the dynamical equations) is concerned. One of them is representative of the whole class, and the symmetry transformations are

[5] Note that this is the basis for the possibility of classification on the grounds of symmetry properties: one of the most important functions of symmetry in science. On this point see the introduction to this volume.

[6] Whence the use, in some of the literature, to define symmetries as transformations that preserve structure. See, in particular, Ismael and van Fraassen, this volume.

the mathematical tools by means of which this is made possible; or, in other words, a reference frame located at any position is representative of all the others, the translational symmetry ensuring the invariance of the physics when passing to another reference frame by means of symmetry transformations. The same is commonly argued in the case of the other global spacetime symmetries. Let us just mention what Eugene Wigner writes in this respect: 'the older laws of invariance', the ones 'which found their perfect, and perhaps final, formulation in the special theory of relativity . . . postulate, in addition to the irrelevance of the absolute position and time of an event, the irrelevance of its orientation and finally, the irrelevance of its state of motion, as long as this remains uniform, free of rotation, and on a straight line' (Wigner, 1967, pp. 4–5).

As already said, in what way the equivalent elements are connected to the presence of irrelevant features in the theory depends on the specific case studied. Moreover, the nature of these 'irrelevancies' is not the same for all types of symmetries, the major difference being between global and local (i.e. dependent on spacetime points) symmetries. In the global case, all the equivalent elements (for example, all the equivalent space locations) have the same physical status ('reality'). But this is controversial in the local case, where, as we shall see in the next section, attributing the same reality to the equivalent elements leads to indeterminism.

It is, however, a fact that, independently of how it is realized, the connection between symmetry, equivalence, and irrelevance is indeed a close one. On this basis, a natural position on the symmetry issue is the one relating the presence of physical symmetries to the presence of irrelevant elements in the physical description. This has an immediate empirical counterpart in terms of non-observability. The idea is that irrelevant theoretical features make no observable difference. Symmetries are thus connected with the presence of non-observable quantities in the physical description, with the corollary that the empirical violation of a symmetry (the paradigmatic case is parity violation in the case of weak interactions) is intended in the sense that 'what was thought to be a non-observable turns out to be actually an observable' (Lee, 1988, p. 178).

The view that 'the root of all symmetry principles lies in the assumption that it is impossible to observe certain basic quantities' (*ibid.*) is quite popular, especially among physicists. It leaves, however, many questions open, the most important of which concerns the nature and origin of the irrelevant/non-observable features.

Dirac, in his 1930 preface to his masterpiece *The Principles of Quantum Mechanics*, gives us a possible clue (p. vii):

> [Nature's] fundamental laws control a substratum of which we cannot form a mental picture without introducing irrelevancies. The formulation of these laws requires the use of the mathematics of transformations.

This may be taken as an illustrious example of the following view: in describing the physical world, because of our inherent epistemic limits in relation to the nature of what we are trying to describe, we introduce irrelevant theoretical elements and this is signalled by the presence of symmetries. This is how far the connection between symmetry and equivalence has brought us until now. But this is still very vague: in what way are these irrelevancies introduced? And what is peculiar to the ones signalled by symmetries? Two ways of addressing these questions are examined separately and then brought together in the concluding section.

4 Symmetry and arbitrariness

Two main points have emerged so far:

- a symmetry gives rise to equivalence classes (the elements that are connected with one another by symmetry transformations form an equivalence class);
- symmetries are related in some way to the presence of irrelevant features in the physical description.

Now, there are two sorts of arbitrariness involved here.

- On the one hand, there is a freedom associated with the fact that a symmetry gives rise to equivalence classes: the freedom to choose one element as representative of the class; which one is completely arbitrary.
- On the other hand, there is a certain arbitrariness in the distinction between what is relevant and what is irrelevant with respect to the physical situation studied. What is relevant or irrelevant may also depend on the level of detail, or the 'scale' (spatial, temporal, etc.), at which the situation is considered.

Let us explore both cases in some more detail, to see how far this takes us.

4.1 Gauge freedom and constraints

The link between symmetry and equivalence is general but, as some of the contributions to this volume clearly show (see especially Martin), there are subtle interpretative questions related to the difference between global and local (or gauge) symmetries. We focus here on the gauge case, for the simple reason that it has greater interest for the purpose of this paper. In the local symmetry case, the freedom to choose one element as representative of the equivalence class ('gauge freedom', in the literature) corresponds to the presence of surplus 'unphysical' degrees of freedom in the theory, an example of what has been called by Redhead 'surplus structure'.[7] There is now a quite intensive discussion in the literature about the significance of these surplus degrees of freedom.[8] The main issue is whether to consider them literally (with indeterminism as a consequence),[9] or to eliminate them in some way,[10] or to search for another type of solution.

A convenient framework for examining the nature and role of the arbitrary features in the gauge case is the *theory of constrained Hamiltonian systems*, going back to Dirac's seminal works in the 1950s. There is in fact an important relation between gauges and constraints, as pointed out also in other contributions to this volume (see especially Earman, Part I). On the one hand, it is shown (by using Noether's second theorem) that all systems with a gauge invariance are necessarily singular systems with constraints.[11] On the other hand, as first conjectured by Dirac in the 1950s, all so-called first-class constraints are demonstrated to be generators of gauge transformations. The presence of arbitrary features in the mathematical framework plays in fact a crucial role in obtaining these results, as will be briefly illustrated here below.

Gauge theories are an example of singular theories, that is theories describing a physical system by more variables than there are physically independent degrees of freedom. In such cases it is easy to show that the general solutions of the Hamiltonian equations of motion with given initial conditions depend on arbitrary functions of time. Let us see briefly how Dirac's analysis proceeds in the simple case of a system with a finite number N of degrees of freedom, with general coordinates q_n $(n = 1, \ldots, N)$ and velocities $\dot{q}_n = dq_n/dt$.[12]

In the singular case, the velocities $\dot{q}_n$ are not uniquely determined in terms of the coordinates q_n and momenta p_n.[13] This means that not all momenta are independent functions of the velocities; that is, there must exist a set of relations of the form

$$\phi_m(q_n, p_n) = 0, \qquad m = 1, \ldots, M,$$

[7] The significance of this surplus structure, first considered in Redhead (1975), is thoroughly analysed in Redhead's contribution to this volume.

[8] See for example Earman (Part I), Martin, and Redhead, this volume, and references therein.

[9] Wallace, this volume, is devoted to clarifying this point.

[10] For example, by moving from the original phase space to a 'reduced space', the points of which are equivalence classes of the original phase space points that are related by symmetry transformations.

[11] According to Noether's second theorem, for a gauge-invariant system – that is, a system invariant under transformations defining a simply connected continuous group, whose parameters are r arbitrary functions of time (or spacetime) – there exist r independent identities of the Euler–Lagrange derivatives of the Lagrange function. These identities are consequences of the gauge invariance and define (Lagrangian) constraints on the system. That gauge-invariant systems are singular systems may be easily seen by analysing these identities.

[12] Here I closely follow Dirac's 1964 *Lectures on Quantum Mechanics* (the classic reference in this regard). More details on what follows in the text can be found in Castellani (2003).

[13] In Lagrangian terms, this is the situation expressed by the singularity of the Hessian matrix of the Lagrange function, that is by the fact that

$$det\left(\frac{\partial^2 L}{\partial \dot{q}^n \partial \dot{q}^{n'}}\right) = 0.$$

called by Dirac the 'primary constraints' of the Hamiltonian formalism.[14] Now, given a set of N phase space degrees of freedom and a set of M primary constraints, time evolution may be generated not only by the canonical Hamiltonian $H_o = p_n\dot{q}_n - L$, but (since we may add to it any linear combination of the ϕs which are zero) by a generalized Hamiltonian of the form

$$H_* = H_o + u_m(q_n, p_n; t)\phi_m(q_n, p_n),$$

where ϕ_m are all primary constraints and u_m are arbitrary functions of time and of the phase space variables. Starting with this expression and carrying out his analysis by imposing all the consistency requirements of the theory, Dirac then arrives at the following final expression for the generalized Hamiltonian:

$$H_* = H + v_a(t)\phi_a,$$

where $H = H_o + U_m\phi_m$ and $\phi_a = V_{am}\phi_m$. The U_m and V_{am} are functions of the phase space variables satisfying to given consistency equations, while the $v_a(t)$ are *arbitrary functions of time* (their number being equal to the number of primary first-class constraints, usually less than the number of all constraints).[15]

The result is thus that, although all consistency requirements have been satisfied, the theory still has arbitrary coefficients which may depend on time, the $v_a(t)$. This means that the general solution of the Hamiltonian equations of motion with given initial conditions depends on arbirary functions of time. What follows from this? According to Dirac, the arbitrariness in the choice of the functions $v_a(t)$ implies that the different trajectories in phase space obtained under time evolution for given initial conditions but for different $v_a(t)$ should be considered as representing the same configuration of the system. It was then conjectured by Dirac (and later demonstrated) that different points in phase space representing the same state of the system are related to one another by gauge transformations that are generated by the first-class constraints of the theory. This result – constraints are generators of gauge transformations – is indeed the upshot of this part of Dirac's analysis.

It is thus made quite precise, in Dirac's analysis, in what the arbitrariness peculiar to the above sort of situation consists, and how the connected surplus theoretical features enter into the theory: in particular, many different phase space points and trajectories representing the same time-dependent configuration of the system. The Hamiltonian treatment à la Dirac shows how they are related to one another by gauge transformations. Now, how should we deal with the equivalent or redundant descriptions we have thus obtained? In agreement with Dirac's position, the

[14] Dirac distinguishes between *primary* and *secondary* constraints (depending on whether their definition is independent of the Lagrangian equations of motion or not), and *first-class* and *second-class* constraints (depending on whether their Poisson brackets with all other constraints is weakly vanishing or not). The really important distinction for the treatment of constrained systems is the second one. For details see Dirac (1964, pp. 14–8).

[15] For details see Dirac (1964, pp. 13–6).

common answer in the physical literature is: by keeping the physically meaningful degrees of freedom only (ideally, by means of a description of the dynamical evolution of the system in terms of the reduced space representing the truly distinct possible configurations of the system). In gauge terms, this corresponds to the common strategy of attributing physical meaning to gauge-invariant quantities only. The underlying idea is that in a theory where the system is described by more variables than the number of independent degrees of freedom, 'the physically meaningful degrees of freedom reemerge as being those invariant under a transformation connecting the variables (gauge transformation)'(Henneaux and Teitelboim, 1992, p. 1). The philosophy, in substance, is the following: for given reasons we introduce extra variables (surplus theoretical features or 'irrelevancies', in Dirac's words) in the theory, and at the same time we bring in a (gauge) symmetry 'to extract the physically relevant context' (*ibid.*).[16]

This is, at first sight, a very strong epistemic attitude towards the meaning of gauge symmetries. I want to suggest a way of justifying a less strongly epistemic approach which nevertheless respects the basic motivations for the above view. The idea is a sort of compromise: symmetries do enter in our way of describing the physical world, but in a way that is significantly 'constrained' by the reality we want to describe. In this regard, particular relevance is to be attributed to the role played by physical scales and related boundary conditions. This is what the concluding part of this paper tries to suggest, by examining the second sort of arbitrariness related to physical symmetries: namely, the degree of arbitrariness in the distinction between what is relevant and what is irrelevant with respect to a given context.

4.2 Symmetries and scales

As we have said, what is relevant or irrelevant with respect to a given context depends not only on the characteristics of the chosen domain (which is obvious), but also on the level of detail at which this domain is considered. Today's physics offers in fact an important suggestion in this sense; that is, the idea at the basis of a recent view of current quantum field theories as *effective theories*: physics can change as one changes the scale considered, at very different ranges of energy scales we can have remarkably different physics. In recent years, in fact, developments in quantum field theory (QFT) and in the application of renormalization group theory techniques have brought a 'change of attitude' in particle physics (Weinberg, 1997, p. 41), which is at the basis of the so-called modern view of QFT: that is, the view

[16] This point of view has been seen by some as problematic in the case of the Aharonov–Bohm effect, where the gauge-invariant quantities are the Yang–Mills fields plus the holonomies. For philosophical reflections on this point see Nounou, this volume.

that 'the most appropriate description of particle interactions in the language of QFT depends on the energy at which the interactions are studied' (Georgi, 1989, p. 446). On this view, current QFTs are understood as *effective field theories*, each effective field theory explicitly referring only to those particles that are actually of importance at the range of energies considered. The key point is the 'decoupling' of the physics at the chosen energy scale from the physics at much higher energies. In the QFT framework, this decoupling of physical phenomena, as well as the changing of the effective physical description as the scale changes, occur according to specific and precise rules. This is due essentially to some important peculiarities of the local quantum field description and, in particular, to the concept of the *renormalization group* and its deep impact on particle physics.

The effective field theory approach suggests that scale considerations might play an important role also when discussing the meaning of physical symmetries. What is relevant or irrelevant is in fact also related to what physical scale we choose. That is, the choice of physical scale imposes a coarse-graining, which in itself creates equivalence classes (the elements of which are all equivalent to one another with respect to the disregarded or 'cut-off' features in the coarse-graining). The cutting-off procedure, by means of which the coarse-graining is effected, thus fixes a distinction between what is being relevant or irrelevant at the chosen scale.

In this way, the arbitrariness implied in the relevant/irrelevant distinction comes to be significantly related to the arbitrariness in the choice of the scale at which the physical situation is studied. The choice of the scale is apparently up to us, so this seems to go well with a purely epistemic attitude towards the meaning of symmetries. But the situation is actually a bit more subtle here. It is true that choosing the scale is up to us, but not *entirely*. The nature of the physical context we want to describe constrains our choice in a significant way. The scale is the one we use in the laboratories, but it is also the one given by the energies typical of the successive states of the universe in its evolution. As is clearly stated in a recent popular book by the physicist Brian Greene (1999, pp. 350–1), 'the significant differences between the forces as we now observe them were all erased by the extremes of energy and temperature encountered in the very early universe. But as the time went by and the universe expanded and cooled, the formalism of quantum field theory shows that this symmetry would have been sharply reduced . . . leading to the comparatively asymmetric form with which we are familiar.'

So, although there is an arbitrariness implied in the choice of scale (and hence in the distinction between what is relevant or irrelevant), this arbitrariness is limited by factors external to us, and this is where the realistic turn to the view discussed in this paper appears. It is a fact that what is a good physical description or approximation in a given regime may not be a good description or approximation in another very different regime. In particular, what are 'unphysical' surplus features in the appropriate description at a determinate regime may become physically relevant features in a very different regime. The following words by the leading theoretical physicist David Gross (1999, p. 58) offer a good and conclusive illustration of what is meant here:

> Today we believe that global symmetries are unnatural... We now suspect that all fundamental symmetries are local gauge symmetries. Global symmetries are either broken, or approximate, or they are the remnants of spontaneously broken local symmetries.

Acknowledgements

Many thanks to Katherine Brading for her invaluable comments on earlier drafts of this paper and thanks also to all the contributors to this volume for the very useful suggestions I have found while reading their papers.

References

Castellani, E. (2003). 'Dirac on gauges and constraints', forthcoming in *International Journal of Theoretical Physics*.

Dirac, P. A. M. ([1930] 1987). *The Principles of Quantum Mechanics*. Oxford: Clarendon Press.

(1964). *Lectures on Quantum Mechanics*. New York: Academic Press.

Georgi, H. M. (1989). 'Effective quantum field theories'. In *The New Physics*, ed. P. Davies, pp. 446–57. Cambridge: Cambridge University Press.

Greene, B. (1999). *The Elegant Universe*. London: Vintage.

Gross, D. (1999). 'The triumph and limitations of quantum field theories'. In *Conceptual Foundations of Quantum Field Theory*, ed. T. Y. Cao, pp. 56–67. Cambridge: Cambridge University Press.

Henneaux, M., and Teitelboim, C. (1992). *Quantization of Gauge Systems*. Princeton, NJ: Princeton University Press.

Lee, T. D. (1988). *Particle Physics and Introduction to Field Theory*. New York: Harwood Academic Publishers.

Perrault, C. (1673). *Les Dix Livres d'Architecture de Vitruve, corrigez et traduits nouvellement en françoys, avec des notes et des figures*. Paris.

Pinter, C. C. (1982). *A Book of Abstract Algebra*. New York: McGraw-Hill.

Redhead, M. (1975). 'Symmetry in intertheory relations'. *Synthese*, **35**, 77–112.

van Fraassen, B. C. (1989). *Laws and Symmetry*. Oxford: Clarendon Press.

Weinberg, S. (1997). 'Changing attitudes and the Standard Model'. In *The Rise of the Standard Model*, ed. L. Hoddeson, L. Brown, M. Riordan, and M. Dresden, pp. 36–44. Cambridge: Cambridge University Press.

Weyl, H. (1952). *Symmetry*. Princeton, NJ: Princeton University Press.

Wigner, E. P. (1967). *Symmetries and Reflections*. Bloomington, IN: Indiana University Press.

Yaglom, I. M. (1988). *Felix Klein and Sophus Lie. Evolution of the Idea of Symmetry in the Nineteenth Century*. Boston-Basel: Birkhäuser.

Index

Printed in the United States
By Bookmasters